Current Topics in Behavioral Neurosciences

Volume 65

Current Topics in Behavioral Neurosciences provides critical and comprehensive discussions of the most significant areas of behavioral neuroscience research, written by leading international authorities. Each volume in the series represents the most informative and contemporary account of its subject available, making it an unrivalled reference source. Each volume will be made available in both print and electronic form.

With the development of new methodologies for brain imaging, genetic and genomic analyses, molecular engineering of mutant animals, novel routes for drug delivery, and sophisticated cross-species behavioral assessments, it is now possible to study behavior relevant to psychiatric and neurological diseases and disorders on the physiological level. The Behavioral Neurosciences series focuses on translational medicine and cutting-edge technologies. Preclinical and clinical trials for the development of new diagnostics and therapeutics as well as prevention efforts are covered whenever possible. Special attention is also drawn on epigenetical aspects, especially in psychiatric disorders.

CTBN series is indexed in PubMed and Scopus.

Christopher Maymon • Gina Grimshaw •
Ying Choon Wu
Editors

Virtual Reality in Behavioral Neuroscience: New Insights and Methods

Springer

Editors
Christopher Maymon
School of Psychology
Victoria University of Wellington
Wellington, New Zealand

Gina Grimshaw
School of Psychology
Victoria University of Wellington
Wellington, New Zealand

Ying Choon Wu
Swartz Center for Computational Neuroscience
University of California, San Diego
San Diego, CA, USA

ISSN 1866-3370 ISSN 1866-3389 (electronic)
Current Topics in Behavioral Neurosciences
ISBN 978-3-031-42994-1 ISBN 978-3-031-42995-8 (eBook)
https://doi.org/10.1007/978-3-031-42995-8

This Springer imprint is published by the registered company Springer Nature Switzerland AG
The registered company address is: Gewerbestrasse 11, 6330 Cham, Switzerland

Paper in this product is recyclable.

Preface

Over the past decade, Virtual Reality (VR) and other extended reality technologies have opened new avenues of research in mind and brain sciences, along with the development of powerful immersive clinical interventions. As a technological medium, VR provides unique opportunities for scientists to transcend the limitations of traditional laboratory settings, while maintaining rigorous experimental control. This innovation is significant, as it enables the exploration of previously impracticable questions that are crucial to advancing our understanding of human cognition, behavior, and brain function. VR allows for the replication of past research findings in contexts that more closely mirror real-world scenarios, allowing the extent to which laboratory findings generalize to ecologically valid situations to be tested. Moreover, VR also makes it possible to test humans in environments that simulate those that are commonly used in animal research, facilitating translational neuroscience. Although predictions vary when it comes to the future success of VR in commercial markets (such as gaming, personal computing, and remote work), when it comes to psychological research, VR is quite likely here to stay. This volume aims to provide a glimpse into some of the novel methods used in VR labs around the world and the wide range of new knowledge that has been created, spanning many of the most prominent domains of cognitive and behavioral neuroscience. We hope these tools and examples will inspire researchers to think about how VR can be used to advance knowledge in yet more domains.

Although the advent of VR research has sparked considerable excitement, the rapid evolution of VR technologies presents a challenge for researchers. The introduction of new VR hardware and software differs from past technological advancements, such as electroencephalography, trans-cranial magnetic stimulation, or high-fidelity eye-tracking systems. All these techniques, which represent high-impact breakthroughs, have mostly transformed either measurement or stimulation methods. However, they could still be integrated into extant research pipelines with relatively little adjustment to the original paradigm. VR, on the other hand, is fundamentally different; for the most part, it does not readily align with existing research procedures. Instead, researchers must develop entirely new protocols and will likely need to pick up skills in game development (or invite a game developer to join the research team). Although these prospects can be daunting, it is our hope that this volume can serve as a guide for new VR researchers trying to navigate the unfamiliar landscape of VR technologies. We hope that this volume will also stimulate discussions about the systems (the hardware and software) used to deliver VR, with the goal of promoting more homogeneity in the VR systems used across studies. Doing so would enable more direct comparisons of findings across different studies and encourage the sharing of research pipelines that can be more easily adopted by other labs.

Unfortunately, it must be noted that much of the practical knowledge in this volume is likely to become outdated in the coming years due to the rapid advancement of VR technologies and the software driving them. For example, in the early stages of conceptualizing, drafting, and editing the first chapters, the HTC Vive Pro Eye headset seemed like the best option for research requiring eye tracking, and we envisioned ourselves to be using this system for many years to come. At the time of this volume's completion, however, the Vive Pro Eye has been discontinued. In its place, more light-weight headsets (with less powerful eye-tracking systems) have been marketed as the headsets of the future. The methods that are useful today may not be as applicable for a researcher using VR in 2030. Despite this caveat, we have endeavored to provide a comprehensive overview of the most cutting-edge VR research as of the early 2020s. Furthermore, while the technology will certainly change, these chapters describe many novel and creative ways that one may use VR to advance knowledge, and we think that these practices will remain central to VR research even as the technology improves.

In this volume, fourteen chapters provide comprehensive views of new methods and insights from VR research informing our understanding of both fundamental and applied topics ranging from vision and memory to imagination, education science, and mental health therapy. *The Promises and Pitfalls of Virtual Reality* (Maymon, Wu & Grimshaw) describes the major benefits, as well as the limitations and challenges, accompanying the choice to use virtual reality in behavioral neuroscience and related fields. *Launching Your VR Neuroscience Laboratory* (Wu, Maymon, Paden & Liu) presents an exhaustive overview of necessary considerations and practical advice for researchers looking to incorporate virtual reality technologies in their laboratory. *Monitoring Brain Activity in VR: EEG and Neuroimaging* (Ocklenburg & Peterburs) describes the benefits and challenges that arise when VR is used alongside neuroimaging and neurostimulation techniques. *Eye Tracking in Virtual Reality* (Anderson, Bischof, & Kingstone) shows how eye movements, head movements, and body movements can be captured in virtual reality, along with new analysis pipelines for determining fixations across spherical displays. *Virtual Reality for Spatial Navigation* (Jeung, Hilton, Berg, Gehrke, & Gramann) describes how human spatial navigation research is evolving alongside virtual reality, which allows for naturalistic locomotion supporting embodied experiences in navigation research paradigms. *Virtual Reality for Vision Science* (Hibbard) discusses the unique contributions that virtual reality makes for evaluating theories about the behavioral goals underpinning visual processing. *VR for Studying the Neuroscience of Emotional Responses* (Andreatta, Winkler, Collins, Gromer, Gall, Pauli, & Gamer) discusses how virtual reality is revolutionizing lab-based emotion induction, opening up new avenues for investigating the neurobiological mechanisms underpinning emotional responses. *VR for Cognition and Memory* (Reggente) presents a comprehensive review of VR research into memory and cognitive processes, highlighting the importance of enhancing ecological validity to explain how VR can shape future cognitive research. *Virtual Reality for Awe and Imagination* (Chirico & Gaggioli) describes how complex emotions, such as awe, can be better understood when we can induce transformative experiences using VR. *Using Extended Reality to Study the Experience of Presence* (Suzuki, Mariola, Schwartzman, & Seth) explores how virtual reality allows researchers to investigate the factors underpinning our subjective sense of reality, i.e., our presence. *Virtual Reality for Learning* (Checa & Bustillo) shows how immersive virtual reality experiences can promote learning across different stages of development. *VR for Pain Relief* (Matamala-Gomez, Donegan, & Swidrak) showcases how virtual reality can effectively modulate pain by inducing an illusion of body ownership. *Virtual Reality for Motor and Cognitive Rehabilitation* (Darekar) describes how new virtual reality clinical interventions are improving rehabilitation of motor and cognitive deficits particularly post stroke or Parkinson's disease. Finally, *Virtual Reality Interventions for Mental Health* (Kothgassner, Reichmann, & Bock) discusses VR therapeutic techniques for the treatment of anxiety disorders, schizophrenia, neurodevelopmental disorders, and eating disorders and highlights the role of gamification in maximizing the success of VR therapies.

We are extremely grateful to the VR researchers who have contributed to this volume, all of whom stand at the forefront of VR research. When viewed together, the chapters in this collection offer a comprehensive snapshot of a rapidly evolving methodology with the capacity to transform our understanding of the mind and advance science beyond findings that can only replicate in laboratory settings.

This project was made possible through grants from the Royal Society Marsden Grant (VUW2005) to co-editor Grimshaw and from the National Science Foundation (DUE-1734883) and the Army Research Laboratory (W911NF2120126) to co-editor Wu.

Wellington, New Zealand — Christopher Maymon
Wellington, New Zealand — Gina Grimshaw
San Diego, CA, USA — Ying Choon Wu

Contents

Part I
Methods in Virtual Reality Research

The Promises and Pitfalls of Virtual Reality

Christopher Maymon, Ying Choon Wu, and Gina Grimshaw

Contents

Abstract Increasingly, Virtual Reality technologies are finding a place in psychology and behavioral neuroscience labs. Immersing participants in virtual worlds enables researchers to investigate empirical questions in realistic or imaginary

C. Maymon (✉) and G. Grimshaw
Victoria University of Wellington, Wellington, New Zealand
e-mail: Christopher.Maymon@vuw.ac.nz; Gina.Grimshaw@vuw.ac.nz

Y. C. Wu
University of California San Diego, San Diego, CA, USA
e-mail: ycwu@ucsd.edu

Curr Topics Behav Neurosci (2023) 65: 3–24
https://doi.org/10.1007/7854_2023_440
Published Online: 17 August 2023

environments while measuring a wide range of behavioral responses, without sacrificing experimental control. In this chapter, we aim to provide a balanced appraisal of VR research methods. We describe how VR can help advance psychological science by opening pathways for addressing many pernicious challenges currently facing science (e.g., direct replication, prioritizing ecological validity). We also outline a range of unique and perhaps unanticipated obstacles and provide practical recommendations to overcome them.

Keywords Ecological validity · Presence · Virtual reality

1 Introduction to the Volume

Imagine a continuum between physical reality at one end, and an entirely digitally mediated reality (for example, as depicted in the *Matrix* films), in which all interoceptive and exteroceptive signals are provided by a simulation, at the other end. Different technologies lie along this physical–virtual continuum, including augmented reality (AR), where digital objects are overlaid onto a video of the real world, and virtual reality (VR), which replaces sensory information about the real world with sensory information corresponding to a virtual environment (Milgram and Kishino 1994). Each technology on the "spectrum of virtuality" may contribute to psychological science; in this volume, we focus on VR and the many unique opportunities that it affords for the advancement of cognitive, affective, and behavioral neuroscience.

To this end, we have collected contributions from experts at the leading edge of VR research around the globe. In this introductory chapter, we describe some of the foundational features of VR and the contributions it can make to behavioral neuroscience research. We also outline some of the unique challenges (theoretical, methodological, and ethical) that arise when using VR in experimental paradigms. By doing so, we hope to help researchers capitalize on the opportunities provided by the technology while circumventing some of the pitfalls that can prevent VR research from delivering on its promise.

The first four chapters of this volume cover fundamental methodologies that combine VR technology with other tools in behavioral neuroscience. In this first chapter, we provide an overview of the technology and its potential applications, and the principles that guide its effective use. This chapter goes hand-in-hand with the following chapter "Launching your VR Neuroscience Laboratory" (Wu et al. 2023), which introduces the technological components of a VR neuroscience laboratory, and provides a collection of practical insights and strategies for different use cases. The next two chapters "Monitoring Brain Activity in VR: EEG and Neuroimaging" (Ocklenburg and Peterburs 2023) and Eye-tracking in Virtual Reality (Anderson et al. 2023) address some of the unique challenges involved in combining VR with these methods.

The following five chaptersdelve deeper into the many ways in which VR is used to uncover novel insights into mind, brain, and behavior. These chapters provide comprehensive overviews of specific research domains, including spatial navigation (Chapter "Virtual Reality for Spatial Navigation"; Jeung et al. 2022), vision science (Chapter "Virtual Reality for Vision Science"; Hibbard 2023), emotion (Chapter "VR for Studying the Neuroscience of Emotional Responses"; Andreatta et al. 2023), memory and cognition (Chapter "VR for Cognition and Memory"; Reggente 2023), awe and imagination (Chapter "VR for Awe & Imagination"; Chirico and Gaggioli 2023), and consciousness (Chapter "Using Extended Reality to Study the Experience of Presence"; Suzuki et al. 2023). Finally, the last four chapters describe new applications of VR to create practical interventions and improve lives. These chapters describe how VR can be used in classroom settings to improve engagement and learning (Chapter "Virtual Reality for Learning"; Checa and Bustillo 2023), distract people from pain using virtual embodiment techniques (Chapter "VR for Pain Relief"; Matamala-Gomez et al. 2023), aid in rehabilitation of motor control and cognitive abilities following a stroke or Parkinson's disease (Chapter "Virtual Reality for Motor and Cognitive Rehabilitation"; Darekar 2023), and provide new diagnostic tools and treatments for psychological disorders such as post-traumatic stress disorder, specific phobia, and depression (Chapter "Virtual Reality Interventions for Mental Health"; Kothgassner et al. 2023).

2 How Psychological Scientists and Virtual Realities Have Worked Together from the Start

Although VR, as a technological medium, has only recently been adopted into psychological science, the practice of immersing participants in some carefully controlled context is fundamental to lab-based psychological research. Historically, these controlled environments (a.k.a. the "psychologist's laboratory"; Danzinger 1994) provide conditions that differ from those in the "real world." By rigorously and thoughtfully controlling the laboratory context, researchers aim to minimize irrelevant sources of variability (i.e., noise) and systematic factors (i.e., confounds) that may influence the dependent measures. The outcome of this process is an experimental design that is *internally* valid, which licenses one to conclude that significant differences in the dependent variables may be confidently attributed to the effect of the independent variable(s). However, the cost of this internal validity has always been external validity – the ability to generalize findings to other contexts.

Much like laboratory contexts, the virtual experiences that are designed today are also approximations (or analogs) of reality, where contextual variables can be carefully considered and controlled. The difference between the psychological scientist and the VR developer is the trade-offs that they need to navigate: while the psychological scientist must balance internal and external validity, the VR developer must balance the sophistication and realism of a given scenario with the computational costs of creating it. However, the VR developer and the psychological scientist share the goal of creating a context that biases participants to produce behaviors that mirror those in the real world's situational homolog. In science, when a study achieves this goal, it is said to be *ecologically* valid; that is, the results of the study will generalize to the real-world context it was meant to represent.

In principle, psychological scientists strive to design studies that are internally *and* ecologically valid. Unfortunately, the conditions necessary to achieve ecological validity seem at times diametrically opposed to the conditions necessary to achieve internal validity. Ecological validity can be achieved when there is verisimilitude between the testing conditions and the conditions of the real world. The trouble is that, often enough, what makes testing conditions naturalistic is precisely the presence of many sources of undesirable variability (Kingstone et al. 2008). The world outside the lab rarely presents internally valid conditions for evaluating theoretically motivated hypotheses, just as laboratory conditions rarely resemble the "real world."

This ecological validity problem is a pernicious one (Holleman et al. 2020). It threatens to reduce the value of every laboratory finding in psychological science – no matter how robust, replicable, or widely cited that finding may be – to nothing more than a "laboratory curiosity" (Gibson 1970). Without ecological validity, findings cannot shed light on real-world phenomena. Lacking any clear solution to this problem, we can forgive scientists of the past for determining that their best course of action was to focus on what they *could* control and prioritize internal validity. The hard problem of ecological validity became a problem for another day. But that day may have arrived; VR could be the solution that finally allows psychological scientists to ask: "do these laboratory findings replicate in naturalistic contexts?"

3 How Immersive Devices Create an Illusion of Presence

We define a VR system as one that uses a head-mounted display (HMD) to fully replace visual information from the "real world" with digitally rendered visual information. This visual input is provided stereoscopically via display screens in front of the participant's eyes, creating a three-dimensional visual landscape. Depending on the system, the simulation may be rendered by a computer tethered to the HMD (via either a cable or wireless receiver), or by a computer inside the HMD itself. A VR system must perform many computations in order to successfully replace a user's sensory experience and generate a compelling simulated environment (Bouchard and Rizzo 2019). These computational demands are not trivial and different companies have leveraged multiple technological innovations to overcome them. For example, whereas Meta's line of Quest headsets overcome certain computational demands using computer vision algorithms, HTC's line of Vive headsets overcome those same demands using infrared light sensors and time-division multiplexing. Regardless of how impressive a simulation may be, the VR system itself is primarily responsible for accomplishing two tasks: (1) capturing human motion and (2) rendering an image from a virtual camera whose position and rotation in the virtual environment is animated by that incoming motion data. When these tasks can be completed at a high enough speed, a VR system is said to be "immersive"; achieving the conditions necessary to induce a stable illusion that the user is looking, not at a series of two-dimensional images, but rather, that they are looking *into* a three-dimensional environment within which they are physically situated. This illusion is referred to as "presence" (Cummings and Bailenson 2016).

Presence is most often defined as the psychological experience of "being there" (Slater and Wilbur 1997; Schubert et al. 2001). Mel Slater attributes "presence" to two distinct yet simultaneous illusions. The first illusion is that the user is in a place (and is therefore called the "place illusion"). The success of the place illusion is primarily a function of the technological properties of the VR system. Features like the visual resolution, frame rate, field of view, and the number of sensory modalities that a system can deliver determine the immersiveness of the virtual environment, resulting in an illusion that the user is embodied within some three-dimensional space. Until recently, only very expensive computers were equipped with graphical processing units capable of meeting the computational demands necessary to reliably induce a place illusion, and the systems that were available were orders of magnitude more expensive than today's VR systems.

Although the place illusion is an essential condition for generating the subjective experience of presence, a second "plausibility" illusion further enhances VR's potential for tapping into human behavior (Slater 2009). It is important to acknowledge that "plausibility" illusion is something of a misnomer, as it does not reflect how much the virtual world looks like a plausible representation of the real world. Instead, the plausibility illusion is experienced when a participant feels that the events in the virtual environment are *actually* happening but are outside of their control. These illusions can be induced in a variety of ways. Borrowing an example from Pan and Slater (2007), imagine that you enter VR and find yourself sitting at a table in a cafe while facing a virtual avatar. If the avatar shifts her posture and gazes around the space naturalistically, and yet she never directs her gaze toward you, then you may feel that you are *physically situated* in the cafe (a place illusion), but you remain orthogonal to the events that occur in that world. If the avatar then makes eye contact with you, you become part of the simulation and experience the plausibility illusion. The distinction between illusions of place and plausibility can also be illustrated using the story of Ebenezer Scrooge from Charles Dickens' *A Christmas Carol.* When the Ghost of Christmas Present allows Scrooge to attend the Cratchit family's party, Scrooge learns that although he can walk amongst the attendees, their behavior indicates that he is not a member of their reality, and that the events he is observing do not refer to him.

3.1 Why Presence Is So Important

Irrespective of the psychological phenomena under investigation, it is pivotal that research participants feel present in the virtual world. The rationale for using VR to study real-world behavior depends on participants feeling present in the virtual environment, just as they do in the physical world. Therefore, it should come as no surprise that presence features in every chapter throughout this volume (and why an entire chapter is devoted to attempts to understand it; Suzuki et al. 2023).

Given its relevance, it is surprising that we do not know much about how presence works, or the mechanisms that give rise to it. We know that a VR system is capable of inducing presence when it incorporates sufficient immersive technological properties (Cummings and Bailenson 2016), and we know that presence can "break" when participants' sensory expectations are violated (Kokkinara and Slater 2014). We also know that simulator sickness disrupts presence (Weech et al. 2019). However, we do not yet know how long it takes to re-induce presence following a "break," and we do not know whether individuals' familiarity using VR technology may facilitate or limit the extent to which people feel present in a virtual environment. We also do not know how other processes impact (and are impacted by) changes in presence.

One reason why presence has proven such an elusive topic of research is because it is quite difficult to measure (Slater et al. 2022). Typically, presence is measured using either self-report ratings while participants are in VR, or questionnaires administered after exiting VR. Commonly used questionnaires include Witmer and Singer's (1998) "Presence Questionnaire," a retrospective state measure that operationalizes presence using subscales measuring realism, possibility to act, and quality of the interface, and the "Immersive Tendencies Questionnaire," a trait measure of an individual's tendency to be immersed in virtual environments and narratives across different types of media. Many attempts have been made to identify reliable objective measures of presence (Van Baren and IJsselsteijn 2004). However, results are mixed. For example, while some earlier research (Meehan et al. 2002, 2003) reported relatively strong correlations between measures of physiological arousal (heart rate and electrodermal activity) and self-reported presence, more recent studies have not observed any relationship (Bailey et al. 2009; Maymon et al. 2023).

Despite these measurement limits, some progress is being made. For example, recent research has demonstrated that presence may be driven by subjective changes in people's emotional experience. In this study conducted in one of our labs (Maymon et al. 2023), we developed a VR simulation inspired by a commercial VR game called *Richie's Plank Experience*. Participants stepped inside an elevator that took them high above a city street. When the elevator door opened, participants saw a wooden plank extending precariously from the elevator and were asked to walk across it. At different points throughout the simulation, participants provided verbal ratings of the extent to which they were experiencing different emotions, as well as how present they felt in the virtual environment. We observed robust increases in measures of physiological arousal (heart rate and electrodermal activity), along with increases in self-reported fear ratings. Interestingly, presence also increased during height exposure, and was related to fear ratings, suggesting that presence may be influenced by particular emotional states.

In a pre-registered follow-up study, we adapted the simulation to include a control condition where the elevator rose but then returned to the ground floor where participants found a wooden plank lying on the pavement. By manipulating the scenario in this way, we were able to control for potential confounds in the design. Specifically, we realized that the environment included some additional immersive features during the height exposure which could explain why presence increased at height (e.g., being able to operate the elevator by pressing a button, feeling a real wooden plank beneath their feet). This control condition was highly useful, as we discovered that while presence *did* significantly increase in both conditions, the magnitude of that increase was much larger in the height condition and only in that condition did we replicate the finding that fear ratings and presence were related. Further, by simultaneously collecting physiological recordings throughout the experiment, we were able to determine which changes in the emotional experience (subjective, physiological, or both) were driving presence. We found that presence was predicted only by changes in self-reported fear and was not related to changes in physiological arousal.

Presence is a tricky construct to study, but it is a worthwhile endeavor. Understanding the mechanisms underpinning presence can enable VR developers and researchers to simulate experiences that correspond more directly to real-world experiences, stimulating more naturalistic and ecologically valid behaviors. Because presence may be causally linked to other cognitive and affective processes, researchers should think carefully not just about the ways in which a simulation is sufficiently immersive, but also about the way in which participants' experience of the content may lead to changes in, for instance, the emotional intensity of the environment, and how those changes may shift presence.

4 The Potential of Virtual Reality for Behavioral Neuroscience Research

Recent years have seen rapid development in the accessibility and practicality of virtual reality for cognitive, affective, and behavioral neuroscience research. State-of-the-art VR systems are affordable and can be powered by desktop computers; headsets are becoming lighter; and new models are equipped with multiple sensors that can track eye movements, facial expressions, and physiological signals. These systems can also interface with conventional neuroscience tools such as EEG or other recording systems. Software is maturing, making it possible for non-specialists to create virtual environments that are suited to their research needs. In the next section of the chapter, we describe several ways that VR systems are changing the research landscape.

4.1 VR Puts Our Participants in Highly Controlled Multisensory Environments

The ability to create alternate realities in which people feel present solves several problems that researchers face when trading off internal for ecological validity. Well-constructed environments can be highly ecologically valid (Parsons 2015), substituting for real environments that would be impossible, impractical, or unethical to produce in laboratories. Virtual environments exist in three dimensions, providing a major advance for vision scientists who have (until now) been largely constrained to the study of visual processing of two-dimensional images presented on a computer screen (see Hibbard 2023). Complex scenarios can be created in which the participant is an active agent, allowing researchers to move away from hypothetical situations in which people estimate how they would behave in a particular situation, and test how they actually do behave (Rosenberg et al. 2013; Schöne et al. 2023). Emotion researchers can create situations that induce genuine fear, or awe, or anger (see Andreatta et al. 2023; Chirico and Gaggioli 2023), and need not rely on pictures, words, or films that induce only pale facsimiles of the real thing. Presence can be enhanced by the addition of other sensory modalities including stereo sound, smells, or tactile feedback via vibrations (e.g., in gloves, or a hand controller) or with real objects (Wu et al. 2023). At the same time, these ecological environments can be highly controlled, allowing researchers to manipulate independent variables while holding all other aspects of the environment constant, and ensuring that all participants experience the same experimental context.

Conversely, VR also allows researchers to import ecologically valid paradigms into the virtual world, increasing their internal validity. For example, Hale and Hamilton (2016) used VR to test the effects of mimicry on prosocial behavior. Earlier studies suggested that mimicry increases prosocial behavior (van Baaren et al. 2004), and that people find mimickers to be more trustworthy (Maddux et al. 2008) and more likable (Kouzakova et al. 2010). These original studies made valiant efforts to achieve ecological validity by employing trained confederates to imitate participants' movements. However, it is unreasonable to expect even professionally trained confederates to precisely replicate the timing of their mimicry equally across all participants and to treat each participant the same way. Hale and Hamilton (2016) replaced the human mimicker with an animated virtual avatar, precisely controlling the timing of movements, while holding potentially confounding social factors like interpersonal warmth constant. The virtual avatar was directly animated by motion data recorded from the participant, which made it a more reliable mimicker than any human partner. Replicating previous findings, participants reported better rapport with avatars who mimicked their actions; however, mimicking avatars were not rated as being more trustworthy or more similar to the participant. These results indicated that previously reported effects may have been the product of confounding social variables introduced when mimicry was performed by real humans, and not caused by mimicry itself.

Another example is provided by the well-validated Trier Social Stress test (Linares et al. 2020). In the conventional version of the task, participants are asked to write a speech which is then evaluated by a panel of judges. In another subtest, participants must perform a series of mental calculations while being evaluated by a stern examiner. The task reliably induces anxiety and stress, but places a large burden on researchers who must enlist the help of a number of confederates, and train them to behave consistently across participants, and across labs. A VR version of the test provides a virtual panel of examiners. Although people on the virtual panel are obviously not "real," the VR version of the task induced stress as indicated by both subjective and endocrine responses (Zimmer et al. 2019). VR therefore reduces the costs of such research, and also makes it possible to fully control the experimental manipulation, making it replicable across participants and across labs.

4.2 *VR Facilitates Translational Neuroscience Research*

Much knowledge in behavioral neuroscience derives from animal research. This is because it is possible to directly manipulate animal brains (e.g., through lesions, the administration of pharmacological agents, through genetic manipulation, through manipulations of stress, hunger, social isolation, etc.). These animals are then tested in well-established behavioral paradigms, allowing researchers to draw causal links between brain and behavior. These brain-behavior relationships can then be extrapolated to humans, often tested with non-invasive technologies such as EEG or fMRI.

One of the barriers to effective translation is that the tasks that are commonly used with laboratory animals are behavioral, while the tasks that are used with humans are often cognitive. For example, in a delay-discounting paradigm, a rat might have to choose (by pressing a lever) between receiving a small reward now or a large reward in the future. Importantly, they receive the reward. In the human analog, participants are asked if they would prefer a small amount of money now, or a large amount of money in the future. But they don't actually receive the reward. Numerous studies have shown that humans do not behave as predictably as rats in these situations, in part because the hypothetical reward decision lacks ecological validity (Vanderveldt et al. 2016). But in VR, it is possible to create a situation for humans that more closely matches the characteristics of animal paradigms, producing reliable discounting functions, and facilitating the translation of research findings from animals to humans (Bruder et al. 2021).

There are many other examples of animal research paradigms that can be implemented for humans in VR. One is the Morris Water Maze (Morris 1984) that is widely used in animal research on the neural substrates of learning and memory. The Virtual Water Maze task is really a collection of paradigms that capture the basic task – navigating to a "safe" place using either landmarks or egocentric trajectory (for a review, see Thornberry et al. 2021). In humans, the task has been combined with single cell recordings in epilepsy patients (e.g., Kunz et al. 2019) and with mobile EEG (e.g., Do et al. 2021), replicating animal studies that link spatial navigation to hippocampal theta activity. Similarly, fear conditioning paradigms, often used in animal research to study the neural mechanisms of fear and anxiety, can be safely and ethically simulated using VR (see Andreatta et al. 2023). VR is well-suited for replicating animal experiments on contextual conditioning, as different contexts (e.g., rooms in a virtual house) can be associated with the conditioned stimulus. A wide range of contexts can be tested by simply altering the simulation, making VR ideal for isolating factors that are associated with conditioning and extinction. These contexts can also be made to resemble real-world contexts (such as a bar, a hospital, etc.), increasing the ecological validity of research findings.

4.3 VR Affords Ecologically Valid Behaviors

The illusion of presence means that people do not just experience the virtual world as real, they also behave within it as they might in physical reality. Traditionally, cognitive and behavioral neuroscientists, particularly those who work with humans, have adopted paradigms that require their participants to remain still. Vision researchers use chin rests, attention researchers constrain eye movements, EEG researchers restrict movement to obtain clean recordings of neural activity. Behavioral responses in these experiments are limited to small movements that can be measured in such circumstances, such as pressing a button or saying a word. But brains (and cognition and emotions) exist in the service of action, and therefore just as it is important to create virtual environments that resemble the real world (i.e., ecologically valid contexts), future research must also afford participants the ability to coordinate action in ways that are similarly unrestricted (i.e., ecologically valid behaviors). Participants in virtual environments are free to move, and the VR system automatically tracks their movements. This freedom to move greatly increases the range of actions that participants can perform, and the dependent variables researchers can collect.

One example of a more complex behavioral response is reaching. In a reach-tracking experiment, participants don't use buttons to indicate a yes/no response. Instead, they reach toward a location in space that corresponds to the response. Because a reach is a continuous action (as opposed to a discrete button press), it is possible to use the reach trajectory to track online conflict resolution and decision-making processes. Reach-tracking can be easily implemented in VR (e.g., Morrison et al. 2023) because the VR system already tracks the movement of the controller. Reach-tracking can then be implemented in a range of virtual environments to determine how contextual factors affect the cognitive processes of interest.

4.4 VR Yields Rich Multivariate Data

VR also affords researchers the ability to collect rich datasets across several systems simultaneously. The VR system itself tracks head position, and further motion tracking can be obtained using hand controllers, foot trackers, or complete motion capture (a.k.a "mocap") suits. Its rich motion tracking capabilities make VR an ideal tool for studying navigation and other aspects of spatial cognition (see Jeung et al. 2022). It is even possible to use treadmills or other devices to extend the range through which participants can move (Hejtmanek et al. 2020). The nature of movements in VR can also be informative. In our plank-walking studies, participants who walked the plank high above the virtual ground took smaller steps and walked more slowly than those in a control condition who walked a plank that appeared to be on the street. These changes in gait were so large as to be apparent to naive observers who watched videos displaying point-light avatars of the participants (Crawford et al. 2023). Other behaviors that might be tracked in VR include head and eye movements (see Anderson et al. 2023), manipulation of virtual (or real) objects, or a wide range of choice behaviors.

In addition to features of body movement, many VR systems on the market today offer a range of tools for detecting and recording diverse metrics related to attention, arousal, emotional state, and more. It is possible, for instance, to capture the participant's field of view on video, providing a real-time record of their experience that can be synchronized to other measures. Headsets can be fitted with eye-trackers that capture eye movements and pupil responses. Wearable sensors can capture a wide range of peripheral responses including cardiac activity, respiration, electrodermal activity, skin temperature, muscle movements, and facial expressions (for more detail, see Wu et al. 2023). Mobile EEG systems with online artifact rejection can be used to capture neural activity during the simulation. Triggers sent from the VR system can be used to synchronize neural and physiological sensors to events in the simulation, allowing for signal averaging and other analytical approaches.

These multiple data streams can be used both to test the effectiveness of a VR scenario and to address fundamental neuroscientific questions about their relationships. For example, a researcher may be interested in determining whether high-arousal fear affects attentional scope. The researcher first needs to reliably manipulate fear. However, fear responses manifest across multiple dimensions (i.e., subjective, physiological, and behavioral channels). The researcher could check that the manipulation worked by (1) asking for subjective ratings at different time points, (2) continuously measuring peripheral physiological signals using wearable technologies, (3) continuously measure root motion in the environment by saving the same positional and rotational coordinates that are used by the game engine to update the virtual camera in real-time, and (4) capture frontal asymmetries that are associated with fear using mobile EEG (El Basbasse et al. 2023). These multiple data streams not only allow the researcher to test that the fear manipulation was successful, they also make it possible to address long-standing theoretical questions about the causal relationships amongst subjective, physiological, and behavioral components of emotion (Cannon 1927; James 1884; Panksepp 1982; Schacter and Singer 1962).

Of course, the sheer amount of data that can be obtained increases the complexity of the research. Continuous recordings across multiple channels create very large datafiles that must be stored and manipulated. Multivariate statistical approaches are necessary, and machine learning algorithms may be more appropriate than conventional statistical analyses in some situations (e.g., Winkler-Schwartz et al. 2019). The synchronization of data streams is also not trivial, although basic principles are not different from those in other neuroscience applications such as EEG or fMRI.

4.5 VR Promotes Direct and Conceptual Replication

Another benefit of VR for addressing current challenges facing psychological science is that simulations can be easily shared with other labs. Doing so has clear advantages for replication studies, and studies that aim to extend research paradigms in ways that closely preserve the testing conditions of an earlier study. Sharing VR simulations between labs can also improve the efficiency of those investigations. Traditionally, researchers are expected to use the "Methods" section from a given empirical study to ascertain the necessary information to replicate the paradigm in their own lab. This practice is often slow and does not ensure that the new stimuli match the stimuli used in the original study. Consequently, if an effect does not replicate, this can spawn debates, not about the robustness of the effect itself, but about whether the new stimuli adequately matched the original.

VR provides a solution to this problem, as the testing conditions of one experiment can be directly matched in replication studies, so long as the simulation is delivered using similar hardware and software. Indeed, sharing simulations can help to meet several key goals outlined in the Association for Psychological Science's strategic plan (Bauer 2023). First, making VR simulations available to labs anywhere in the world encourages researchers to examine the robustness of a given effect across diverse study populations. Second, the ever-increasing affordability and availability of VR technology means that teams with limited funding for research will be able to afford the technology necessary to present even the most impressive VR simulations, thereby promoting a more globally representative psychological science. Finally, Bauer (2023) highlights the need for an increased focus on *authenticity* in psychological science, encouraging a shift toward a science that concerns itself with the extent to which findings can explain behavior in the real world, even if that shift incurs a cost to a study's internal validity. VR not only provides a way to achieve these goals, but it can also do so without sacrificing internal validity. Therefore, we think that VR will play a major role in how psychological science overcomes the replication crisis and will become an increasingly common methodology in cognitive and behavioral neuroscience laboratories.

4.6 The Impact of VR on Research Ethics: Challenges and Opportunities

VR engenders new concerns about the ethical treatment of research participants. A number of adverse experiences can occur when a person enters VR, including eye strain, headaches, dizziness, and nausea (i.e., simulator sickness; Weech et al. 2019). Additionally, VR poses a risk of bodily injury. Because participants cannot see objects in the real world when they are in VR, any objects left in the VR "play area" present a hazard. Fall risks are particularly dangerous for a participant in VR because they also have no way of knowing where the walls are. Therefore, they have no way of knowing whether they may be stumbling toward a wall or the edge of a table as they try to regain their footing.

Currently, no "gold standard" exists for mitigating these risks. As with any psychological research, researchers must ensure that the experiment does not cause physical or psychological harm to participants. In our labs, we have developed protocols for minimizing these risks. For example, in our research into fear using height exposure (Maymon et al. 2023), we ensure that experimenters stand beside the participant while they walk across the wooden plank, and that experimenters are hypervigilant to signs that they may lose their balance and be prepared to catch the participant if they should lose their balance. Experimenters are also trained to respond to instances when a participant reports feeling nauseous or dizzy, by erring on the side of caution and ending the experiment. In our experience, simulator sickness does not subside while the participant remains in VR. Rather, participants who try to "muscle through" those symptoms and remain in VR will usually experience worsening symptoms and will take longer to recover after exiting VR.

Interestingly, VR may also provide a means of overcoming particular ethical restrictions. The idea here is that some in vivo research paradigms that have been deemed unethical, may be ethically permissible when participants experience the paradigm in an environment that they know is not real. Perhaps the most famous example of this comes from Slater et al. (2006) who developed a VR analog of Milgram's obedience study. The original procedure involved the use of deception, whereby a participant was instructed by an experimenter to deliver increasingly dangerous electric shocks to another person (the learner). Participants were led to believe that the learner was just another participant in the experiment, and that if the outcome of a coin toss were different, the participant would have been the one receiving the shocks. The world was astonished when Milgram (1963, 1974) reported that 65% of participants proceeded to deliver shocks that were clearly marked as lethal, despite the learner screaming in protest. Of course, no shocks were being delivered, and participants were (eventually) made aware that the learner was an actor pretending to be shocked. Nevertheless, participants later reported experiencing considerable psychological distress as a result of their actions during the experiment. Since then, direct empirical replications have been impossible.

Slater et al. (2006) developed a VR scenario whereby the confederate receiving the shocks was a computer-generated avatar. Although the avatar was programmed to behave as if they were experiencing pain following each shock, participants were regularly reminded during the experiment that the avatar was not real and could not experience pain. Therefore, this replication study did not involve the use of deception. Nevertheless, participants who saw and heard the avatar suffering, showed patterns of subjective, physiological, and behavioral responses consistent with experiencing stress and caring for the well-being of the virtual avatar. Interestingly, despite knowing full well that the avatar was not experiencing pain, 27% of participants asked to stop the experiment before reaching the end of the study.

With this VR paradigm, it is now possible to evaluate the proposed mechanisms underpinning obedience to authority figures, in ways that are perhaps less ethically dubious. In a follow-up study from the same lab, Gonzalez-Franco et al. (2018) used the same paradigm to critically examine Milgram's "agentic state" interpretation for his original findings. According to Milgram, people were willing to cause harm to the confederate because people are inclined to focus on efficiently doing what is commanded of them by an authority figure, even at the expense of harmful consequences. If it is the case that participants are submitting to authority without considering the harm to the confederate, then participants should be equally likely to obey when the avatar is not visible. In this study, Gonzalez-Franco et al. (2018) demonstrated that when an avatar was present, participants spontaneously adapted their tone to emphasize the correct response, in an apparent attempt to help the avatar avoid receiving a shock. Furthermore, they found that participants who identified more with scientific pursuits displayed more concern for the avatar, were more likely to provide help and (perhaps due to having attempted to provide help) exhibited less stress.

This demonstration has inspired others to imagine ways to study behaviors in VR that would be unethical to study otherwise. However, VR should not be viewed as negating ethical scrutiny, and there is ongoing debate about the kinds of criteria that ethics committees should apply when reviewing VR research. Ramirez (2019) proposes an *equivalence* principle, stating that "If it would be wrong to subject a person to an experience, then it would be wrong to subject a person to a virtually-real analog of that experience. As a simulation's likelihood of inducing virtually-real experiences in its subject increases, so too should the justification for the experimental protocol." Returning to the plank-walk, it would certainly be unethical to put participants in physical danger by walking a real plank suspended at extreme heights (or to require participants to escape from a burning building; Bernardini et al. 2023). VR analogs remove the physical risk, but not the fear. Ethics committees must therefore evaluate the risks arising from experiences that manipulate emotional states or beliefs using VR, independent of the physical risks that would be present in the real world.

5 Experimental Research Using VR: Recommendations for Best Practices

VR brings with it a variety of novel and unapparent challenges. In this section, we provide five recommendations from our own experiences adopting VR technology into our neuroscience laboratories.

5.1 *Recommendation 1: Anticipate That People May Differ in Their Response to Being in VR by Adding a Baseline Measure of the Dependent Variable*

VR studies are often designed such that participants, drawn from a common population, are randomly assigned to one of two simulations which the researcher predicts will be associated with differential patterns of responding. The researcher may improve this experimental design by adding a within-subjects variable, allowing them to demonstrate that these randomly assigned groups are indeed similar on relevant measures *prior* to entering VR. If researchers observe an interaction effect in this 2 × 2 mixed model design, they can more confidently attribute that difference to the effect of their manipulation. However, it is often worth extending the levels of the within-subjects variable further to include a measurement taken after participants have entered VR but before the events differ as a function of the independent variable.

This practice can be particularly helpful when VR is used to demonstrate effects of emotional change. In such studies, it is recommended that participants enter VR and are first immersed in an "emotionally neutral" environment where they can acclimate to this new simulation. Studies interested in manipulating emotional states should collect baseline measures both before participants enter VR (to account for individual differences at the trait level), as well as during the neutral environment (to account for individual differences in participants' response to being immersed in VR) before proceeding to the environment where groups differ as a function of condition. This neutral environment can also be used to help participants learn the rules of the new reality, like whether and how they can move through the environment, which objects they can interact with, and how. When a study aims to measure naturalistic behavioral responses (like which of two buttons a participant reaches toward or how quickly participants can navigate a maze) providing ample opportunity to interact in the virtual world prior to the task can ensure that participants are similarly familiar with being embodied in the virtual environment.

5.2 *Recommendation 2: Avoid Tasks That Require Participants to Remain in VR for Long Periods of Time*

When designing a VR study, it is important to take into consideration how long participants will be expected to remain inside the headset. This is especially pertinent for cognitive researchers who may rely on tasks that require participants to complete a large number of trials in an experimental session. While it may be reasonable to ask participants to remain in front of a computer for an hour or more, this regimen is not reasonable for those in VR. Being in VR for long periods of time (i.e., more than 20 min) can have a number of deleterious effects on participants' impression of the virtual environment, their physical comfort, and their task performance (Souchet et al. 2022). Over longer periods of time, participants' neck muscles may fatigue from supporting the weight of the headset. VR headsets also tend to generate heat and this may cause participants to become overheated and dizzy. Additionally, if that heat causes participants to sweat in the headset, this sweat can become condensation across the lenses, compromising visual clarity. Some of these difficulties may be overcome by future technological innovations, but presently, the best course of action for ensuring the best experience and data quality is to design studies that keep participants in VR only as long as is necessary.

5.3 *Recommendation 3: Conduct Rigorous Pilot Studies of New VR Scenarios to Help Identify Potential Nuisance Variables*

By design, VR studies introduce more contextual nuance relative to traditional laboratory studies. Although VR allows every aspect of a virtual environment to be controlled, knowing how to set each potentially relevant parameter is no trivial task. Furthermore, in our experience, it is quite difficult to predict which features of a virtual environment may have an undesirable impact on participants. We have found that the best practice for identifying potential nuisance variables is by conducting pilot studies. Additionally, it is often helpful to plan these pilot studies in such a way that participants can provide open-ended feedback about their impressions of the environment and especially about what in the environment may have attracted their attention.

Pilot studies can test several assumptions about participants' experience in a virtual environment. For example, imagine that you create a VR scenario in which participants are expected to press a button on a wall using their hand controller. It may seem obvious that participants would realize that they could interact with the button; however, participants are often surprised to learn that they can operate a button the same way they would outside of VR, by reaching their hand toward it until the button has been depressed. It is also common to encounter the opposite situation, where a participant incorrectly assumes that objects in VR can be interacted with as if they were physically real. For example, in our early fear studies (Maymon et al. 2023), we were surprised to observe that some participants reached out to grab the edges of the doors of the elevator, only to remember that there was no wall there to grasp. We responded to this observation by widening the elevator so that the edge of the door was further away. Finally, pilot testing can help identify where there may be disruptions in the performance of the VR system during the experiment. VR technology can be disrupted by even momentary errors in motion tracking or if some event causes an increase in the real-time computational demands. These disruptions typically result in a visual glitch (where the headset may default to presenting a gray screen or may render images from an incorrect point-of-view) or a lag in the framerate, which can break presence, interrupt data collection, and cause simulator sickness. Therefore, it is important that researchers are also recording when and where technological glitches may be occurring so that they may be addressed before collecting data for the main experiment.

5.4 *Recommendation 4: Don't Try to Go It Alone*

The creation of even seemingly straightforward virtual environments capable of inducing presence can be a daunting and unfamiliar technological challenge for cognitive and behavioral neuroscientists. It is one thing to expect that researchers can pick up some skill in MATLAB to program tasks with relatively few parameters; it is an altogether different thing to expect researchers to simulate context in a way that compels participants to behave naturalistically. Moreover, in multi-modal paradigms, detecting heterogeneous streams of data and synchronizing them with virtual world events typically requires at least some degree of iterative design and testing. When it comes to developing the virtual environment, it is best to reach out to artists and game developers working with VR who can translate your idea into a quality environment. Consulting with developers and artists can also reveal alternative ways to create a scene that may be more efficient, higher quality, and more affordable, than how you may envision creating the environment.

5.5 *Recommendation 5: Recognize When Using VR Is Not Appropriate*

Scientists should think critically about whether an idea for a VR study makes appropriate use of VR. It is certainly not the case that all behavioral research would be improved simply by embedding a study in VR. It helps to ask what specific value VR adds to the interpretation of your data. This is also a helpful exercise for ensuring that the way you are using VR makes the best use of the technology. One of the biggest challenges when designing psychological experiments using VR is ensuring that the virtual environment is actually capable of representing the intended real-world situation. While VR technology has made significant advancements in recent years, the technology struggles to fully replicate some sensory experiences, such as touch and smell, which may contribute meaningfully to the experience that a study aims to simulate (Vasser and Aru 2020). Even visual experiences in VR may not sufficiently replicate visual experiences in the real world. The field of view in virtual reality is typically constrained (Ragan et al. 2015), and the frame rate may not be sufficient to simulate real motion. Finally, it is important to consider whether your specific research question may contraindicate the use of VR. For example, VR may not be well-suited for investigating special populations such as those suffering from psychosis, depersonalization/derealization disorder, or specific phobia. In such cases, it is important to assume that the VR experience may be appraised as real, which may have lasting impacts on the individual. For more information about contraindications of VR, see Kothgassner et al. 2023.

6 Conclusions

VR is making waves throughout psychological science, and we have only just begun to appreciate its unique potential to improve scientific pursuits across nearly every domain. When utilized appropriately, VR could help researchers overcome some of the most pernicious challenges facing the field. Virtual environments can be both precisely controlled and naturalistic, providing a solution to the otherwise intractable compromise between internal and ecological validity. VR can also improve the accuracy and efficiency of replication studies and unlocks new methods for translating animal research paradigms for human participants. By leveraging positional tracking data collected by the VR system alongside wireless wearable recording devices, VR studies can permit researchers to pre-register stronger, more specific predictions across a multitude of dependent measures. In this chapter, we have endeavored to provide a balanced account that tempers the technology's many promises with some of VR's new and in some cases unavoidable challenges. The many promises of VR are achieved when researchers carefully consider whether VR is appropriate for their use case, consult with specialists to develop quality simulations, and conduct rigorous pilot studies.

References

Anderson NC, Bischof WF, Kingstone A (2023) Eye tracking in virtual reality. In: Maymon C, Wu Y, Grimshaw G (eds) Virtual reality in behavioural neuroscience: new insights and methods. Springer

Andreatta M, Winkler MH, Collins P, Gromer D, Gall D, Pauli P, Gamer M (2023) VR for studying the neuroscience of emotional responses. In: Maymon C, Wu Y, Grimshaw G (eds) Virtual reality in behavioural neuroscience: new insights and methods. Springer

Bailey R, Wise K, Bolls P (2009) How avatar customizability affects children's arousal and subjective presence during junk food–sponsored online video games. Cyberpsychol Behav 12(3):277–283. https://doi.org/10.1089/cpb.2008.0292

Bauer PJ (2023) Responding to the Association for Psychological Science strategic plan, 2022–2027. Psychol Sci 34(1):3–7. https://doi.org/10.1177/09567976221133816

Bernardini G, Lovreglio R, Quagliarini E, D'Orazio M (2023) Can active and passive wayfinding systems support fire evacuation in buildings? Insights from a virtual reality-based experiment. J Build Eng 74:106778

Bouchard S, Rizzo AS (2019) Applications of virtual reality in clinical psychology and clinical cognitive neuroscience – an introduction. In: Rizzo A, Bouchard S (eds) Virtual reality for psychological and neurocognitive interventions. Springer, pp 1–13. https://link.springer.com/10.1007/978-1-4939-9482-3

Bruder LR, Scharer L, Peters J (2021) Reliability assessment of temporal discounting measures in virtual reality environments. Sci Rep 11(1):7015

Cannon WB (1927) The James-Lange theory of emotions: a critical examination and an alternative theory. Am J Psychol 39(1/4):106–124

Checa D, Bustillo A (2023) Virtual reality for learning. In: Maymon C, Wu Y, Grimshaw G (eds) Virtual reality in behavioural neuroscience: new insights and methods. Springer

Chirico A, Gaggioli A (2023) Virtual reality for awe and imagination. In: Maymon C, Wu Y, Grimshaw G (eds) Virtual reality in behavioural neuroscience: new insights and methods. Springer

Crawford MT, Maymon CN, Miles N, Blackburne K, Tooley MD, Grimshaw GM (2023) Emotion in motion: perceiving fear in the behaviour of individuals from minimal motion capture displays. https://doi.org/10.31219/osf.io/bdnyw

Cummings JJ, Bailenson JN (2016) How immersive is enough? A meta-analysis of the effect of immersive technology on user presence. Media Psychol 19(2):272–309

Danzinger K (1994) Constructing the subject: historical origins of psychological research. Cambridge University Press, Cambridge

Darekar A (2023) Virtual reality for motor and cognitive rehabilitation. In: Maymon C, Wu Y, Grimshaw G (eds) Virtual reality in behavioural neuroscience: new insights and methods. Springer

Do TTN, Lin CT, Gramann K (2021) Human brain dynamics in active spatial navigation. Sci Rep 11(1):13036

El Basbasse Y, Packheiser J, Peterburs J, Maymon C, Güntürkün O, Grimshaw G, Ocklenburg S (2023) Walk the plank! Using mobile electroencephalography to investigate emotional lateralization of immersive fear in virtual reality. R Soc Open Sci 10(5):221239

Gibson JJ (1970) On the relation between hallucination and perception. Leonardo 3:425–427

Gonzalez-Franco M, Slater M, Birney ME, Swapp D, Haslam SA, Reicher SD (2018) Participant concerns for the learner in a virtual reality replication of the Milgram obedience study. PLoS One 13(12):e0209704

Hale J, Hamilton A (2016) Testing the relationship between mimicry, trust and rapport in virtual reality conversations. Sci Rep 6:35295. https://doi.org/10.1038/srep35295

Hejtmanek L, Starrett M, Ferrer E, Ekstrom AD (2020) How much of what we learn in virtual reality transfers to real-world navigation? Multisens Res 33(4–5):479–503

Hibbard PB (2023) Virtual reality for vision science. In: Maymon C, Wu Y, Grimshaw G (eds) Virtual reality in behavioural neuroscience: new insights and methods. Springer

Holleman GA, Hooge ITC, Kemner C, Hessels RS (2020) The 'real-world approach' and its problems: a critique of the term ecological validity. Front Psychol 11:721. https://doi.org/10.3389/fpsyg.2020.00721

James W (1884) What is an emotion? Mind 9(34):188–205

Jeung S, Hilton C, Berg T, Gehrke L, Gramann K (2022) Virtual reality for spatial navigation. In: Maymon C, Wu Y, Grimshaw G (eds) Virtual reality in behavioural neuroscience: new insights and methods. Springer

Kingstone A, Smilek D, Eastwood JD (2008) Cognitive ETHOLOGY: a new approach for studying human cognition. Br J Psychol 99:317–340. https://doi.org/10.1348/000712607x251243
Kokkinara E, Slater M (2014) Measuring the effects through time of the influence of visuomotor and visuotactile synchronous stimulation on a virtual body ownership illusion. Perception 43(1): 43–58
Kothgassner OD, Reichmann A, Bock MM (2023) Virtual reality interventions for mental health
Kouzakova M, van Baaren R, van Knippenberg A (2010) Lack of behavioral imitation in human interactions enhances salivary cortisol levels. Horm Behav 57(4–5):421–426
Kunz L, Wang L, Lachner-Piza D, Zhang H, Brandt A, Dümpelmann M et al (2019) Hippocampal theta phases organize the reactivation of large-scale electrophysiological representations during goal-directed navigation. Sci Adv 5(7):eaav8192
Linares NN, Charron V, Ouimet AJ, Labelle PR, Plamondon H (2020) A systematic review of the trier social stress test methodology: issues in promoting study comparison and replicable research. Neurobiol Stress 13:100235
Maddux WW, Mullen E, Galinsky AD (2008) Chameleons bake bigger pies and take bigger pieces: strategic behavioral mimicry facilitates negotiation outcomes. J Exp Soc Psychol 44(2):461–468
Matamala-Gomez M, Donegan T, Świdrak J (2023) VR for pain relief. In: Maymon C, Wu Y, Grimshaw G (eds) Virtual reality in behavioural neuroscience: new insights and methods. Springer
Maymon CN, Crawford MT, Blackburne K, Botes A, Carnegie K, Mehr SA et al (2023) The presence of fear: how subjective fear, not physiological arousal, shapes the experience of presence. https://doi.org/10.31219/osf.io/2dx7n
Meehan M, Insko B, Whitton M, Brooks FP Jr (2002) Physiological measures of presence in stressful VEs. ACM Trans Graph (Tog) 21(3):645–652. https://doi.org/10.1145/566654.566630
Meehan M, Razzaque S, Whitton MC, Brooks FP Jr (2003) Effect of latency on presence in stressful VEs. In: Null. IEEE, p 141
Milgram S (1963) Behavioral study of obedience. J Abnorm Soc Psychol 67:371–378
Milgram S (1974) Obedience to authority. McGraw-Hill
Milgram P, Kishino F (1994) A taxonomy of mixed reality visual displays. IEICE Trans Inform Syst 77:1321–1329
Morris R (1984) Developments of a water-maze procedure for studying spatial learning in the rat. J Neurosci Methods 11(1):47–60
Morrison S, Erb CD, Egbert MD, Gurr M, Nicholson-Brown A, Lee D (2023) Exploring the dynamics of cognitive control using virtual reality. In: Proceedings of the annual meeting of the cognitive science society
Ocklenburg S, Peterburs J (2023) Monitoring brain activity in VR: EEG and neuroimaging. In: Maymon C, Wu Y, Grimshaw G (eds) Virtual reality in behavioural neuroscience: new insights and methods. Springer
Pan X, Slater M (2007) A preliminary study of shy males interacting with a virtual female. In: Presence: the 10th annual international workshop on presence
Panksepp J (1982) Toward a general psychobiological theory of emotions. Behav Brain Sci 5(3): 407–422
Parsons TD (2015) Virtual reality for enhanced ecological validity and experimental control in the clinical, affective and social neurosciences. Front Hum Neurosci 9:660. https://doi.org/10.3389/fnhum.2015.00660
Ragan ED, Bowman DA, Kopper R, Stinson C, Scerbo S, McMahan RP (2015) Effects of field of view and visual complexity on virtual reality training effectiveness for a visual scanning task. IEEE Trans Vis Comput Graph 21(7):794–807
Ramirez EJ (2019) Ecological and ethical issues in virtual reality research: a call for increased scrutiny. Philos Psychol 32(2):211–233
Reggente N (2023) VR for cognition and memory. In: Maymon C, Wu Y, Grimshaw G (eds) Virtual reality in behavioural neuroscience: new insights and methods. Springer

Rosenberg RS, Baughman SL, Bailenson JN (2013) Virtual superheroes: using superpowers in virtual reality to encourage prosocial behavior. PLoS One 8(1):e55003

Schacter S, Singer J (1962) Cognitive, social, and physiological determinants of emotional state. Psychol Rev 69(5):379–399

Schöne B, Kisker J, Lange L, Gruber T, Sylvester S, Osinsky R (2023) The reality of virtual reality. Front Psychol 14:1093014

Schubert T, Friedmann F, Regenbrecht H (2001) The experience of presence: factor analytic insights. Presence 10:266–281. https://doi.org/10.1162/105474601300343603

Slater M (2009) Place illusion and plausibility can lead to realistic behaviour in immersive virtual environments. Philos Trans Royal Soc B Biol Sci 364(1535):3549–3557

Slater M, Wilbur S (1997) A framework for immersive virtual environments (FIVE): speculations on the role of presence in virtual environments. Presence 6:603–616

Slater M, Antley A, Davison A, Swapp D, Guger C, Barker C et al (2006) A virtual reprise of the Stanley Milgram obedience experiments. PLoS One 1(1):e39

Slater M, Banakou D, Beacco A, Gallego J, Macia-Varela F, Oliva R (2022) A separate reality: an update on place illusion and plausibility in virtual reality. Front Virt Real 3:914392

Souchet AD, Philippe S, Lourdeaux D, Leroy L (2022) Measuring visual fatigue and cognitive load via eye tracking while learning with virtual reality head-mounted displays: a review. Int J Human–Comput Interact 38(9):801–824

Suzuki K, Mariola A, Schwartzman DJ, Seth AK (2023) Using extended reality to study the experience of presence. In: Maymon C, Wu Y, Grimshaw G (eds) Virtual reality in behavioural neuroscience: new insights and methods. Springer

Thornberry C, Cimadevilla JM, Commins S (2021) Virtual Morris water maze: opportunities and challenges. Rev Neurosci 32(8):887–903

Van Baren J, IJsselsteijn W (2004) Measuring presence: a guide to current measurement approaches (OmniPres project IST 2001-39237). Retrieved from Eindhoven University of Technology, IST FET OMNIPRES project: http://www8.informatik.umu.se/~jwworth/PresenceMeasurement.pdf

Van Baaren RB, Holland RW, Kawakami K, Van Knippenberg A (2004) Mimicry and prosocial behavior. Psychol Sci 15(1):71–74

Vanderveldt A, Oliveira L, Green L (2016) Delay discounting: pigeon, rat, human – does it matter? J Exp Psychol: Animal Learn Cogn 42(2):141

Vasser M, Aru J (2020) Guidelines for immersive virtual reality in psychological research. Curr Opin Psychol 36:71–76

Weech S, Kenny S, Barnett-Cowan M (2019) Presence and cybersickness in virtual reality are negatively related: a review. Front Psychol 10:158

Winkler-Schwartz A, Bissonnette V, Mirchi N, Ponnudurai N, Yilmaz R, Ledwos N et al (2019) Artificial intelligence in medical education: best practices using machine learning to assess surgical expertise in virtual reality simulation. J Surg Educ 76(6):1681–1690

Witmer BG, Singer MJ (1998) Measuring presence in virtual environments: a presence questionnaire. Presence Teleoper Virtual Environ 7:225–240. https://doi.org/10.1162/105474698565686

Wu YC, Maymon C, Paden J, Liu W (2023) Launching your VR neuroscience laboratory. In: Maymon C, Wu Y, Grimshaw G (eds) Virtual reality in behavioural neuroscience: new insights and methods. Springer

Zimmer P, Buttlar B, Halbeisen G, Walther E, Domes G (2019) Virtually stressed? A refined virtual reality adaptation of the trier social stress test (TSST) induces robust endocrine responses. Psychoneuroendocrinology 101:186–192

Launching Your VR Neuroscience Laboratory

Ying Choon Wu, Christopher Maymon, Jonathon Paden, and Weichen Liu

Contents

Abstract The proliferation and refinement of affordable virtual reality (VR) technologies and wearable sensors have opened new frontiers in cognitive and behavioral neuroscience. This chapter offers a broad overview of VR for anyone interested in leveraging it as a research tool. In the first section, it examines the fundamental functionalities of VR and outlines important considerations that inform the development of immersive content that stimulates the senses. In the second section, the focus of the discussion shifts to the implementation of VR in the context of the neuroscience lab. Practical advice is offered on adapting commercial, off-the-shelf devices to a researcher's specific purposes. Further, methods are explored for recording, synchronizing, and fusing heterogeneous forms of data obtained through the VR system or add-on sensors, as well as for labeling events and capturing game play. The reader should come away with an understanding of fundamental

Y. C. Wu (✉), J. Paden, and W. Liu
University of California San Diego, San Diego, CA, USA

C. Maymon
Victoria University of Wellington, Wellington, New Zealand

Curr Topics Behav Neurosci (2023) 65: 25–46
https://doi.org/10.1007/7854_2023_420
Published Online: 13 June 2023

considerations that need to be addressed in order to launch a successful VR neuroscience research program.

Keywords EEG · Eye-tracking · Head-mounted display · Multi-modal biosensing · Research platform · Virtual reality

1 Introduction

Creating a research platform that leverages the capabilities of virtual reality (VR) demands diverse expertise to a greater degree than most typical scientific endeavors centered on the human mind, brain, and behavior. While at first glance, the technical challenges of such an undertaking may seem a barrier to entry, they also create opportunities for cross-disciplinary collaboration that can drive innovation. Indeed, VR platforms can support seamless merging of experimental research with therapy, art, gaming, educational tools, and other applications. A VR-based installation, for example, can be designed to serve as both an art piece and a material for a study on esthetic experience. For this reason, one should give special advance consideration to the audience that is the target of one's project. The broad appeal and functionality of VR can readily transcend the boundaries of the specific scientific community that may be the most obvious target of one's research question; and the impact of one's study can be substantially enhanced by availing oneself of a broader range of stakeholders and strategic partnerships.

This chapter reviews key features of diverse VR systems that impact comfort and useability. It also examines several factors that are important for immersion and the cultivation of presence, which is the feeling of being physically located in a virtual space (Barbot and Kaufman 2020). While commercial game developers may seek to evoke a strong sense of presence that increases the immersive draw of a given game, researchers should consider what aspects of presence are most important for their objectives. In a study of acrophobia, for instance, it may be desirable to manipulate many sensory elements of the virtual experience – including the sounds, sights, and sensations associated with heights – in order to successfully elicit a fear of heights. On the other hand, to study aspects of spatial navigation, a virtual environment that simply replicates a sense of depth and optic flow may be sufficient. This chapter offers an overview of immersive technologies for VR – how they work and how they can be implemented in a research setting. Further, diverse approaches are examined to creating immersive content ranging from replications of real-world spaces to fantastic fictional worlds.

An additional set of vital considerations that all VR researchers must address is the types of data that they intend to record and how they plan to analyze them. Many VR systems can be either purchased off the shelf with integrated capabilities or are compatible with after-market sensors that detect and record many channels of information about a user's behavior in the game world, including head position and orientation, head movements, hand movements, gaze points, and more. Moreover, full body motion capture of the limbs and torso can be achieved with auxiliary trackers. Further, with ongoing advances in wireless and wearable biosensing technologies and signal processing techniques, it is now feasible to measure heterogeneous activities of the central, peripheral, and autonomic nervous systems in tandem with data acquired through the VR system. Indeed, the ability to integrate virtual experience with these peripheral recording systems makes VR particularly valuable in behavioral neuroscience research. However, these exciting advances also bring many challenges. This chapter offers practical advice on building an optimal research platform, acquiring and synchronizing heterogeneous modalities of data, labeling events, and capturing game play. The reader is invited to explore subsequent chapters of this volume to learn more about the state of the art in neuroimaging (chapter "Monitoring Brain Activity in VR: EEG and Neuroimaging") and eye-tracking (chapter "Eye-Tracking in VR") in VR-based research paradigms.

2 Behind the Magic of Virtual Reality

2.1 Visuo-Spatial Cues

Perhaps the most well-known approach to achieving an immersive, 360° VR experience in three dimensions is through a head-mounted display (HMD) – which is essentially a goggle-style display system that fits over the orbital region of the face and is held in place via head straps. Although all visual input is presented binocularly via a flat display system that rests only a few inches from the user's eyes, depth cues are simulated through stereopsis – that is, slightly offset perspectives of the same image are presented to each eye, usually through separate video displays or the same display with dual feeds. Fresnel plastic lenses serve to support more comfortable viewing conditions and cause images to appear situated further away than their actual position on the display.

Positional and orientational tracking allows the HMD system to update visual input as one moves through space and looks in different directions, affording the illusion of being situated in the virtual setting. Positional tracking can be accomplished by a variety of methods. Some systems (like the HTC Vive (https://www.vive.com/)) use "outside in" tracking, which involves the use of base stations positioned at opposite corners of the VR area emitting infrared (IR) lasers that are detected by photosensors on the headset and handheld controllers. By using the timing of photosensor activation, the system obtains positional estimations and can adjust the viewing angle of the virtual camera, rendering an image to each lens while accounting for the user's motion. On the other hand, standalone devices, such as the Meta Quest (https://www.meta.com/) series, use an "inside-out" system. Here, outward-facing cameras are located on the headset and generate a real-time map of the physical environment. Using this map, the headset continuously estimates the position of the user within the physical environment using computer vision algorithms like feature detection. These "inside-out" positional estimations are then combined with parallel streams of angular velocity and acceleration from inertial measurement units (IMUs) in the headset and controllers, contributing to the ongoing estimation of the headset's rotation and position.

Cave Automatic Virtual Environment (CAVE) systems (Manjrekar et al. 2014) rely on the same basic mechanisms as an HMD, such as stereopsis and positional tracking, but use displays scaled to the size of a room. The walls, ceiling, and even the floor of a CAVE are lined with flat panel displays, which can be seamlessly adjoined when bezel-free screens are used. Passive, anaglyph interlacing, or active shutter approaches are employed to achieve a stereoscopic 3D effect. In both cases, the user wears specialized glasses. Positional tracking can be achieved through sensors mounted to the glasses or attached elsewhere to the user's body. Notably, CAVEs can support multiple simultaneous users, but the environment can only update relative to a single user's position. More information on these and other vision science topics can be found in chapter "VR for Vision Science" of this volume.

Factors Impacting the Quality of an HMD-Based Visual Experience

Screen Resolution

Because the visual display is viewed at very close proximity to the user's eye, low screen resolution will cause images to appear pixelated. At least 4 K resolution (3,840 × 2,160) is recommended, but successful VR-based research has been carried out at lower resolutions.

Field of View (FOV)

FOV is measured in degrees and represents the range of ocular visibility. The binocular human FOV approaches 180°, with around 120° overlapping for stereoscopic vision (Read 2021). Popular VR headsets such as the Meta Quest and the HTC Vive currently support FOVs between 90° and 110°. Although an HMD with a wider FOV can approximate true-to-life viewing conditions more accurately, there are also trade-offs with resolution – as a wider FOV will yield a lower pixel density.

Interpupillary Distance (IPD)

To avoid eye strain, it is important that the optical center of the HMD lenses aligns with interpupillary distance. In some headsets, the user's IPD is measured through integrated infrared eye cameras, and a calibration routine assists the user in attaining optimal lens spacing using an adjustment knob. Other headsets allow you to manually shift the lenses into different spacing settings based on IPD ranges, and users are expected to ascertain their IPD on their own.

Comfort and Fit

A poorly fitting HMD can result in shifting or sliding of the headset, muscle strain, light bleed, and other distractions and discomforts. Most HMDs are

(continued)

secured through a combination of lateral and coronal head straps or rigid plastic bands that can be adjusted to different head sizes via Velcro or an adjustment wheel. After-market accessories, such as masks or attachable covers, can improve the effectiveness and hygienics of the facial interface, which is the soft foam lining around the edges of the open portion of the headset that presses against the face.

Users Who Wear Glasses

While contact lenses are the preferred method of vision correction while wearing an HMD, glasses can be accommodated in various ways. In some headsets, such as versions of the Vive, the distance between the lens and the face can be lengthened via side knobs so that the headset can fit over most eye-wear frames. In other cases, spacers may be added between the facial interface and the headset, effectively extending the distance between the lenses and the face. Prescription lens inserts may also be purchased for many popular brands of headsets, obviating the need to wear glasses.

2.2 *Engaging with Single- and Multi-Player Virtual Spaces*

In most HMD VR systems, the perimeter of a play area – that is the physical space in which users engage in the virtual world – must be defined at the outset. When a user approaches the edges of the play area, a boundary warning grid can appear, and stepping outside of this boundary can result in exiting the virtual world. In cases when the extent of a virtual world exceeds the boundaries of a play area, a number of different techniques exist to support locomotion and other forms of movement and travel. Examples of solutions that involve minimal degrees of freedom include riding in a vehicle on a track or entering a portal, which can directly transport users from one fixed location to another. On the other hand, controllers can be used to achieve self-directed locomotion over large regions of space. Teleporting, for instance, usually involves aiming a controller or simply looking at a desired location in the scene and clicking a button to advance one's position to that location. Teleporting is often combined with walking and allows users to traverse much greater distances more quickly than would be possible through walking alone. Controllers can also be used to achieve various forms of world pulling, such as skiing, rock climbing, swimming, and ladder climbing, and to steer virtual vehicles. Finally, VR-integrated omni-treadmills support continuous naturalistic walking without controllers, and new vehicle simulation devices (like the NOVA by Eight360 (https://www.eight360.com/)) support synchronous 360° rotation while operating a vehicle in VR.

Another category of actions supported in VR is the manipulation of virtual objects and substances. Many popular systems, such as Meta Quest and Vive, can track movements and configurations of users' hands in real time, supporting naturalistic interactions with a virtual environment based on stereotyped hand gestures.

Alternatively, hand controllers can serve as a proxy for users' hands. Continuously tracking the controllers' position and rotation and mapping gestures to available buttons/levers allows users to interact with virtual objects in many ways, such as holding, moving, dropping, throwing, and so forth. In situations where physical space may be limited or when virtual objects are distant, it is also possible to program hand controllers to select and pull objects through space toward. Interactions like these can even be driven by users' gaze, which can be estimated on the basis of head orientation via a ray projected from the center of the headset. When the ray intersects with the collider of a virtual object, that object may be "selected." Additionally, Unity and Unreal game engines, both of which are software frameworks that support video game development, offer libraries that support recognition of spoken or written keywords and phrases, allowing users to effect changes in their environments or on non-player characters through simple commands (e.g., uttering "Lights on," or "Sleep").

Although the head and the hands are the primary effectors in most VR experiences, full body tracking is also feasible and is currently used in applications such as virtual martial arts. Typically, wireless sensors are strapped to the limbs and torso to obtain continuous estimates of joint position and rotation, from which a whole-body model can be computed via forward and inverse kinematics and updated in real time. Notably, a representation of the user's body can be visible to the user as well as to other users in multi-player contexts. VR technology has reached a point where multiple people can interact in the same virtual environment, whether they are physically situated in the same space or not. This feature makes VR a desirable space to investigate social interactions, as it allows researchers to manipulate variables in ecologically valid ways, without sacrificing experimental control (Pan and Hamilton 2018). In an early study, Bailenson et al. (2003) immersed participants in a virtual environment with an avatar, varying the avatar's gender, gaze behavior, and whether it was controlled by another human or by a computer. The researchers measured the distance participants maintained from these avatars when the avatar approached the participant. Behavioral measures like these may constitute more ecologically valid evidence relative to self-report measures about the extent to which participants attributed sentience to the avatar.

Analogously, Altspace VR, a social platform supporting live virtual events and gatherings (McVeigh-Schultz et al. 2018), has been used in research conducted by the first author and colleagues examining compassion fatigue in health care workers (Wu et al. in prep). The study involved meditation and compassion cultivation activities held in a virtual courtyard and studio, followed by a simulated clinic session with a "patient" (whose avatar was animated by an actor joining the virtual session from a different location). Electroencephalographic (EEG) and electrocardiographic (ECG) data were recorded, as well as behavioral measures reflecting participants' subjective empathy and compassion. This work was part of a larger longitudinal study examining the impact of virtual meditation booster sessions on compassion cultivation training.

2.3 *Sensory Engagement and Immersion*

VR represents a seductive yoking of paradoxical capabilities. On the one hand, it offers ever increasing realism in its appeal to sensory experience. On the other, it offers the unique opportunity to divorce the senses from reality through forays into the fantastic and impossible. On the realism side of this dichotomy, it is now possible to feel simulated gun shots, desert winds, or rain drops and to smell burning tire rubber or brewing coffee while gaming. On the fantastic side, it is possible to create novel experiences by changing the physics governing different forms of sensory input. A player can experience the capability of changing properties of ambient sounds through body movement, for instance, or can try out the perspective afforded by a viewpoint that is only 5 in. from the ground – or 50 feet in the air. It is also possible to isolate unitary modalities of sensory experience. For instance, listening with eyes closed to a soundscape (Schafer 1993) of a tropical rainforest at dawn via 360° audio recording can stimulate mental imagery and allow listeners to become immersed in their imagined environments in the absence of any other sensory input that would normally occur if one were actually present in the represented setting.

Current solutions to haptic and olfactory stimulation can be grouped roughly into portable, wearable, and contact-free approaches. For instance, a portable solution to creating simple force cues can be accomplished by programming the hand controllers to vibrate upon colliding with certain types of virtual surfaces. Other examples include using some of the basic outfitting of an oxygen bar to deliver different scents or air temperature and humidity sensations directly to the nostrils, simulating changes in the ambient environment. Additionally, long or short bursts of forced air directed toward the face via mounted nozzles can simulate different patterns of airflow (Rietzler et al. 2017).

Wearables offer more elaborate solutions. Gloves, sleeves, vests, masks, and even full body suits can deliver complex sensory stimulation in synchronization with specific virtual events. These devices are capable of simulating a variety of touch and force sensations, as well as feelings of heat, cold, wind, water, and sound through vibratory motors, haptic transducers, micro heaters and coolers, and even direct electrostimulation – e.g., using Teslasuits (Caserman et al. 2021). Multi-sensory masks that mesh with popular VR headsets are able not only to supply haptic stimulation to the face, but also hundreds of distinct smells. It is worth noting that some of these sensory feedback systems only interface with specific gaming or training applications, whereas others are compatible with major game engines and offer Application Programming Interfaces (APIs) and Software Development Kits (SDKs) to support development and customization.

As an alternative to wearables or portable devices, tactile sensations can be induced through mid-air haptic technologies, which modulate volleys of ultrasonic energy in order to produce discernible pressure on the surface of the skin when in the vicinity of the device. Currently, these systems work best with the hands and are capable of simulating different sensations of texture and motion, such as the feeling of trickling water or rising bubbles. These devices are intended for use in a variety of settings, including interactive kiosks; however, developers have successfully integrated mid-air haptics into VR applications.

Whereas synthesizing olfactory, haptic, and other tactile sensations often requires some type of add-on gear, immersive sound experiences can be realized through spatial audio functions that are commonly available to the VR development community and supported by most HMD systems. Spatial audio can be understood as a rendering process such that a sound source is locked to a point in space rather than a person's head (Moore 1983). Thus, even though players may be wearing headphones attached to their HMD, sounds will become quieter as they move away from its source, and louder as they draw closer. In addition, ongoing work on path-traced acoustics allows developers to model changes in a sound's acoustic energy depending on properties such as the size, shape, and materials of the virtual environment (Beig et al. 2019). Thus, a narrow, enclosed space with many reflective surfaces will yield different acoustic qualities than an open outdoor space or one filled with sound-absorbing materials such as carpets and drapes. Likewise, a sound coming from around a corner will have different properties from one coming from a direct, un-occluded source.

Notably, as an alternative to simulating various kinds of sensory feedback produced by real-world stimuli, it is also possible to integrate real-world stimuli as physical props within a virtual experience. For instance, in recent work by the second author (Maymon et al. 2023), participants balanced across a real wooden plank which corresponded to a matching VR plank appearing to extend precariously over a city street from a height of 80 stories. Similarly, another research group added real textured surfaces within their play area in order to enhance the feeling that participants were inside a virtual cave (Kisker et al. 2021).

2.4 Creating Immersive Content

Creating your own virtual content necessitates selecting the most suitable game engine or platform to support development. At the time of writing this chapter, Unity (https://unity.com/) and Unreal Engine (https://www.unrealengine.com/) are the most broadly used game engines, each delivering advanced, out-of-the-box solutions. However, alternatives also exist, such as Vizard (https://www.worldviz.com/vizard-virtual-reality-software), which offers analytical tools that are specifically designed to support research, and A-Frame (Neelakantam and Pant 2017), which is a web-based framework. Additionally, live, mixed reality platforms offer a framework tailored to creating social VR-based paradigms.

It is a widely held view that Unity is a better choice for novice programmers. Unity supports scripting in C#, which presents a lower barrier to entry than programming in Unreal, which uses C++. Further, because Unity has attracted a large, active community of users, as well as numerous industry partners, abundant documentation and instances of prior work are available. Unity also offers a fairly intuitive user interface, along with team collaboration tools. On the other hand, the Unreal game engine is known for more powerful rendering and much higher resolution graphics. It supports multi-player stacks and servers natively, whereas Unity can require peripheral libraries for multiple players. Unreal also offers node-based visual programming (Blueprints). It is worth noting that for both Unity and Unreal, pre-existing assets – that is, digital objects, sounds, images, animations, and other content that comprise the virtual world – as well as code, plugins, and other functionality, can all be obtained either for free or at a cost. In this way, many basic environments can be created simply by placing assets and specifying their properties in the editor.

When designing a VR study, the choice of which platform to use depends on the content that one wishes to create or use. Many different approaches requiring varying degrees of customization and development can be adopted. Obviously, presenting 360° images or videos from an existing library requires less specialized functionality than creating a novel, believable, and fully interactive world. In the first case, the visual content is captured from the real world using a 360° camera, whereas the latter case is much more complex. Customized assets must be created and programmed with the interactive features that support the desired experience. For instance, if one wishes for some objects to clatter and bounce when they are dropped, these properties must be programmatically associated with those objects. On the other hand, it is also possible to use procedural assets generated natively by a game engine or draw from existing asset libraries or packages. This approach requires considerably less investment of resources than creating custom assets from scratch. For those whose research needs require the development of a customized experience from scratch, it is advisable to avail oneself of tools that support modeling, animating, and texturing, such as Blender (https://www.blender.org/), Maya (https://www.autodesk.com/products/maya/), Houdini (https://www.sidefx.com/products), Tilt Brush (https://www.tiltbrush.com/), and many others.

While it is often desirable to construct new experimental content that is customized to a specific research question, creating optimized and visually immersive virtual environments may require a dedicated team of creators – which may be prohibitively expensive when setting up a new study. One alternative approach that researchers may wish to consider is to "mod" an existing VR game. Modding is a process whereby users can write their own code and embed them into the file structure of a particular game. The result is a version of that game that is modified by that code. Modding is available only for those simulations whose production studio releases development files (e.g., Minecraft, Skyrim). If, however, a researcher can find an existing simulation that includes a scenario that could be minimally altered to be suitable for experimentation (perhaps you want to remove on-screen distractions like the character's health bar or add in a particular recording of your experiment instructions), then one can save considerable time and budget by developing the code necessary to shape an existing simulation, rather than to create a new virtual environment from scratch.

Fig. 1 Point cloud representation of the Star of India obtained through photogrammetry (courtesy of Scott McAvoy on Open Heritage 3D)

In addition to creating assets using 3d software, it is also possible to leverage 3d capture technologies that draw upon the real world as input (Fig. 1). *Photogrammetry* is a process that involves constructing virtual 3d objects that users can walk around and view from all angles using simple photographs as inputs. This approach has become widely adopted for building at all scales – from scanning small handheld objects all the way up to large geo-regions, including cities.

Photogrammetry largely runs on *structure from motion* (SfM) techniques (Özyeşil et al. 2017), which involve estimating 3d structure from a sequence of 2d images (photographs or video). They have become a standard in software such as Agisoft Metashape (https://www.agisoft.com/), Reality Capture (https://www.capturingreality.com/), and VisualSFM (C. Wu 2011) – to name a few – and are common practice in diverse fields, including forensics and cyber archeology. The output of SfM algorithms can be stored in many forms compatible with VR practices, though the two most typical formats are LAS/LAZ point clouds or meshes. Meshes often required post processing in some form of traditional 3D software such as Blender or Maya to clean up floating points, fix holes, or modify texture maps.

An alternative approach to real-world capture is LiDAR scanning (Light Detection and Ranging), which is a laser-pulse sensing technology that can map the topography of objects and the environment. It yields point cloud representations that can be imported into modeling platforms, such as VR Sketch, so that textures and other features can be added before placement in a virtual scene. Because LiDAR lasers can penetrate foliage, it is heavily favored in contexts in which vegetation can obscure the true contours of an area (Chase et al. 2017).

3 VR as a Research Tool

On the one hand, VR is an attractive research tool that promises an unprecedented degree of controlled stimulation coupled with the possibility of capturing an exceptionally complex, multimodal matrix of continuous data representing diverse, temporally synchronized aspects of behavior and physiology. HMD systems track the positions of the controllers and the player's head in the virtual environment, as well as head orientation. Additional sensors added to the body and hands (e.g., haptic gloves) can provide a finely detailed, dynamic model of limb movement. Further, systems with eye-tracking technologies and ray-casting algorithms can offer continuous estimates of the user's gaze in the virtual environment.

In addition to the integrated sensors that come built-in to many VR systems, it is also possible to leverage wearable biosensing technologies that measure EEG, ECG, respiration, temperature, photoplethysmography (PPG), and more. Many of these devices are sufficiently lightweight and unobtrusive to be worn concomitantly with an HMD – and in this way, a wide range of actions and events, such as looking toward a particular location, walking, teleporting, and picking up or coming into contact with objects can be parsed and analyzed in conjunction with simultaneously recorded physiological data that reflect modulations of activities in the central and peripheral nervous systems. MRI-compatible VR systems have been developed as well (Adamovich et al. 2009).

On the other hand, as a corollary to these exciting possibilities are several hurdles that a VR-based research paradigm must overcome above and beyond the issues of game development and user experience covered in the previous section. Most commercially available VR systems do not come research-ready out of the box. To obtain access to the desired data, a license may be required or customized API or driver. Further, precautions are necessary to ensure participant safety and comfort, and to ensure that the various sensors used in a study do not interfere with one another. Groundwork must also be laid to synchronize heterogeneous modalities of data that are collected – either in real time or offline. Finally, the problem of analyzing data obtained during engagement in immersive VR is non-trivial. The past decades of cognitive neuroscience research have largely been devoted to paradigms that involve motionless volunteers viewing 2D displays and pressing buttons in response to experimentally relevant events. However, when recording EEG, ECG, or other time-series modalities that are distorted by body movement, strategies for removing or correcting motion artifacts are crucial for studies using immersive VR. Analogously, approaches for parsing saccades and fixations under conditions of head and body movement in 3D space are critical for eye-tracking work performed in immersive contexts. More information on EEG and eye-tracking in VR can be found in chapters "Monitoring Brain Activity in VR: EEG and Neuroimaging" and "Eye-Tracking in VR", respectively, of this volume.

3.1 Safety and Comfort in VR

A few basic precautions are advised in order to reduce the risk of injury or discomfort. First and foremost, in the case of room-scale activities, the play space should be clear of hazards that could cause users to trip, fall, or otherwise injure themselves. While wearing a headset, users should be monitored to ensure that they do not become entangled in cables or collide with walls or other stationary objects on the perimeter of the play space. Secondly, steps should be taken to ensure proper fit of the headset and adjustment of the IPD in order to avoid blurry images, ocular problems, and uncomfortable shifting or sliding of the headset during movement. Thirdly, common negative "side effects" associated with VR usage, such as motion sickness or headache, may be mitigated to some degree by the design of the virtual environment (e.g., by minimizing exposure to intense light or reducing optic flow) (Nichols and Patel 2002). Further, limiting the length of exposure (15 min is a conservative time window) and ensuring adequate breaks are important measures to minimizing negative experiences (Kaimara et al. 2022). Finally, consideration should be given to the needs of special populations, such as those with impaired balance or limited vision. Most head-mounted VR systems are not recommended by the manufacturers for children below the age of 12 or 13.

Because headsets will likely be worn by many different users, maintaining the hygienic standards of equipment also deserves some forethought. High-end solutions include sanitizing systems that are specialized for headsets and that rely on short wave ultraviolet light (UV-C). A less expensive approach involves replacing foam facial interfaces with silicone or vinyl ones that can easily be cleaned with a disinfectant wipe. Disposable face covers or breathable, washable fabric ones are also available on the market.

3.2 Choosing a Suitable HMD

What to Expect from Off-the-Shelf Head-Mounted VR Hardware

It goes without saying that all head-mounted VR systems involve some sort of head-mounted display. Some systems, such as the Varjo XR3, which implements hand tracking and inside-out positional tracking, do not require additional hardware. Other systems include hand controllers and base stations or other forms of positional trackers. Many VR systems require tethering to a PC or gaming laptop, such as the HP Fury. Tethering can be accomplished via a physical cable or in some cases, wirelessly through after-market adapters. Some systems, such as the Meta Quest series, are standalone – that is, they operate without external computing support.

(continued)

Space Requirements

The amount of physical play space needed for head-mounted VR depends on the kind of interaction that is expected. Viewing a 360° video or experience a simple scene, such as riding a cart on a track, might be suitable for seated participants. Likewise, it may be possible to navigate large virtual environments while walking on a treadmill or sitting in a chair and using the teleport function. On the other hand, if natural walking or other forms of locomotion are required of participants, then a room-scale VR setup with at least 6.5 by 5 feet of open space free of obstacles and hazards is advised.

Integrated Sensors

Off-the-shelf systems support positional tracking of the headset and controllers. Other integrated sensing capabilities vary widely from model to model, ranging from hand and body tracking, eye gaze tracking, facial expression capture, and monitoring of brain and cardiac activities. It should be noted that in some cases, licenses or specialized APIs are necessary to access these data.

Requisite Technical Expertise

Most off-the-shelf systems are designed to work right of the box, requiring only basic technical abilities. Further, some platforms offer game editors that allow simple VR experiences to be created without coding. On the other hand, scripting and other operations necessary to support complex interactions require advance knowledge of the game engine's native programming language. Additionally, more advanced computational skills and knowledge may be required to visualize data from peripheral sensors during engagement in VR, generate event markers in real time, and synchronize and record data streams with event markers.

A central distinction to consider in selecting a VR system hinges around tethered versus untethered models (Table 1). A tethered system is an HMD that is connected physically to a PC through a cord – either USB-c, USB-3, HDMI, or a combination of the 2 (USB, HDMI). It uses the PC that it is connected to for computation, data, power, and rendering. In some cases, a wireless adapter kit may be added to eliminate the need for a physical cable-to-PC connection – but the PC is still required for computation and rendering. On the other hand, untethered systems – often referred to as "standalone" – are endowed with their own fully embedded processor, communications, OS, and power. These devices can range from simple cellphone-based 360° viewers, to the modern gaming headsets of the Meta Quest series and Vive Focus, which feature 6 DoF (degrees of freedom), inside-out tracking, and more. These systems provide convenience for novice VR users and serve well as prototyping devices for people on a budget.

Many HMDs feature integrated sensors above and beyond standard IMUs that track head position and orientation. Infrared eye-tracking, for instance, can support analysis of eye movement, gaze, and pupillary response (PR) of individuals in

Table 1 Pros and cons of tethered versus untethered head-mounted VR

Tethered pros	Tethered cons
• Usually guaranteed high performance • Greater FOV (110°) • Higher resolution (depending on PC rig) – supporting ultra high-resolution displays • Higher throughput for multi-threaded work • Customizable for external peripheries (Wi-Fi, hardline connections) and eye-tracking (90+ Hz) • Faster internet speed potential through fiber optics • Usually a high-performance Rig (HPR) can support multiple types of HMD systems and makers • The tether can serve as a boundary or safety line to control distance • No battery to recharge	• Costly setup • A lot of equipment • Not easily portable • Tether can be cumbersome in some experiences
Untethered pros	**Untethered cons**
• Cheaper • Significantly more portable • Typically lighter weight and easier to wear • Faster to deploy for user testing	• Less power and performance • Not as much capability for peripheral or external device integration • Narrow field of view (avg 90° FOV) • Lower resolution • Needs regularly recharging • Often comes with unique user login and side support registration apps or user agreements • Slower Wi-Fi connection

VR. The HTC Vive Pro Eye series, HP Reverb G2 Omnicept edition (https://www.hp.com/), and Varjo series (https://varjo.com/), among others, all feature integrated eye trackers. Additionally, after-market hardware, such as the Pupil Core from Pupil Labs (https://pupil-labs.com/), is compatible with some VR devices, such as the Vive.

Current head-mounted VR systems – particularly high-end ones – also tend to support face, body, or hand tracking through integrated cameras – or compatible aftermarket add-ons. These capabilities allow users' facial expressions, as well as the position and orientation of their hands, fingers, and body to be tracked and modeled in the virtual world. Intriguingly, some systems, such as the HP Reverb G2 Omnicept, combine input derived from multiple sensors simultaneously, including face cameras, eye trackers, and a PPG sensor in the headset in order to compute online estimates of a user's cognitive load (Siegel et al. 2021).

Wireless, dry EEG sensors have also been directly incorporated into VR HMDs. The nGoggle (Fig. 2), for instance, features a bank of electrodes over the occipital area of the head that can measure EEG activities associated with visual processing of stimuli presented in the headset (Nakanishi et al. 2017). The DSI-VR300 (https://wearablesensing.com/dsi-vr300/), which is commercially available, offers active electrodes and a somewhat broader distribution of electrodes over parietal and

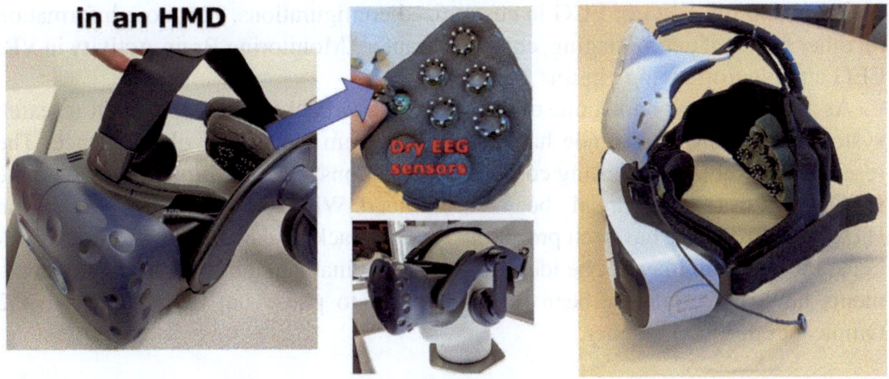

Fig. 2 Customized VR (HTC Vive, left and nGoggle, right) with integrated EEG electrodes

Table 2 Comparing custom builds with add-on sensors to pre-built VR sensing packages

Advantages to custom builds	Advantages to pre-built systems
• Modular setups assure ideal fit for any subject. Pre-built setups are usually tailored only to average size adults • More flexibility in optimizing sensor positions, sampling rates, active versus passive electrodes, wet versus dry electrodes, and more • Individual parameters can be easily modified for new experiments (e.g., switching from an 8- to a 32-channel EEG system while keeping the same VR hardware or upgrading the VR headset to a different one that has built-in eye-tracking)	• Overall cost may be lower because sensors are integrated in the VR headset • Proprietary data protocols may facilitate (but may also hinder) access to the data • Integrated sensors may be more comfortable given the snug fit that is required of the headsets for an immersive VR experience

frontal portions of the head in order to support embedded real-time P300 analysis in brain–computer interface (BCI) applications.

We feel it is important to mention, however, that EEG, PPG, ECG, and other physiological signals have been successfully recorded by the first and second authors of this chapter during engagement in VR-based paradigms using add-on rather than integrated sensors. Indeed, if done correctly, add-ons can provide a more suitable and higher performing solution than currently available pre-built setups because add-on systems are typically optimized for capturing data tailored to specific research needs, whereas pre-built systems typically prioritize user experience and ease of implementation. Table 2 outlines the advantages of both of these approaches. Currently, a variety of mobile EEG options are available, including the Brain Systems Live Amp (https://www.brainproducts.com/), the mBrainTrain Smarting Pro (https://mbraintrain.com/), and the versatile, compact Mentalab ExplorePlus (https://mentalab.com/). It is worth noting that the ExplorePlus is currently the world's smallest research grade system (Niso et al. 2023) and flexibly supports both wet and dry EEG or ECG in customized configurations. For more information on other forms of neuroimaging, consider chapter "Monitoring Brain Activity in VR: EEG and Neuroimaging" in this volume.

As a precautionary note, one of the primary drawbacks of building a VR research system using consumer-grade hardware stems from the risk of obsolescence. The researcher is subject to shifting corporate inclinations. As of writing this chapter, the HTC Vive Pro Eye has already been discontinued. While the upcoming release of an HTC Vive Pro 2 Eye has been promised, the eye-tracking capabilities of this second-generation system may not be identical to the original version. Likewise, announcements have unfortunately been made of plans to phase out the HP Reverb G2 Omnicept edition altogether.

3.3 Eye-tracking in VR

With the growing availability of HMDs that feature mobile eye trackers, new possibilities have opened for studying naturalistic, unconstrained gaze behaviors in 3D space (see also chapter "Eye-Tracking in VR" for an in-depth discussion about eye-tracking and head tracking in virtual environments). Through continuous tracking of data about the user's head location and orientation within the virtual world, the head-mounted system can establish the distance between the user and objects and other features of the virtual world at all times, making it possible to compute gaze points as 3D vectors using relative eye position co-ordinates. A gaze intersection point (GIP) is the nearest point in the virtual environment that is intersected by the vector-based gaze ray. Thus, in virtual paradigms, it is possible to study eye movements and gaze both through traditional approaches that estimate changes in visual angle and point of regard based on information about the relative position of the eye in the head, as well as through the analysis of GIPs.

Ongoing work in the primary author's lab suggests that caution must be exercised when analyzing eye movement in contexts that also involve unconstrained head movement, as vestibular-ocular reflex (VOR) can confound the differentiation of fixations from saccades. VOR triggers compensatory eye movements that allow an individual to maintain gaze on a fixed object while moving his or her head. During head movements that trigger VOR, the rate of change of the position of the eye-in-head during a fixation can actually exceed the rate of change of eye position that occurs during small saccades (Tatler et al. 2019), making it extremely difficult to discern the onset of a fixation using algorithms that rely on pupil movement velocity. In other words, even though a person's gaze may have already reached a target, the persistence of low amplitude eye movement may hinder precise determination of fixation onset.

This possibility is supported by Fig. 3, which plots fixation-related potentials (FRPs) derived over CPz in a classic P300 oddball paradigm involving either a saccade only (left) or a saccade and a head turn (right) to the standard or deviant target. In the saccade only condition, fixation-locking yields a robust P300 effect

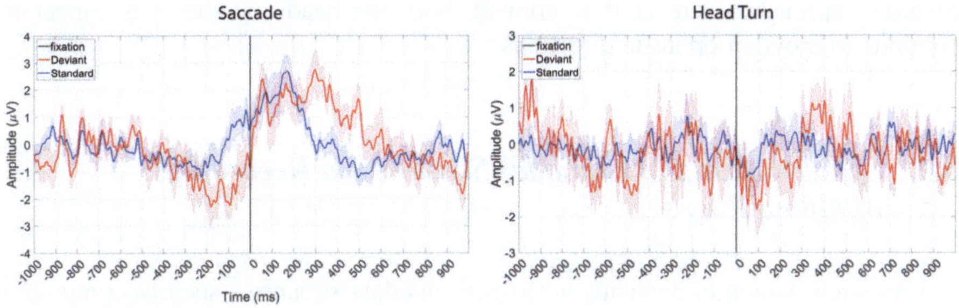

Fig. 3 Fixation-related potentials computed in response to high (red) or low (blue) probability visual targets. Participants made a saccade to the target (left) or a head turn and a saccade (right). Time zero is fixation onset

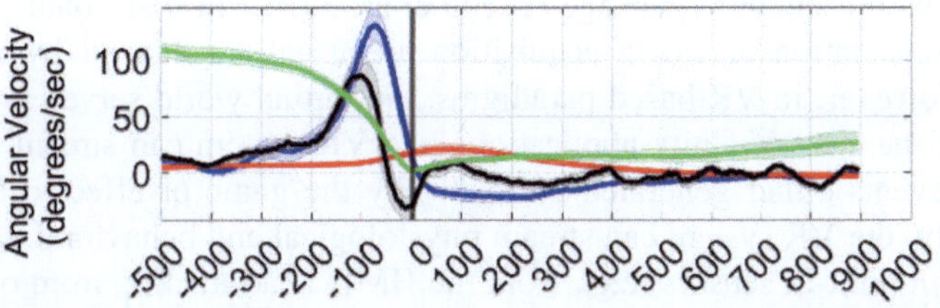

Fig. 4 Angular velocity of head (red) and eye (blue) movement during a head turn and saccade to a peripheral target. Gaze velocity is plotted in black. The green line indicates the rate of change of the distance of gaze points from the target

from approximately 250 to 550 ms post-fixation. It also reveals sensitivity to standards versus deviants before fixation onset. On the other hand, in the head turning condition, the oddball response is smaller and more difficult to distinguish from the response to standards, and no reliable effects before fixation onset can be discerned. These mushier results in the head turning condition can likely be attributed to two primary factors – namely, imprecise fixation-locking due to VOR, as well as muscle artifacts introduced into the raw EEG during head movement (Wu et al. 2023).

As an alternative to classifying fixations solely on the basis of pupil velocity, it may be possible to use information from the GIP stream to understand when a fixation is occurring on a target of interest. Algorithms can be developed that detect when the GIP is in contact with a target irrespective of pupil movement velocity. Figure 4, for instance, plots the angular velocity of head movement (red line), eye movement (blue line), and gaze (black line), as well as the rate of change in distance of each GIP co-ordinate relative to the target (green line), during a head turn to a peripheral visual target. Time zero is the moment when the GIP first overlaps with the target (causing the distance from the target to register as zero). This time point is presumably the onset of the fixation on the target. Notably, while the rate of change of gaze approaches zero at this moment, both the head and the eyes appear to continue to move in opposite directions.

3.4 Synchronizing Multimodal Signals and Recordings of Game Play

If a research paradigm demands heterogeneous data streams – such as continuous physiological measurements coupled with eye-tracking and motion capture data – the challenge of aligning these diverse time series in a common temporal frame is non-trivial. The primary author's research group uses Lab Streaming Layer (LSL) – a freely available API that supports the streaming and recording of multimodal data over a local network and integration of these data streams with event codes generated by a stimulus presentation package (Ojeda et al. 2014). In traditional experiments, stimulus presentation and data acquisition are usually managed by independent systems. However, in VR-based paradigms, the virtual world serves as the "stimulus" and in the case of Unity applications, the VR system can stream event codes related to events either generated internally by the game or effected by the user. Additionally, the VR system can stream physiological and behavioral data obtained from its own built-in sensors (e.g., from the HMD head tracker, from other peripheral IMU trackers, from eye-tracking systems, and so forth) to be synchronized via LSL with time-series data obtained from other sensors that are not integrated with the VR system.

Examples of event codes streamed to LSL by the authors include trigger pulls to pick up virtual objects, ratings of flow during an in-game survey, and the names of virtual objects fixated during game play. Many other events could be created based on participants' location in and interaction with the virtual environment – including body movements, eye movement, and speech. Through a dynamic library file available in the LSL package, it is possible to define streams associated with these event types and then specify updating functions that push event information to the stream at regular intervals. Each stream is recorded as a vector of values (e.g., event codes) and associated timestamps. Because LSL allows heterogeneous data streams to be aligned within a common time scale, it is possible to synchronize event streams during offline analysis with other simultaneously recorded physiological and behavioral data acquired during the experiment. This approach can support event-related analyses, including event-related potentials (ERPs), FRPs, task-evoked pupillary response (TEPR), and more.

To relate time-series data to the player's experience in the virtual world, it may be useful to record not only event markers, but also a video capture of the game play (see Sect. 3.5). A fairly simple approach to synchronizing a screen capture with the other data streams involves creating a visible marker on the screen that is activated by a recordable event. For instance, a flash of color could appear in the corner in response to a trigger pull, that is recorded as a separate LSL stream. In this way, one can align various time-series data recordings with the known time of the trigger pull and the known time of the color flash in the screen capture.

3.5 Capturing Video

Video capture of a VR session falls into 3 unique categories – namely, traditional screen capture, virtual camera capture, and mixed reality capture. Traditional screen capture involves recording the scene that is cast to the flat 2d desktop monitor. It typically reflects the user's egocentric perspective. Screen capture tools have become ubiquitous and are deployed on almost all operating systems (e.g., Quicktime on Mac, Win+G on Windows). Many third-party apps are also available such as Nvidia GeForce Experience (https://www.nvidia.com/enus/geforce/geforce-experience/) and Open Broadcast Software (OBS) (https://obsproject.com/) – an open-source video audio mixer software for parsing content from multiple streams into a single directed format that can then be recorded or streamed in real time to external sources. This platform would be useful in cases where you may have multiple virtual cameras and want to stitch them all together in real-time.

Alternatively, if one wishes to capture activities occurring at a specific location in the virtual environment or from an allocentric point of view, virtual cameras can be placed into the game programmatically and operated as viewports into the game. With them you can follow the player, free flying (as if controlling a drone). Static cameras can also be stationed in the 3D world space, making it possible to capture a 360° view from a fixed location. This technique can also be used to replay the point of view of the user and can contain different information from scene casting to the desktop monitor. Virtual cameras have become mainstays for broadcasting in-game events to the outside world and are supported by most game engines. Third-party plugins can be obtained as well.

The last method to be examined in this section is mixed reality capture. It involves capturing the physical, real-world body of the player mixed within the digital world. This technique uses a combination of virtual cameras and real-world cameras calibrated together to place the real person within the confines of the game he or she is playing. Traditional green screen photography and virtual cameras are used to mix worlds. Pervasive in Hollywood filming today, mixed reality capture is more resource intensive than the previously described methods and generally requires a specialized space big enough to mask the backgrounds out and additional equipment to manage the virtual camera interface with a physical device. There are many tools and systems that can now be used to this end, including MixCast (https://mixcast.me/) by Intel and Reality Mixer, which can be obtained through Apple as an iOS-based app that relies on iPads or iPhones (though a beta version that interfaces with certain VR headsets is also available through Steam). Microsoft Mixed Reality Capture Studios is another resource, supporting volumetric video and hologram creation. It is built in natively to the Microsoft Mesh platform and works both with VR and Hololens2.

4 Conclusion

This chapter underscores the fact that accomplishing behavioral neuroscience in VR is a cross-disciplinary team effort that mixes the roles of experimentalists, developers, and engineers. However, VR is a versatile medium that can be adapted to different research objectives and the capabilities of different research teams. Off-the-shelf solutions exist in the form of plug-and-play head-mounted systems with integrated sensors, as well as asset libraries, game engines, and customizable game platforms. Alternatively, it is possible to develop specialized content from scratch and modify existing, commercially available VR hardware to suit specific experimental needs.

As outlined in the first section of this chapter, developing a VR-based research platform requires attention to the kinds of interactivity and immersive sensory content – such as visual, auditory, haptic, and olfactory stimulation – that will be put to play in the virtual experience that is part of the research paradigm. As outlined in the second section, one's research goals can guide the selection of hardware and possibly add-on sensors. Further, a variety of open-source solutions can be implemented to overcome the challenge of capturing and synchronizing diverse streams of data. As a final note, advances in signal processing methods and theoretical frameworks for interpreting experimental outcomes are still catching up with the pace of research possibilities opened by VR technologies. The reader is invited to explore the other chapters of this book to gain further insight into solutions that have already been successfully implemented in response to these challenges.

Acknowledgment This chapter was supported by grant #IIS-2017042 from the National Science Foundation and and #W911NF-21-2-0126 from the Army Research Laboratory to the first author, and by a grant from the Royal Society of New Zealand Marsden fund (VUW-2005) which supported the second author.

References

Adamovich SV, August K, Merians A, Tunik E (2009) A virtual reality-based system integrated with fmri to study neural mechanisms of action observation-execution: a proof of concept study. Restor Neurol Neurosci 27(3):209–223. https://doi.org/10.3233/RNN-2009-0471

Bailenson JN, Blascovich J, Beall AC, Loomis JM (2003) Interpersonal distance in immersive virtual environments. Pers Soc Psychol Bull 29(7):819–833. https://doi.org/10.1177/0146167203029007002

Barbot B, Kaufman JC (2020) What makes immersive virtual reality the ultimate empathy machine? Discerning the underlying mechanisms of change. Comput Hum Behav 111:106431. https://doi.org/10.1016/j.chb.2020.106431

Beig M, Kapralos B, Collins K, Mirza-Babaei P (2019) An introduction to spatial sound rendering in virtual environments and games. Comput Games J 8(3–4):199–214. https://doi.org/10.1007/s40869-019-00086-0

Caserman P, Krug C, Göbel S (2021) Recognizing full-body exercise execution errors using the Teslasuit. Sensors 21(24). https://doi.org/10.3390/s21248389

Chase ASZ, Chase DZ, Chase AF (2017) Lidar for archaeological research and the study of historical landscapes. In: Masini N, Soldovieri F (eds) Sensing the past, vol 16. Springer, Cham, pp 89–100. https://doi.org/10.1007/978-3-31950518-3_4

Kaimara P, Oikonomou A, Deliyannis I (2022) Could virtual reality applications pose real risks to children and adolescents? A systematic review of ethical issues and concerns. Virtual Reality 26(2):697–735. https://doi.org/10.1007/s10055-021-00563-w

Kisker J, Lange L, Flinkenflügel K, Kaup M, Labersweiler N, Tetenborg F et al (2021) Authentic fear responses in virtual reality: a mobile EEG study on affective, behavioral and electrophysiological correlates of fear. Front Virtual Reality 2. https://doi.org/10.3389/frvir.2021.716318

Manjrekar S, Sandilya S, Bhosale D, Kanchi S, Pitkar A, Gondhalekar M (2014) CAVE: an emerging immersive technology – a review. In: 2014 UKSim-AMSS 16th international conference on computer modelling and simulation. IEEE, pp 131–136, https://doi.org/10.1109/UKSim.2014.20

Maymon CN, Crawford MT, Blackburne K, Botes A, Carnegie K, Mehr S, Meier J, Murphy J, Myles N, Robinson K, Tooley M, Grimshaw GM (2023) The presence of fear: subjective fear, not arousal, drives the experience of presence. [Unpublished manuscript]

McVeigh-Schultz J, Márquez Segura E, Merrill N, Isbister K (2018) What's it mean to "be social" in VR? Mapping the social VR design ecology. In: Proceedings of the 19th international ACM SIGACCESS conference on computers and accessibility – DIS '18. ACM Press, New York, pp 289–294. https://doi.org/10.1145/3197391.3205451

Moore FR (1983) A general model for spatial processing of sounds. Comput Music J 7(3):6. https://doi.org/10.2307/3679589

Nakanishi M, Wang Y-T, Jung T-P, Zao JK, Chien Y-Y, Diniz-Filho A et al (2017) Detecting glaucoma with a portable brain-computer interface for objective assessment of visual function loss. JAMA Ophthalmol 135(6):550–557. https://doi.org/10.1001/jamaophthalmol.2017.0738

Neelakantam S, Pant T (2017) Introduction to A-frame. In Learning Web-based virtual reality. Apress, Berkeley, pp 17–38, https://doi.org/10.1007/978-1-4842-2710-7_4

Nichols S, Patel H (2002) Health and safety implications of virtual reality: a review of empirical evidence. Appl Ergon 33(3):251–271. https://doi.org/10.1016/S00036870(02)00020-0

Niso G, Romero E, Moreau JT, Araujo A, Krol LR (2023) Wireless EEG: a survey of systems and studies. Neuroimage 269:119774. https://doi.org/10.1016/j.neuroimage.2022.119774

Ojeda A, Bigdely-Shamlo N, Makeig S (2014) MoBILAB: an open source toolbox for analysis and visualization of mobile brain/body imaging data. Front Hum Neurosci 8:121. https://doi.org/10.3389/fnhum.2014.00121

Özyeşil O, Voroninski V, Basri R, Singer A (2017) A survey of structure from motion. Acta Numerica 26:305–364. https://doi.org/10.1017/S096249291700006X

Pan X, de Hamilton AFC (2018) Why and how to use virtual reality to study human social interaction: the challenges of exploring a new research landscape. Br J Psychol 109(3): 395–417. https://doi.org/10.1111/bjop.12290

Read JCA (2021) Binocular vision and stereopsis across the animal kingdom. Annu Rev Vis Sci 7: 389–415. https://doi.org/10.1146/annurev-vision-093019-113212

Rietzler M, Plaumann K, Kränzle T, Erath M, Stahl A, Rukzio E (2017) Vair: simulating 3D airflows in virtual reality. In: Proceedings of the 2017 CHI conference on human factors in computing systems. ACM, New York, pp 5669–5677. https://doi.org/10.1145/3025453.3026009

Schafer RM (1993) The soundscape: our sonic environment and the tuning of the world. Simon and Schuster

Siegel EH, Wei J, Gomes A, Oliviera M (2021) HP Omnicept cognitive load database (HPO-CLD)–developing a multimodal inference engine for detecting real-time mental workload in VR

Tatler BW, Hansen DW, Pelz JB (2019) Eye movement recordings in natural settings. In: Klein C, Ettinger U (eds) Eye movement research: an introduction to its scientific foundations and applications. Springer, Cham, pp 549–592. https://doi.org/10.1007/978-3-030-20085-5_13
Wu C (2011) VisualSFM: a visual structure from motion system. http://www.cs.washington.edu/homes/ccwu/vsfm
Wu YC, Carrillosulub E, Vieten C (in prep) Cultivating empathic resonance through virtual reality
Wu YC, Chang C, Liu W, Stevenson C, Cohen Hoffing R, Thurman S, Jung T-P (2023) Free moving gaze-related electroencephalography in mobile virtual environments. Presented at the vision science society annual meeting. J Vis

Monitoring Brain Activity in VR: EEG and Neuroimaging

Sebastian Ocklenburg and Jutta Peterburs

Contents

Abstract Virtual reality (VR) is increasingly used in neuroscientific research to increase ecological validity without sacrificing experimental control, to provide a richer visual and multisensory experience, and to foster immersion and presence in study participants, which leads to increased motivation and affective experience. But the use of VR, particularly when coupled with neuroimaging or neurostimulation

---

S. Ocklenburg (✉)
Department of Psychology, Faculty for Life Sciences, MSH Medical School Hamburg, Hamburg, Germany

ICAN Institute for Cognitive and Affective Neuroscience, MSH Medical School Hamburg, Hamburg, Germany

Faculty of Psychology, Institute of Cognitive Neuroscience, Biopsychology, Ruhr University Bochum, Bochum, Germany
e-mail: Sebastian.ocklenburg@medicalschool-hamburg.de

J. Peterburs
Institute of Systems Medicine & Department of Human Medicine, MSH Medical School Hamburg, Hamburg, Germany

Curr Topics Behav Neurosci (2023) 65: 47–72
https://doi.org/10.1007/7854_2023_423
Published Online: 13 June 2023

techniques such as electroencephalography (EEG), functional magnetic resonance imaging (fMRI), or transcranial magnetic stimulation (TMS), also yields some challenges. These include intricacies of the technical setup, increased noise in the data due to movement, and a lack of standard protocols for data collection and analysis. This chapter examines current approaches to recording, pre-processing, and analyzing electrophysiological (stationary and mobile EEG), as well as neuroimaging data recorded during VR engagement. It also discusses approaches to synchronizing these data with other data streams. In general, previous research has used a range of different approaches to technical setup and data processing, and detailed reporting of procedures is urgently needed in future studies to ensure comparability and replicability. More support for open-source VR software as well as the development of consensus and best practice papers on issues such as the handling of movement artifacts in mobile EEG-VR will be essential steps in ensuring the continued success of this exciting and powerful technique in neuroscientific research.

Keywords Electroencephalography (EEG) · Functional magnetic resonance imaging (fMRI) · Methodological challenges · Replicability · Virtual reality

1 Introduction

It has been suggested that using virtual reality (VR) setups for stimulus presentation can lead to enhanced ecological validity and experimental control in many sub-disciplines of neuroscience (Parsons 2015). Typically, VR setups can provide a richer and more stimulating visual experience than the static and simplistic stimuli used in many neuroscientific experiments. This leads to increased presence, i.e., the perceptual illusion of "being there in the virtual space", motivation, and affective experience (Bohil et al. 2011). Using VR, a wide variety of behaviors, ranging from mere passive viewing of objects and naturalistic visual scenes to active exploration of expansive virtual environments, to decision-making, learning, memory, and even interaction with (virtual) characters can be assessed, with negligible risks for the participants. The types of virtual environments are equally diverse, ranging from naturalistic to futuristic, from simplistic to complex, while still offering a maximum of experimental control. Many neuroscientific techniques, such as functional magnetic resonance imaging (fMRI), severely limit the range of movements participants can perform. This hampers the investigation of certain cognitive functions such as spatial navigation, especially also when considering the constraints imposed by classic stimulus presentation techniques. VR can provide the opportunity to conduct research on such functions in a more ecologically valid way than 2D stimulus presentation.

Rapid technological advances over the past few years have rendered VR technology more convenient and accessible, more affordable, and more versatile. As a result, VR is increasingly used in combination with neuroimaging techniques, EEG, and non-invasive brain stimulation (NIBS). However, this increase contrasts with a lack of standard protocols for data collection and established toolboxes for data analysis. Many researchers have built individual VR neuroimaging systems and created pipelines for data analysis based on their unique requirements. Moreover, even though more software is made available to create virtual environments for use in neuroscientific studies, a certain degree of VR coding expertise may be needed to interface them with the imaging technology and to extract meaningful data.

Since the benefits of VR for use in neuroscientific studies outweigh the challenges, its use will increase in the next decade. Many cognitive and/or behavioral neuroscientists who are already familiar with EEG, fMRI, or NIBS may find themselves interested in VR but hesitant due to little knowledge about its practical and methodological possibilities. The present chapter will examine current approaches to creating virtual environments, recording, pre-processing, and analyzing electrophysiological (stationary and mobile EEG) and neuroimaging data (fMRI and functional near-infrared spectroscopy, fNIRS) recorded during engagement with virtual environments. This chapter will also discuss approaches to synchronizing these data with other data streams generated by various events.

2 Different Approaches to Creating Virtual Environments

The main difficulties in applying VR in neuroscience experiments often lie in the tedious implementation as it is typically time consuming, and researchers may lack expertise regarding computational aspects, especially when it comes to synchronizing or harmonizing different data streams. It is not surprising that previous studies report a range of approaches to creating virtual environments from quite simplistic to increasingly complex. Early work has frequently made use of video games for computers or game consoles. This likely required considerable coding skills and effort to adapt the existing virtual environment to a given study's needs. Researchers have therefore increasingly made use of dedicated VR software, and some software solutions have even been designed specifically for research use (e.g., Mueller et al. 2012). Table 1 provides a list of studies that combined VR with EEG, focusing on the specific VR software and hardware that was used. Studies that combined VR with fMRI or functional near-infrared spectroscopy (fNIRS) are detailed in Table 2. It strikes the eye that regarding software, the Unity game engine and Vizard have been the most popular tools for creating virtual environments in more recent studies.

Table 1 Overview of studies combining virtual reality with electroencephalographic (EEG) recording

Study	VR software	VR hardware	Recording	EEG system
Bischof and Boulanger (2003)	Mandala	NEC MultiSync LCD screen	Oscillations	Stationary; NeuroScan NuAmps
Burns and Fairclough (2015)	Playstation 3 video game	Playstation 3 & head-mounted display or LCD TV screen	Event-related potentials	Stationary; BioSemi ActiveTwo
Dey et al. (2019)	Unity 3D	HTC Vive headset	Oscillations	Mobile; BrainProducts LiveAmp
Ehinger et al. (2014)	WoldViz Vizard	NVIS nVisor SX60 head-mounted display & position tracker	Oscillations	Mobile; unspecified
El Basbasse et al. (2023)	Unity game engine & steam VR	HTC Vive headset	Oscillations	Mobile; BrainProducts LiveAmp
Hofmann et al. (2021)	Russian VR coasters	HTC Vive headset & headphones	Oscillations	Mobile; BrainProducts LiveAmp
Kober and Neuper (2012)	Virtual reality projector Cube3D^2	2 m x 2 m projection screen	Event-related potentials	Stationary; brain products BrainAmp, BrainVision analyzer
Kober and Neuper (2011)	Virtual reality projector Cube3D^2	Shutter glasses for stereoscopic imaging	Oscillations, event-related synchronization/ desynchronization	Stationary; brain products BrainAmp, BrainVision analyzer
Lapborisuth et al. (2019)	NEDE	Oculus Rift DK2	Event-related potentials	Stationary; BioSemi ActiveTwo
Liang et al. (2018)	Unity game engine & Cyberith Virtualizer	HTC Vive headset & Cyberith Virtualizer omnidirectional Treadmill	Oscillations	Mobile; BrainProducts MOVE
Lin et al. (2007a)		Stewart dynamic motion platform + real car & 7 projectors	Event-related spectral perturbation (ERSP)	Stationary; NeuroScan NuAmps
Lin et al. (2007b)		Motion platform & projectors	Event-related potentials	Unspecified
Ma et al. (2022)	Unity 3D	Pico neo 2 eye, balance board, kinematic sensors	Oscillations	Unspecified ANT neuro system

(continued)

Table 1 (continued)

Study	VR software	VR hardware	Recording	EEG system
de Oliveira et al. (2018)	Nintendo® Wii (virtual reality gaming interface)	Wii balance board & 47-in. TV screen	Oscillations	Mobile; Emotiv EPOC®
Peterson et al. (2018)	Unity 5	Oculus Rift DK 2 & Microsoft Kinect V2	Event-related activity	(Stationary system used as mobile system) BioSemi active two
Schubring et al. (2020)	Unity 3D	HTC Vive	Oscillations	Stationary; electrical geodesics, Net Station acquisition
Slobounov et al. (2015)	VTC open GL	Samsung 3D TV screen & CrystalEyes stereo glasses	Oscillations	Stationary; NeuroScan SynAmps
Stolz et al. (2019)	Vizard	Oculus Rift DK 2	Event-related potentials	Mobile; Emotiv EPOC®
Török et al. (2014)	VIRCA	CAVE automatic virtual environment	Event-related potentials	Mobile; BrainProducts MOVE
Tremmel et al. (2019)	Unity	HTC Vive headset & VIVE controllers	Event-related activity	Mobile; Guger Technologies g. MOBIlab
Tromp et al. (2018)	Vizard software	NVIS nVisor SX60 head-mounted display	Event-related potentials	Unspecified
Wang et al. (2020)	Unity game engine	HTC Vive headset & Vive trackers	Oscillations	(Stationary system used as mobile system) BioSemi active two

3 VR Hardware Options

The variety in approaches to creating virtual environments is paralleled by the variety of VR hardware options. Larger (TV) screens integrated immersive projection systems, and head-mounted displays have all been used for stimulus presentation. EEG can be quite flexibly combined with all these options, although head-mounted systems may need to be modified to not interfere with electrode placement. This is also true for fNIRS. As can be seen in Table 1, recent EEG studies tend to use head-mounted displays, and, interestingly, increasingly also mobile rather than stationary EEG systems. Positioning and space constraints in the scanner bore substantially limit the range of VR hardware options in fMRI studies. Therefore, most researchers have opted for screens or mirror projections or specific MR-compatible goggles. These hardware setups thus are not drastically different

Table 2 Overview of studies combining virtual reality with functional magnetic resonance imaging or functional near-infrared spectroscopy

Study	VR software	VR hardware
VR-fMRI studies		
Adamovich et al. (2009)	Virtools, VRPack, VRPN	5DT data glove 16 MRI
Baumgartner et al. (2008)	NoLimits roller coaster simulation & presentation software	VisuaStim XGA goggles & headphones
Burgess et al. (2001)	PC video game (Duke Nukem 3D)	24 cm x 16 cm screen, keypad
Calhoun et al. (2002)	PC video game (need for speed II)	Modified game pad controller
Carvalho et al. (2006)	STISIM drive simulator	Hand-held controller, steering wheel, foot pedals
Jang et al. (2005)	IREX®	TV screen, video camera, cyber glove
King et al. (2006)	Unreal engine, 3D studio max, Microsoft visual C++	Unspecified
Mellet et al. (2010)	Virtools 3.0	CrystalEyes stereo shutter glasses & joystick (in pre-scanning training outside of the scanner)
Mueller et al. (2012)	C++, 3ds max, vision game engine	Video projection & mirror
You et al. (2005)	IREX®	TV screen, video camera, cyber glove
VR-fNIRS studies		
Holper et al. (2010)	Torque 3D	Computer screen, 3D digital compasses for arm position, data gloves
Landowska et al. (2018)	Unity3D	Oculus Rift DK2, Octave (immersive projection system)
Putze et al. (2019)	Unity, LSL4Unity	HTC Vive
Seraglia et al. (2011)	GlovePie	Custom-built VR helmet based on bike helmet, with LCDs attached to the front, tracker for head position, Nintendo Wiimote controller

from non-VR stimulus presentation. With regard to additional data streams and/or response recording, researchers have used keyboard or keypads, game console controllers, specific controllers such as steering wheels or foot pedals. Furthermore, recording of peripheral physiological responses such as heart rate, breathing rate, skin conductance responses, and more is feasible (e.g., El Basbasse et al. 2023).

In the following sections of this chapter, we will review specific studies that combined VR with EEG (Sect. 4), fMRI and fNIRS (Sect. 5), and neurostimulation (Sect. 6).

4 Using EEG to Monitor Electrophysiological Brain Responses in VR

EEG is a widely used neuroscientific technique that measures the brain's electrophysiological activity at rest or while a task is performed (Luck 2014). Participants wear an elastic cap outfitted with several measurement electrodes (typically 32 to 64, but setups with higher and lower numbers of electrodes are available) that measure voltage changes at the scalp to estimate electric activity in the brain. Most EEG research over the last decades has been performed with stationary, wired EEG systems that confine data collection to psychological laboratories (Metzen et al. 2021). However, as technology advances and EEG systems get smaller and smaller, the number of studies using mobile, often wireless, EEG systems has been increasing (Delaux et al. 2021). Both stationary and mobile EEG have been combined with different types of VR setups for stimulus presentation. The following sections will provide an overview of the various technical approaches used, starting with stationary EEG systems, followed by mobile setups.

4.1 *Stationary EEG: Oscillations*

One of the major ways to analyze EEG data is to use transformations of the raw signal to analyze the so-called oscillations in different frequency bands (Herrmann et al. 2016) that have been linked to various cognitive processes and states of consciousness. The most common oscillations investigated in EEG research are alpha, beta, gamma, delta, and theta oscillations (Herrmann et al. 2016).

Many EEG-VR studies have focused on different types of oscillations as dependent variables. For example, an early study from 2007 investigated EEG dynamics during motion sickness (Lin et al. 2007a). The researchers used a 360° 3D VR scene of a driving environment while participants were seated in a car on a six degree-of-freedom (DOF) motion platform. They measured *event-related spectral perturbation* (ERSP), a parameter that reflects dynamic changes in amplitude of the broadband EEG spectrum as a function of time relative to an event of interest (Makeig 1993). Results showed that most subjects experienced an 8–10 Hz power increase parietally and over motor areas to motion sickness-related phenomena. Of note, the same setup was also used to assess drivers' cognitive responses during VR driving using event-related potentials (ERPs; see below).

More recently, a study compared alpha and beta oscillations in a VR environment programmed with the game engine Unity3D and presented using a Vive VR headset (HTC Corporation, Taoyuan City, Taiwan) to presentation on a 2D TV screen (Schubring et al. 2020). The authors found that both alpha and beta activity was larger in the VR condition than in the 2D condition and concluded that enhanced immersion in VR setting could potentiate EEG oscillations. Alpha oscillations were also a focus point in a recent study by Ma et al. (2022). Here, participants stood on a balance board while wearing a virtual reality VR headset, immersed in a virtual environment that created multisensory conflict. Frontal and occipital alpha power was measured to assess top-down control of visual processing and suppression of irrelevant visual information.

In an older study, 3D presentation did not involve a head-mounted display but a TV screen while participants were wearing stereo glasses (Slobounov et al. 2015). Here, higher frontal-midline theta was observed during the encoding phase of a navigation task for 3D compared to 2D presentation. Theta oscillations were also investigated in an earlier EEG-VR navigation study (Kober and Neuper 2011). Here, participants performed a navigation task presented on a 2 m-by-2 m projection screen under a Cube3D2 VR projector (Digital Image, Overath, Germany). The authors found that processing of spatial cues induced cortical theta activity in comparison with a baseline condition, an effect that was more pronounced in female than in male participants. Of note, a link between theta activity and navigation in 3D environments had already been supported by an early study using screen-based presentation of VR (Bischof and Boulanger 2003).

4.2 Stationary EEG: Event-Related Potentials

ERPs are the second most utilized way to analyze EEG data. ERPs are stereotype positive or negative components in the EEG signal that occur after a specific sensory (e.g., reading a word), cognitive (e.g., making an error), or motor (e.g., pressing a button) event (Luck 2014). Assessment of ERPs typically involves repeatedly presenting the same stimulus or asking the participant to repeatedly perform the same response. As a result, experimental paradigms can be quite lengthy in order for researchers to obtain enough trials to ensure reliable assessment of ERPs with acceptable levels of noise. Moreover, millisecond-exact timing is essential for reliable analysis, and typically the so-called triggers are sent from the stimulus presentation software to the EEG recording computer to timestamp the relevant event in the EEG datafile for later data analysis. Sending triggers typically involves additional cables and in case of some EEG system additional hardware ("trigger boxes"), which is not needed when only recording oscillations. These two factors make the implementation of ERP research in dynamic VR worlds that involve active movement of the participant more challenging than the assessment of oscillations. Nevertheless, a few EEG-VR studies have assessed ERPs. One early study from 2007 used a screen-based VR driving environment while the participant was located on a six degree-of-freedom (DOF) motion platform (Lin et al. 2007b). Participants were tested using a traffic-light events task, and EEGLAB software (Delorme and Makeig 2004) was used to assess the P300 component in the ERP. The P300 was also assessed in a more recent study that used a head-mounted display (Oculus Rift DK2, Oculus VR, Menlo Park, USA) to display a VR target detection task in which subjects moved through a three-dimensional maze (Lapborisuth et al. 2019). The paradigm was programmed using NEDE (naturalistic experimental design environment), an open-source scripting suite for developing experiments in 3D virtual environments (Jangraw et al. 2014).

Other EEG-VR studies have focused on earlier ERP components than the P300. In a study on video gaming participants played the video game "WipeoutHD Fury" on a Playstation 3 gaming system (Sony Liverpool Studios, Liverpool, UK) using either a head-mounted display or a standard LCD TV (Burns and Fairclough 2015). An acoustic oddball paradigm was presented over the soundtrack of the game using E-Prime v2.0 software (Psychology Software Tools, Sharpsburg, USA). The authors found that screen presentation mode did not affect various ERP components including the late slow wave, the P1 and the late negative component. A similar setup (i.e., auditory paradigm presented during navigation an unrelated visual VR environment) was used in another study on auditory ERPs (Kober and Neuper 2012). Here, participants navigated through a virtual city that was presented using a 2 m-by-2 m projection screen and a Cube3D[2] VR projector (Digital Image, Overath, Germany). They had to rate their personal experience of presence in the VR environment while hearing frequent standard acoustic stimuli or infrequent deviant tones. Results revealed that early auditory ERP components were not affected by presence in VR, but increased presence in VR was associated with decreased late negative slow wave amplitudes. Since late negative slow wave amplitudes have been linked to central stimulus processing and allocation of attention, these results can be interpreted to reflect strong allocation of attention to the VR environment in individuals experiencing strong presence, which decreases the attentional resources available for processing VR-irrelevant stimuli.

Language-related ERPs such as the N400 have also been investigated in VR. One study (Tromp et al. 2018) had participants wear the EEG beneath an NVIS nVisor SX60 head-mounted display (NVIS, Reston, USA) in order to study language processing in a noisy naturalistic environment. Participants performed a speech-based mismatch paradigm in a virtual restaurant that had been programmed using Vizard software version 4.08 (WorldViz, Santa Barbara, USA). They recorded N400 ERP components in response to a mismatch between a virtual partner's speech (I ordered the *fish*) and the item on their plate. Typical N400 responses were observed, demonstrating the validity of VR-EEG methods for conducting cognitive neuroscience in complex settings with high ecological validity.

4.3 Mobile EEG and VR

Mobile EEG is a relatively recent evolution of stationary EEG. Two major developments were crucial for its development: 1) technical advances that reduced the size of the amplifier, typically the bulkiest part of the EEG setup, and 2) the development of reliable wireless transmission of EEG signals, allowing for a reduction or elimination of movement-hindering cables when using EEG (Lau-Zhu et al. 2019).

A typical mobile EEG setup for VR research is shown in Fig. 1. The 32-electrode EEG cap (MES Forschungssysteme GmbH, Wörthsee, Germany) in this picture has

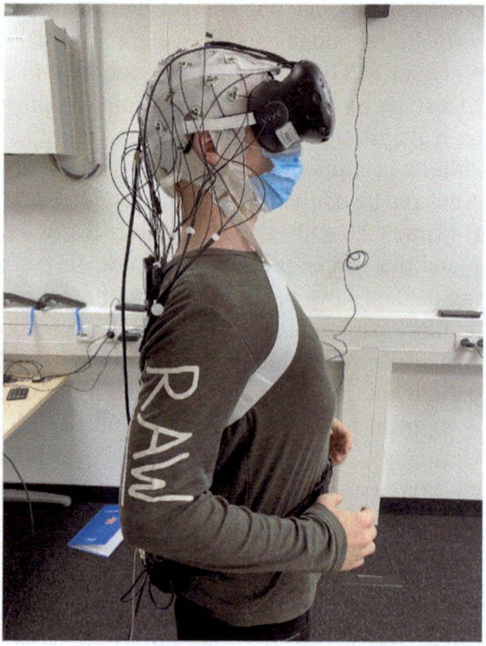

Fig. 1 A volunteer wearing a 32-electrode mobile EEG system with an EEG cap that has been modified to fit a VR headset

been modified to fit a Vive VR headset (HTC Corporation, Taoyuan City, Taiwan) by removing the original straps from the VR headset and outfitting the EEG cap with custom-made sewn-on Velcro straps located between electrode sites to minimize interference with electrophysiological recordings. Custom modification of the EEG cap was performed by the company that produced it but could in theory also be performed by researchers themselves if somewhat skilled in sewing. The EEG amplifier is tucked away in a little fabric bag located at the lower back of the EEG cap; it is therefore not visible in the picture. Chest straps ensure that movement of the setup is minimized while the subject is moving around during a task.

There are several different options for mobile EEG systems available that vary considerably regarding the number of electrodes, size, price, and other parameters. Several papers have focused on the use of such mobile EEG systems in VR research.

In an early paper from 2014, the authors conducted a comparison between wireless and wired EEG recordings in a virtual reality lab (Török et al. 2014). They used a 62-channel Brain Products MOVE mobile EEG system (Brain Products GmbH, Gilching, Germany) and recorded data from one participant while he was walking around or sitting in a cave automatic virtual environment (CAVE) (Cruz-Neira et al. 1992), and while various possible sources of electrical noise were turned on and off. The authors found that electrical noise was present in the mobile EEG data, but mostly above the frequency ranges of brain-related EEG signals. In the second experiment, the authors used a simple visual experiment in the CAVE and also a traditional screen-based setup in an EEG lab. EEG was recorded wired and wireless, and the authors assessed event-related potentials associated with early visual processing such as the N1. The shape of the N1 was not affected by the type of transmission, but N1 latency was decidedly longer in the CAVE than in the traditional setup (keep in mind that this study was in only one participant, so no statistics were calculated).

While a CAVE is a VR setup that uses projection of visual input on the walls of the room in which the participant is located, most mobile EEG studies used head-mounted displays to generate VR. One early mobile EEG study on spatial navigation in VR (Ehinger et al. 2014) used an NVIS nVisor SX60 head-mounted display (NVIS, Reston, USA) in combination with a position tracker system. Participants and technical setup were located on a cart that was moved along a predefined pathway. The authors assessed alpha band suppression during turning movements.

Similar combinations of head-mounted displays and mobile EEG systems have also been used in other studies. For example, Dey et al. (2019) combined a 32-channel LiveAmp mobile EEG system (Brain Products GmbH, Gilching, Germany) with a Vive VR headset (HTC Corporation, Taoyuan City, Taiwan) to investigate the EEG correlates of cognitive training in VR. The same combination was also used in a recent study on the decoding of emotional arousal from the EEG signal during VR rollercoaster rides (Hofmann et al. 2021). In this study, no VR paradigm was created specifically for the study but a commercial rollercoaster ride software available on the gaming platform Steam was used ("Russian VR Coasters" by 191 Funny Twins Games, Ekaterinburg, Russia). For researchers without extensive VR programming expertise, using commercial software like this is a good option to still be able to conduct mobile EEG-VR research. Due to the wide range of available commercial VR software, this idea may also be useful in other fields of psychological research. For example, the authors of this book chapter have used commercially available horror gaming software in preliminary, unpublished VR research.

The Vive head-mounted display has also been used in combination with a Brain Products MOVE mobile EEG system (Brain Products GmbH, Gilching, Germany) in a recent mobile EEG study on the dissociation of frontal-midline delta-theta and posterior alpha oscillations in a navigation task (Liang et al. 2018). Another study combined the Vive with the BioSemi Active II EEG system (BioSemi, Amsterdam, The Netherlands) which originally is a stationary system that can also be used in mobile settings (Wang et al. 2020).

One particularly interesting application of mobile EEG-VR research is research on high heights exposure and related fear – a topic that is almost impossible to investigate with EEG outside of an VR environment due to both technical and ethical concerns. One study assessed EEG and other markers of physiological stress during VR beam walking at high heights (Peterson et al. 2018). In this study, participant wore an Oculus Rift DK2 head-mounted display (Oculus VR, Irvine, USA) and a 136-channel BioSemi Active II EEG system (BioSemi, Amsterdam, The Netherlands), as well as heart rate and electrodermal activity sensors. The virtual environment consisted of a virtual balance beam located either on the ground or 15 m above ground. It was created using Unity 5 software (Unity Technologies, San Francisco, USA). Subjects moved around the VR environments using gestures that were tracked by a Microsoft Kinect V2 (Microsoft, Redmond, USA). The authors found that during virtual beam walking above ground, source-localized EEG peak amplitude of the anterior cingulate significantly decreased. As this brain area is important for sensing loss of balance, the authors concluded that their VR task provided a realistic experience.

Of note, some studies have used an Emotiv EPOC® system (Emotiv Systems, Sydney, Australia) in mobile EEG-VR research (de Oliveira et al. 2018; Stolz et al. 2019). The EPOC® is a small, commercially available low-density EEG system that has the advantage of being considerable cheaper than the typical mobile EEG system. A study investigating the data quality that can be obtained with this system concluded that good quality, single-trial EEG data can be obtained with the EPOC® (Debener et al. 2012).

4.4 *Mobile EEG and VR: Movement Artifacts*

As EEG systems get smaller and smaller and newer VR systems allow room-scale tracking, paradigms that allow participants to move freely in VR will become increasingly common in the future. However, free movements can increase noise in the EEG data by generating movement artifacts. Figure 2 shows EEG raw data recorded with a LiveAmp mobile EEG system (Brain Products GmbH, Gilching, Germany). Periods of increased movements are clearly distinguishable from periods of standing still.

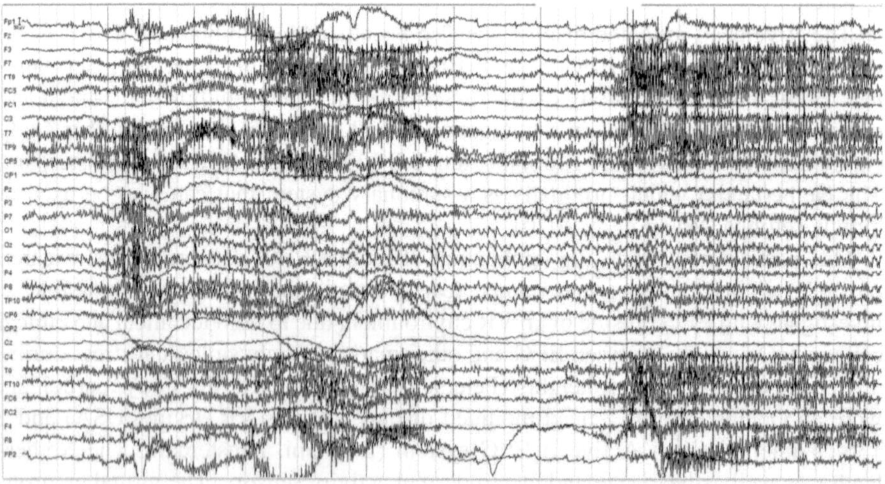

Fig. 2 Raw EEG signal obtained with a mobile EEG system. The participant alternated between moving and standing still. Increased noise during periods of movement is clearly visible in the raw EEG signal

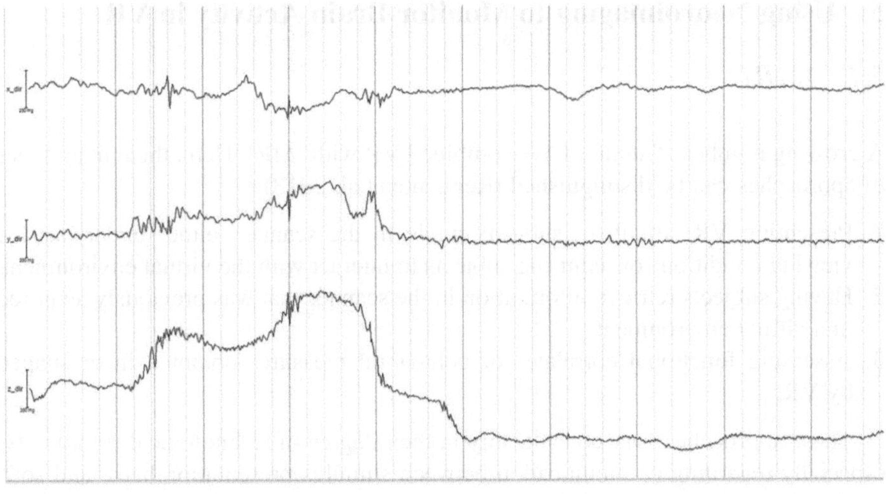

Fig. 3 Raw movement signal from the accelerometer of a mobile EEG system. The participant alternated between moving and standing still. Increased movements are clearly visible in the first half of the recording

As such movement-related artifacts may impair EEG data quality, some mobile EEG system include or allow for accelerometers to measure movements in the X, Y, and Z directions so that they can be assessed in later analysis. This then typically results in 3 additional channels in the recording (see Fig. 3).

Unfortunately, there are currently no established standards on how to deal with movement artifacts in EEG paradigms that utilize free movement in VR or physical space. One option is to collect critical data while participants are stationary, in between periods of movement. However, advances in signal processing and statistical techniques may make it possible to extract clean EEG signals during movement. Tremel et al. (2019) had participants perform a VR n-back task that involved putting colored balls into treasure chests. The authors used a technique called warp correlation filter to filter movement artifacts out of the EEG signal. This technique had previously been used successfully to clean mobile EEG data in a non-VR paradigm (Gwin et al. 2010). These approaches will make it possible for researchers to use a wider (and more ecologically) valid range of behavioral responses in their research.

Other approaches such as using accelerometer data as co-variates in statistical analysis of EEG data or using independent component analysis (ICA) as implemented in most EEG analysis software packages are also feasible. However, best practice guidelines for how to optimally deal with movement artifacts in mobile EEG-VR setups are currently still missing and more systematic research into this question is clearly needed.

5 Using Neuroimaging to Monitor Brain Activity in VR

5.1 *fMRI*

A growing number of studies have combined VR with fMRI. Here, three main types of approaches can be distinguished (Reggente et al. 2018):

1. Presenting VR stimuli to subjects while in the scanner, either under passive viewing conditions or requiring subjects to interact with the virtual environment.
2. Having subjects retrieve information in the scanner that was previously acquired in a virtual environment.
3. Assessing functional correlates of behavioral measures obtained in or shaped by VR.

These approaches vary considerably in their degree of technological complexity, especially regarding communication between stimulus presentation hard- and software and the MR scanner. Unfortunately, best practice guidelines or handy manuals on how to implement VR in fMRI research are yet lacking, and while researchers have found a range of different solutions, especially early publications often lack appropriate detail when reporting the methods, substantially limiting reproducibility and comparability of the findings.

Early work has frequently made use of video games for computers or game consoles, which likely required considerable coding skills and effort to adapt the existing virtual environment to the researchers' needs. For instance, in one study the authors modified the PC video game Duke Nukem 3D (3D Realms Entertainment, Apogee Software Ltd., Garland, TX) to create a virtual town for subjects to move through using four keys on a response pad while receiving objects from virtual people in different locations (Burgess et al. 2001). During scanning, subjects completed a forced choice recognition test in the VR environment. The use of VR for stimulus presentation in this study allowed memory testing for different aspects of a lifelike event and its context. Data analysis was event-related, with event codes logged for the responses in the memory test. Event codes for other parts of the VR stimulation were not required.

VR has also been used to investigate brain activations during simulated driving using the game Need for Speed II™ (Electronic Arts, Redwood City, CA) (Calhoun et al. 2002), as well as an interactive driving simulator (STISIM M500W, Systems Technology Inc., Hawthorne, CA) that allows for custom-designed roadway environments and situations (Carvalho et al. 2006). Here, the authors extended the standard technical setup (which included a hand-held controller and a projector outside the scanner room and behind the scanner that projected to a translucent screen that the subjects could see in a mirror attached to the head coil) with a steering wheel located just outside the scanner bore and foot pedals located at the subject's feet. These devices were connected to computers outside the scanner room via a shielded cable to enable continuous sampling of information on steering and pedal activity.

Moreover, a highly immersive virtual roller coaster scenario using the commercial software "NoLimits Roller Coaster Simulation" (www.nolimitscoaster.com) has been employed in fMRI research (Baumgartner et al. 2008). Of note, stimulus presentation was controlled by *Presentation*® software (Neurobehavioral Systems, Berkeley, CA), which is widely used in experimental psychology and neuroscience for its powerful interface for physiological measures such as EEG, MEG, or fMRI. Stimuli were presented by means of stereoscopic MR-compatible goggles (VisuaStim XGA; Resonance Technology, Northridge, California) which have become increasingly common in MRI setups, and headphones were used for auditory stimulation.

Virtual environments in fMRI research were also constructed using the Unreal engine (Epic Games Inc., Raleigh, NC) (e.g., King et al. 2006). These researchers made additions using 3D Studio Max (Autodesk, San Rafael, CA) and Microsoft Visual C++, the latter of which to enable communication via the computer's parallel port. In this study, subjects had to either shoot or bandage a virtual human or an aggressive humanoid, and an event-related fMRI design was used to compare brain activations for context-appropriate and context-inappropriate aggressive or compassionate responses.

The Unreal engine, along with other game engines such as Unity (Unity Technologies, San Francisco, CA) or SteamVR (Valve, Kirkland, WA), is among the most popular software for VR experiments. Recent work has indicated that even though this software is not typically engineered for use in neuroscientific experiments and may not come with built-in features for data collection and output, possibly compromising precision and accuracy of time critical measurements, it does achieve high levels of precision and accuracy both concerning stimulus duration and response time measurement, at least when used in combination with the commonly used HTC Vive (Wiesing et al. 2020).

A commercial VR stimulation solution specifically developed for neurobehavioral and functional MRI studies was VR World 2 (Baumann et al. 2003), which was intended to provide an accessible drag-and-drop concept for building virtual environments. It required a standard computer with a high-performance graphics card running Windows, and the software offered a range of options for implementing simultaneous physiological monitoring or interfacing with MR scanners or EEG systems. Unfortunately, VR World 2 is no longer available as a standalone software. Other researchers have developed open-source solutions, for example the software Reactor-Man which was specifically designed to provide accurate stimulus and event control as typically required in psychological experiments (Wolter et al. 2007). Software like this could therefore prove to be particularly useful for fMRI studies with event-related designs.

Some researchers have relied on third parties to develop their virtual environments (Mellet et al. 2010). In this work, the authors presented an immersive virtual replication of their lab to study the representation of topographical knowledge learned in VR. Of note, these researchers also used commercial software 3DVIA Virtools 3.0 (now Dassault Systèmes SE, Vélizy-Villacoublay, France) for stimulus presentation.

In some studies, the virtual environments are a VR version of a classic tasks used in Experimental Psychology. For example, Beck et al. created a line bisection task that depicted objects (a toilet paper roll and a test tube) in realistic settings (on a metal holder or a wooden shelf), and subjects had to indicate whether these objects were centered or not (Beck et al. 2010).

VR systems have also been used in fMRI scanners to assess motor behavior and concurrent brain activity (Adamovich et al. 2009), which presents particular challenges given the space constraints. In this study, subjects first observed the movements of virtual hand avatars and subsequently imitated them while viewing the avatars as animated by their own movements. The technical setup included an MRI-compatible right hand 5DT Data Glove 16 MRI (Fifth Dimension Technologies, 5DT Data Glove 16 MRI, http://www.5dt.com) equipped with fiberoptic sensors (Ascension Flock of Birds 6° of freedom sensors; Ascension Technology Corp., http://www.ascension-tech.com) to measure hand movement. Subjects wore the glove in the magnet, and communication between the glove and the stimulation computer was realized by signals transmitted via fiberoptic cables that were digitized and fed into the computer's serial port. The virtual environment was created using Virtools (Dassault Systèmes, Virtools Dev 3.5, 2006: http://www.virtools.com) with the VRPack plugin that communicates with the open-source VRPN (Virtual Reality Peripheral Network) (Taylor et al. 2001). Using peripherals like gloves and foot pedals increases the range of movements (or simulated movements) that can be studied in the scanner.

A commercial solution for interfacing kinematic data with VR particularly for interactive rehabilitation applications is the IREX® virtual reality system (GestureTek Health, Toronto, ON). Two studies (Jang et al. 2005; You et al. 2005) used VR as an intervention and compared pre- and post-VR training motor activation in stroke patients, characterizing both VR-induced neuroplastic changes as evident in fMRI data and associated motor recovery on the behavioral level.

A VR stimulus application based on C++ that allows fast and easy presentation of VR environments in fMRI studies without additional expert knowledge has been developed (Mueller et al. 2012). Unlike many other platforms, this application also supports real-time data analysis (as, for instance, needed in real-time fMRI experiments) due to a bidirectional communication interface based on an integrated Transmission Control Protocol (TCP). Of note, real-time fMRI analysis in their VR-fMRI experiments was realized using the commercial software Turbo BrainVoyager 3.0 (Brain Innovation BV, Maastricht, NL). The virtual environment itself comprised two components: (1) a virtual scene or 3D model created using the commercial professional design software 3ds Max (Autodesk, San Rafael, CA), and (2) a game engine framework that calculates interactions between the subject and the 3D model and simulates human behavior and realistic environmental conditions. In addition, the generation of event-markers and interactive elements was supported by a virtual event system. The latter aspects were realized using Vision Game Engine (Havok, Dublin, Ireland), a popular tool in the computer gaming industry. While it is a commercial software product, the original developer Trinigy (Eningen, Germany) allowed its usage for non-commercial and scientific purposes for free. All relevant components of the Vision Game Engine were embedded C++ source code programmed with Microsoft Visual Studio 2008, Microsoft.NET Framework 2.5., and PhysX, version 2.8.1 (NVIDIA, Santa Clara, CA), which is freely available.

5.2 *fNIRS*

fNIRS combines the spatial acuity of fMRI with the (relative) freedom of movement of EEG, rendering the technology of particular interest for research into mechanisms of neurorehabilitation. Along these lines, a VR-fNIRS neurorehabilitation system based on a VR environment (built with the Torque multi-user 3D gaming environment; GarageGames, Oregon, USA) that contained a virtual limb performing a grasping task was developed (Holper et al. 2010). The VR stimulation largely followed the procedures outlined in earlier work (Eng et al. 2007) in which representations of a subject's own arms and hands were displayed on a large computer screen (94 cm diagonal) and controlled by arm position trackers (3D digital compasses; Honeywell) and data gloves. Of note, by using a screen for VR presentation, this procedure circumvented the main challenge when combining VR with fNIRS: (in)compatibility of the sensor cap with common VR headsets. Some researchers have developed creative solutions to this problem. For example, Seraglia et al. (2011) built a customized VR helmet by modifying a bike helmet with cut outs for the fNIRS sensors and optical fibers over parietal and occipital regions (Seraglia et al. 2011). They attached the LCDs from a V8 Head Mount Display (Virtual Research Systems, Aptos, CA) to the front of the helmet and added a Velcro belt in the back to counterbalance the weight. This belt itself was secured to the participant's back using a thoracic belt. A tracker for head position was also attached to the helmet above the LCDs. Participants interacted with the VR using a Nintendo Wiimote controller that operates as a normal mouse when coupled with the software GlovePie (http://glovepie.org/). In the virtual line bisection task programmed by the researchers, the Wiimote moved a small red dot to simulate a laser pointer, and participants could record the pointer's position by pressing the A button.

In one study, fNIRS was recorded from central and parietal sites using a 27-channel optical topography system (Brite24, Artinis Medical Systems) in a real and a virtual environment (Ulsamer et al. 2020). This fNIRS system is rather plug-and-play and uses a soft neoprene head cap that poses less interference with head-mounted displays. Subjects in this study wore the HTC Vive Pro (HTC, Xindian, New Taipei, Taiwan). Several other recent studies have also used versions of the HTC Vive in combination with fNIRS (Lamb et al. 2018; Kim et al. 2019; Putze et al. 2019). Putze et al. (2019) had subjects perform a VR version of the n-back task, including physical interaction using the standard VIVE controller. The task was implemented using the Unity framework, with Lab Streaming Layer middleware (https://github.com/sccn/labstreaminglayer) and LSL4Unity plugin used for synchronization (https://github.com/xfleckx/LSL4Unity). Most notably, the authors have made the software for their experiment available in an Open Science Framework (OSF) repository (https://osf.io/yhtz8/).

Landowska et al. (2018) tested different ways to present virtual environments in combination with fNIRS: the Oculus Rift DK2 by Oculus VR as a head-mounted display, and Octave, an elaborate, stationary surround immersive projection technology (IPT) system (https://www.salford.ac.uk/our-facilities/octave). Octave provides an octagonal IPT space of approximately 32.6 m^2 in which immersive projection is delivered via surrounding wall and floor displays. It also comprises a 32-channel wave field synthesis acoustic system with 264 speakers to support an authentic acoustic experience. The authors assessed signal-to-noise ratios, freedom of movement, and motion artifacts as well as hemodynamic response from the prefrontal cortex in IPT. The results indicated that fNIRS is compatible with both approaches and most of the movements they allow participants to make.

6 Neurostimulation Techniques and VR

We conclude this chapter by highlighting a few ways that VR can be combined with non-invasive brain stimulation techniques such as transcranial magnetic stimulation (TMS) (Lefaucheur 2019) or transcranial direct current stimulation (tDCS) (Chase et al. 2020). These techniques allow researchers to assess causal relationships between brain function and cognition and hold promise for many applied research domains such as spatial navigation or vehicle driving which may particularly benefit from the use of more ecologically valid settings.

Some studies have used anodal or cathodal transcranial direct current stimulation (tDCS). This is commonly applied for 20 min immediately prior to performance of an experimental task. The use of electrodes attached to the scalp therefore does not always have to be coordinated with the use of head-mounted displays for VR delivery. For example, Ferrucci et al. (2019) applied anodal or sham tDCS to the cerebellum after subjects had navigated a virtual environment and learned the location of an object in the encoding phase of their task. 30 min after stimulation, subjects had to retrieve the object from a different starting point in the virtual environment. In this study, the experimental task was programmed using the software Unity 3D (www.unity3d.com), VR delivery was realized by using an Oculus Rift DK2, and subjects used a joypad to move around in the virtual city.

An older study by Beeli et al. (2008) applied cathodal, anodal, and sham tDCS to the right dorsolateral prefrontal cortex while subjects were watching a virtual roller coaster ride in order to modulate the experience of presence. The same commercial rollercoaster simulation software (http://www.nolimitscoaster.com) used by Baumgartner et al. (2008) in combination with fMRI was also used in this study. Psychophysiological measurements such as electrodermal activity (EDA) and the electro-myogram (EMG) were also recorded.

Of note, Asbee and Parsons (2021) and Brunyé et al. (2019) provide a small meta-analysis and review, respectively, of studies combining tDCS and VR. While these publications offer guidance to researchers regarding the effectiveness of this type of brain stimulation and the types of questions that can be addressed combining these methods, information pertaining to methodological and technical aspects is limited.

Transcranial magnetic stimulation (TMS) is an alternative technique that is also typically applied either repetitively over a period of time prior to task performance or concurrently during task performance, with single or double pulses time-locked to stimuli or responses. The TMS coil is positioned over the scalp (often guided by a structural scan of the participant's brain) and fixed either with a stand of some sort or held by the experimenter. In any case, the coil usually does not spatially interfere with standard head-mounted displays.

One study (Bassolino et al. 2018) combined TMS applied to the motor cortex to stimulate the hand corticospinal representation with visual hand feedback provided in VR, in order to investigate the induction of embodiment for the virtual hand. In other words, they integrated TMS and VR to elicit the Rubber Hand Illusion. Subjects received single TMS pulses over motor cortex (or vertex, as a control condition) at specific stimulation intensities. The virtual hand was presented using an Oculus Rift (Oculus VR, Menlo Park, CA, USA) head-mounted display and in-house software designed for creating and running experiments in VR (ExpyVR, EPFL, http://lnco.epfl.ch/expyvr). This software also logged the subjects' answers and generated triggers for the TMS pulses that were transmitted by a Magstim monophasic stimulator (Magstim Co., Whitland, UK) with a laptop-parallel-adapter-card.

While the technical setup used by Bassolino et al. (2018) includes sophisticated in-house software that is not available to all researchers in the field, Talkington et al. (2015) have shared a protocol that synchronously samples data generated by TMS, electromyography, and 3D motion capture in the context of virtual reality stimulus presentation and feedback. The authors describe overall connectivity, intersystem signaling, as well as temporal synchronization of the recorded data. Synchronization is mainly accomplished by a customizable circuit that consists of readily available off-the-shelf components which require minimal electronics assembly skills. Most importantly, the publication is supplemented by a helpful video guide that can be found at http://www.jove.com/video/52906/ .

7 Conclusion

Neuroscience researchers are increasingly using virtual reality in creative ways to advance our understanding of the relationship between brain and behavior. Early research was often very methodological – aimed at addressing the viability of combining VR with neuroscientific research paradigms and establishing fundamental effects. We expect future research to unleash the potential of VR to yield new insights that are not possible with conventional laboratory methods. However, challenges remain. It is evident that there still is a lack of consensus regarding methodological "gold standards" for the use of VR in EEG and neuroimaging, but there are some promising developments. The use of now redundant software and hardware in older studies renders comparisons between early work and more recent studies and their results difficult. This problem is more pronounced in VR neuroscience research than in many other fields of psychology and neuroscience, due to the fast turnover of technologies in VR and the need for highly specialized interfaces between VR and neuroscientific tools. Particularly the VR software development scene is highly dynamic, and smaller, independent companies and their products are often bought by bigger corporations. As a result, some software might be discontinued or become unavailable. On the other hand, new tools are developed at a high rate, and increasingly so with researchers as targeted audience. For example, Vizard (WorldViz, Santa Barbara, CA) is a comprehensive VR software platform that allows Python-based development of virtual environments. Importantly, it offers connectivity with a range of research-grade hardware devices such as data gloves, head-mounted VR displays, body motion and eye trackers, or biofeedback monitors. It also has certain data management features built-in, e.g., processing data or streaming data out in real time.

Open-source software products such as OpenSimulator (www.opensimulator.org), a versatile and powerful multi-platform, multi-user 3D application server, substantially increase the accessibility of VR applications. However, it must be noted that open-source solutions tend to require more computational expertise to adapt them to the researcher's needs. Nevertheless, virtual environments created by open-source (or commercial) software these days can be customized to meet the needs of a wide variety of tasks and to offer sufficient experimental control. In addition, (hardware) equipment available in most scanner suites, e.g., stereoscopic MR-goggles or joysticks/joypads, further add to VR's accessibility and help pave the way for advanced research and even clinical applications.

Another challenge with some of the previous VR neuroimaging work is insufficiently detailed reporting of procedures which sometimes impairs reproducibility. Nonetheless, similar approaches have now been applied in several studies, as evident in this review, reflecting the emergence of a methodological foundation to build upon in future studies. Taken together, we are convinced that given the strong benefits of VR, e.g., research paradigms in more naturalistic environments with high ecological validity (de Gelder et al. 2018), it will see increased use in neuroscientific research over the next decade. More support for open-source VR software as well as the development of consensus and best practice papers on issues like how to handle movement artifacts in mobile EEG-VR will be essential steps in ensuring the continued success of this exciting and powerful technique in neuroscientific research.

Acknowledgments The authors thank Yasmin El Basbasse for help with literature research.

References

Adamovich SV, August K, Merians A, Tunik E (2009) A virtual reality-based system integrated with fmri to study neural mechanisms of action observation-execution: a proof of concept study. Restor Neurol Neurosci 27(3):209–223. https://doi.org/10.3233/RNN-2009-0471

Asbee J, Parsons TD (2021) Effects of transcranial direct current stimulation on cognitive and affective outcomes using virtual stimuli: a systematic review. Cyberpsychol Behav Soc Netw. https://doi.org/10.1089/cyber.2020.0301

Bassolino M, Franza M, Bello Ruiz J, Pinardi M, Schmidlin T, Stephan MA et al (2018) Non-invasive brain stimulation of motor cortex induces embodiment when integrated with virtual reality feedback. Eur J Neurosci 47(7):790–799. https://doi.org/10.1111/ejn.13871

Baumann S, Neff C, Fetzick S, Stangl G, Basler L, Vereneck R, Schneider W (2003) A virtual reality system for neurobehavioral and functional MRI studies. Cyberpsychol Behav 6(3): 259–266. https://doi.org/10.1089/109493103322011542

Baumgartner T, Speck D, Wettstein D, Masnari O, Beeli G, Jäncke L (2008) Feeling present in arousing virtual reality worlds: prefrontal brain regions differentially orchestrate presence experience in adults and children. Front Hum Neurosci 2:8. https://doi.org/10.3389/neuro.09.008.2008

Beck L, Wolter M, Mungard NF, Vohn R, Staedtgen M, Kuhlen T, Sturm W (2010) Evaluation of spatial processing in virtual reality using functional magnetic resonance imaging (FMRI). Cyberpsychol Behav Soc Netw 13(2):211–215. https://doi.org/10.1089/cyber.2008.0343

Beeli G, Casutt G, Baumgartner T, Jäncke L (2008) Modulating presence and impulsiveness by external stimulation of the brain. Behav Brain Funct 4:33. https://doi.org/10.1186/1744-9081-4-33

Bischof WF, Boulanger P (2003) Spatial navigation in virtual reality environments: an EEG analysis. Cyberpsychol Behav 6(5):487–495. https://doi.org/10.1089/109493103769710514

Bohil CJ, Alicea B, Biocca FA (2011) Virtual reality in neuroscience research and therapy. Nat Rev Neurosci 12(12):752–762. https://doi.org/10.1038/nrn3122

Brunyé TT, Hussey EK, Fontes EB, Ward N (2019) Modulating applied task performance via transcranial electrical stimulation. Front Hum Neurosci 13:140. https://doi.org/10.3389/fnhum.2019.00140

Burgess N, Maguire EA, Spiers HJ, O'Keefe J (2001) A temporoparietal and prefrontal network for retrieving the spatial context of lifelike events. Neuroimage 14(2):439–453. https://doi.org/10.1006/nimg.2001.0806

Burns CG, Fairclough SH (2015) Use of auditory event-related potentials to measure immersion during a computer game. Int J Hum-Comput Stud 73:107–114. https://doi.org/10.1016/j.ijhcs.2014.09.002

Calhoun VD, Pekar JJ, McGinty VB, Adali T, Watson TD, Pearlson GD (2002) Different activation dynamics in multiple neural systems during simulated driving. Hum Brain Mapp 16(3): 158–167. https://doi.org/10.1002/hbm.10032

Carvalho KN, Pearlson GD, Astur RS, Calhoun VD (2006) Simulated driving and brain imaging: combining behavior, brain activity, and virtual reality. CNS Spectr 11(1):52–62. https://doi.org/10.1017/S1092852900024214

Chase HW, Boudewyn MA, Carter CS, Phillips ML (2020) Transcranial direct current stimulation: a roadmap for research, from mechanism of action to clinical implementation. Mol Psychiatry 25(2):397–407. https://doi.org/10.1038/s41380-019-0499-9

Cruz-Neira C, Sandin DJ, DeFanti TA, Kenyon RV, Hart JC (1992) The CAVE: audio visual experience automatic virtual environment. Commun ACM 35(6):64–72. https://doi.org/10.1145/129888.129892

de Gelder B, Kätsyri J, de Borst AW (2018) Virtual reality and the new psychophysics. Br J Psychol 109(3):421–426. https://doi.org/10.1111/bjop.12308

de Oliveira SMS, Pimenta de Medeiros CS, Pacheco TBF, Bessa NPOS, Silva FGM, Tavares NSA et al (2018) Electroencephalographic changes using virtual reality program: technical note. Neurol Res 40(3):160–165. https://doi.org/10.1080/01616412.2017.1420584

Debener S, Minow F, Emkes R, Gandras K, de Vos M (2012) How about taking a low-cost, small, and wireless EEG for a walk? Psychophysiology 49(11):1617–1621. https://doi.org/10.1111/j.1469-8986.2012.01471.x

Delaux A, de Saint Aubert J-B, Ramanoël S, Bécu M, Gehrke L, Klug M et al (2021) Mobile brain/body imaging of landmark-based navigation with high-density EEG. Eur J Neurosci. https://doi.org/10.1111/ejn.15190

Delorme A, Makeig S (2004) EEGLAB: an open source toolbox for analysis of single-trial EEG dynamics including independent component analysis. J Neurosci Methods 134(1):9–21. https://doi.org/10.1016/j.jneumeth.2003.10.009

Dey A, Chatburn A, Billinghurst M (2019) Exploration of an EEG-based cognitively adaptive training system in virtual reality. In: 2019 IEEE conference on virtual reality and 3D user interfaces (VR). IEEE, Osaka, pp 220–226

Ehinger BV, Fischer P, Gert AL, Kaufhold L, Weber F, Pipa G, König P (2014) Kinesthetic and vestibular information modulate alpha activity during spatial navigation: a mobile EEG study. Front Hum Neurosci 8:71. https://doi.org/10.3389/fnhum.2014.00071

El Basbasse Y, Packheiser J, Peterburs J, Maymon C, Güntürkün O, Grimshaw G, Ocklenburg S (2023) Walk the Plank! Using mobile EEG to investigate emotional lateralization of immersive fear in virtual reality. R Soc Open Sci 10:221239. https://doi.org/10.1098/rsos.221239

Eng K, Siekierka E, Pyk P, Chevrier E, Hauser Y, Cameirao M et al (2007) Interactive visuo-motor therapy system for stroke rehabilitation. Med Biol Eng Comput 45(9):901–907. https://doi.org/10.1007/s11517-007-0239-1

Ferrucci R, Serino S, Ruggiero F, Repetto C, Colombo D, Pedroli E et al (2019) Cerebellar transcranial direct current stimulation (tDCS), leaves virtual navigation performance unchanged. Front Neurosci 13:198. https://doi.org/10.3389/fnins.2019.00198

Gwin JT, Gramann K, Makeig S, Ferris DP (2010) Removal of movement artifact from high-density EEG recorded during walking and running. J Neurophysiol 103(6):3526–3534. https://doi.org/10.1152/jn.00105.2010

Herrmann CS, Strüber D, Helfrich RF, Engel AK (2016) EEG oscillations: from correlation to causality. Int J Psychophysiol 103:12–21. https://doi.org/10.1016/j.ijpsycho.2015.02.003

Hofmann SM, Klotzsche F, Mariola A, Nikulin V, Villringer A, Gaebler M (2021) Decoding subjective emotional arousal from EEG during an immersive virtual reality experience. eLife 10. https://doi.org/10.7554/eLife.64812

Holper L, Muehlemann T, Scholkmann F, Eng K, Kiper D, Wolf M (2010) Testing the potential of a virtual reality neurorehabilitation system during performance of observation, imagery and imitation of motor actions recorded by wireless functional near-infrared spectroscopy (fNIRS). J Neuroeng Rehabil 7(1):57. https://doi.org/10.1186/1743-0003-7-57

Jang SH, You SH, Hallett M, Cho YW, Park C-M, Cho S-H et al (2005) Cortical reorganization and associated functional motor recovery after virtual reality in patients with chronic stroke: an experimenter-blind preliminary study. Arch Phys Med Rehabil 86(11):2218–2223. https://doi.org/10.1016/j.apmr.2005.04.015

Jangraw DC, Johri A, Gribetz M, Sajda P (2014) NEDE: an open-source scripting suite for developing experiments in 3D virtual environments. J Neurosci Methods 235:245–251. https://doi.org/10.1016/j.jneumeth.2014.06.033

Kim G, Buntain N, Hirshfield L, Costa MR, Makana CT (2019) Processing racial stereotypes in virtual reality: an exploratory study using functional near-infrared spectroscopy (fNIRS). In: Schmorrow DD, Fidopiastis CM (eds) Augmented cognition, vol 11580. Springer. Lecture Notes in Computer Science, Cham, pp 407–417

King JA, Blair RJ, Mitchell DGV, Dolan RJ, Burgess N (2006) Doing the right thing: a common neural circuit for appropriate violent or compassionate behavior. Neuroimage 30(3):1069–1076. https://doi.org/10.1016/j.neuroimage.2005.10.011

Kober SE, Neuper C (2011) Sex differences in human EEG theta oscillations during spatial navigation in virtual reality. Int J Psychophysiol 79(3):347–355. https://doi.org/10.1016/j.ijpsycho.2010.12.002

Kober SE, Neuper C (2012) Using auditory event-related EEG potentials to assess presence in virtual reality. Int J Hum-Comput Stud 70(9):577–587. https://doi.org/10.1016/j.ijhcs.2012.03.004

Lamb R, Antonenko P, Etopio E, Seccia A (2018) Comparison of virtual reality and hands on activities in science education via functional near infrared spectroscopy. Comput Educ 124:14–26. https://doi.org/10.1016/j.compedu.2018.05.014

Landowska A, Royle S, Eachus P, Roberts D (2018) Testing the potential of combining functional near-infrared spectroscopy with different virtual reality displays—oculus rift and oCtAVE. In: Jung T, Dieck MC (eds) Augmented reality and virtual reality. Springer. Progress in IS, Cham, pp 309–321

Lapborisuth P, Faller J, Koss J, Waytowich NR, Touryan J, Sajda P (2019) Investigating evoked EEG responses to targets presented in virtual reality. In: Annual international conference of the IEEE engineering in medicine and biology society. IEEE Engineering in Medicine and Biology Society, pp 5536–5539. https://doi.org/10.1109/EMBC.2019.8856761

Lau-Zhu A, Lau MPH, McLoughlin G (2019) Mobile EEG in research on neurodevelopmental disorders: opportunities and challenges. Dev Cogn Neurosci 36:100635. https://doi.org/10.1016/j.dcn.2019.100635

Lefaucheur J-P (2019) Transcranial magnetic stimulation. In: Handbook of clinical neurology, vol 160, pp 559–580. https://doi.org/10.1016/B978-0-444-64032-1.00037-0

Liang M, Starrett MJ, Ekstrom AD (2018) Dissociation of frontal-midline delta-theta and posterior alpha oscillations: a mobile EEG study. Psychophysiology 55(9):e13090. https://doi.org/10.1111/psyp.13090

Lin C-T, Chuang S-W, Chen Y-C, Ko L-W, Liang S-F, Jung T-P (2007a) EEG effects of motion sickness induced in a dynamic virtual reality environment. In: Annual international conference of the IEEE engineering in medicine and biology society. IEEE Engineering in Medicine and Biology Society, pp 3872–3875. https://doi.org/10.1109/IEMBS.2007.4353178

Lin C-T, Chung I-F, Ko L-W, Chen Y-C, Liang S-F, Duann J-R (2007b) EEG-based assessment of driver cognitive responses in a dynamic virtual-reality driving environment. IEEE Trans Biomed Eng 54(7):1349–1352. https://doi.org/10.1109/TBME.2007.891164

Luck SJ (2014) An introduction to the event-related potential technique, A Bradford book series, 2nd edn. MIT Press, Cambridge. Available online at https://search.ebscohost.com/login.aspx?direct=true&scope=site&db=nlebk&db=nlabk&AN=798344

Ma, L., Marshall, P.J. & Wright, W.G. (2022). The impact of external and internal focus of attention on visual dependence and EEG alpha oscillations during postural control. J NeuroEngineering Rehabil 19, 81. https://doi.org/10.1186/s12984-022-01059-7

Makeig S (1993) Auditory event-related dynamics of the EEG spectrum and effects of exposure to tones. Electroencephalogr Clin Neurophysiol 86(4):283–293. https://doi.org/10.1016/0013-4694(93)90110-H

Mellet E, Laou L, Petit L, Zago L, Mazoyer B, Tzourio-Mazoyer N (2010) Impact of the virtual reality on the neural representation of an environment. Hum Brain Mapp 31(7):1065–1075. https://doi.org/10.1002/hbm.20917

Metzen D, Genç E, Getzmann S, Larra MF, Wascher E, Ocklenburg S (2021) Frontal and parietal EEG alpha asymmetry: a large-scale investigation of short-term reliability on distinct EEG systems. Brain Struct Funct. https://doi.org/10.1007/s00429-021-02399-1

Mueller C, Luehrs M, Baecke S, Adolf D, Luetzkendorf R, Luchtmann M, Bernarding J (2012) Building virtual reality fMRI paradigms: a framework for presenting immersive virtual environments. J Neurosci Methods 209(2):290–298. https://doi.org/10.1016/j.jneumeth.2012.06.025

Parsons TD (2015) Virtual reality for enhanced ecological validity and experimental control in the clinical, affective and social neurosciences. Front Hum Neurosci 9:660. https://doi.org/10.3389/fnhum.2015.00660

Peterson SM, Furuichi E, Ferris DP (2018) Effects of virtual reality high heights exposure during beam-walking on physiological stress and cognitive loading. PloS One 13(7):e0200306. https://doi.org/10.1371/journal.pone.0200306

Putze F, Herff C, Tremmel C, Schultz T, Krusienski DJ (2019) Decoding mental workload in virtual environments: a fNIRS study using an immersive n-back task. In: 2019 41st annual international conference of the IEEE engineering in medicine and biology society (EMBC). IEEE, Berlin, pp 3103–3106

Reggente N, Essoe JK-Y, Aghajan ZM, Tavakoli AV, McGuire JF, Suthana NA, Rissman J (2018) Enhancing the ecological validity of fMRI memory research using virtual reality. Front Neurosci 12., Article 408:408. https://doi.org/10.3389/fnins.2018.00408

Schubring D, Kraus M, Stolz C, Weiler N, Keim DA, Schupp H (2020) Virtual reality potentiates emotion and task effects of alpha/beta brain oscillations. Brain Sci 10(8). https://doi.org/10.3390/brainsci10080537

Seraglia B, Gamberini L, Priftis K, Scatturin P, Martinelli M, Cutini S (2011) An exploratory fNIRS study with immersive virtual reality: a new method for technical implementation. Front Hum Neurosci 5:176. https://doi.org/10.3389/fnhum.2011.00176

Slobounov SM, Ray W, Johnson B, Slobounov E, Newell KM (2015) Modulation of cortical activity in 2D versus 3D virtual reality environments: an EEG study. Int J Psychophysiol 95(3): 254–260. https://doi.org/10.1016/j.ijpsycho.2014.11.003

Stolz C, Endres D, Mueller EM (2019) Threat-conditioned contexts modulate the late positive potential to faces-A mobile EEG/virtual reality study. Psychophysiology 56(4):e13308. https://doi.org/10.1111/psyp.13308

Talkington WJ, Pollard BS, Olesh EV, Gritsenko V (2015) Multifunctional setup for studying human motor control using transcranial magnetic stimulation, electromyography, motion capture, and virtual reality. J Vis Exp 103. https://doi.org/10.3791/52906

Taylor RM, Hudson TC, Seeger A, Weber H, Juliano J, Helser AT (2001) VRPN. In: Green M, Shaw C, Wang W (eds) Proceedings of the ACM symposium on virtual reality software and technology – VRST '01. The ACM symposium. Baniff, Alberta, Canada, 11/15/2001–11/17/2001. ACM Press, New York, p 55

Török A, Sulykos I, Kecskes-Kovacs K, Persa G, Galambos P, Kobor A et al (2014) Comparison between wireless and wired EEG recordings in a virtual reality lab: Case report. In: 2014 5th IEEE conference on cognitive Infocommunications (CogInfoCom). IEEE, Vietri sul Mare, pp 599–603

Tremmel C, Herff C, Sato T, Rechowicz K, Yamani Y, Krusienski DJ (2019) Estimating Cognitive Workload in an Interactive Virtual Reality Environment Using EEG. Front Hum Neurosci 13:401. https://doi.org/10.3389/fnhum.2019.00401

Tromp J, Peeters D, Meyer AS, Hagoort P (2018) The combined use of virtual reality and EEG to study language processing in naturalistic environments. Behav Res Methods 50(2):862–869. https://doi.org/10.3758/s13428-017-0911-9

Ulsamer P, Pfeffel K, Müller NH (2020) Brain activation in virtual reality for attention guidance. In: Zaphiris P, Ioannou A (eds) Learning and collaboration technologies. Human and technology ecosystems, vol 12206. Springer. Lecture Notes in Computer Science, Cham, pp 190–200

Wang W-E, Ho RLM, Gatto B, van Der Veen SM, Underation MK, Thomas JS et al (2020) A novel method to understand neural oscillations during full-body reaching: a combined EEG and 3D virtual reality study. IEEE Trans Neural Syst Rehabil Eng 28(12):3074–3082. https://doi.org/10.1109/TNSRE.2020.3039829

Wiesing M, Fink GR, Weidner R (2020) Accuracy and precision of stimulus timing and reaction times with unreal engine and SteamVR. PloS One 15(4):e0231152. https://doi.org/10.1371/journal.pone.0231152

Wolter M, Armbruester C, Valvoda JT, Kuhlen T (2007) High ecological validity and accurate stimulus control in VR-based psychological experiments

You SH, Jang SH, Kim Y-H, Hallett M, Ahn SH, Kwon Y-H et al (2005) Virtual reality-induced cortical reorganization and associated locomotor recovery in chronic stroke: an experimenter-blind randomized study. Stroke 36(6):1166–1171. https://doi.org/10.1161/01.STR.0000162715.43417.91

Eye Tracking in Virtual Reality

Nicola C. Anderson, Walter F. Bischof, and Alan Kingstone

Contents

Abstract This chapter explores the current state of the art in eye tracking within 3D virtual environments. It begins with the motivation for eye tracking in Virtual

N. C. Anderson (✉), W. F. Bischof, and A. Kingstone
University of British Columbia, Vancouver, BC, Canada
e-mail: nicola.anderson@ubc.ca; wfb@ualberta.ca; alan.kingstone@ubc.ca

Curr Topics Behav Neurosci (2023) 65: 73–100
https://doi.org/10.1007/7854_2022_409
Published Online: 30 January 2023

Reality (VR) in psychological research, followed by descriptions of the hardware and software used for presenting virtual environments as well as for tracking eye and head movements in VR. This is followed by a detailed description of an example project on eye and head tracking while observers look at 360° panoramic scenes. The example is illustrated with descriptions of the user interface and program excerpts to show the measurement of eye and head movements in VR. The chapter continues with fundamentals of data analysis, in particular methods for the determination of fixations and saccades when viewing spherical displays. We then extend these methodological considerations to determining the spatial and temporal coordination of the eyes and head in VR perception. The chapter concludes with a discussion of outstanding problems and future directions for conducting eye- and head-tracking research in VR. We hope that this chapter will serve as a primer for those intending to implement VR eye tracking in their own research.

Keywords Eye movements · Head movements · Virtual reality

1 Why Track Eyes in VR?

VR is a good design choice for human performance experiments for a number of reasons, but tracking the eyes in VR provides several key advantages over traditional computer-based or mobile eye tracking research that we touch on throughout this chapter. Most notably, VR eye tracking allows for the simultaneous tracking of the eyes and other head and body movements with respect to a common reference frame (see Sect. 5.1), allowing for the precise calculation and dissociation of the relative contributions of these different movements to more general attentional control (see Sect. 6). Note that throughout this chapter, we use the terms looking, eyes, and gaze synonymously.

Much of what we know about visual attention and eye movement control is derived from studies that require people to look at images presented on a computer monitor while their head is restrained. There is, however, growing recognition that eye movements measured in the lab when an observer's head movements are discouraged or restrained are not representative of how people move their eyes in everyday life when their head is free to move (e.g., Backhaus et al. 2020; Hooge et al. 2019; Kingstone et al. 2008; Land and Hayhoe 2001; Risko et al. 2016; 't Hart et al. 2009). For example, Foulsham et al. (2011) asked participants to watch first-person video clips of someone walking across campus. While there was some bias to look in the center of the video, their gaze (i.e., the direction of their eyes) was spread over the whole scene, looking at objects and the people that the walker encountered. When the same participants physically walked across campus with a mobile eye tracker, they often focused on the path, and the eyes remained relatively centered in the visual field as defined by their head orientation (Foulsham and Kingstone 2017). In other words, when the head was free to move, people tended to move their head in order to redirect their gaze to objects and people. In the lab, Solman et al. (2017) have shown that when participants are required to look at a scene through an asymmetric window that is yoked to their eyes, eye movements target regions within the window. However, when the window is yoked to an observer's head movements, the head moves to reveal new information outside the window, presumably so that the eyes can then examine visual information within the new window. These studies suggest that when people are allowed to move their head, they do so, and that the head acts to reveal new information for the eyes to exploit.

In the analysis of head movements, we have learned that the relative timing of eye and head movements may suggest whether attentional selection is reflexive or volitional (Doshi and Trivedi 2012; Freedman 2008; Zangemeister and Stark 1982). For instance, when the eyes move before the head, these are unplanned, reflexive movements (usually to a suddenly presented stimulus such as a flash of lightning) and involve shifts of less than 45° (Barnes 1979; Flindall et al. 2021). When the head leads the eyes, however, these are thought to be large, planned, purposeful movements, often to a known target location. These conclusions are mainly based on experiments where participants respond to simple light displays or targets on a screen but it has also been shown in more naturalistic settings, where, for example, when we prepare to cross a street or when we prepare a lane change while driving (Doshi and Trivedi 2012).

In VR, when people are asked to view scenes in 360°, the attention system must coordinate eye movements with other head and body movements to explore the full range of the visual field (if this is desired). When looking at 360° panoramic scenes, observers spend the majority of the time exploring along the horizon (Rai et al. 2017; Sitzmann et al. 2018), using their head and other body movements to extend the field of view for the eyes (Bischof et al. 2020). When viewing landscape and fractal panoramic scenes that are rotated (for example, 45° clockwise), the head tends to rotate in a similar manner in order to bring the scenes closer to their canonical upright position for the eyes (Bischof et al. 2020), converging with other evidence suggesting that the head acts in service of the eyes to extend the range of possible viewable locations (Solman et al. 2017; Solman and Kingstone 2014).

On the other hand, studies in VR have taught us that the eyes, head (and body) may move in ways that diverge from what we might expect. When observers are asked to search 3D environments, the effective field of view, or visual span, is much larger than reported in studies using smaller images on a computer monitor (David et al. 2021). In a study where observers viewed large, flat (i.e., non-panoramic) landscape and fractal scenes, it was found that the head and eyes responded differently to scene rotations, where the eyes seemed to be more sensitive to rotation than the head (Anderson et al. 2020). In other work with panoramic scenes, it has been shown that the head is less affected by the masking of central or peripheral scene information than the eyes (David et al. 2022). In addition, in many of these works, the extent that participants move their head (and body) varies largely between individuals (Anderson et al. 2020; Bischof et al. 2020; Jacobs et al. 2020). This variation mirrors earlier observations by researchers who looked at the extent that observers prefer to recruit their head in response to peripheral target acquisition –

resulting in some observers being dubbed "movers" and others "non-movers" (e.g., Delreux et al. 1991; Fuller 1992a, 1992b; Goldring et al. 1996; Stahl 2001). Taken together, these works provide varying degrees of evidence that head and eye movements may diverge in their control strategies, leading researchers to speculate that the head may be under more deliberate, cognitive control (David et al. 2022), or utilize different spatial reference frames (Anderson et al. 2020). Importantly, these and other studies in VR have uncovered interesting findings and generated novel questions about the complex dynamics between eye, head, and other movements in fields of view that extend beyond the standard computer monitor.

2 VR and Eye Tracking: Hardware

In this section we review major hardware and software used for eye tracking in VR. We first present the hardware used or stimulus presentation, namely head-mounted displays and projection displays, followed by the presentation of eye-tracker and head-tracker hardware.

2.1 Stimulus Hardware

VR stimuli can be presented in one of the two major configurations, head-mounted displays and projection displays. In *head-mounted displays* (HMDs), stimuli are presented using a head-mounted viewer with two displays, one for each eye (see Fig. 1). An eye tracker can be mounted inside the viewer, and the position and orientation of the HMD can be tracked with multiple methods depending on the headset used.

When setting up a VR eye tracking lab, some consideration needs to be taken to how participants physically interact with the headset. Most headsets with built-in eye tracking are (to date) still tethered to a computer via a cable (as in the HTC Vive). In

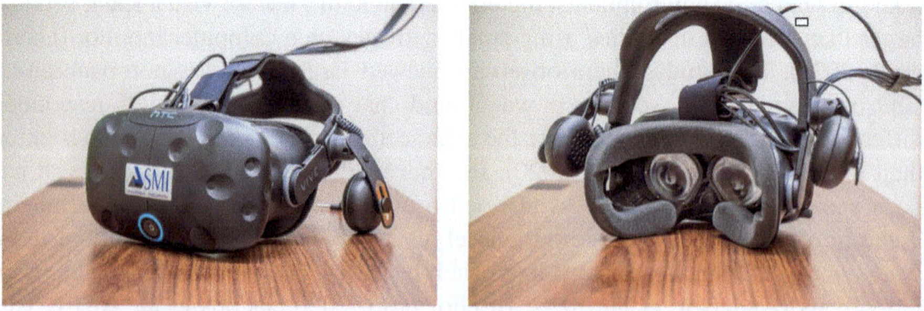

Fig. 1 (Left) Outside view of HTC Vive HMD. (right) Inside view of HTC Vive HMD with SMI eye tracker. Photos taken by Jacob Gerlofs

our experience, participants are aware of any cables from the headset or response apparatus (i.e., keyboard) that are attached to the computer. This may affect their ability or tendency to move freely in the VR environment (if this is desired). For example, in recent published and unpublished work (Bischof et al. 2020; Jacobs et al. 2020) where participants responded on a tethered keyboard while looking at 360° scenes, participants rarely, if ever, rotated the full 360°, despite being seated in a swivel chair. One way to mitigate this is to mount the cables to the ceiling and use VR controllers or wireless keyboards for manual responses. It remains an interesting open question whether it makes a difference in how participants attend to VR scenes if they are seated (as in the vast majority of our studies) or standing (as in, for example, David et al. 2020, 2021).

In *projection displays*, sometimes referred to as caves, the stimuli are projected onto 1–6 walls with the observer standing in the middle (e.g., Visbox, Inc., 2022). Eye movements of observers can be recorded using a remote eye tracker attached to a screen or using a head-mounted eye tracker. In the latter case, head movements need to be tracked to convert gaze direction in head-centered coordinates into an environmental stimulus-centered (i.e., allocentric) framework. Alternatively, the gaze direction can be converted to allocentric coordinates using markers attached to the walls of the cave.

2.2 Eye-Tracker Hardware

The goal of eye tracking is often to determine the direction of gaze in the allocentric coordinate system of the presented virtual world. This chapter focuses on the use of head-mounted displays for VR. In this case, gaze determination involves two parts: determining gaze direction in head-centered coordinates (e.g., Pupil Invisible – Eye Tracking Glasses for the Real World – Pupil Labs 2022; Sensomotoric 2017) and determining the position and orientation of the head using a head tracker (see below) or via scene camera motion analytics (Ohayon and Rivlin 2006).

2.3 Head-Tracker Hardware

Several systems are available for tracking the head position and direction. In most commercial VR systems (e.g., the HTC Vive), the motion tracker is a system component. In other systems like those using projection displays, this must be achieved using an independent head tracker, for example with visual sensing (V120 n.d.) or with an inertial tracker composed of accelerometers and gyroscopes (Blue Trident IMU | Inertial Sensor by Vicon | Biomechanic Tracking 2022).

3 VR and Eye Tracking: Software

There are three major software systems for creating VR worlds, Unity, Unreal Engine, and Vizard. We review each system briefly, but since our own experience is based on Unity, we later provide a detailed example using that system. Not every VR eye tracker is compatible with each of these systems (e.g., Pupil Labs does not yet have a plugin for Unreal Engine), so it is worth making note of the availability and support for different software prior to purchasing any VR eye tracking hardware.

3.1 Unity

Unity is a game engine developed by Unity Technologies in 2005. The engine has been extended to support a large range of platforms, including desktops, mobiles, and VR platforms. It is very popular for mobile game development and used for games such as Pokémon Go or Call of Duty. It is considered easy to use for beginners and can be used for creating 3D and 2D worlds. Creating a virtual world consists of setting the world up using an interactive development environment and expanding and controlling it using C# or Java scripting language.

There are a number of toolboxes that help in the development of Unity-based experiments, such as UXF (Brookes et al. 2020), USE (Watson et al. 2019), bmlTUX (Bebko and Troje 2020), and VREX (Vasser et al. 2017). Later in this chapter we present a basic and straightforward approach to developing experiments with Unity that focuses on the measurement and use of eye and head movements in experiments (see Sect. 4). Nevertheless, these could potentially be used in conjunction with any of the above tools.

3.2 Unreal Engine

Unreal Engine is a game engine developed by Epic Games in 1998. The engine was first developed for first-person shooter games, but has since been used in an expanding range of 3D games and has been adopted by the film and television industry. The unreal engine is scripted in C++ and supports a wide range of desktop, mobile, console, and virtual reality platforms. The Unreal Engine has only limited support for popular eye trackers, for example the Tobii or the SMI eye trackers.

3.3 Vizard

Worldviz has developed a Python toolkit for developing VR applications, which supports multiple stimulus devices, eye and body trackers, input devices, such as gloves, haptic devices, and controllers (Vizard | Virtual Reality Software for Researchers 2022). Figure 2 shows a small example of a Vizard program, which includes setting up a VivePro eye tracker.

3.4 *Stimuli*

One of the primary advantages of VR is the flexibility of the virtual world, where participants can be immersed in a fully rendered 3D replication of the world to any number of more simplified (or fantastical) environments. This has some implications for eye tracking that are worth noting. For example, in some of our recent work, we were interested in how image (and screen) rotation might have affected saccade direction biases. In traditional, desktop-based setups, it has been shown that rotating scenes affects the predominance of horizontal and vertical saccades (Foulsham et al.

```
import viz
import vizact

viz.setMultiSample(4)
viz.fov(60)
viz.go()
# Set up the VivePro eye tracker
VivePro = viz.add('VivePro.dle')
eyeTracker = VivePro.addEyeTracker()
# Create an empty array to put some pigeons in.
pigeons = []
# Go through a loop six times.
for eachnumber in range(6):
    # Create pigeon
    newPigeon = viz.addAvatar('pigeon.cfg')
    # Place the new pigeon on the x-axis.
    newPigeon.setPosition([eachnumber, 0, 5])
    #Add the new pigeon to the "pigeons" array.
    pigeons.append(newPigeon)
# Move the view to see all pigeons
viz.MainView.move([2.5,-1.5,1])
# Get gaze matrix in local HMD coordinate system
gazeMat = eyeTracker.getMatrix()
# Transform gaze matrix into world coordinate system using main view
gazeMat.postMult(viz.MainView.getMatrix())
# Intersect world gaze vector with scene
line = gazeMat.getLineForward(1000)
info = viz.intersect(line.begin, line.end)
if info.valid:
    print('User is looking at', info.point)
```

Fig. 2 Simple Vizard example program

Fig. 3 Example participant view of a rotated scene in Anderson et al. (2020)

2008; Foulsham and Kingstone 2017). However, in VR, it is possible to rotate not only the image itself, but also the entire screen it is projected on – a real-world equivalent would be rotating the entire computer monitor, not just the image content (Anderson et al. 2020). In this case, we were interested in how observers looked at the scenes themselves (and not so much whether they looked at the rest of the virtual scene). Therefore, eye movements were reported based on their 2D position on the plane in the virtual world, very similar to how they would be reported in a traditional, computer-monitor-based eye tracking study (see Fig. 3).

In more complicated situations, researchers might be interested in how participants search for objects in a fully rendered 3D scene (e.g., David et al. 2021). In this case, the researchers were interested in the objects the participants looked at, as well as general search measures such as scanning, verification, and search time. For fully immersive 360° panoramic scenes, gaze position might be reported with respect to the scene sphere. Each of these scenarios has different demands from the data processing and analysis that need to be kept in mind. We discuss the reference frames commonly used in VR below (Sect. 5.1).

4 Eye and Head Movements in 360° Scenes: An Example in Unity

In this section, we present our work on eye and head movements in 360° panoramic scenes, providing a concrete example and "how-to" information for developing experiments in Unity. We also include details that might not be in our published work (for example, experiment control flow principles in C#). Note that several excellent experiment builder type programs have been in development for Unity (Sect. 3.1), but here we aim to provide a working example of a basic Unity project that includes eye tracking. For the purposes of this example, we assume that the reader has a basic understanding of the Unity Editor and project setup, as explained in the Beginner Unity tutorials. A good place to start might be the "Unity Essentials pathway in Unity Learn" (Unity Essentials 2022).

In this example study, participants were asked to move their eyes and head to look at a random selection of 80 indoor and outdoor 360° panoramic scenes and remember them for a later memory test. Participants looked at the scenes using a head-mounted display with a built-in SMI eye tracker, and each scene was displayed for 10 s. A uniform gray scene with a black fixation cross presented directly ahead of the participant was displayed between images, and participants were instructed to look at this fixation cross and press a key to initiate the next trial. At the beginning of every 20th trial, the eye tracker was calibrated using a 9-point display specific to the SMI software.

4.1 Unity 3D Environment

The Unity project architecture of our example experiment is quite straightforward. The scene consists of a few basic elements (see Fig. 4) defined below:

- Camera: The camera is a Unity object provided by SMI. It represents the headset and moves in the space when the headset is moved or worn by participants. The camera provided by SMI determines not only the point of view, but it also manages the eye tracking. Eye tracker related functionality can be set and changed via an SMI-specific script, which is attached to an otherwise standard camera object.
- Scene Sphere: This is a 3D sphere game object with a shader and sphere collider attached. The shader is somewhat special, such that textures applied to the surface of the sphere (the panoramic scenes) are visible not from the outside, but the inside, to allow observers in the middle of the sphere to see the scenes. The collider provides the physical properties needed for this object to interact with other physical aspects of the project.
- Directional Light: A light source is required for illuminating the inside of the sphere. That light source is positioned below the camera and aligned with the forward direction of the camera.
- Experiment: The empty game object "Experiment" has a script attached with the name Experiment.cs. The bulk of the work in the experiment is done in this script, key components of which are described in greater detail below. Importantly, game objects that are modified or referred to in this script are attached to this script in the Unity environment (for example, the SceneSphere and DirectionalLight objects shown in the Inspector window of Fig. 4).

Fig. 4 Example of a simple experimental setup for a Unity project. Objects in the scene are listed in the "Hierarchy" window (**a**), which consists of the camera (in this case, a camera combined with an SMI eye tracker), an EventSystem, a sphere (SceneSphere) with a special shader and a collider, a directional light, a GUI object, and an empty game object called "Experiment." The current scene (the spherical stimulus as seen from the outside) is shown in (**b**) and the initial display (for entering experiment information) is shown in (**c**). The "Inspector" window (**d**) shows the details of the "Experiment" game object, which has a transform indicating where this invisible empty game object is in the virtual space, as well as a script (also named "Experiment") attached to it. The bottom of the display lists assets that can be included in the experiment (**e** and **f**)

- GUI object: This object is responsible for interacting with the experimenter to record participant and stimulus information.
- Event system: This is used for sending events to other objects based on input, for example, a keyboard or an eye tracker. The Event System consists of a few components that work together to send events.

4.2 Experimental Control Flow

One of the things typically taken care of by experiment builders such as PsychoPy (Peirce et al. 2019) and OpenSesame (Mathôt et al. 2012) is experimental control flow, that is, the movement from one trial to the next, and the management of different experimental blocks. In Unity, experimental control is complicated by the fact that Unity is in charge of program execution and is calling the user-defined Experiment script at more or less regular intervals. Essentially, the script that takes care of experiment control flow is called during each headset frame refresh. For this reason, the script has to keep track of the state of execution to guarantee an orderly execution. This is described in detail in the next section.

4.3 Script Experiment.cs

The Experiment.cs script is attached to the "Experiment" Unity object and is responsible for most of the "real work" done in the experiment. This script has links to relevant objects in the scene, such as the light source and the stimulus sphere, and it initiates and modifies their states. The script is written in C# and has two main functions, Start() and Update().

The Start() function is called at the beginning of the experiment and contains code that must be executed at the beginning of the experiment (e.g., initializing the variables that are constant for the entire experiment such as participant information and accessing the stimulus sphere material and setting it to a variable for later use). This script reads in a control file with descriptive information about each trial. Each row in this file represents a trial and includes information about the image name, what type of image it is (in our example, either an indoor or outdoor scene), and further information associated with the trials (see Table 1):

The Update() function is responsible for the bulk of the experimental control flow code. It is called from Unity approximately 50–70 times per second. For this reason, the code must keep track of the state of the experiment, such that, on every call to Update(), the script continues at the correct place. To achieve this, we use a C# programming structure called a switch statement (essentially a series of if... else if statements), where a code block gets executed based on a match to a list of possible states called experimentPhase (see Fig. 5). Then we only need to keep track of the state experimentPhase between calls to Update(). Each phase state takes care of a particular part of the experiment such as calling for a calibration (in our example, this occurs every 20 trials), showing a stimulus, or waiting for a participant response.

Each case in the switch case statement has links to others, so that for example, after the "waitForParticipant" state, which waits for the participant to press a specific key on the keyboard, the variable "phase" is assigned to the "stimulus" case and the code for controlling what happens during stimulus presentation is executed. This basic structure can be adapted to suit many types of simple experimental design. For

Table 1 Example control file containing trial information

Uid	Image	Image type	Task
1	indoor_pano_aaabgnwpmpzcgv	Indoor	Encoding
2	indoor_pano_aaabnctbyjqifw	Indoor	Encoding
3	indoor_pano_aaacisrhqnnvoq	Indoor	Encoding

```
enum ExperimentPhase { preparation, calibration, waitForCalibration,
        waitForParticipant, stimulus, finished };

switch (phase) {

        case ExperimentPhase.preparation:

                // code run prior to starting experiment
                phase = ExperimentPhase.calibration;
        break;

        case ExperimentPhase.calibration:

                // code to run calibration coroutine
                phase = ExperimentPhase.waitForParticipant;
        break;

        case ExperimentPhase.waitForCalibration:

                // code to show fixation screen while waiting for calibration to run
                phase = ExperimentPhase.waitForParticipant;
        break;

        case ExperimentPhase.waitForParticipant:

                // wait for participant to indicate they're ready for the trial to start
                if (ParticipantResponse()) {
                        phase = ExperimentPhase.stimulus;
                }
        break;

        case ExperimentPhase.stimulus:

                // code to run trial sequence coroutine
        break;

        case ExperimentPhase.finished:

                // code to run at end of experiment
                QuitExperiment();

        break;
}
```

Fig. 5 Switch statement used to keep track of what state the experiment is in on every given call of Update()

example, including an experimental block structure requires the addition of a phase that checks the trial number after each trial is run. If it matches the number of trials in a block, an instruction screen specific to that block is run. How the trial changes across blocks, in this case, is handled in the stimulus phase with an if statement checking which block is currently being run. The block information itself could be located in the .csv file containing the trial information.

4.4 Eye Tracking Implementation

The SMI eye tracker is able to record eye movements at a rate of about 250 Hz, but the Unity system invokes the Update() function at a rate of only about 50–70 Hz. For this reason, if eye tracking is controlled through the Unity system, eye tracking speed is substantially reduced. In our example, we run the eye tracker independently of the Unity system and at the maximum possible speed (i.e., at 250 Hz) via multithreading. This is accomplished by creating and starting a gaze tracking function in a new thread, i.e., in a program segment that runs independently of the rest of the program. The function called in this thread GazeTrack() creates two lists, gazeTime that contains the timestamp output from a custom SMI function, and gazeDirection, which indicates the current gaze direction. At the end of each trial, the data in these lists are then passed to a function called WriteGazeData(), which transforms the data appropriately and writes them to the output file.

As shown in Fig. 6, gaze direction is returned by the SMI system as a ray emanating from the center of the headset. This must be translated to spherical coordinates (longitude and latitude) using functions that compute the intersection of the gaze ray with the stimulus sphere. Note that due to characteristics of Unity, hits can only be detected from the outside, so the ray must be sent out and then reversed in direction in order to hit the outside of the sphere. Alternatively, this could be computed directly using information on the location and orientation of the headset with respect to the center of the sphere, the gaze direction, and the radius of the sphere.

Other data, in particular headset position-related data obtained through Unity, is recorded at about 50–70 Hz. In WriteGazeData() from Experiment.cs the two data streams can be combined and synchronized, provided that precise timing information has been recorded for gaze and the other variable. In our example, this is done via linear interpolation (see Experiment.cs).

5 Data Handling

In this section, we outline some of the data handling issues that are unique to VR.

5.1 Reference Frames

Before diving into the specifics of eye, head, and gaze analysis in VR, it is worth clarifying the reference frame we are dealing with (see Hessels et al. 2018). Most readers may be familiar with the popular desktop-based eye tracking technology, where observers are required to sit in a chin rest at a set distance away from a computer monitor. In this situation, the head is fixed, and eye movements are reported with respect to the computer monitor, usually in some form of pixel location or degrees of visual angle. In other words, the reference frame for eye movements is the screen, which typically encompasses approximately 30–50° visual angle, depending on the particular setup.

```
IEnumerator StartTrialDuration(string imageName, float duration)
{
	...
	recordGaze = true;

	Thread gazeThread = new Thread(GazeTrack);
	gazeThread.Start();

	SetSphereImage(imageName);

	yield return new WaitForSeconds(duration);

	recordGaze = false;
	WriteGazeData();
	currentTrialNumber++;

	...
}

...

void GazeTrack()
{
	Vector3 rayCast;

	while (recordGaze)
	{
		timeStamp = SMI.SMIEyeTrackingUnity.Instance.smi_GetTimeStamp();

		if (timeStamp > oldTimeStamp)
		{
			if (nGazeReadings < maxGazeReadings)
			{
				rayCast =
SMI.SMIEyeTrackingUnity.Instance.smi_GetCameraRaycast();
				gazeTime[nGazeReadings] = timeStamp;
				gazeDirection[nGazeReadings] = rayCast;
				nGazeReadings += 1;
			}
			oldTimeStamp = timeStamp;
		}
	}
}
```

Fig. 6 Example code for implementing eye tracking in C# via multithreading

In mobile eye tracking experiments, the eyes are tracked by one or two cameras pointed toward the eyes, while the scene is recorded in a forward-facing camera and the head and body are free to move naturally. The reference frame in this case is the scene camera, with eye position reported with respect to their location in the scene camera, ranging from around 60–120° visual angle. In this case the head is free to move, so the reported eye coordinates are essentially head-based. Head movements can be estimated from the movements of the scene camera (e.g., Dufaux and Konrad 2000; Ohayon and Rivlin 2006), while gaze position is typically hand coded, or it can be extracted with the use of reference markers placed in the world and detected via software (for e.g., using Pupil Labs, Kassner et al. 2014).

In VR experiments, similarly to mobile eye tracking, the eyes are tracked by cameras mounted within the HMD. The head and body are often free to move naturally, although the weight of the headset, participant pose (standing vs. sitting), and potential tethering can influence participants' range of motion. One significant advantage of this setup is that eye and head movements can be tracked with respect to a common reference frame. This means that things like eye eccentricity in the head coordinate system, as well as the contributions of head movements to gaze position can be calculated precisely (more on this below). Note, however, that without special equipment, neck movements cannot be differentiated from chair and torso movements. For the sake of simplicity, in the example we present below, we use the position of the HMD in space as a proxy for head movements alone.

5.2 Fixation and Saccade Detection

Unlike more standard, desktop, and mobile-based eye tracking, to date, VR eye tracking implementations do not come with analysis programs such as Eyelink's DataViewer (SR Research Ltd.) or Pupil Lab's Pupil Player (Core – Pupil Player, 2022) that automatically parse gaze data into blinks, fixations, and saccades. In this section, we review event detection in gaze analysis that can be used in VR with a focus on the detection of fixations and saccades, while other ocular events such as smooth pursuit, micro-saccades, or blinks are ignored (see, for example, Holmqvist and Andersson 2017). Note that the SMI eye tracker used in our example in Sect. 6 automatically omits blinks from the recorded data. Figure 7 shows an example output from an event detection algorithm for a single trial in our example dataset.

There are two fundamentally different approaches to gaze analysis. The first approach starts with the detection of fixations, and saccades are defined as

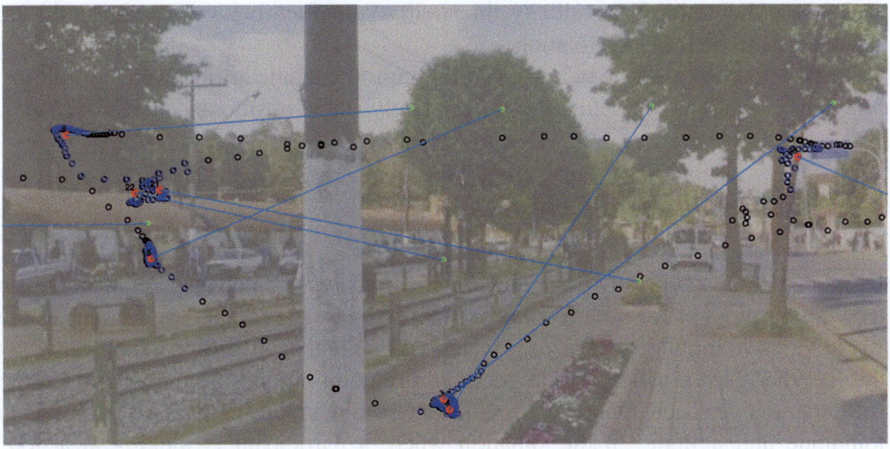

Fig. 7 Example visualization of the IDT algorithm (zoomed in for clarity). Black dots are gaze samples, red dots are fixations, and blue dots are the gaze samples that contribute to each nearby fixation. Green dots are averaged head positions during a given fixation

differences between successive fixations, whereas the second approach starts with the detection of saccades, and fixations are defined as stable points between saccades. A popular method for the detection of fixations is the Dispersion-Threshold (IDT) algorithm (Komogortsev et al. 2010; Salvucci and Goldberg 2000), which assumes that the dispersion of gaze points within a fixation is relatively small (in our studies typically 2.5–3°) and that the duration of fixations exceeds a minimum duration (in our studies typically 80 ms). Specifically, the IDT algorithm proceeds as follows:

1. Initialize a window of gaze points to cover duration threshold. Drop the data points if the dispersion of the gaze points exceeds the dispersion threshold.
2. Add further gaze points to the window as long as the dispersion of the gaze points does not exceed the dispersion threshold.
3. Define the fixation position as the centroid of gaze points.
4. Remove the gaze points of the fixation and start again from step 1.

An alternative method for fixation detection relies on gaze vector velocities, where in step 2, gaze points are added to the window as long as the velocity of successive gaze points does not exceed the velocity threshold. For both, the IDT algorithm and the velocity algorithm, saccades are defined as differences between successive fixations.

The second approach begins with the detection of saccades, and fixations are defined as stable points between saccades. The detection of saccades is based on the assumption that motion above a velocity threshold is assumed to be (part of) a saccade. Specifically, the algorithm proceeds as follows:

1. Calculate all gaze velocities between successive gaze points.
2. Detect peak velocities (which are assumed to define the middle of a saccade).
3. Add velocities immediately before the peaks and immediately after the peaks as long as they exceed a velocity threshold. Velocities below that threshold are assumed to be part of a fixation.
4. Peak velocities must be below a certain limit to exclude artifacts, such as blinks.
5. Finally, fixations are defined as the relatively stable positions between saccades.

5.3 Using Spherical Coordinates

Event detection is typically done on gaze positions represented as pixel locations on a computer monitor (or with respect to some standard coordinate frame, such as the world video in mobile eye tracking systems). In VR the situation is complicated by the fact that gaze positions could occur at any point around the participants in the fully immersive space. One way to represent gaze positions in such a situation is in longitude and latitude. In the situation where a participant is looking at a 360° panoramic scene, these could be calculated with respect to the scene sphere (as in the example presented in Sect. 4). However, in a fully immersive 3D environment, this has typically been done by computing gaze positions in longitude and latitude

Fig. 8 Example gaze points P and Q projected onto a virtual unit sphere surrounding the VR headset. The red line represents the angular distance between the fixations, which is calculated using the great circle distance

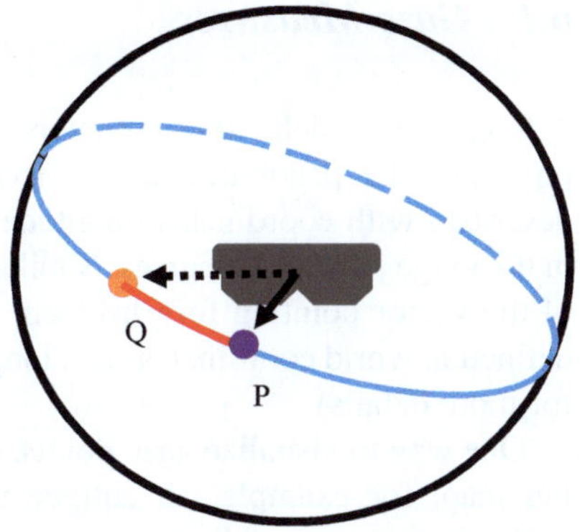

with respect to an imaginary unit sphere surrounding the headset (see Fig. 8). Event detection is then done using these spherical coordinates (David et al. 2020, 2021), and other gaze measures, such as what 3D object a participant was looking at, can be obtained via raycasting or by utilizing physical interactions of the gaze ray and objects in the virtual world. For a detailed example of how to extract gaze ray information from a virtual sphere, see Sect. 4.4.

There are a few key points to keep in mind when doing event detection in a fully immersive VR environment where circular statistics must be taken into account (Batschelet 1981; Bischof et al. 2020 Appendix 1). One must pay particular attention to how distances between successive gaze positions are calculated. This has implications for the dispersion threshold (usually represented in degrees visual angle) as well as the saccade amplitudes (distances between fixations). In a fully immersive 360° world, these distances must be calculated using the great circle distance, otherwise known as the orthodromic distance, which is defined as the shortest distance between two points on a sphere (see Fig. 8).

6 Data Analysis

In this section, we focus on the analysis of gaze and head movements of observers in a 360° panoramic virtual environment. We use the data from our example project from Sect. 4, however, many of the measures we outline below can be adapted to other situations in VR that are not specific to panoramic scenes. To recap, the virtual environment is produced by projecting images on the interior of a sphere with the observer's head at the center of the sphere. In our examples, the environment is static, with all points of the environment at the same distance from the observer, and it contains no moving objects that can be tracked. In the following sections, we present first the basics of gaze analysis and the analysis of saccades and head movements. Finally, we present an analysis of eye-in-head measures and the spatial and temporal relationship between eye and head movements.

6.1 Gaze Measures

A gaze point is defined as the intersection of the gaze vector with the virtual sphere on which the panoramas are projected. Azimuth and elevation of this point are described with coordinates longitude in the range [−180,180] degrees and latitude in the range [−90,90] degrees. Similarly, we define the head point as the intersection of the vector pointing forward from the face with the virtual sphere, and it is also defined in world coordinates (i.e., longitude and latitude of the panorama, see Sect. 5 for more details).

One way to visualize gaze points, or sets of gaze points, is to project them onto a flat map, for example, an equirectangular (or equidistant) projection map. This projection maps meridians into to vertical straight lines of constant spacing, introducing distortions near the poles compared to the equator. This is illustrated in Fig. 9.

An analysis of typical fixation patterns in spherical displays shows that there seems to be a preference for fixations along the equator of the virtual sphere, a tendency that is referred to as equator bias (see Fig. 10 (left)). There are multiple causes that may contribute to the equator bias. First, if participants inspect the panorama with neck extension and flexion in a resting state and the eyes are centered in the head coordinate system, then there is a natural preference for fixations along the horizon. Second, an analysis of typical panorama images shows that on average, edge density is strongest along the equator (see Fig. 10 (right)), which may be due to the fact that there is simply more content along the horizon in typical panoramic scenes (as in, for e.g., Torralba et al. 2006). Figure 10 (right) was constructed by computing the edge images of a large number of panorama images and averaging them.

Edges occur in regions where there is a strong change in gray-level of the panorama images. To generalize the edge density map, one can compute the entropy of local neighborhoods in the panorama images. This is done by computing gray-level histograms in a grid overlaid over the image, then computing the entropy of the

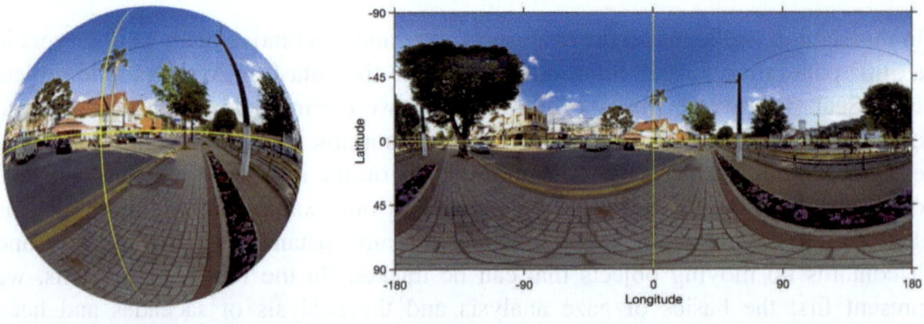

Fig. 9 (Left) Stimulus sphere. (Right) Equirectangular (equidistant) map of the stimulus. The yellow lines indicate the equator and the center meridian. Note the distortions near the north and south poles. In analyzing fixation patterns, these distortions must be taken into account

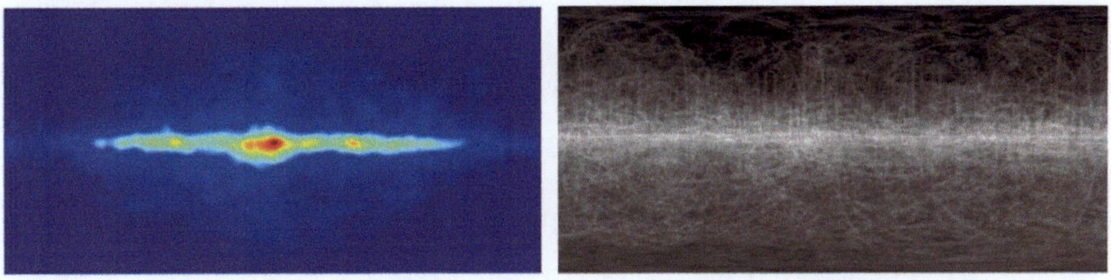

Fig. 10 (Left) Fixation heatmap. (Right) Edge density averaged over many images

histograms, and finally averaging the entropy images over many panorama images. In this analysis, regions with large entropy are assumed to "attract" attention, leading to peaks in the fixation heatmaps. As a further generalization of local entropy maps, one can compute the saliency maps of panoramic images. In Engineering and Computer Science (e.g., De Abreu et al. 2017; Sitzmann et al. 2018), there is substantial work on the saliency of panoramic images, but that work always includes fixation patterns for predicting saliency, while we are interested in predicting fixation patterns from saliency based on image properties.

For other general gaze movement measures that might be of interest in panoramic scenes, see Bischof et al. (2020) and David et al. (2022).

6.2 Analysis of Saccades

Saccade patterns in panoramic scenes are closely related to the fixation distributions. Given the large spread of fixations along the horizon of the images, it is plausible that saccade directions also align with the scene horizons. This is illustrated in Fig. 11 (left), which shows a polar histogram of saccade directions. The histogram shows that most saccades were made along the horizon direction of the panoramas. In addition, Fig. 11 (right) also shows a histogram of the saccade amplitudes. On average, in free viewing of natural panoramic scenes saccades are typically on the order of 10–20° visual angle. For a more detailed analysis of saccade characteristics in panoramic scenes, see David et al. (2022).

6.3 Head Analysis

Head movements are defined by head pitch, which is achieved through neck extension and flexion, head yaw, which is achieved through lateral neck rotation, and head roll, which is achieved through lateral bending of the neck. A head point is defined as the intersection of the vector pointing forward from the face with the virtual panorama sphere surrounding the observer. It is defined in world coordinates (i.e., longitude and latitude of the panorama). Head movements with a VR viewer are

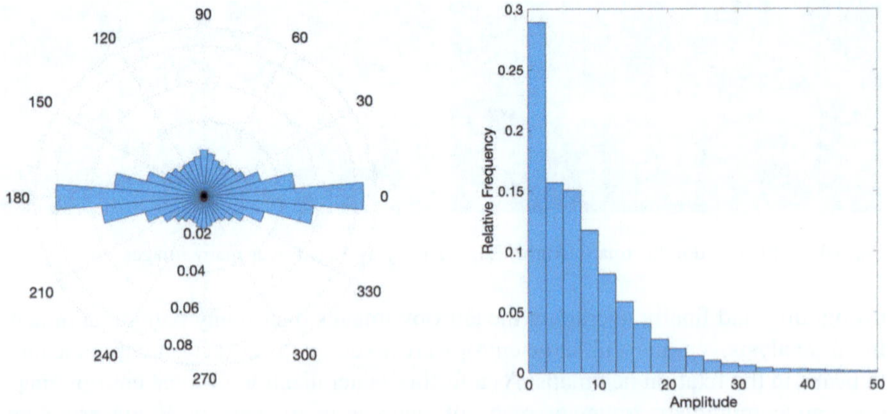

Fig. 11 (Left) Polar histogram of saccade directions. Most saccades are aligned with the horizon of the panorama images. (Right) Histogram of saccade amplitudes

not ballistic in the way that eye movements are. For this reason, there are, in contrast to gaze, no natural demarcations for head shifts and head fixations.

Gaze and head positions are illustrated in Fig. 12. One way to analyze head movements is to examine head point patterns independent of gaze. To relate head movement information to gaze, one can link the two using common timing information. Alternatively, one can compute the average of head points during a fixation (referred to, in this chapter, as head average) and then directly relate the head averages to fixations. In our experience, the latter approach has been the most straightforward for relating head movements to fixations, however, care must be taken when interpreting head positions in this way.

Comparable to gaze, one can plot a heatmap of head averages for sets of images. Figure 13 shows a head average heatmap for the same images used in Fig. 10. Given that gaze positions deviate only by a moderate angle from the head position (see Sect. 6.4.1 below), it is not surprising that the head points are also concentrated along the horizon.

6.4 Eyes

We define eyes-in-head as the gaze direction in a head-centered coordinate system. The eye hit point, or eye point, is defined as the difference between gaze point and head point and is also expressed in world coordinates (i.e., in longitude and latitude), with the origin (longitude and latitude equal to 0°) at the head point. Figure 13 (right) illustrates the eye heatmap corresponding to the gaze heatmap in Fig. 10 (left) and the head heatmap in Fig. 13 (left). Note that in computing the eyes-in-head map, care must be taken to account for the distortions near the north and south poles because

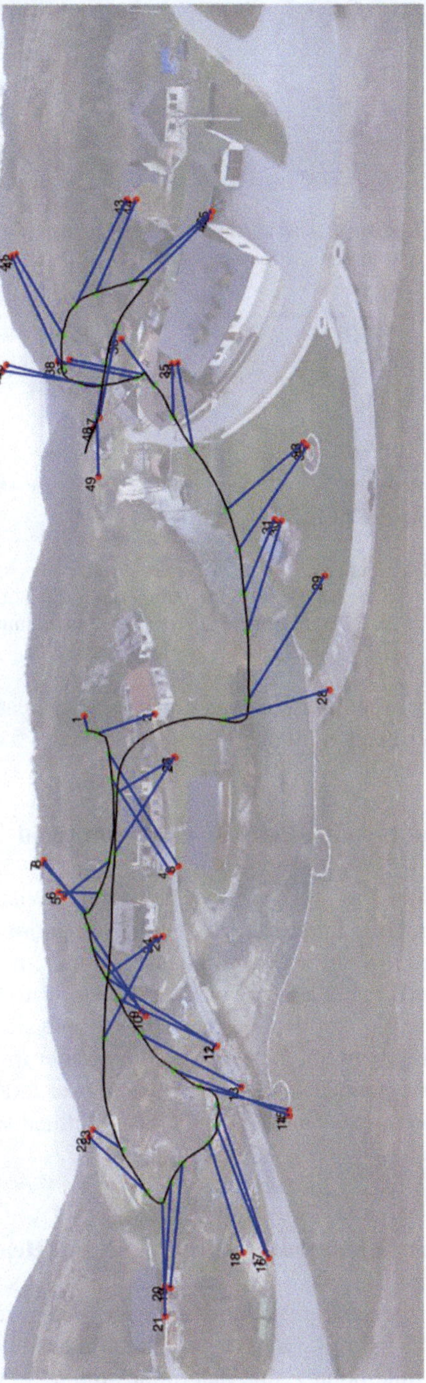

Fig. 12 Example panorama map with gaze and head positions. The red circles indicate gaze fixations, the black line shows the head positions, the green circles indicate head averages during the fixations, and the blue lines connect fixations with the corresponding head averages. These lines thus represent distances between gaze and head and are determined by the eye direction in a head-centered coordinate system

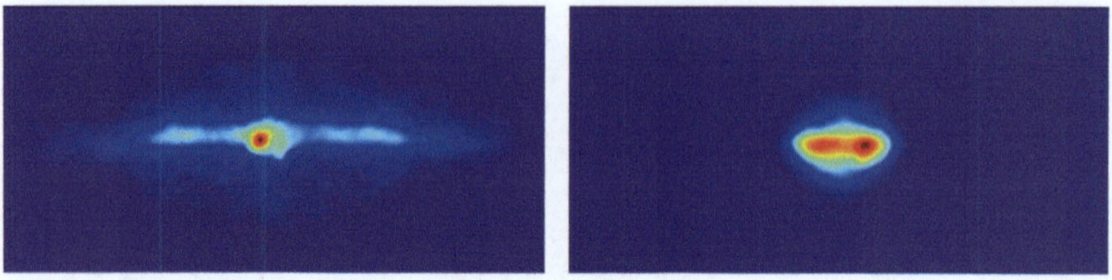

Fig. 13 (Left) Heatmap of head averages; (right) Heatmap of eye-in-head positions. Both heatmaps were computed for the same set of images as in Fig. 10

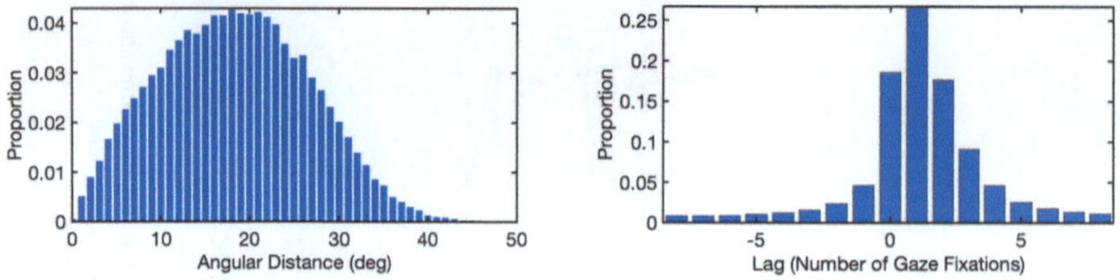

Fig. 14 (Left) Distribution of distances between head points and gaze points. (Right) Histogram of gaze-head lags, with the lag expressed in number of gaze fixations. If gaze has a positive lag, then gaze is leading. If gaze has a negative lag, then gaze is trailing

the distance between meridians is much smaller near the poles than at the equator. This is achieved using circular statistics (see Sect. 5.3).

6.4.1 Spatial Relation Between Gaze and Head

As illustrated in Fig. 13 (right), gaze points deviate only moderately from head points. Typically, in unrestrained viewing of panoramic scenes, gaze deviates from the head direction only by a moderate amount, is somewhat ahead of the head movement, and covers a larger area of the visual field. The latter helps to preserve energy because moving the eyes requires less physical effort than moving the head. The relationship between gaze and head is further illustrated in Fig. 14 (left), which shows the histogram of distances between gaze fixations and head averages. This corresponds to eye eccentricity in the head-defined visual field. As seen in Fig. 14, most eye eccentricities are in the range 10–25°.

6.4.2 Temporal Relation Between Gaze and Head

To determine the temporal relationship between gaze fixations and head positions, one proceeds as follows: Given a gaze fixation g_i and a set of head fixations h_j before and after the time point i, one determines the h_j^{min} with minimum distance. If the

h_j^{min} occurs before g_i, then it is concluded that head is leading gaze, otherwise it is concluded that head is lagging behind gaze. The results of this minimum distance analysis for a set of panorama images are shown in Fig. 14 (right), which indicates that most minimum distances occur at a positive lag (i.e., the gaze positions are most often leading the head positions). The lag peak is around 1–2 fixations, suggesting that gaze leads the head by approximately 200 ms.

6.5 Observations on Eye and Head Movement Behavior While Looking at 360° Scenes

The studies mentioned throughout the chapter and the data reported in this section revealed several important characteristics of the interplay between eye and head movements in 360° panoramic scenes. First, not unsurprisingly, the head and eyes tend to follow along the horizon of 360° scenes, where most head and gaze positions are found along the horizon and most saccades are made along the horizon (at least in cases where a notable horizon exists, Bischof et al. 2020). Second, we found consistently that the head tends to follow the eyes, indicating (based on earlier research, Doshi and Trivedi 2012; Freedman 2008; Zangemeister and Stark 1982) that viewing 360° scenes in VR follows a pattern of reflexive orienting, most likely with the eyes and head responding to image cues. This observation stands in contrast to studies where the head played a more central role by moving to reveal new information to the eyes (Solman et al. 2017). Third, the eyes tend to stay relatively close to the center of the head-defined visual field, consistent with the results obtained with mobile eye tracking (e.g., Foulsham et al. 2011; 't Hart et al. 2009).

More speculatively, we have noticed that there are typically substantial individual differences in the amount that an observer moves their head during visual explorations. While some observers move their head extensively, others keep their head very still. This distinction between head "movers" and "non-movers" has been found repeatedly in the kinematic literature (e.g., Delreux et al. 1991; Fuller 1992a, b; Goldring et al. 1996; Stahl 2001). Taken together, these works provide evidence that head and eye movements diverge in their control strategies, leading researchers to speculate that the head may be under more deliberate, cognitive control (David et al. 2022) or are sensitive to different spatial reference frames (Anderson et al. 2020).

7 Open Questions and Future Directions

Implementing eye tracking into experiments in VR can be challenging, but we hope that this chapter provides researchers with practical and actionable advice on how to begin. At the start of this chapter, we made a case for why it is important to track eye movements in VR, and in this section, we leave readers with a sense of the types of questions that VR can help us answer about visual cognition and behavior:

1. What is the nature of the relationship between eye, head, and other body movements in supporting human cognition?

   This question is not only a kinematic one, where researchers may be interested in exactly how and when the cognitive system recruits different effectors like the head and body to support ongoing thought and behavior, but it also can provide clues to how these systems relate to everyday, realistic environments. VR sits between the extremely constrictive computer-based eye-tracker situations (where most research has been conducted) and the unrestrained but more natural approach taken with mobile eye trackers by researchers interested in the more complex but less controlled everyday situations. VR provides a way to precisely measure human movements in naturalistic situations, while simultaneously controlling the external inputs to the system. One can imagine that by simplifying the VR environment similar to computer-based studies (e.g., Folk et al. 1992; Henderson 2016; Silvis and Donk 2014; Theeuwes 1994; van Zoest et al. 2004), combined with added head and body movement information, it would be possible to tackle some of the fundamental, unanswered questions involving eye and head movements (see, for example, Flindall et al. 2021).
2. What are the consequences of studying human cognition in a situation where the head is restrained?

   This question refers to the growing realization among researchers that restricting head movements in a chin rest, and pre-selecting the stimuli that observers are allowed to view, may bias the type of data one obtains and the conclusions one reaches. First, participants must explore the stimuli using eye movements alone. In contrast, our studies have shown that the eyes and head work in conjunction to explore the visual world. Second, in a paradigm with peripheral masking, David et al. (2021) have demonstrated that the previously reported visual span of 6° visual angle obtained with head-fixed studies may be a gross underestimation. It should be emphasized that the head movements in our studies involved only changing the head orientation (pitch, yaw, roll), but not changing the position of the head in space.
3. What are the similarities and differences between perceiving and acting on virtual items and versus those that are real?

   This question is motivated by the growing recognition that the way people see and act on objects that are virtual may often engage very different forms of cognition and behavior, as well as different brain systems, than those that are real (e.g., Dosso and Kingstone 2018; Freud et al. 2018; Gallup et al. 2019). Determining when one's findings are specific to virtual stimuli versus when they generalize to real-world situations is of fundamental and profound importance, and one that VR research promises to unlock in the future.

It is our hope and intention that the material presented in this chapter will provide researchers with the basic skills necessary to begin to engage in this exciting research enterprise. We have focused on elements that are critical to conduct research in VR that focuses on the movement of the head and eyes, and how they relate to one another as well as the external environment. The power of VR is that one can create, manipulate, and control the environment that an individual is immersed within, ranging from the simple environments that are routinely used in lab experiments to much more complex real-world situations, to creating environments that are, literally, not of this world! The challenge, of course, is how one can make sense of the data that one obtains in such studies. We hope that the material presented in this chapter will enable researchers to pursue their own research questions in a manner that is both theoretically exciting and empirically tractable.

References

Anderson NC, Bischof WF, Foulsham T, Kingstone A (2020) Turning the (virtual) world around: patterns in saccade direction vary with picture orientation and shape in virtual reality. J Vis 20(8):21–21

Backhaus D, Engbert R, Rothkegel LOM, Trukenbrod HA (2020) Task-dependence in scene perception: head unrestrained viewing using mobile eye-tracking. J Vis 20(5):3–3. https://doi.org/10.1167/jov.20.5.3

Barnes GR (1979) Vestibulo-ocular function during co-ordinated head and eye movements to acquire visual targets. J Physiol 287(1):127–147. https://doi.org/10.1113/jphysiol.1979.sp012650

Batschelet E (1981) Circular statistics in biology. Academic Press, New York, p 388

Bebko AO, Troje N (2020) bmlTUX: design and control of experiments in virtual reality and beyond. PsyArXiv. https://doi.org/10.31234/osf.io/arvkf

Bischof WF, Anderson NC, Doswell MT, Kingstone A (2020) Visual exploration of omnidirectional panoramic scenes. J Vis 20(7):23. https://doi.org/10.1167/jov.20.7.23

Blue Trident IMU | Inertial Sensor by Vicon | Biomechanic Tracking (2022) Vicon. Retrieved March 30, 2022, from https://www.vicon.com/hardware/blue-trident/

Brookes J, Warburton M, Alghadier M, Mon-Williams M, Mushtaq F (2020) Studying human behavior with virtual reality: the unity experiment framework. Behav Res Methods 52(2):455–463. https://doi.org/10.3758/s13428-019-01242-0

Core – Pupil Player (2022) Pupil Labs. Retrieved March 31, 2022, from https://docs.pupil-labs.com

David EJ, Beitner J, Võ ML-H (2020) Effects of transient loss of vision on head and eye movements during visual search in a virtual environment. Brain Sci 10(11):841. https://doi.org/10.3390/brainsci10110841

David EJ, Beitner J, Võ ML-H (2021) The importance of peripheral vision when searching 3D real-world scenes: a gaze-contingent study in virtual reality. J Vis 21(7):3–3

David EJ, Lebranchu P, Perreira Da Silva M, Le Callet P (2022) What are the visuo-motor tendencies of omnidirectional scene free-viewing in virtual reality? J Vis 22(4):12. https://doi.org/10.1167/jov.22.4.12

De Abreu A, Ozcinar C, Smolic A (2017) Look around you: saliency maps for omnidirectional images in VR applications. 2017 Ninth international conference on quality of multimedia experience (QoMEX), pp 1–6

Delreux V, Abeele SV, Lefevre P, Roucoux A (1991) Eye–head coordination: influence of eye position on the control of head movement amplitude. Brain Space:38–48

Doshi A, Trivedi MM (2012) Head and eye gaze dynamics during visual attention shifts in complex environments. J Vis 12(2):9–9

Dosso JA, Kingstone A (2018) Social modulation of object-directed but not image-directed actions. PLoS One 13(10):e0205830. https://doi.org/10.1371/journal.pone.0205830

Dufaux F, Konrad J (2000) Efficient, robust, and fast global motion estimation for video coding. IEEE Trans Image Process 9(3):497–501

Flindall J, Sara A, Kingstone A (2021) Head and eye movements are each facilitated by the offset of a central fixation point in a virtual gap paradigm. Exp Brain Res 239(1):117–126. https://doi.org/10.1007/s00221-020-05905-9

Folk CL, Remington RW, Johnston JC (1992) Involuntary covert orienting is contingent on attentional control settings. J Exp Psychol Hum Percept Perform 18:1030–1030

Foulsham T, Kingstone A, Underwood G (2008) Turning the world around: patterns in saccade direction vary with picture orientation. Vision Res 48(17):1777–1790

Foulsham T, Walker E, Kingstone A (2011) The where, what and when of gaze allocation in the lab and the natural environment. Vision Res 51(17):1920–1931

Foulsham T, Kingstone A (2017) Are fixations in static natural scenes a useful predictor of attention in the real world. Can J Exp Psychol / Revue canadienne de psychologie expérimentale 71 (2):172–181

Freedman EG (2008) Coordination of the eyes and head during visual orienting. Exp Brain Res 190(4):369

Freud E, Macdonald SN, Chen J, Quinlan DJ, Goodale MA, Culham JC (2018) Getting a grip on reality: grasping movements directed to real objects and images rely on dissociable neural representations. Cortex 98:34–48. https://doi.org/10.1016/j.cortex.2017.02.020

Fuller J (1992a) Comparison of head movement strategies among mammals. The headneck sensory motor system. Oxford University Press, New York, pp 101–112

Fuller J (1992b) Head movement propensity. Exp Brain Res 92(1):152–164

Gallup AC, Vasilyev D, Anderson N, Kingstone A (2019) Contagious yawning in virtual reality is affected by actual, but not simulated, social presence. Sci Rep 9(1):294. https://doi.org/10.1038/s41598-018-36570-2

Goldring JE, Dorris MC, Corneil BD, Ballantyne PA, Munoz DR (1996) Combined eye-head gaze shifts to visual and auditory targets in humans. Exp Brain Res 111(1):68–78

Henderson JM (2016) Gaze control as prediction. Trends Cogn Sci. http://www.sciencedirect.com/science/article/pii/S1364661316301929

Hessels RS, Niehorster DC, Nyström M, Andersson R, Hooge IT (2018) Is the eye-movement field confused about fixations and saccades? A survey among 124 researchers. R Soc Open Sci 5(8): 180502

Holmqvist K, Andersson R (2017) Eye tracking: a comprehensive guide to methods. Paradigms and measures

Hooge IT, Hessels RS, Niehorster DC, Diaz GJ, Duchowski AT, Pelz JB (2019) From lab-based studies to eye-tracking in virtual and real worlds: conceptual and methodological problems and solutions

Jacobs O, Anderson NC, Bischof WF, Kingstone A (2020) Into the unknown: head-based selection is less dependent on peripheral information than gaze-based selection in 360-degree virtual reality scenes. PsyArXiv. https://doi.org/10.31234/osf.io/2qtcw

Kassner M, Patera W, Bulling A (2014) Pupil: an open source platform for pervasive eye tracking and mobile gaze-based interaction. Proceedings of the 2014 ACM international joint conference on pervasive and ubiquitous computing: adjunct publication, pp 1151–1160

Kingstone A, Smilek D, Eastwood JD (2008) Cognitive ethology: a new approach for studying human cognition. Br J Psychol 99(3):317–340

Komogortsev OV, Gobert DV, Jayarathna S, Gowda SM (2010) Standardization of automated analyses of oculomotor fixation and saccadic behaviors. IEEE Trans Biomed Eng 57(11): 2635–2645

Land MF, Hayhoe M (2001) In what ways do eye movements contribute to everyday activities? Vision Res 41(25):3559–3565

Mathôt S, Schreij D, Theeuwes J (2012) OpenSesame: an open-source, graphical experiment builder for the social sciences. Behav Res Methods 44(2):314–324

Ohayon S, Rivlin E (2006) Robust 3d head tracking using camera pose estimation. 18th international conference on pattern recognition (ICPR'06), vol 1, pp 1063–1066

Peirce J, Gray JR, Simpson S, MacAskill M, Höchenberger R, Sogo H, Kastman E, Lindeløv JK (2019) PsychoPy2: experiments in behavior made easy. Behav Res Methods 51(1):195–203. https://doi.org/10.3758/s13428-018-01193-y

Pupil Invisible – Eye tracking glasses for the real world – Pupil Labs (2022) Retrieved March 30, 2022, from https://pupil-labs.com/products/invisible/

Rai Y, Gutiérrez J, Le Callet P (2017) A dataset of head and eye movements for 360 degree images. Proceedings of the 8th ACM on multimedia systems conference, pp 205–210

Risko EF, Richardson DC, Kingstone A (2016) Breaking the fourth wall of cognitive science: real-world social attention and the dual function of gaze. Curr Dir Psychol Sci 25(1):70–74

Salvucci DD, Goldberg JH (2000) Identifying fixations and saccades in eye-tracking protocols. Proceedings of the 2000 symposium on eye tracking research and applications, pp 71–78

SensoMotoric (2017) SensoMotoric Instruments. [Apparatus and software]. https://en.wikipedia.org/wiki/SensoMotoric_Instruments

Silvis JD, Donk M (2014) The effects of saccade-contingent changes on oculomotor capture: salience is important even beyond the first oculomotor response. Atten Percept Psychophys 76(6):1803–1814

Sitzmann V, Serrano A, Pavel A, Agrawala M, Gutierrez D, Masia B, Wetzstein G (2018) Saliency in VR: how do people explore virtual environments? IEEE Trans Vis Comput Graph 24(4): 1633–1642. https://doi.org/10.1109/TVCG.2018.2793599

Solman GJ, Foulsham T, Kingstone A (2017) Eye and head movements are complementary in visual selection. R Soc Open Sci 4(1):160569

Solman GJ, Kingstone A (2014) Balancing energetic and cognitive resources: memory use during search depends on the orienting effector. Cognition 132(3):443–454. https://doi.org/10.1016/j.cognition.2014.05.005

Stahl JS (2001) Eye-head coordination and the variation of eye-movement accuracy with orbital eccentricity. Exp Brain Res 136(2):200–210

't Hart BM, Vockeroth J, Schumann F, Bartl K, Schneider E, König P, Einhäuser W (2009) Gaze allocation in natural stimuli: comparing free exploration to head-fixed viewing conditions. Vis Cogn 17(6–7):1132–1158. https://doi.org/10.1080/13506280902812304

Theeuwes J (1994) Endogenous and exogenous control of visual selection. Perception 23(4): 429–440

Torralba A, Oliva A, Castelhano MS, Henderson JM (2006) Contextual guidance of eye movements and attention in real-world scenes: the role of global features in object search. Psychol Rev 113(4):766–786. https://doi.org/10.1037/0033-295X.113.4.766

Unity Essentials (2022) Unity learn. Retrieved March 31, 2022, from https://learn.unity.com/pathway/unity-essentials

V120 (n.d.) Duo – an optical tracking system in a single, plug-and-play package. OptiTrack. Retrieved March 30, 2022, from http://optitrack.com/cameras/v120-duo/index.html

van Zoest W, Donk M, Theeuwes J (2004) The role of stimulus-driven and goal-driven control in saccadic visual selection. J Exp Psychol Hum Percept Perform 30(4):746

Vasser M, Kängsepp M, Magomedkerimov M, Kilvits K, Stafinjak V, Kivisik T, Vicente R, Aru J (2017) VREX: an open-source toolbox for creating 3D virtual reality experiments. BMC Psychol 5(1):4. https://doi.org/10.1186/s40359-017-0173-4

Visbox, Inc. (2022) Retrieved March 30, 2022, from http://www.visbox.com/
Vizard | Virtual Reality Software for Researchers (2022) Retrieved March 30, 2022, from https://www.worldviz.com/vizard-virtual-reality-software
Watson MR, Voloh B, Thomas C, Hasan A, Womelsdorf T (2019) USE: an integrative suite for temporally-precise psychophysical experiments in virtual environments for human, nonhuman, and artificially intelligent agents. J Neurosci Methods 326:108374. https://doi.org/10.1016/j.jneumeth.2019.108374
Zangemeister WH, Stark L (1982) Types of gaze movement: variable interactions of eye and head movements. Exp Neurol 77(3):563–577. https://doi.org/10.1016/0014-4886(82)90228-X

Part II
VR to Study the Mind

Virtual Reality for Spatial Navigation

Sein Jeung, Christopher Hilton, Timotheus Berg, Lukas Gehrke, and Klaus Gramann

Contents

S. Jeung
Department of Biological Psychology and Neuroergonomics, Technische Universität Berlin, Berlin, Germany

Kavli Institute for Systems Neuroscience, Norwegian University of Science and Technology, Trondheim, Norway

Max Planck Institute for Human Cognitive and Brain Sciences, Leipzig, Germany

C. Hilton, T. Berg, and L. Gehrke
Department of Biological Psychology and Neuroergonomics, Technische Universität Berlin, Berlin, Germany

K. Gramann (✉)
Department of Biological Psychology and Neuroergonomics, Technische Universität Berlin, Berlin, Germany

Center for Advanced Neurological Engineering, University of California, San Diego, CA, USA
e-mail: klaus.gramann@tu-berlin.de

Curr Topics Behav Neurosci (2023) 65: 103–130
https://doi.org/10.1007/7854_2022_403
Published Online: 14 December 2022

Abstract Immersive virtual reality (VR) allows its users to experience physical space in a non-physical world. It has developed into a powerful research tool to investigate the neural basis of human spatial navigation as an embodied experience. The task of wayfinding can be carried out by using a wide range of strategies, leading to the recruitment of various sensory modalities and brain areas in real-life scenarios. While traditional desktop-based VR setups primarily focus on vision-based navigation, immersive VR setups, especially mobile variants, can efficiently account for motor processes that constitute locomotion in the physical world, such as head-turning and walking. When used in combination with mobile neuroimaging methods, immersive VR affords a natural mode of locomotion and high immersion in experimental settings, designing an embodied spatial experience. This in turn facilitates ecologically valid investigation of the neural underpinnings of spatial navigation.

Keywords Embodiment · Mobile brain–body imaging · Multisensory integration · Reference frame · Spatial navigation · VR

1 Spatial Navigation as an Embodied Experience

> "I will take the Ring", he said, "though I do not know the way."—J.R.R. Tolkien, The Fellowship of the Ring

As well-demonstrated in the famous book and movie trilogy The Lord of the Rings, where Frodo is tasked with journeying from one end of Middle-earth to the other, figuring out where the goal is and how to get there is often one of the essential components of solving real-life tasks that take place in physical space. Spatial navigation, the process of forming, updating, and retrieving the knowledge of where to go, comprises a multitude of complex physical and cognitive processes such as perception of relevant spatial cues, motor execution, decision-making, and memory (Montello 2005).

A variety of tasks can be construed as spatial navigation; yet, there is a gradient of how *embodied* the behavior is. For instance, looking at the map of Middle-earth to draw an optimal route between the Shire and Mordor is a rather abstract, or *disembodied* form of spatial navigation. On the other hand, when the same route is navigated on foot, every step toward Mordor, being the farthest away from home than one's ever been, will evoke embodied, more vivid physical and cognitive experiences.

The latter form of spatial navigation can be seen as more *naturalistic*, as movement through physical space is of more fundamental evolutionary relevance. The ability to learn the causal structures in the environment and to deal with uncertainties of dynamically changing scenes and objects is important for efficient interaction with the world (Klippel 2003). We constantly predict the influx of sensory data based on probabilistic analyses of previously experienced contingencies between actions and perceptual events (Clark 2013; Friston 2010; Rao and Ballard 1999). This is inherently tied to the body's capacity to act on the environment, rendering the action-perception cycle of cognition into an embodied process (Friston 2012). Hence, research efforts attempting to explain ecologically valid navigation behavior should consider the embodied experience with respect to environmental affordances (Gramann 2013).

During spatial navigation as an embodied experience, cues from various sensory modalities contribute to forming a convergent source of spatial information (Millar 1994; Schinazi et al. 2016; Chen et al. 2017) and active body motion affects the quality of spatial representations (Simons and Wang 1998; Wang and Simons 1999). We discuss below how the choice of strategies for spatial navigation is influenced by the types of sensory inputs and environmental affordances and how these can lead to different measured signals in the brain. To some extent, stationary and non-immersive virtual reality (VR) can account for these aspects. However, the use of mobile, immersive VR can help experiments better approximate spatial navigation as it happens in real life, i.e., with a full range of sensory inputs and affordances.

2 Strategies for Spatial Navigation

Different pictures can come to mind when one thinks of their latest episode of navigating through space. The way to the goal may have been guided by a navigation aid such as a map or GPS, a landmark visible from afar, or simply a sequence of familiar scenes in memory. Depending on the types and quality of information at hand and action choices available, the exact strategy for spatial navigation can vary from individual to individual, and from situation to situation. Again, we take Frodo's journey as an example, to show how the act of moving from one place to another can be broken down into a variety of tasks that are all framed as spatial navigation. While some of the journey takes place in locations with which he is familiar, such as the Shire, much of it is through novel environments requiring the full range of navigation mechanisms available to Frodo and the fellowship.

Initially, Frodo must avoid the road and travel through the wilderness, relying on his path integration abilities, defined as constant updating of position and orientation during motion (Mittelstaedt and Mittelstaedt 1982), to guide him through the forests. Various types of idiothetic information (i.e., information originating from self-motion) such as proprioception and optic flow contribute to this process (Chrastil et al. 2019). Proprioceptive information about self-movement and body position originates from sensors in our muscles and joints (Chance et al. 1998), as do motor efference cues (Sheeran and Ahmed 2020). The vestibular system, part of the inner ear, provides information about the acceleration of the head in all three dimensions (Goldberg et al. 2012) to compute the relative change in orientation and position of the navigator with respect to landmarks and other sources of information – even in the absence of vision.

Path integration is useful in open environments where direct trajectories can be taken to achieve navigation goals. However, such abilities are less helpful in the Mines of Moria, when the fellowship instead falls back on the "always follow your nose" approach to wayfinding, which involves the continuation of a straight trajectory until a cue for action guides them otherwise (a well-documented strategy for route navigation; Klippel 2003; Meilinger et al. 2012, 2014). In addition, allothetic information (i.e., information originating from the external environment) perceived through multiple senses is used to guide navigation. While vision plays a special role for human navigation (Ekstrom 2015), other senses contribute significantly to our updating of position and orientation (Campos et al. 2012; Lappe et al. 2007). This includes olfaction (Wu et al. 2020; Hamburger and Knauff 2019, Samwise Gamgee: "What's that horrible stink? I wonder if there's a nasty bog nearby?"), and acoustic information for estimating distances (Dodsworth et al. 2020; Kolarik et al. 2016; Geronazzo et al. 2016, Boromir sounding The Horn of Gondor), or as a cue to support the visual system (Werkhoven et al. 2014; Rossier et al. 2000).

As proximity to his goal location, Mount Doom, increases, Frodo is able to use a well-known landmark as a beacon to guide navigation – the Eye of Sauron. Information about one's position relative to the environment is predominantly perceived through the visual sense. Salient features of the built and natural environment such as noticeable buildings, a lake or park, intersections and other features can serve as landmarks that indicate specific spatial locations and orienting points. Thus, landmarks aid navigation behavior and allow for re-orienting if the navigator becomes lost (Chan et al. 2012). When Frodo finds his route to Mount Doom blocked by the impassable Black Gate, Smeagol is on hand to provide his knowledge represented as a cognitive map (Epstein et al. 2017; Tolman 1948) to plan a novel detour via the pass of Cirith Ungol.

Frodo's journey required him to integrate information from all of his senses, and his existing spatial knowledge, to solve a range of navigation problems. In most cases, this means that both allothetic and idiothetic sources are combined to successfully orient and navigate in the environment and the more reliable source dynamically determines the navigation strategy (Nardini et al. 2008; Fetsch et al. 2009). Of course, even Frodo failed on occasion, becoming hopelessly lost in the hills of Emyn Muil!

3 Neural Basis of Spatial Navigation

For solving a navigation task, the brain uses all types of available resources to reach the goal, resulting in multiple areas being active in parallel (Chrastil 2013). Thus, when designing and interpreting a study that makes a statement about the functional role of a brain region in spatial navigation, attention has to be paid to the types of spatial cues as well as action choices available to the participants. To show how different modalities of sensory information and actions of the agent shapes the representations of space in the brain, we briefly review findings about three brain regions that are particularly well-investigated in relation to spatial navigation: the parietal cortex, the medial temporal network including the hippocampus, and the retrosplenial complex (RSC).

3.1 The Parietal Cortex and Multisensory Integration

Spatial representations, like many other types of information represented in the brain, make use of signals received from the sensory organs specialized on modality-specific input such as the retina or the cochlea. Once filtered through the thalamus (with the exception of olfaction), the sensory information is forwarded to brain regions that further process specific attributes from low-level to higher-level aspects of the stimulus, still confined within that sensory modality. The modality-specific sensory information (vision, sound, etc.) is then integrated in the parietal cortex, situated between the occipital and the frontal cortices, to form a multi-modal representation that is behaviorally relevant (Brang et al. 2013; Whitlock 2017).

In the context of spatial navigation, it is suggested that the sensory information integrated into the parietal area represents egocentric information (viewpoint-dependent self-environment representations), interacting with the RSC where allocentric information (viewpoint-independent environment-environment representations) is incorporated to create global mappings (see Clark et al. (2018) for review). Importantly, based on its role in integrating sensory information from multiple modalities that result from movement in the environment, the posterior parietal cortex (PPC) plays a role in the embodied experience of space. As reviewed by Whitlock (2017), historic clinical case studies demonstrated that damage to the PPC is linked to deficits in the generation of body image, peripersonal space, and the ability to act upon one's surrounding (Bálint 1909; Critchley 1953; Head and Holmes 1911). One striking example is Bálint's syndrome, where bilateral damage to the PPC causes the inability to move a hand toward a target object guided by vision, among other symptoms (Bálint 1909). Parietal damage is also associated with the inability to locate or recognize one's own body (Critchley 1953) or to update cognitive heading in the absence of corresponding proprioceptive and vestibular information (Seubert et al. 2008). As these findings are closely linked to embodied experience of space, body movement through physical space is a critical element in experimental investigation of how the parietal cortex is involved in spatial navigation.

3.2 The Medial Temporal Network

At the core of the medial temporal network supporting spatial navigation is the hippocampus. The most notable function of the hippocampus in spatial navigation is the representation of the navigator's allocentric position in space. This is exemplified by "place cells" that fire whenever the animal is in a specific location irrespective of its orientation (O'Keefe et al. 1998). Numerous reports on impairments in spatial cognition in patients with hippocampal damage demonstrate the involvement of the hippocampus in processing allocentric information, as shown in the meta-analysis by Kessels et al. (2001).

Directly adjacent to the hippocampus in the medial temporal network lies the entorhinal cortex, which hosts a special type of neurons that are called "grid cells." These cells fire in a regular manner commensurate with movement through space and provide a metric for spatial representations (Hafting et al. 2005). As the population activity from sources localized to the entorhinal cortex is observable in humans using noninvasive methods such as fMRI (Doeller et al. 2010; Kunz et al. 2015) and MEG (Staudigl et al. 2018), grid cell research holds the key to testing models of spatial information processing in the medial temporal network in a healthy population.

Neural populations in the hippocampus display prominent theta oscillation in the frequency range between 4 and 10 Hz in rodents (Buzsáki 2005). The power and frequency of the theta activity show linear correlations with speed of locomotion as well as spatial learning in freely navigating rodents (Young et al. 2021). In humans, this rhythmic activity shows a different profile, being weaker and also slower (Jacobs 2014). It is possible that this observed difference between human and rodent theta oscillation stems from different experimental setups and the presence of idiothetic information for the animal subjects.

Indeed, body-based signals play an important role in firing of cells in the medial temporal network. For example, vestibular lesions led to the disruption of hippocampal theta rhythm in rodents (Russell et al. 2006). Similarly, when vestibular input was selectively impaired in rodents, theta oscillations in the entorhinal cortex were abolished and grid cell firing was interrupted (Jacob et al. 2014). In mice, the entorhinal cortex is shown to encode a variety of self-motion signals including head movement (Mallory et al. 2021). Thus, the medial temporal network integrates idiothetic as well as allothetic information about movement in space as the basis for an allocentric representation.

3.3 The Retrosplenial Complex

Associated with allocentric and egocentric spatial representations, respectively, the medial temporal and parietal networks closely interact with each other via the RSC (Vann et al. 2009; Byrne et al. 2007). Often seen as part of the retrosplenial-parietal network, the RSC has dense anatomical connectivity with both parietal and medial temporal regions (Kobayashi and Amaral 2003) and thus has been proposed to play a role in conversion between the allocentric and egocentric representations from parietal and temporal brain areas (Becker and Burgess 2000; Vann et al. 2009; Baumann and Mattingley 2021). Note that some integration of ego- and allocentric information may take place within the parietal cortex (Whitlock et al. 2008; Save and Poucet 2009). Clark (2013) suggests that spatial information is organized in a gradient from the parietal cortex to RSC, with an increasing representation of allocentric information in the RSC and more specialized encoding of egocentric information in the parietal cortex.

In a model proposed by Byrne et al. (2007), the RSC hosts the circuit for translating between ego- and allocentric coordinates, allowing each location in an environment to have a single representation regardless of where it is viewed from. The translation between the different frames of reference in the RSC is achieved using signals from "head direction cells" (Taube et al. 1990) housed throughout the limbic system. These cells provide information about changes in individual heading (based on idiothetic information) that allow integration of the navigator's directional changes with the hippocampal positional system (Muller et al. 1996). While heading changes can be computed based on optic flow (Gramann et al. 2006, 2010), using external landmarks further allows to calibrate and update the path integration system during navigation and is used whenever stable landmarks are perceived (Etienne and Jeffery 2004).

The above-described neural systems underlying spatial navigation support complex, multisensory processes that are based on full-body movement through the environment. Neuroscientific research on human spatial navigation, however, is restricted by several factors. The portability of brain imaging devices is one of the most limiting aspects (Gramann et al. 2011, 2014), leading to a prevalence of stationary experimental protocols. This in turn limits the multimodal sensory experience of space restricting the interpretation of the neural dynamics associated with this kind of navigation protocols.

4 Non-immersive Virtual Reality Setups for Spatial Navigation

To capture the many facets of spatial navigation both on the behavioral and neural levels, we would ideally observe animals as they naturally move through space. However, to induce lifelike spatial navigation in humans in a laboratory setting is a non-trivial task, due to our larger body size. Such research traditionally faces barriers driven by the need to incorporate relevant spatial scales into experimental paradigms (Wolbers and Wiener 2014; Spiers and Maguire 2006) and the restrictions of traditional brain imaging methods. The adoption of VR as a common tool over the past decades in navigation research was a step forward in addressing some of these challenges. Below we discuss the advantages provided by using *non-immersive* (in contrast to *immersive*) VR, defined as a methodology that utilizes a laboratory desktop setup in which participants are stationary and explore virtual environments via an input device such as keyboard and joystick.

4.1 Free Manipulation of Space

In spatial cognition research it is particularly challenging to exert control over real-world environments in order to systematically manipulate experimental variables. Such challenges originate with the obvious difficulty of altering the architecture of spaces (e.g., Bécu et al. 2020). Additionally, real-world environments contain many confounding features which can impact the behavior of participants. Environments in VR on the other hand can be created according to the experimenters' precise needs, such as in route navigation paradigms which contain virtual mazes with visually identical intersections and corridors, for the systematic variation of landmark identities and locations (e.g., Janzen (2006); Wiener et al. (2013); Grzeschik et al. (2020). Waller and Lippa (2007) used such a paradigm to determine whether landmarks can be used as associative cues for navigation (e.g., turn left at the church) and as beacons (e.g., move toward the church) by systematically manipulating the placement of landmark objects relative to route heading (also see Wiener et al. 2013).

4.2 Compatibility with Neuroimaging

Many early navigation studies were limited in the extent to which the neural correlates of navigation behavior could be (non-invasively) investigated in humans. The rapid advancement of neuroimaging technologies over the last decades allowed for new insights into the neural networks and brain dynamics underlying navigation, such as the discovery of larger hippocampal formations in experienced London taxi drivers, as navigation experts, compared to bus drivers as control participants that simply follow the same routes every day (Maguire et al. 2000). With the addition of VR, such studies can now be conducted with online, functional recording of brain activity for greater insights, such as with the virtual taxi driving game used by Spiers and Maguire (2007). Using functional magnetic resonance imaging (fMRI), they found a correlation between metric goal distance (i.e., allocentric navigation) and activity in the entorhinal complex.

This neural code for allocentric space representation could be dissociated from the representation of egocentric directions to goal locations that correlated with activity in the parietal cortex. Indeed, several findings originally constrained to investigations without neuroimaging data, such as the importance of decision point locations in the environment for navigation strategies (Blades et al. 1992; Golledge 1995), have been since revisited in VR with the addition of neuroscientific methods (e.g., Janzen and Van Turennout (2004) demonstrating selective para-hippocampal encoding of objects located at decision points).

4.3 Challenges with Transferring Animal Paradigms to Human Studies

A key challenge from the aforementioned barriers of controlling space and compatibility with neuroimaging is the transfer of paradigms from animal to human research. Early insights into the neural underpinnings of the navigation system came from nonhuman animal research. A primary example is the Morris Water Maze (MWM; Morris 1981) task, which is arguably one of the most valuable spatial learning tasks to date (Thornberry et al. 2021) as it allows for investigating different navigation strategies, ranging from egocentric path replication to allocentric goal-directed navigation, and the related neural dynamics. The core component of this task is to submerge the animal, typically a rodent, in a pool of opaque water so that they would search for a hidden platform to free themselves (see Hodges 1996; Bolding and Rudy 2006; Vorhees and Williams 2006 for reviews on MWM procedures). Allothetic information is provided to the animals in the form of distal landmarks surrounding the pool to allow them to find the platform without any proximal visual cues inside the pool. Combining information about distal landmark configurations, self-motion, and local metric information about the distance between the platform and the pool border enables a variety of strategies to solve the task (Schoenfeld et al. 2017).

While some studies have used real-world human analogs of the MWM (Laczó et al. 2010), this comes with clear financial, ethical, and logistical barriers alongside the difficulty with studying brain activity in such tasks (since traditional imaging techniques do not allow for movement, and do not mix well with water). As a solution, VR has allowed researchers to create virtual versions of tasks from animal research for use with motion-restricted human subjects at lower cost and higher precision. Human virtual MWM tasks have revealed considerable cross-species generalization of benchmark spatial memory effects. These include maze learning and memory performance over time (Schoenfeld et al. 2017), the role of the hippocampus in place learning (Goodrich-Hunsaker et al. 2010), performance deficits for cognitively aged humans (Zhong et al. 2017), and the impact of traumatic brain injury (Tucker et al. 2018).

4.4 Beyond Real-World Space

An unusual method to understand the cognitive representation of space in humans has been to create environments that do not conform to the principles of real space. Such environments fall into two categories: (1) extreme reductionist environments that provide only sparse spatial information and (2) non-euclidean spaces that introduce virtual spatial configuration that are physically impossible, such as "wormholes" (locations in the environment that connect two spatially distant locations to be temporally close).

Extreme reductionist environments, such as the star field, allow for the targeting of specific components of the navigation system. In the star field, an infinite empty plane exists with only floating lights ("stars") to induce perception of self-motion. Here, the environment does not display any landmarks or geometric features to guide action, and navigators must thus rely on the visual flow to navigate by the stars (Goeke et al. 2013). As another example, in an fMRI by Wolbers et al. (2007), a path integration task was presented using a visually sparse virtual environment where only the optic flow of the floor texture was available as a cue, to reveal differential engagement of the hippocampus, the human motion complex, and the medial prefrontal cortex in self-motion processing.

Non-euclidean environments have been used in non-immersive VR to understand how real-world physics is reflected in the way humans represent space. One debate in the field of navigation is whether space is represented in a common, euclidean, reference frame or whether more local chunks of knowledge are formed independently of others. Of course, since all space encountered by humans conform to euclidean postulates, it is impossible to dissociate behavior from a euclidean representation in the real world. To dissociate behavior and euclidean representations, Warren et al. (2017) created a virtual environment that contained "wormholes," thus making it non-euclidean. They found that participants exhibited behavior consistent with a non-euclidean representation of space, such as using the wormholes to take shortcuts between locations, and even reported that all but one of their participants failed to notice the environmental abnormality (see also Tversky 1992; Kluss et al. 2015; Muryy and Glennerster 2021 for other non-euclidean space studies). In this way, VR not only makes studying spatial navigation behavior more logistically straightforward and cost-efficient, but it also expands the possibilities for experimental paradigms beyond what is possible in the real world.

4.5 Enhanced Replicability

The features of physical space vary substantially around the globe, and thus real-world navigation tasks in one geographic region may not be fully comparable with those used in other geographic regions. Such conflicts can occur due to differences in global structures of built environments (e.g., streets arranged in grids for US cities vs irregular street alignments in European cities; Coutrot et al. 2020) or due to visual differences in architecture. One solution to this has been the creation of environments in VR that reflect real locations, such as Virtual Tübingen (Van Veen et al. 1998) – a virtual 3D model of the German town of Tübingen – which has been used by many research groups thus standardizing the impact of the environment across studies. Similarly, some virtual environments and corresponding tasks have been shared for use openly, such as, among others, the Reference Frame Proclivity Test (Gramann et al. 2005; Goeke et al. 2015), Virtual Silcton (Weisberg et al. 2014), a route-learning testing suite (Wiener et al. 2020), and the Landmarks task (Starrett et al. 2021), which can act as standardized assessments of spatial learning.

5 From Non-immersive to Immersive VR for Spatial Navigation Research

Considering the stationary experimental setups used in combination with non-immersive VR and the embodied nature of spatial navigation, a legitimate question arises: how much do the findings from non-immersive VR, mostly lacking physical motion, generalize to behavior as it occurs in day-to-day life? Fortunately, many of the processes investigated in non-immersive studies translate well to navigation skills in real life (Thornberry et al. 2021; Hejtmanek et al. 2020; Richardson et al. 1999). However, not all questions about real-life spatial navigation can be answered using non-immersive VR, not only due to the lacking proprioceptive and vestibular inputs but also due to constraints on the types of interactions that it supports.

5.1 Locomotion Interfaces, Sensory Immersion, and Embodied Spatial Navigation

Physical navigation is an interplay between the navigator's action and perception of the resultant information, received from several sensory channels and converging as the basis for a robust spatial representation (Gramann 2013). For example, each rotation is associated with proprioceptive and vestibular feedback that is automatically used to update the navigator's orientation in space (Farrell and Robertson 1998).

To describe the inseparable connection between action and perception, Gibson introduced the term *affordance* in his 1979 seminal work on the ecological approach to perception (Gibson 1979). According to Gibson, affordances are possibilities for action that a given environment, object, or interactive technology offers a person with given physical abilities. Thus, the higher the number of sensory channels that are available for interaction, the more a given virtual environment resembles the

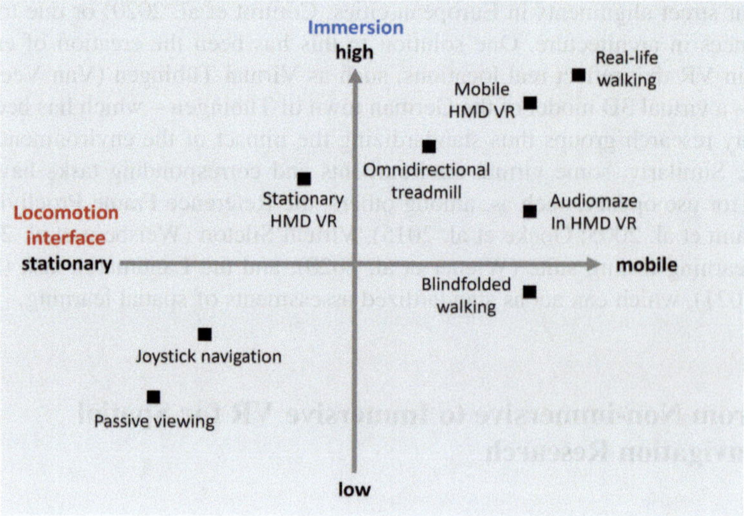

Fig. 1 Examples of experimental setups for researching the neuroscience of spatial navigation, placed in a hardware design space. For illustrative purposes, the 2D space is defined by the two dimensions locomotion interface and sensory immersion. On the horizontal axis, locomotion interface ranges from setups with little physical motion other than eye movements, termed "passive viewing", to mobile setups (mobile HMD VR, blindfolded walking). The mobile setups can all be grouped as MoBI methods, where locomotion is realized via naturalistic body motion. However, the degree of mobility in MoBI and real life are not equated due to persisting spatial limitations (e.g., confined tracking range) imposed by VR or neuroimaging hardware. On the vertical axis, sensory immersion ranges from setups with few interaction channels such as passive viewing with low-resolution visual stimuli and joystick control (bottom left quadrant) to setups incorporating body motion as well as high-resolution visual stimuli, such as mobile HMD VR. To demonstrate the differing degrees of sensory immersion within setups with comparable mobility, blindfolded walking with no visual input is contrasted with audiomaze (with auditory spatial cues, Miyakoshi et al. 2021) and the usual application of mobile HMD VR with high-resolution visual input

real-world. For example, controlling the viewpoint in a virtual environment through head rotations is a unique feature of immersive VR that uses head-mounted displays (HMDs), because such control requires motion tracking of head rotations. In VR with HMD, users can interact with the virtual environment through head movements, *affording* an interaction that is grounded in real-world experience.

For spatial navigation research, we highlight two core aspects of the embodied navigation experience that are difficult to investigate using non-immersive VR: (1) *natural* locomotion, and (2) high sensory immersion. Although not representing orthogonal axes, these two factors jointly alienate the experiences of participants in VR from real-life spatial navigation. Figure 1 depicts locomotion interface and sensory immersion as the two axes defining the hardware space used for investigating the neural basis of spatial navigation in humans. Artificial, non-natural, modes of locomotion and low sensory immersion entails a lack of body-based idiothetic signals, limited affordances, and a potential conflict between the virtual and physical spaces.

Locomotion interface can be seen as a type of environmental affordance, namely the available *means* for exploration offered by VR technology. As Taube et al. (2013) point out in their critical review on navigation in VR with fMRI, physical movement is a crucial component in experimental works on spatial navigation. In non-immersive VR, i.e., when images are presented on a desktop screen, the stimulation affords participants to explore the picture through directed eye movements (see Fig. 1, left). In a more interactive type of VR, an environment may afford to be explored through operation of input devices with concurrent directed eye movements (Wiener et al. 2012). Here, goal-directed sensory fusion occurs, where a causal relation between proprioceptive feedback from the finger joints during key press control is integrated with the visual sensory information of changing scene perception. Already at this point, *active* behavior is observed where participants exert control to sample *novel* sensory information (see Fig. 1, "joystick navigation"). At the rightmost end of the axis are setups with full body motion including physical walking and head turning, achieved by fully mobile implementations of VR (Do et al. 2021; Gehrke and Gramann 2021) and live navigation studies such as Wunderlich and Gramann (Wunderlich and Gramann 2021). These studies exemplify Mobile Brain/Body Imaging (MoBI; Makeig et al. 2009; Gramann et al. 2011, 2014; Jungnickel and Gramann 2016) setups, where brain activity is recorded in participants actively moving through physical spaces.

The axis of sensory immersion complements the principal dimension of locomotion interface. Here, we follow the definition of Slater (2009), who defines the level of sensory immersion by the number of sensory channels affording interaction. In VR, interaction channels are added via the HMD, with which participants are able to change their viewpoint via rotation and translation of their head, or even the whole body. Inclusion of sensory immersion as the second component accounts for setups that use additional sensory modalities surpassing the visual domain. Designing immersive environments, for example by adding olfactory stimulation to trigger temperature illusions (Brooks et al. 2020) or adding physical forces to the environment (Lopes et al. 2015), significantly influence the computational demands and in turn neural mechanisms for successful task completion (Gehrke et al. 2019, 2022).

Note that this formulation serves to place non-immersive and immersive VR in a continuous space, rather than presenting them as discrete categories. The setups described in the previous section on non-immersive VR are presented on a flat screen and the participants are asked to remain stationary. However, there is still some variation in how much interaction is possible in non-immersive VR setups (e.g., does the participant only passively view the stimuli or are they allowed to self-pace the presentation, or even simulate translation by key press?). Further, there are setups that fall somewhere between non-immersive and fully immersive VR, such as the Cave Automated Virtual Environment (CAVE, Cruz-Neira et al. 1992), which does not use an HMD but creates an immersive visual experience. Additionally, the axes of locomotion interface and sensory immersion are not orthogonal, because the level of sensory immersion increases as participants become more mobile.

5.2 Better Approximation of Real Life in Immersive VR

Thanks to the more naturalistic mode of locomotion and higher sensory immersion, immersive VR creates a better approximation of real-life experiences compared to non-immersive VR. The cognitive load that is created by the difficulty of interacting with an unrealistic experimental setup can create artifacts in the recorded brain activity. This does not only contribute to the level of random noise but may also qualitatively alter the process subserving spatial navigation. For instance, recruitment of the hippocampal formation in spatial memory may be affected by the cognitive load on working memory (Jeneson et al. 2011). While the highest level of ecological validity is achieved in real-world environments (Wunderlich and Gramann 2021; Hejtmanek et al. 2020), immersive VR experiments provide an alternative that is more efficient to implement, better controlled and documented, and easier to reproduce.

There are several studies investigating performance in various spatial navigation tasks with different levels of immersion. Waller et al. (1998) showed that virtual environments that induce more lifelike experiences facilitated the transfer of spatial knowledge learned in VR to a real-world situation. Ijaz et al. (2019) suggest a positive effect of immersive versus non-immersive VR on spatial navigation memory in older adults, with higher landmark recall scores, less navigational mistakes, and higher level of presence with no significantly increased stress experience compared to a desktop condition. Further research by Parong et al. (2020) showed how spatial learning is mediated by the feeling of presence in an immersive setting.

5.3 Inclusion of Body-Based Cues

One of the factors that underlie these benefits of immersive VR is the inclusion of body-based signals. Non-immersive VR primarily simulates visual experiences among other sensory information. For this reason, Taube et al. (2013) have questioned the ecological validity of fMRI navigation studies using in-scanner VR. The imaging method requires the participants to be motionless, which contradicts the requirements of navigation processes that rely on physical locomotion accompanied with motor, vestibular, and proprioceptive systems (Gramann 2013).

As a step forward, some setups enable participants to move their legs on a treadmill placed in front of a screen (Lövdén et al. 2012; Schellenbach et al. 2010; Lövdén et al. 2005). This way, the translative motion of the lower limbs in natural locomotion can be approximated. However, the rotation components of lifelike locomotion are not accounted for. This is addressed by the use of immersive HMD VR in combination with omni-directional treadmills (Hejtmanek et al. 2020; Bellmund et al. 2020), where the full spectrum of rotation is supported. This also allows the participant to rotate their head and body independently. The addition of proprioception and vestibular input in immersive HMD VR facilitates higher engagement and more accurate heading computation (Gramann et al. 2021). Finally, there are fully mobile immersive VR setups where the translative vestibular input is replicated by means of natural walking in the lab space (Do et al. 2021; Gehrke and Gramann 2021). As seen here, the locomotion interface used in a setup is a major determinant of which body-based signals are included in the participant's experience. See Sect. 6 for a more detailed overview of VR setups for spatial navigation, categorized by locomotion interfaces.

5.4 Embodied Affordances

The two axes, locomotion interface and sensory immersion (Fig. 1), define the actions afforded by the environment and significantly affect the spatial navigation networks under consideration. Clark (2015) sees affordance as determined not only by the properties of the environment, but also by the agent's physical structure, capacities, and skills. Crucially, the multisensory experience of self-motion, combining idiothetic cues and referencing the self to allothetic cues, depends on environmental affordances (Delaux et al. 2021; Do et al. 2021). By maintaining a model of the body and the space around it (Riva et al. 2019), immersive VR supports embodied experience of the space and subsequently a set of affordances that reflect the physicality of the navigating agent. For example, a person's height in HMD VR will be reflected in the height of the virtual camera and alter the visual input, whereas in desktop VR the camera height is commonly fixed across different participants.

In humans, Djebbara et al. (2019) showed that spatial-architectural affordances directly influence the neural mechanisms underlying perception of space and the concurrent motor action preparation and execution. Such findings shed light on the necessity for *natural* movement affordances mimicking paradigms used in real-life settings or in animal research.

5.5 Reduced Conflict Between Reference Frames

A task presented in non-immersive VR can lead to simultaneous retention of conflicting spatial representations. In most flat-screen setups where the display device is distant from the participant, they can directly perceive that their body exists outside of the presented virtual environment. Further, the mismatch between the imagined locomotion and the lack of idiothetic signals that would normally accompany such locomotion leads to a conflict. Although the same space can be represented from multiple perspectives and through varying levels of abstraction, this can significantly alter the way spatial information is processed by the brain. The Reference frame proclivity test (Gramann et al. 2005; Goeke et al. 2015) is an example of how the response behavior differs drastically due to the fact that participants do not always align their own heading in the physical space to follow the virtual camera. Specifically, participants choose different return trajectories after moving through space depending on if their heading representation is aligned with their physical self in space, or the virtual heading on screen. Taube et al. (2013) pointed out that the supine position of the participants inside the MR-scanner can induce a conflict between the signal from the vestibular organ and the imagined upright position in VR, leading them to retain multiple reference frames simultaneously. Therefore, findings about reference frame processing in non-immersive VR may not always generalize to the way humans perceive and interact with space in real life.

6 Locomotion Interfaces in Immersive VR for Spatial Navigation

The use of immersive VR facilitates research of spatial navigation as an embodied phenomenon. Its usefulness, however, depends on the research question and neuroimaging method at hand and should be evaluated on a case-by-case basis. To this end, it is helpful to turn to a systematic framework to assess the transferability of individual paradigms to immersive VR, such as the "VR-Check" (Krohn et al. 2020). The VR-Check framework evaluates VR applications on 10 main dimensions, including technical feasibility and immersive capacities, for example. However, the combination of the topic of spatial navigation and neuroimaging further complicates the situation. Technological limitations often lead to an exclusive choice between naturalistic task and *concurrent* neuroimaging. On the one hand, immersive stimulation with natural locomotion makes it difficult to record concurrent brain activity, due to sensitivity of the neurophysiological recording to motion artifacts or non-portability of the hardware. On the other hand, capturing the brain activity during the task often limits locomotion to a static interface.

In certain situations, immersive VR is not free from these limitations. For instance, as discussed in Taube et al. (2013), head motion is inevitably restricted in an MR scanner. When no head motion is required, it is possible to use real-time fMRI in combination with fMRI-compatible HMDs (see Wiederhold and Wiederhold 2008 for a discussion on its potential use). The use of in-scanner VR mainly serves to enhance the visual experience and the interaction via other body parts. For example, Limanowski et al. (2017) successfully tracked hand motion during in-scanner VR.

Separating the neuroimaging and VR recording sessions is a workaround for exploiting the advantage of both VR technology and the neuroimaging method. This approach can be applied to all imaging modalities but is especially useful for fMRI studies. Huffman and Ekstrom (2019), for instance, manipulated the availability of body-based cues during navigation using an omnidirectional treadmill, by selectively allowing the participants to use physical motion or joystick control to translate and rotate in HMD VR. A subsequent test of spatial knowledge in-scanner revealed comparable performance across the two learning modalities. Similarly, lesion studies using fully mobile VR are another important way to gain insights about the underlying neural processes in navigation. For example, Parslow et al. (2005) investigated participants with hippocampal lesions in immersive VR and identified selective impairment of allocentric spatial memory.

Still, there is no doubt that real-time investigation of processes that happen *during* navigation is central to our knowledge on embodied spatial navigation. With the development of mobile brain-imaging methods, there have been attempts to combine neuroimaging with some degree of concurrent physical motion achieved through the use of immersive, mobile VR. As movement through space is at the core of embodied experience of space, it is meaningful to categorize findings from the studies investigating cognitive neuroscience of human spatial navigation, among other domains, by the types of locomotion interfaces applied.

6.1 Neuroimaging in Stationary VR with Unrestricted Head Motion

Neuroscientific investigation of spatial navigation behavior is faced with the conflict between acquiring neuroimaging data containing minimal motion artifacts and the need to elicit life-like experiences through physical space. One way to settle this conflict is to keep the participants largely stationary, but at the same time allow them to look around by moving their heads. The neuroimaging method used in this type of setup should be tolerant to some degree of head motion.

With the introduction of optically pumped magnetometers (OP-MEG) that enhance the wearability of MEG hardware (Boto et al. 2018), MEG studies now allow for less constrained head motion during recording. Roberts et al. (2019) demonstrated the potential of combining OP-MEG with VR, by measuring the oscillatory and visually evoked signals during the use of an HMD. Combined with the capability of OP-MEG to access deeper structures in the brain such as the hippocampus (Barry et al. 2019), it is expected to become an important tool in spatial navigation research.

Nonetheless, this kind of VR setup does neither support the use of vestibular cues resulting from translation nor does it allow use of information derived from natural limb motion (e.g., proprioception arising from taking steps). In addition, learning to use the provided mode of control can be a task of its own. For example, Ruddle et al. (2013) reported the ease of navigating via physical walking in immersive VR compared to joystick navigation. The authors showed that participants in immersive VR performed better when both rotation and translation were achieved through body motion, compared to when the translation input was provided by a joystick.

6.2 Semi-Mobile Neuroimaging in Immersive VR

The size of physical space needed for designing a naturalistic experience is dependent on the specific task. For example, in an EEG study by Gramann et al. (2021), the behavior of interest was full-body rotation, which did not require translation of the body through the environment. Here, the activity in the RSC differed substantially for physical rotations (full body rotation) as compared to visual flow rotations (provided via a display), demonstrating the importance of vestibular information in updating of orientation.

However, when the research question relates to exploration of large-scale environments, physical restrictions imposed by the hardware system and the lab space becomes more relevant. When participants need to move their body within a limited range due to wires attached to the measurement devices or the lack of navigable space, one can opt to employ a treadmill or similar devices that allow participants to move and use the accompanying proprioceptive feedback. These interfaces constrain the location of the body to the limited area on the treadmill but allow for walking motion through a large-scale environment.

In their behavioral studies, Bellmund et al. (2020) and Hejtmanek et al. (2020) used omni-directional treadmills with immersive VR to synchronize the locomotion of the virtual camera with the lower limb motion of the participant in a large-scale virtual environment. A similar setup was used in an EEG study by Liang et al. (2018), where they observed the increase of frontal-midline theta oscillations during physical motion (see Fig. 2, panels A and B).

One caveat of this approach is that the idiothetic cues may be contradictory, especially when it comes to the translation component of natural locomotion: the limb motion on a treadmill triggers proprioceptive cues signaling translation, but the vestibular system signals that the location of the participant is fixed. This prevents the hardware combination from addressing the interaction dynamics between the idiothetic cues from different sources.

6.3 Fully Mobile Neuroimaging in VR

The combination of MoBI and immersive VR is a promising approach that has been proven to be powerful in investigating brain dynamics in actively behaving participants (Jungnickel and Gramann 2016). A portable or even wireless neuroimaging system is a prerequisite for implementation of experimental paradigms in fully mobile setups, as shown in Fig. 1, panel C. At the moment, only EEG and functional near-infrared spectroscopy (fNIRS) meet the criteria for fully mobile brain imaging, thanks to the relatively light weight of the amplifiers and sensors.

EEG comes with limited spatial but high temporal resolution of the recorded neural dynamics. Thus, EEG is an excellent method to track the neural dynamics associated with fast-paced cognitive processes. In their EEG study, Delaux et al.

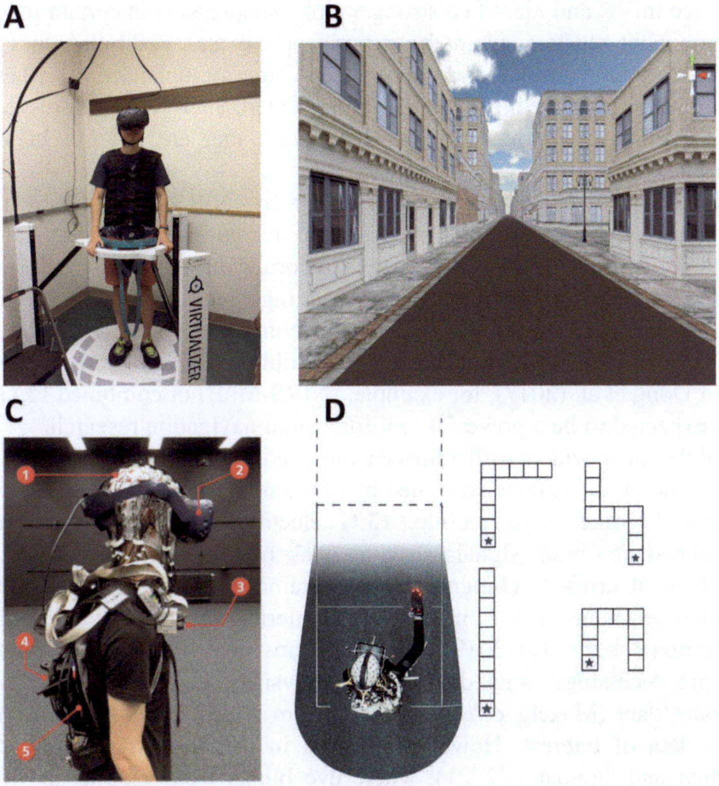

Fig. 2 Examples of locomotion interfaces used with mobile EEG. (**a**) Liang et al. (2018) used an omnidirectional treadmill (Cyberith GmbH, Herzogenburg, Austria) to allow participants to move their lower limbs to control their movement through the environment in immersive VR. Participants were equipped with a wireless EEG recording system. (**b**) A snapshot of the virtual environment used in Liang et al. (2018). (**c**) In their navigation study, Delaux et al. (2021) used a fully wireless MoBI setup with immersive VR, with the following hardware components: (1) EEG cap (128 channels, Brain Products, Gilching, Germany); (2) VIVE Pro VR HMD (HTC Corporation, Taoyuan, Taiwan); (3) Wifi transmitter for EEG data (Move system, Brain Products); (4) Additional motion capture tracker (VIVE tracker); and (5) Backpack computer running the virtual environment (Zotac PC, ZOTAC Technology Limited, Fo Tan, Hong Kong). (**d**) A birds-eye view of a participant in Gehrke et al. (2018) with a fully wireless MoBI setup. A visually sparse environment was presented in immersive VR, as shown in the schematic illustration of the invisible mazes used in the experiment

(2021) used MoBI in immersive VR to reveal contribution of sensorimotor areas during landmark-based navigation (see Fig. 2, panel C). The combination of MoBI and immersive VR facilitates a flexible manipulation of different sensory channels, as seen in Miyakoshi et al. (2021), Gehrke et al. (2018), and Gehrke and Gramann (2021), where auditory stimuli and sparse visual stimuli were presented as spatial cues while participants equipped with EEG and HMD were navigating using full-body motion (Fig. 2, panel D). Ehinger et al. (2014) tested path integration performance in VR and identified stronger alpha suppression in certain independent EEG component clusters when the proprioceptive and vestibular cues were in conflict. Additionally, they found an increase in anterior alpha activity when vestibular but no proprioceptive information was presented. Such works address important aspects of sensory experience during navigation that are left out in stationary experiments.

The condition of portability is also fulfilled by fNIRS systems which provide limited temporal resolution but good spatial resolution for superficial cortical sources. Takeuchi et al. (2016), for instance, recorded the activity from the prefrontal cortex in a multitasking situation during walking. See Lancia et al. (2017) for a review of potential application of fNIRS for spatial navigation research, focused on the mobility aspect of it. Considering the compatibility of fNIRS with immersive VR as seen in Dong et al. (2017), for example, fNIRS-MoBI or combined EEG-fNIRS-MoBI is expected to be a powerful tool for spatial navigation research.

One of the most evident difficulties encountered by MoBI in immersive VR is the increased amount of artifacts generated by physical motion as well as the HMD. As participants become more mobile, EEG electrodes inevitably pick up more motion-related non-brain signals such as neck muscle activity, eye movements, and mechanical artifacts (Jungnickel and Gramann 2016). Using HMD VR in OP-MEG experiments creates magnetic field interference, causing a drop in signal-to-noise ratio (Roberts et al. 2019). These problems are usually addressed at the stage of data preprocessing, using data-driven analysis techniques suitable for multi-dimensional data (Makeig et al. 2009) that aim at eliminating non-brain signals from the data of interest. However, as seen in a real-life navigation study by Wunderlich and Gramann (2021), where eye blinks were used as the feature of interest, these non-brain signals may not necessarily be irrelevant to the process of interest and instead be additional sources of information.

7 Conclusion

Spatial navigation at its essence is deeply intertwined with the physicality of the surrounding space. It is an embodied process that does not only involve various cognitive processes but also motor actions that we recruit in order to better perceive and act in the environment. By using immersive VR, we can get closer to the phenomenon as it is experienced in real life, while maintaining the degree of experimental control provided by non-immersive VR.

Wearable HMDs used in VR allow a wide range of locomotion interfaces to be used in the simulated space. More naturalistic locomotion with motion tracking enables participants to interact with the space using more sensory channels. Naturalistic modes of locomotion and high sensory immersion lead to the inclusion of body-based cues, embodied affordances, and a reduction in potential conflicts between multiple reference frames. These factors serve to go beyond the abstract experiences presented by non-immersive VR, to better approximate navigation behavior in real life.

Ranging from rather stationary applications to MoBI setups that allow for full body motion, there is a variety of locomotion interfaces that can be used together with immersive VR. Concurrent neuroimaging in fully mobile setups is not as easy to employ as stationary neuroimaging due to the need for mobile, lightweight hardware and artifacts associated with active movement. Depending on the research question, an immersive VR setup with only the head or limb motion within a restricted area can be an excellent alternative to a fully wireless MoBI setup. However, full MoBI combined with immersive VR provides a plethora of unique advantages for spatial navigation research by engaging the whole body of the participant in spatial navigation as a tangible interaction between them and the world.

In the end, if we were to study the navigation behavior of Frodo and the fellowship during their journey across Middle Earth, it would be a great injustice to do so in a manner that restricted the role of their large hobbit feet, short dwarven legs, or pointy elf ears, or to alienate them from the lucid experience of being in the environments they traversed, from the dense Fangorn Forest to the open wastes of Mordor. The same is true when studying human spatial cognition, with the methodological and technological advances that we covered in this chapter providing great opportunities for the future of spatial navigation and neuroscience research.

References

Bálint R (1909) Psychic paralysis of gaze, optic ataxia, and spatial disorder of attention. Eur Neurol 25:67–81

Barry DN, Tierney TM, Holmes N, Boto E, Roberts G, Leggett J, Bowtell R, Brookes MJ, Barnes GR, Maguire EA (2019) Imaging the human hippocampus with optically-pumped magnetoencephalography. NeuroImage 203:116192

Baumann O, Mattingley JB (2021) Extrahippocampal contributions to spatial navigation in humans: a review of the neuroimaging evidence. Hippocampus, 1–18. https://doi.org/10.1002/hipo.23313

Becker S, Burgess N (2000) Modelling spatial recall, mental imagery and neglect. Adv Neural Inf Proces Syst 13

Bécu M, Sheynikhovich D, Tatur G, Agathos CP, Bologna LL, Sahel JA, Arleo A (2020) Age-related preference for geometric spatial cues during real-world navigation. Nat Hum Behav 4:88–99

Bellmund JL, De Cothi W, Ruiter TA, Nau M, Barry C, Doeller CF (2020) Deforming the metric of cognitive maps distorts memory. Nat Hum Behav 4:177–188

Blades M et al (1992) Developmental differences in the ability to give route directions from a map. J Environ Psychol 12:175–185

Bolding K, Rudy JW (2006) Place learning in the Morris water task: making the memory stick. Learn Mem 13:278–286

Boto E, Holmes N, Leggett J, Roberts G, Shah V, Meyer SS et al (2018) Moving magnetoencephalography towards real-world applications with a wearable system. Nature 555(7698):657–661

Brang D, Taich ZJ, Hillyard SA, Grabowecky M, Ramachandran VS (2013) Parietal connectivity mediates multisensory facilitation. NeuroImage 78:396–401

Brooks J, Nagels S, Lopes P (2020) Trigeminal-based temperature illusions. In: Proceedings of the 2020 CHI conference on human factors in computing systems, pp 1–12

Buzsáki G (2005) Theta rhythm of navigation: link between path integration and landmark navigation, episodic and semantic memory. Hippocampus 15:827–840

Byrne P, Becker S, Burgess N (2007) Remembering the past and imagining the future: a neural model of spatial memory and imagery. Psychol Rev 114:340

Campos JL, Butler JS, Bülthoff HH (2012) Multisensory integration in the estimation of walked distances. Exp Brain Res 218:551–565

Chan E, Baumann O, Bellgrove MA, Mattingley JB (2012) From objects to landmarks: the function of visual location information in spatial navigation. Front Psychol 3:304

Chance SS, Gaunet F, Beall AC, Loomis JM (1998) Locomotion mode affects the updating of objects encountered during travel: the contribution of vestibular and proprioceptive inputs to path integration. Presence 7:168–178

Chen X, McNamara TP, Kelly JW, Wolbers T (2017) Cue combination in human spatial navigation. Cogn Psychol 95:105–144

Chrastil ER (2013) Neural evidence supports a novel framework for spatial navigation. Psychon Bull Rev 20:208–227. https://doi.org/10.3758/s13423-012-0351-6

Chrastil ER, Nicora GL, Huang A (2019) Vision and proprioception make equal contributions to path integration in a novel homing task. Cognition 192:103998

Clark A (2013) Whatever next? Predictive brains, situated agents, and the future of cognitive science. Behav Brain Sci 36:181–204. https://doi.org/10.1017/S0140525X12000477

Clark A (2015) Surfing uncertainty: prediction, action, and the embodied mind. Oxford University Press

Clark BJ, Simmons CM, Berkowitz LE, Wilber AA (2018) The retrosplenial-parietal network and reference frame coordination for spatial navigation. Behav Neurosci 132:416–429

Coutrot A, Manley E, Yesiltepe D, Dalton RC, Wiener JM, Holscher C, Hornberger M, Spiers HJ (2020) Cities have a negative impact on navigation ability: evidence from 38 countries. Biorxiv

Critchley M (1953) The parietal lobes. Hafner, New York

Cruz-Neira C, Sandin DJ, DeFanti TA, Kenyon RV, Hart JC (1992) The cave: audio visual experience automatic virtual environment. Commun ACM 35:64–73

Delaux A, Aubert JBS, Ramanöel S, Bécu M, Gehrke L, Klug M, Chavarriaga R, Sahel JA, Gramann K, Arleo A (2021) Mobile brain/body imaging of landmark-based navigation with high-density EEG. Eur J Neurosci. https://doi.org/10.1111/ejn.15190

Djebbara Z, Fich LB, Petrini L, Gramann K (2019) Sensorimotor brain dynamics reflect architectural affordances. Proc Natl Acad Sci 116:14769–14778

Do TTN, Lin CT, Gramann K (2021) Human brain dynamics in active spatial navigation. Sci Rep 11:1–12

Dodsworth C, Norman L, Thaler L (2020) Navigation and perception of spatial layout in virtual echo-acoustic space. Cognition 197:104185

Doeller CF, Barry C, Burgess N (2010) Evidence for grid cells in a human memory network. Nature 463:657–661

Dong D, Wong LK, Luo Z et al (2017) Assessment of prospective memory using fnirs in immersive virtual reality environment. J Behav Brain Sci 7:247

Ehinger BV, Fischer P, Gert AL, Kaufhold L, Weber F, Pipa G, König P (2014) Kinesthetic and vestibular information modulate alpha activity during spatial navigation: a mobile EEG study. Front Hum Neurosci 8:71

Ekstrom AD (2015) Why vision is important to how we navigate. Hippocampus 25:731–735

Epstein RA, Patai EZ, Julian JB, Spiers HJ (2017) The cognitive map in humans: spatial navigation and beyond. Nat Neurosci 20:1504

Etienne AS, Jeffery KJ (2004) Path integration in mammals. Hippocampus 14:180–192

Farrell MJ, Robertson IH (1998) Mental rotation and automatic updating of body centered spatial relationships. J Exp Psychol Learn Mem Cogn 24:227

Fetsch CR, Turner AH, DeAngelis GC, Angelaki DE (2009) Dynamic reweighting of visual and vestibular cues during self-motion perception. J Neurosci 29:15601–15612

Friston K (2010) The free-energy principle: a unified brain theory? Nat Rev Neurosci 11:127–138

Friston K (2012) Embodied inference and spatial cognition. Cogn Process 13

Gehrke L, Gramann K (2021) Single-trial regression of spatial exploration behavior indicates posterior EEG alpha modulation to reflect egocentric coding. Eur J Neurosci 54(12):8318–8335

Gehrke L, Iversen JR, Makeig S, Gramann K (2018) The invisible maze task (IMT): interactive exploration of sparse virtual environments to investigate ActionDriven formation of spatial representations. In: Creem-Regehr S, Schöning J, Klippel A (eds) Spatial cognition XI. Springer, Cham, pp 293–310

Gehrke L, Akman S, Lopes P, Chen A, Singh AK, Chen HT, Lin CT, Gramann K (2019) Detecting visuo-haptic mismatches in virtual reality using the prediction error negativity of event-related brain potentials. In: Proceedings of the 2019 CHI conference on human factors in computing systems – CHI'19. ACM Press, New York, 427:1–427:11

Gehrke L, Lopes P, Klug M, Akman K, Gramann K (2022) Neural sources of prediction errors detect unrealistic VR interactions. J Neural Eng 19(3):036002. https://doi.org/10.1088/1741-2552/ac69bc

Geronazzo M, Avanzini F, Fontana F (2016) Auditory navigation with a tubular acoustic model for interactive distance cues and personalized head-related transfer functions. J Multimodal User Interf 10:273–284

Gibson JJ (1979) The ecological approach to visual perception. Mifflin and Company, Houghton

Goeke C, Kornpetpanee S, Köster M, Fernández-Revelles AB, Gramann K, König P (2015) Cultural background shapes spatial reference frame proclivity. Sci Rep 5:1–13

Goeke CM, König P, Gramann K (2013) Different strategies for spatial updating in yaw and pitch path integration. Front Behav Neurosci 7:5

Goldberg JM, Wilson VJ, Angelaki DE, Cullen KE, Broussard DM, Fukushima K, Buttner-Ennever J, Minor LB (2012) The vestibular system: a sixth sense. Oxford University Press

Golledge RG (1995) Path selection and route preference in human navigation: a progress report. In: International conference on spatial information theory. Springer, pp 207–222

Goodrich-Hunsaker NJ, Livingstone SA, Skelton RW, Hopkins RO (2010) Spatial deficits in a virtual water maze in amnesic participants with hippocampal damage. Hippocampus 20:481–491

Gramann K (2013) Embodiment of spatial reference frames and individual differences in reference frame proclivity. Spatial Cogn Comput 13:1–25

Gramann K, Müller HJ, Eick EM, Schönebeck B (2005) Evidence of separable spatial representations in a virtual navigation task. J Exp Psychol Hum Percept Perform 31:1199

Gramann K, Müller H, Schönebeck B, Debus G (2006) The neural basis of ego-and allocentric reference frames in spatial navigation: evidence from spatiotemporal coupled current density reconstruction. Brain Res 1118:116–129

Gramann K, Onton J, Riccobon D, Mueller HJ, Bardins S, Makeig S (2010) Human brain dynamics accompanying use of egocentric and allocentric reference frames during navigation. J Cogn Neurosci 22:2836–2849

Gramann K, Gwin JT, Ferris DP, Oie K, Jung TP, Lin CT, Liao LD, Makeig S (2011) Cognition in action: imaging brain/body dynamics in mobile humans. Rev Neurosci 22:593–608

Gramann K, Ferris DP, Gwin J, Makeig S (2014) Imaging natural cognition in action. Int J Psychophysiol 91:22–29

Gramann K, Hohlefeld FU, Gehrke L, Klug M (2021) Human cortical dynamics during full-body heading changes. Sci Rep 11:1–12

Grzeschik R, Hilton C, Dalton RC, Konovalova I, Cotterill E, Innes A, Wiener JM (2020) From repeating routes to planning novel routes: the impact of landmarks and ageing on route integration and cognitive mapping. Psychol Res 85(6):1–13

Hafting T, Fyhn M, Molden S, Moser MB, Moser EI (2005) Microstructure of a spatial map in the entorhinal cortex. Nature 436:801–806

Hamburger K, Knauff M (2019) Odors can serve as landmarks in human wayfinding. Cogn Sci 43: e12798

Head H, Holmes G (1911) Sensory disturbances from cerebral lesions. Brain 34:102–254

Hejtmanek L, Starrett M, Ferrer E, Ekstrom AD (2020) How much of what we learn in virtual reality transfers to real-world navigation? Multisens Res 33:479–503

Hodges H (1996) Maze procedures: the radial-arm and water maze compared. Cogn Brain Res 3: 167–181

Huffman DJ, Ekstrom AD (2019) A modality-independent network underlies the retrieval of large-scale spatial environments in the human brain. Neuron 104:611–622.e7

Ijaz K, Ahmadpour N, Naismith SL, Calvo RA (2019) An immersive virtual reality platform for assessing spatial navigation memory in predementia screening: feasibility and usability study. JMIR Mental Health 6:e13887

Jacob PY, Poucet B, Liberge M, Save E, Sargolini F (2014) Vestibular control of entorhinal cortex activity in spatial navigation. Front Integr Neurosci 8:38

Jacobs J (2014) Hippocampal theta oscillations are slower in humans than in rodents: implications for models of spatial navigation and memory. Philos Transact R Soc B Biol Sci 369:20130304

Janzen G (2006) Memory for object location and route direction in virtual large-scale space. Q J Exp Psychol 59:493–508

Janzen G, Van Turennout M (2004) Selective neural representation of objects relevant for navigation. Nat Neurosci 7:673–677

Jeneson A, Mauldin KN, Hopkins RO, Squire LR (2011) The role of the hippocampus in retaining relational information across short delays: the importance of memory load. Learn Mem 18:301–305

Jungnickel E, Gramann K (2016) Mobile brain/body imaging (mobi) of physical interaction with dynamically moving objects. Front Hum Neurosci 10:306

Kessels RP, de Haan EH, Kappelle LJ, Postma A (2001) Varieties of human spatial memory: a meta-analysis on the effects of hippocampal lesions. Brain Res Rev 35:295–303

Klippel A (2003, September) Wayfinding choremes. In: International conference on spatial information theory. Springer, Berlin, pp 301–315

Kluss T, Marsh WE, Zetzsche C, Schill K (2015) Representation of impossible worlds in the cognitive map. Cogn Process 16(1):271–276

Kobayashi Y, Amaral DG (2003) Macaque monkey retrosplenial cortex: II. Cortical afferents. J Comp Neurol 466:48–79

Kolarik AJ, Moore BCJ, Zahorik P, Cirstea S, Pardhan S (2016) Auditory distance perception in humans: a review of cues, development, neuronal bases, and effects of sensory loss. Atten Percept Psychophys 78:373–395

Krohn S, Tromp J, Quinque EM, Belger J, Klotzsche F, Rekers S, Chojecki P, De Mooij J, Akbal M, McCall C, Villringer A, Gaebler M, Finke C, ThöneOtto A (2020) Multidimensional evaluation of virtual reality paradigms in clinical neuropsychology: application of the VR-check framework. J Med Internet Res 22

Kunz L, Schröder TN, Lee H, Montag C, Lachmann B, Sariyska R, Reuter M, Stirnberg R, Stöcker T, Messing-Floeter PC et al (2015) Reduced grid-cell–like representations in adults at genetic risk for alzheimer's disease. Science 350:430–433

Laczó J, Andel R, Vyhnalek M, Vlcek K, Magerova H, Varjassyova A, Tolar M, Hort J (2010) Human analogue of the Morris water maze for testing subjects at risk of alzheimer's disease. Neurodegener Dis 7:148–152

Lancia S, Mammarella S, Bianco D, Quaresima V (2017) Is wireless functional nearinfrared spectroscopy (fnirs) 3d neuroimaging feasible to map human navigation in the real-world? In: International conference on spatial information theory. Springer, pp 73–78

Lappe M, Jenkin M, Harris LR (2007) Travel distance estimation from visual motion by leaky path integration. Exp Brain Res 180:35–48

Liang M, Starrett MJ, Ekstrom AD (2018) Dissociation of frontal-midline delta, theta and posterior alpha oscillations: A mobile EEG study. Psychophysiology 55:e13090

Limanowski J, Kirilina E, Blankenburg F (2017) Neuronal correlates of continuous manual tracking under varying visual movement feedback in a virtual reality environment. NeuroImage 146:81–89

Lopes P, Ion A, Baudisch P (2015) Impacto: simulating physical impact by combining tactile stimulation with electrical muscle stimulation. In: Proceedings of the 28th annual ACM symposium on user interface software & technology, pp 11–19

Lövdén M, Schellenbach M, Grossman-Hutter B, Krüger A, Lindenberger U (2005) Environmental topography and postural control demands shape aging-associated decrements in spatial navigation performance. Psychol Aging 20:683

Lövdén M, Schaefer S, Noack H, Bodammer NC, Kühn S, Heinze HJ, Düzel E, Bäckman L, Lindenberger U (2012) Spatial navigation training protects the hippocampus against age-related changes during early and late adulthood. Neurobiol Aging 33:620–6e9

Maguire EA, Gadian DG, Johnsrude IS, Good CD, Ashburner J, Frackowiak RS, Frith CD (2000) Navigation-related structural change in the hippocampi of taxi drivers. Proc Natl Acad Sci 97: 4398–4403

Makeig S, Gramann K, Jung TP, Sejnowski TJ, Poizner H (2009) Linking brain, mind and behavior. Int J Psychophysiol 73:95–100

Mallory CS, Hardcastle K, Campbell MG, Attinger A, Low IIC, Raymond JL, Giocomo LM (2021) Mouse entorhinal cortex encodes a diverse repertoire of self-motion signals. Nat Commun 12: 671

Meilinger T, Franz G, Bülthoff HH (2012) From Isovists via mental representations to behaviour: first steps toward closing the causal chain. Environ Plan B Plan Design 39:48–62

Meilinger T, Riecke BE, Bülthoff HH (2014) Local and global reference frames for environmental spaces. Q J Exp Psychol 67:542–569

Millar S (1994) Understanding and representing space theory and evidence from studies with blind and sighted children. Oxford University Press

Mittelstaedt H, Mittelstaedt ML (1982) Homing by path integration. In: Avian navigation. Springer, pp 290–297

Miyakoshi M, Gehrke L, Gramann K, Makeig S, Iversen J (2021) The AudioMaze: an EEG and motion capture study of human spatial navigation in sparse augmented reality. Eur J Neurosci 54(12):8283–8307

Montello DR (2005) Navigation. Cambridge University Press

Morris RG (1981) Spatial localization does not require the presence of local cues. Learn Motiv 12: 239–260

Muller RU, Ranck JB Jr, Taube JS (1996) Head direction cells: properties and functional significance. Curr Opin Neurobiol 6:196–206

Muryy A, Glennerster A (2021) Route selection in non-euclidean virtual environments. PLoS One 16:e0247818

Nardini M, Jones P, Bedford R, Braddick O (2008) Development of cue integration in human navigation. Curr Biol 18:689–693

O'Keefe J, Burgess N, Donnett JG, Jeffery KJ, Maguire EA (1998) Place cells, navigational accuracy, and the human hippocampus. Philos Transact R Soc London Ser B Biol Sci 353: 1333–1340

Parong J, Pollard KA, Files BT, Oiknine AH, Sinatra AM, Moss JD, Passaro A, Khooshabeh P (2020) The mediating role of presence differs across types of spatial learning in immersive technologies. Comput Hum Behav 107:106290

Parslow DM, Morris RG, Fleminger S, Rahman Q, Abrahams S, Recce M (2005) Allocentric spatial memory in humans with hippocampal lesions. Acta Psychol 118:123–147

Rao RP, Ballard DH (1999) Predictive coding in the visual cortex: A functional interpretation of some extra-classical receptive-field effects. Nat Neurosci 2:79–87

Richardson AE, Montello DR, Hegarty M (1999) Spatial knowledge acquisition from maps and from navigation in real and virtual environments. Mem Cogn 27:741–750
Riva G, Wiederhold BK, Mantovani F (2019) Neuroscience of virtual reality: from virtual exposure to embodied medicine. Cyberpsychol Behav Soc Netw 22:82–96
Roberts G, Holmes N, Alexander N, Boto E, Leggett J, Hill RM, Shah V, Rea M, Vaughan R, Maguire EA et al (2019) Towards OPM-MEG in a virtual reality environment. NeuroImage 199:408–417
Rossier J, Haeberli C, Schenk F (2000) Auditory cues support place navigation in rats when associated with a visual cue. Behav Brain Res 117:209–214
Ruddle RA, Volkova E, Bülthoff HH (2013) Learning to walk in virtual reality. ACM Transact Appl Percept (TAP) 10:1–17
Russell NA, Horii A, Smith PF, Darlington CL, Bilkey DK (2006) Lesions of the vestibular system disrupt hippocampal theta rhythm in the rat. J Neurophysiol 96:4–14
Save E, Poucet B (2009) Role of the parietal cortex in long-term representation of spatial information in the rat. Neurobiol Learn Mem 91:172–178
Schellenbach M, Lövdén M, Verrel J, Krüger A, Lindenberger U (2010) Sensorimotor-cognitive couplings in the context of assistive spatial navigation for older adults. GeroPsych 23(2):69
Schinazi VR, Thrash T, Chebat D (2016) Spatial navigation by congenitally blind individuals. WIREs Cogn Sci 7:37–58
Schoenfeld R, Schiffelholz T, Beyer C, Leplow B, Foreman N (2017) Variants of the Morris water maze task to comparatively assess human and rodent place navigation. Neurobiol Learn Mem 139:117–127
Seubert J, Humphreys GW, Müller HJ, Gramann K (2008) Straight after the turn: the role of the parietal lobes in egocentric space processing. Neurocase 14:204–219
Sheeran WM, Ahmed OJ (2020) The neural circuitry supporting successful spatial navigation despite variable movement speeds. Neurosci Biobehav Rev 108:821–833
Simons DJ, Wang RF (1998) Perceiving real-world viewpoint changes. Psychol Sci 9:315–320
Slater M (2009) Place illusion and plausibility can lead to realistic behaviour in immersive virtual environments. Philos Transact R Soc B Biol Sci 364:3549–3557
Spiers HJ, Maguire EA (2006) Thoughts, behaviour, and brain dynamics during navigation in the real world. NeuroImage 31:1826–1840
Spiers HJ, Maguire EA (2007) A navigational guidance system in the human brain. Hippocampus 17:618–626
Starrett MJ, McAvan AS, Huffman DJ, Stokes JD, Kyle CT, Smuda DN et al (2021) Landmarks: a solution for spatial navigation and memory experiments in virtual reality. Behav Res Methods 53(3):1046–1059
Staudigl T, Leszczynski M, Jacobs J, Sheth SA, Schroeder CE, Jensen O, Doeller CF (2018) Hexadirectional modulation of high-frequency electrophysiological activity in the human anterior medial temporal lobe maps visual space. Curr Biol 28:3325–3329
Takeuchi N, Mori T, Suzukamo Y, Tanaka N, Izumi SI (2016) Parallel processing of cognitive and physical demands in left and right prefrontal cortices during smartphone use while walking. BMC Neurosci 17:1–11
Taube JS, Muller RU, Ranck JB (1990) Head-direction cells recorded from the postsubiculum in freely moving rats. I. Description and quantitative analysis. J Neurosci 10(2):420–435
Taube JS, Valerio S, Yoder RM (2013) Is navigation in virtual reality with fMRI really navigation? J Cogn Neurosci 25:1008–1019
Thornberry C, Cimadevilla JM, Commins S (2021) Virtual Morris water maze: opportunities and challenges. Rev Neurosci 32(8):887–903
Tolman EC (1948) Cognitive maps in rats and men. Psychol Rev 55:189
Tucker LB, Velosky AG, McCabe JT (2018) Applications of the Morris water maze in translational traumatic brain injury research. Neurosci Biobehav Rev 88:187–200
Tversky B (1992) Distortions in cognitive maps. Geoforum 23:131–138

Van Veen HA, Distler HK, Braun SJ, Bülthoff HH (1998) Navigating through a virtual city: using virtual reality technology to study human action and perception. Futur Gener Comput Syst 14: 231–242

Vann SD, Aggleton JP, Maguire EA (2009) What does the retrosplenial cortex do? Nat Rev Neurosci 10:792–802

Vorhees CV, Williams MT (2006) Morris water maze: procedures for assessing spatial and related forms of learning and memory. Nat Protoc 1:848–858

Waller D, Lippa Y (2007) Landmarks as beacons and associative cues: their role in route learning. Mem Cogn 35:910–924

Waller D, Hunt E, Knapp D (1998) The transfer of spatial knowledge in virtual environment training. Presence Teleop Virt 7:129–143

Wang RF, Simons DJ (1999) Active and passive scene recognition across views. Cognition 70(2): 191–210

Warren WH, Rothman DB, Schnapp BH, Ericson JD (2017) Wormholes in virtual space: from cognitive maps to cognitive graphs. Cognition 166:152–163

Weisberg SM, Schinazi VR, Newcombe NS, Shipley TF, Epstein RA (2014) Variations in cognitive maps: understanding individual differences in navigation. J Exp Psychol Learn Mem Cogn 40:669

Werkhoven P, van Erp JB, Philippi TG (2014) Navigating virtual mazes: the benefits of audiovisual landmarks. Displays 35:110–117

Whitlock JR (2017) Posterior parietal cortex. Curr Biol 27:R691–R695

Whitlock JR, Sutherland RJ, Witter MP, Moser MB, Moser EI (2008) Navigating from hippocampus to parietal cortex. Proc Natl Acad Sci 105:14755–14762

Wiederhold BK, Wiederhold MD (2008) Virtual reality with fMRI: a breakthrough cognitive treatment tool. Virtual Reality 12:259–267

Wiener JM, Hölscher C, Büchner S, Konieczny L (2012) Gaze behaviour during space perception and spatial decision making. Psychol Res 76:713–729

Wiener JM, de Condappa O, Harris MA, Wolbers T (2013) Maladaptive bias for extrahippocampal navigation strategies in aging humans. J Neurosci 33:6012–6017

Wiener JM, Carroll D, Moeller S, Bibi I, Ivanova D, Allen P, Wolbers T (2020) A novel virtual-reality-based route-learning test suite: assessing the effects of cognitive aging on navigation. Behav Res Methods 52:630–640

Wolbers T, Wiener JM (2014) Challenges for identifying the neural mechanisms that support spatial navigation: the impact of spatial scale. Front Hum Neurosci 8:571

Wolbers T, Wiener JM, Mallot HA, Büchel C (2007) Differential recruitment of the hippocampus, medial prefrontal cortex, and the human motion complex during path integration in humans. J Neurosci 27:9408–9416

Wu Y, Chen K, Ye Y, Zhang T, Zhou W (2020) Humans navigate with stereo olfaction. Proc Natl Acad Sci 117:16065–16071

Wunderlich A, Gramann K (2021) Eye movement-related brain potentials during assisted navigation in real-world environments. Eur J Neurosci 54(12):8336–8354

Young CK, Ruan M, McNaughton N (2021) Speed modulation of hippocampal theta frequency and amplitude predicts water maze learning. Hippocampus 31:201–212

Zhong JY, Magnusson KR, Swarts ME, Clendinen CA, Reynolds NC, Moffat SD (2017) The application of a rodent-based Morris water maze (mwm) protocol to an investigation of age-related differences in human spatial learning. Behav Neurosci 131:470

Virtual Reality for Vision Science

Paul B. Hibbard

Contents

Abstract Virtual reality (VR) allows us to create visual stimuli that are both immersive and reactive. VR provides many new opportunities in vision science. In particular, it allows us to present wide field-of-view, immersive visual stimuli; for observers to actively explore the environments that we create; and for us to

P. B. Hibbard (✉)
Department of Psychology, University of Essex, Colchester, UK
e-mail: phibbard@essex.ac.uk

Curr Topics Behav Neurosci (2023) 65: 131–160
https://doi.org/10.1007/7854_2023_416
Published Online: 2 February 2023

understand how visual information is used in the control of behaviour. In contrast with traditional psychophysical experiments, VR provides much greater flexibility in creating environments and tasks that are more closely aligned with our everyday experience. These benefits of VR are of particular value in developing our theories of the behavioural goals of the visual system and explaining how visual information is processed to achieve these goals. The use of VR in vision science presents a number of technical challenges, relating to how the available software and hardware limit our ability to accurately specify the visual information that defines our virtual environments and the interpretation of data gathered in experiments with a freely moving observer in a responsive environment.

Keywords Augmented reality · Head mounted displays · Immersion · Presence · Virtual reality · Vision science · 3D Displays

1 Introduction

This chapter discusses the significant opportunities for advancing our understanding of vision made possible by virtual reality (VR). It presents the benefits and drawbacks of using VR to study visual perception, focussing not only on the current state of the art, but also on the inherent differences between VR and other display technologies and methodological approaches. Since VR can mean many different things in different contexts, I begin with a working definition of VR for the purposes of this discussion. Lanier (2017) provides many thought-provoking definitions, some of which are especially relevant for our purposes. In particular, his fourth definition provides the most useful framing of VR for exploring its role as a tool for vision science:

> The substitution of the interface between a person and the physical environment with an interface to a simulated environment.
> (Lanier 2017, page 47)

In practice, I define the visual component of VR as a display that is both *immersive* and *responsive*. The typical example hardware is a head-mounted display, although many of the same principles and practical considerations apply to other implementations, such as a CAVE, in which the user is surrounded by large display screens at a distance of several metres (Cruz-Neira et al. 1992). A head-mounted display provides a binocular, wide field-of-view screen which displays the virtual environment, while excluding any visual input from the real physical environment. This is combined with 6°-of-freedom motion tracking of the headset, so that its position and orientation in 3D space are known and can be used to update the images presented to the observer in response to their movement (Fig. 1). In addition, tracking of the observer's hand positions or more extensive position tracking of their body is required to produce an environment and display that update in response

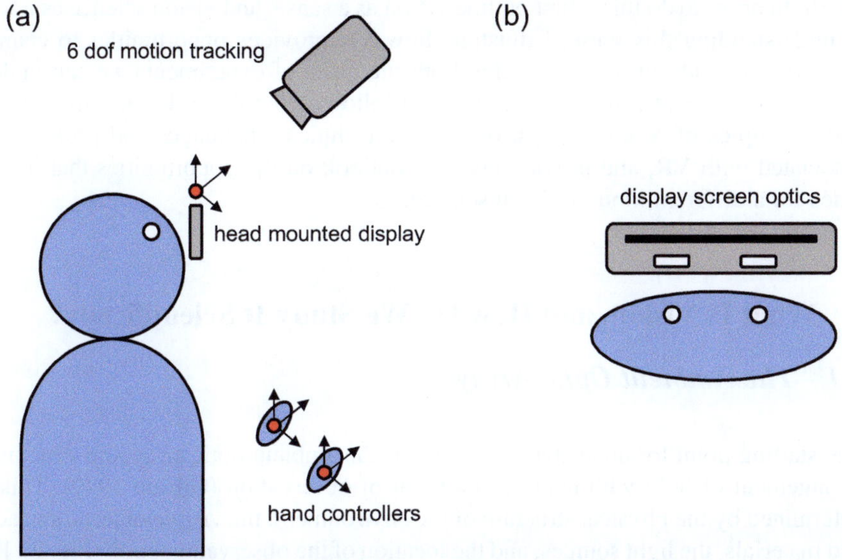

Fig. 1 The essential components of VR for visual display. (**a**) The observer wears a head-mounted display and typically interacts with the environment via two hand-held controllers. The motion of the headset, and the hand controllers, is tracked to provide 6°-of-freedom estimates of their location and orientation in 3D space. (**b**) Within the head-mounted display, the display screen is placed close to the observer's eyes. This creates a wide, binocular field of view, and occludes the view of the outside world. The screen optics ensure that the image is formed at a comfortable distance

to the user's actions. Together, these properties can be used to create an environment that is immersive (providing the optic array of the virtual environment, while excluding the optic array of the physical environment) and responsive (changing in near-to-real time in response to the user's actions).

Head-mounted VR systems achieve this immersive experience by combining a number of key components. First, the virtual environment must be created. This may be built in a games engine or other software platform and defines the components of the environment and their behaviour. These include, for example, the objects and other characters in the environment, their behaviour in response to simulated physics, and possible interactions with the user. It also includes components such as lighting and atmospheric effects that determine the appearance of the environment for the user. As the user is a part of the environment, tracking of their own motion is also required so that their view of the environment can be updated, and to allow interactions such as picking up and moving objects. The observer's view of the environment is then rendered and presented on binocular display screens with a wide field of view. As these screens are placed very close to the user's eyes, the headset also includes optics that allow the displayed images to be clearly and comfortably perceived. This creates an effective image plane at a fixed distance of a metre or so from the observer.

With this definition in mind, we can explore the novel opportunities provided by VR for the scientific study of our sense of vision (Scarfe and Glennerster 2015,

2019). In order to do this, I first outline vision as a sense, and vision science as a way of understanding this sense. I illustrate how VR provides opportunities to change how we can study vision, expanding both the types of experiments we can undertake, and the theoretical understanding that follows from these. I will illustrate this with examples of recent studies, outline the technical challenges and possibilities associated with VR, and provide my own outlook on the opportunities that it provides for a different kind of vision science.

2 What Is Vision, and How Do We Study It Scientifically?

2.1 The Ambient Optic Array

The starting point for understanding vision is the ambient optic array, the structured arrangement of light with respect to a point of observation (Gibson 1979). This is determined by the physical structure of the environment: the visible objects, surfaces and materials, the light sources, and the location of the observation point (Fig. 2). For each direction from the observation point, the spectral intensity of light depends on

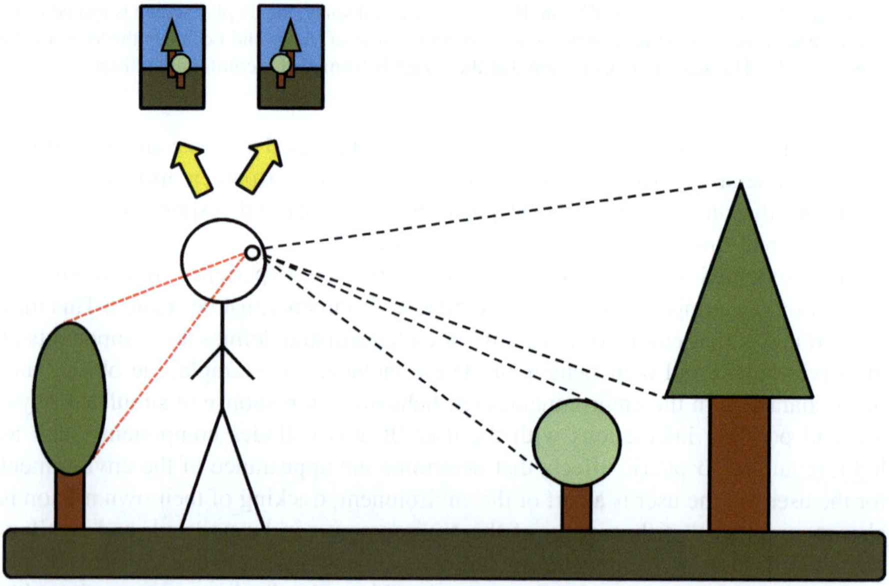

Fig. 2 The ambient optic array and the retinal images. The ambient optic array is defined at each point in space. In this figure the optic array is shown from the observer's point of view. This includes visual information arriving at this point from all locations. However, the viewer can only see those components lying within their field of view (such as the dashed black lines) but not points lying outside this (such as the dotted red lines). The projection of the optic array through the observer's eyes creates the two retinal images

the illumination and reflectance properties of the physical structures along that line of sight. As the observer moves, the ambient optic array is sampled from different locations. This definition of the starting point of vision does not depend on any properties of the observer, but is a definition of the information available from the world in the form of the structured array of light. In particular, the optics of the eye or camera and the imaging surface, which together may be involved in creating a focussed image from the optic array, are not part of the definition (Rogers 2021). As most people have two eyes, we sample the ambient optic array from two points simultaneously. As we move around the environment, the locations from which it is sampled change.

2.2 The Retinal Image

Each of our eyes contains a lens and a cornea, which together focus the light to create an image from the optic array on the retinal surface. This means that, at each moment in time, the two retinal images provide us with a partial sample of the ambient optic array from two locations. The information in each image can be defined as a function of visual direction, rather than the location on the retina to which that direction projects. The visual direction to each visible point is defined as the orientation of the line connecting that point to the centre of the eye. Since there is a one-to-one mapping between visual direction and retinal location, each point on the retina can be associated with a single visual direction (Koenderink 1984; Rogers 2021).

In creating a visual display for VR, our goal is to position a display screen so as to create a desired pair of retinal images, and thus a pair of samples of the ambient optic array that correspond not to the physical surroundings of the user, but to the virtual environment that we wish them to experience. VR thus provides us with the ability to create the experience of whatever virtual visual environment we want. From the perspective of vision science, it provides a means of experimentally controlling the ambient optic array of the observer. This control of the entirety of the visual input, in a way that mimics the ambient optic array in the physical world, and its ability to change in response to the movement and actions of the observer differ fundamentally from the more restricted opportunities available on more traditional display screens with a fixed location relative to the observer.

2.3 The Visual System

The human visual system consists of the eyes and a large number of cortical and subcortical areas in the brain. This system includes the photoreceptors in the retina which perform the task of transduction, creating a neural response to the incoming stimulus; the optic nerve, transmitting this signal to the brain; and the areas of the brain involved in visual processing. It is the task of vision science to understand the role of the visual system in making use of the information provided by the optic array.

2.4 *What Is Vision Science?*

This is the scientific study of visual perception. It assesses the processes by which we use visual information, how vision influences and affects our behaviour and experience, and the means by which this is achieved. While this is a broad undertaking, requiring multiple, complementary approaches, in its simplest terms it provides three types of understanding (Koenderink 2019). The first is the physiology of the way that the visual system responds to light. The second is how this information is used to control our behaviour. The third is the phenomenological experience of seeing.

2.4.1 Responses of the Visual System to the Ambient Optic Array

This endeavour covers topics such as the spectral response properties of retinal photoreceptors (Schnapf et al. 1988); the receptive fields of neurons in the lateral geniculate nucleus (Cleland et al. 1971) and primary visual cortex (Hubel and Wiesel 1962); the retinotopic maps formed in the classical visual areas (Zeki 1978); nonlinear aspects of visual responses (Wilson 1980; Heeger 1992); the response properties of higher-level cortical areas, such as the fusiform face area (Kanwisher and Yovel 2006); and the organisation of visual areas into visual processing streams (Livingstone and Hubel 1987; Milner and Goodale 2006). It also covers our more abstract conceptualisations such as filters and channels (Braddick et al. 1978; Graham 1989; Zhaoping 2014) and the relationship between the responses of populations of neurons and our perceptual judgements (Parker and Newsome 1998).

2.4.2 The Use of Visual Information to Control Behaviour

Vision provides us with information that allows us to make better decisions and to execute actions with skill and dexterity. It allows us to recognise objects and people; to understand the emotional state of others, to recognise them as individuals, and to interact with them socially; to move around our environment safely, and to pick up and manipulate objects. To understand the role of vision therefore is to provide an account not only of how information is detected, encoded, and transmitted, but also of how it is employed in all these ways and more. These tasks and behaviours themselves need to be part of our theories.

2.4.3 The Phenomenological Experience of Seeing

The most obvious fact of vision, to a sighted observer, is not the physiological responses of their visual system, nor the importance of these responses in enacting behaviour, but the immediate conscious experience of seeing that places them in the here and now. As with understanding the use of vision in the control of behaviour, virtual reality provides the opportunity for the controlled, measurable assessment of visual experience as it relates to our everyday behaviour. The potential for a traditional psychophysical experiment to provide generalisable insights on this topic is limited. The full reality of our perception, as individuals inhabiting and immersed in a three-dimensional world cannot be extrapolated from experiments in which immobilised observers make forced-choice responses about the nature of abstract patterns presented on a display screen.

3 The Role of Display Technology in Shaping Vision Science

As our available display technologies have developed, the kinds of empirical understanding that are possible from laboratory-based experiments have grown. In understanding the opportunities afforded by virtual reality, it is instructive to consider it alongside other technologies. It is also important to remember that VR does not make these other technologies obsolete; rather, it adds to the kinds of experiments that it is possible to undertake.

Prior to the ubiquitous adoption of computer-generated, screen-presented stimuli in vision research, experiments typically used optical apparatus to create and present stimuli (Koenderink 1999). These would include light sources, prisms, mirrors, beam-splitters, filters, and such like. These would be arranged with great precision using optical benches and rails, to present the stimulus to an observer whose head position was held firmly in place using a bite-bar. Oscilloscopes, and then graphics cards and computer monitors, offered increased flexibility, complexity, and ease with which images could be created and displayed. This means that, if an image can be specified as an array of numbers representing the intensity in the red, green, and blue channels, it can be displayed, subject to the resolution and dynamic range limits of the screen (Pelli 1997). This has been especially facilitated by the development of software platforms that take care of the interactions with the computer hardware (Brainard 1997; Peirce 2007), allowing the experimenter to concentrate on the specification of the stimuli. These software and hardware technologies have greatly facilitated the creation and display of stimuli. These stimuli can be ordered in terms of the degree to which they take account of the observer's point of view.

3.1 Simple Visual Patterns

Some visual stimuli are defined with no reference to a real or imagined object beyond the retinal image. They exist purely as visual patterns on the screen, and examples include random dot patterns (Julesz 1971; Braddick 1974), gratings (Braddick 1981), plaids (Adelson and Movshon 1982), and Gabor patches (Daugman 1985). They are not intended to be interpreted as projections of physical objects, but are designed to test the way in which visual information is encoded and used for perceptual judgements. They are used to address questions such as our acuity to spatial alignment or orientation (Westheimer 1972; Heeley et al. 1997), whether shape and depth can be perceived purely on the basis of binocular disparities (Julesz 1971), or how neural responses are interpreted to determine the perceived direction of motion of an image (Adelson and Movhson 1982; Newsome et al. 1989).

3.2 Pictures of Physical Objects

The same display devices can also present stimuli that are images of real or fictional objects, scenes, or people. This can be achieved either through the use of photographs or videos captured using a camera or through the rendering of stimuli intended to be images of objects and scenes with specified physical properties. Creating such stimuli requires an understanding of, and ability to implement, the properties of projective geometry, lighting, and image formation.

If we assume the location of the optical point of observation within the scene, then it is a simple process to determine the visual direction from that point to any 3D location in space. An example here could be a stereoscopically presented stimulus in which the left and right images are intended as the views of an object with a particular 3D location, shape, and size. Each point on the object is projected onto the appropriate location in the left and right images, according to its defined location relative to the optical centres of the left and right eye (Johnston 1991; Johnston et al. 1993).

In creating these images, we need to specify the location of the observer, and for them then to be presented so that the retinal image faithfully reflects the experience of viewing the scene from that position. This requires accurate positioning of the observer's eyes and spatial calibration of the display screen. If this is not done, then we create a situation in which the observer is looking at a picture in which their point of observation and that used in creating the stimuli do not necessarily agree. It has been observed that much of vision science, through using stimuli presented on 2D display screens, is a science of how we see pictures, and not of how we see real objects in the three-dimensional world (Wade 2013).

3.3 *The Ambient Optic Array Sampled from a Specific Viewpoint*

It is however possible to position the observer and the images so that they accurately recreate the retinal images that would be experienced from a particular 3D scene viewed from the observer's location. This can be achieved with great precision using a bite-bar to keep the observer's head fixed and a sighting device to ensure that this fixed location is in the correct position relative to the screen. Once this position is known, a mapping between visual directions from the optical centres and pixel locations on the screen can be determined, for example by positioning a physical grid in front of the displays. This mapping can then be used to accurately transform the images and position them on the display screens (Backus et al. 1999). If photographs are used as stimuli, it is also necessary to calibrate the camera images (Hibbard 2008). If all the appropriate steps are followed, then it is possible to create stimuli that form known retinal images for the observer, replicating the images that would be formed in an equivalent three-dimensional scene. This approach is used, for example, in studies of how binocular information is used in the perception of 3D structure (Backus et al. 1999; Hillis et al. 2004; Watt et al. 2005).

3.4 *The Ambient Optic Array of a Freely Moving Observer, Interacting with a 3-Dimensional World*

The most direct way in which this can be achieved, of course, is through conducting experiments directly in the real world using three-dimensional scenes and objects as stimuli (Gibson 1950). This approach is very uncommon, save for some notable and valuable exceptions. Studies of inattention blindness have, for example, used real, staged environments, in which actors may be swapped during interactions with participants, demonstrating a surprising lack of awareness of this change (Simons and Levin 1998). Studies of the accuracy of distance perception have also been performed extensively outside of the laboratory (Plumert et al. 2005). Research in the natural environment using mobile eye tracking also provides an understanding of how observers sample the optic array when performing everyday tasks (Foulsham et al. 2011). Such studies are valuable since they provide evidence of how we see, and act, in the natural environment, rather than in response to pictures on a display screen (Koenderink 1999; Wade 2013).

Experiments in the natural environment present a variety of practical challenges, however. The first is that they are very labour intensive. The situation must be set up in the same way for each participant. This requires the actors, the props, and the interactions between these and the participant to be arranged and controlled as similarly as is possible on every repetition. Even if this can be achieved, there will always be a limit to the amount of experimental control that is possible. However carefully the scene is set, there will always be some differences which cannot be controlled. It can therefore be very difficult to replicate these studies with any precision. Another problem is that the data needed to fully describe the trial – the shape, size, and location of all objects in the scene, and the movement of the participant and other people throughout the trial will typically not be recorded in fine detail, even if video and audio recordings of the scene are made.

In some experimental designs, it is possible to use quite complex natural visual stimuli, while varying only a single parameter of interest. This provides some of the advantages of research in natural environments, although without the ability to study natural behaviour. For example, in studies of the perception of distance it is possible to vary distance specified by convergence while keeping all other visual information unaltered (Tresilian et al. 1999; Mon-Williams et al. 2000). This has the advantage of providing precise experimental control while studying perception in a full-cue, natural environment rather than using typically sparse laboratory stimuli – thus at a natural 'operating point' (Koenderink 1998). It is limited, however, in that while the visual stimuli may contain all the complexity of a typical scene, the tasks used in experiments are highly constrained.

3.5 The Ambient Optic Array of a Freely Moving Observer, Interacting with a 3-Dimensional World

Many of the advantages of performing experiments in the natural environment can be achieved in VR, whilst maintaining greater control over the stimuli presented and the data collected. VR thus provides an ideal environment in which to undertake experiments involving a freely moving observer interacting with a 3-dimensional world (Scarfe and Glennerster 2015). In comparison with complex real-world stimuli, the major advantages of VR are control, automation, and data capture. The exact same environment can be presented on each trial, specified by the experimenter down to the precise location of every point on every object, the nature of the lighting, and the behaviour of all the components. Once created, no further input from the experimenter is required. In addition, a detailed description of the scene, and the user's movements, can be recorded. This recording can specify the virtual 3D scene, the projected 2D images, and the relationship between the two. The latter, for example, means that we have access to the ground truth associated with each pixel in the display, so that properties such as the 3D location to which that point corresponds and the object to which it belongs can all be recorded (Goutcher et al. 2021).

Some of the challenges of conducting experiments in the real world remain when they are performed in virtual reality, however. For example, because the stimulus depends on the user's motion within – and interaction with – the environment, each trial will be different for every observer, and different on every repetition, even when the specified environment is identical.

4 New Tasks for New Stimuli

Experiments in vision science are defined not just by the kinds of stimuli that are presented, but also by the tasks set for the observer, and the behavioural data that are collected. Perhaps the simplest example is the forced-choice discrimination task, in which the observer is presented with two stimuli and asked to judge whether they are the same or different. Brindley (1960) defined this as a 'type A' psychophysical task (Brindley 1960). Type A tasks have the strong theoretical advantage of a clear linking hypothesis which allows us to make inferences about the physiological responses to the stimuli presented (Brindley 1960; Morgan et al. 2013). If an observer can discriminate reliably below two stimuli in a type A task, then there must be some difference in the way that the brain responds to them. Other psychophysical tasks ('type B' tasks) require the observer to report on the appearance of the stimulus (for example, its orientation or direction of motion) and thus depend on the observer's judgement criteria as well as the neural encoding of the stimulus.

For type B tasks, there is often an agreed 'correct' answer. When using an abstract visual pattern, this will be a property of the image itself, for example the orientation of the grating, or the direction of motion of the random dot pattern. In other cases, whether or not a response is correct depends on both the observer's judgement and the experimenter's definition of the distal stimulus of which the proximal retinal images are projections. For example, the experimenter determines that a particular size and direction of binocular disparity is the projection onto the retinas of a particular depth separation (Johnston 1991), that an image texture results from a particular slant and tilt of a planar surface (Hillis et al. 2004), or the 'veridical' 3D shape depicted in a single 2D image (Pizlo 2010). In all these cases, the 'experimenter's share', the beholder's share (Koenderink et al. 2001), and the actual stimulus presented are all essential components determining whether or not a response is correct, or the way in which it is wrong.

Beyond forced-choice psychophysical experiments, researchers may also seek to understand how visual information is used in the control of behaviour, such as maintaining a desired direction of heading (Warren and Hannon 1988; Rushton et al. 1998), or reaching to pick up an object (Servos et al. 1992; Watt and Bradshaw 2000; Bradshaw et al. 2004; Melmoth and Grant 2006). In these cases, the experimental data are the measures of the observer's behaviour itself, and the interpretation of the data can depend on a number of theoretical assumptions. For example, in studies of reaching and grasping, the maximum speed of motion and the maximum size of the grip aperture are taken as measures of the apparent distance and size of the object to be picked up, respectively (Servos et al. 1992).

When interpreting such natural behavioural responses, as opposed to forced-choice categorisations, it is important to appreciate that they are not direct measures of 'pure' perception, but are also determined by the requirements of the action to be completed. For example, peak speed of motion and peak grip aperture will incorporate margins of error, dependent on the participant's confidence in their perceptual estimates, so as not to bump into or knock over the target (Keefe et al. 2019). There is thus rarely a simple one-to-one mapping that can be established between estimated perceptual parameters and behavioural measures. This greatly complicates the interpretation of data in comparison with forced-choice tasks, as for example when comparing the degree to which visual illusions affect perception and action, since there is no single measure that can be meaningfully compared across different tasks (Franz et al. 2000).

Virtual reality provides great opportunities in expanding the type of behavioural data that can be used in experiments, beyond simple forced-choice tasks, while maintaining a high level of experimental control. The benefit of less constrained responses is the increased range of theoretical questions that can be posed, but this also presents much more challenging problems in the interpretation of data.

5 A Vision Science of Natural Environments and Natural Tasks

VR allows us to immerse a participant in the ambient optic array associated with a virtual environment; for this environment to be altered in response to the participant's behaviour; for complex naturalistic behaviour to be undertaken in this environment; and for full details of the structure of the environment, and the participant's actions, to be recorded. Together, these properties of VR expand the possibilities for experiments which address different areas of vision science, and different types of theoretical questions about vision, than traditional screen-based, forced-choice experiments.

Traditional psychophysical techniques tend to be of most relevance to understanding how information is encoded, rather than how this is used in everyday vision. Typical models of visual encoding will include channels composed of perceptual filters or templates, non-linearities, sources of noise, and a final decision-making stage (Dosher and Lu 1999). Because these models are generally used to account for performance in forced-choice categorisation tasks, they do not address how the information delivered by these channels might be used in the control of action, or in more complex cognitive tasks. Whether or not this presents a significant limitation depends on the nature of the visual processing under investigation.

Several rather different characterisations of the purpose of vision have been presented. One class, exemplified by the concept of 'inverse optics', sees the task of visual processing as reconstructing a representation of the three-dimensional environment on the basis of the information available from the 2D retinal images (Adelson and Pentland 1996; Pizlo 2001; Mamassian et al. 2002). In such models, the goal is the creation of a general-purpose visual representation, which is then accessed for all subsequent tasks (Marr 1982). This might, for example, specify the distance to the seen point in each visual direction (Descartes 1637/1965; Sedgwick 2021), this distance plus the surface orientation at each point (Marr and Nishihara

1978), or some other representation that specifies the three-dimensional structure of the environment, and our location within it.

Under this characterisation, it would be possible to account for how this process of inverse optics is achieved, without needing to concern ourselves with the nature of the tasks for which it is subsequently used. Conversely, if the encoding of visual information is not separable from the ways in which it is used, then it would be necessary to use complex, naturalistic tasks in order to understand perception.

Another important consideration is the way in which visual perception depends upon and makes assumptions about the structure of the environment. Gibson's theory of ecological optics outlines a number of arguments for why typical everyday environments, and typical everyday activities within these environments, should be used in vision research.

The first consideration is that our environment is structured in ways that shape the nature of the visual information that is available and in ways that determine how this is then used in perception. It is expected that the encoding of visual information will be tuned to these regularities, so as to be optimised for the typical natural environment (Olshausen and Field 1996; Simoncelli and Olshausen 2001). The distribution and material properties of objects and surfaces in our environment determine the statistical and structural properties of images. These include the luminance (Field and Brady 1997; Balboa et al. 2001; Rogers 2021), binocular disparity (Hibbard 2007, 2008; Sprague et al. 2015), colour (Chiao et al. 2000), and motion (van Hateren 1993; Dong and Atick 1995a, b) characteristics of typical natural scenes. The use of virtual reality to create naturalistic scenes as stimuli in psychophysical studies allows us to replicate these characteristics more fully than can be achieved using typical display screens with a smaller field of view in which stimuli are not updated to reflect the observer's movement.

One example is the way that different sources of information are used in the perception of distance and depth. Optimal cue combination models specify that cues should be combined through a weighted-averaging process, with the weights determined by the relative reliabilities of different cues (Landy et al. 1995). Laboratory experiments with carefully specified stimuli have been successful in testing the predictions of these models (Ernst and Banks 2002; Hillis et al. 2004; Watt et al. 2005; Keefe et al. 2011). However, such models and experiments by themselves do not provide any information about the relative importance of cues in the natural environment. To do this requires some understanding of the reliabilities of these cues in typical everyday scenes (Nagata 1991; Cutting and Vishton 1995; Hibbard 2021), and to test the predictions in this context requires us to perform experiments using such scenes. VR is valuable in this context in allowing us to perform psychophysical tests of the accuracy of 3D perception in typical everyday scenes, while manipulating the information that is available to observers (Hornsey et al. 2020; Hornsey and Hibbard 2021; Hartle and Wilcox 2021). Hornsey and Hibbard (2021) used this approach to quantify the contributions of various depth cues to the accuracy of distance and size judgements. They showed that binocular and pictorial cues both contribute to improved precision in size judgements, as predicted by theoretical models, and that binocular cues are important not only in near space, but also beyond distances of 10 m.

Gibson (1950) proposed that the structure of the environment shapes the way that we make use of information in perception. He proposed a 'grounded' theory of perception, reflecting the fact that our natural environment is composed of surfaces rather than unconnected points. These surfaces create structure in the visual input and may also determine the goals of perception; for example, we may judge the distance of objects relative to their background context, rather than their distance from the observer through empty space (Glennerster and McKee 1999; Glennerster et al. 2002; Petrov and Glennerster 2004; He and Ooi 2000; Sedgwick 2021). For these influences of surfaces, or other features of the environment on our perception to be captured, it is important that they are included in the stimuli used in experimental studies. The full extent of the structure of the natural environment can readily be incorporated when conducting experiments outdoors (Gibson 1950) or in VR, but not using traditional screen displays.

Gibson also emphasised the importance of thinking of perception as a process of active exploration, rather than passive observation. That is, we create the visual input that we experience by the way that we move in our environment and sample the ambient optic array. Again, this process of active exploration can only be incorporated into empirical studies by conducting them in the physical environment or in VR. Active exploration of the world creates optic flow (Lee 1980; Rogers 2021). As we move from one point to another, we experience different samples of the ambient optic array. The structure of surfaces in the environment means that these samples are related in predictable ways, creating higher-level patterns of motion which provide reliable information about the structure of the environment. Not only does this redundancy allow us to encode information efficiently (Olshausen and Field 1996; Simoncelli and Olshausen 2001), but it can also inform us directly about the structure of the environment, and our actions, without this needing to be mediated by simpler representations. Examples of this type of structure that have been proposed include gradients of motion and the orientation of surfaces; dynamic occlusion and the presence and depth order of surfaces; and the relationship between the focus of expansion of motion and the observer's direction of heading (Rogers 2021).

The way in which we make use of visual information in everyday tasks also needs to be considered to properly understand how visual information is processed. Gibson proposed that we directly access the information that supports our actions in the world (Gibson 1979; Sedgwick 2021). Other formulations of the way in which visual information is used by observers include van Uexküell's sensorimotor loop (Koenderink 2019), O'Regan and Noe's sensorimotor account of vision (O'Regan and Noe 2001), and the interface theory of perception (Hoffman et al. 2015), in which the goal of perception is not to create veridical representations, but to guide adaptive behaviours. Empirical testing of these theories therefore depends on the appropriate use of typical everyday tasks, an endeavour for which VR is ideally suited (Tarr and Warren 2002; Bhargava et al. 2020). This has been applied, for example, in understanding important behaviours such as reaching and grasping

(Hibbard and Bradshaw 2003; Klinghammer et al. 2016; Kopiske et al. 2019) and navigation (Tarr and Warren 2002; Muryy and Glennerster 2021).

VR has also been used to empirically evaluate the nature of our perception of space, notably addressing the question of whether we maintain a stable 3D representation of the visual world while we navigate within it. Glennerster and colleagues have, for example, shown that we can fail to notice large changes in the scale of our environment (Glennerster et al. 2006); that judgements of distances between multiple pairs of points are not necessarily mutually consistent within this environment (Svarverud et al. 2012); that our ability to point to previously seen targets is not consistent with a single, stable 3D representation (Vuong et al. 2019; Scarfe and Glennerster 2021); and that we successfully use navigation strategies that are inconsistent with a Euclidean representation of space (Muryy and Glennerster 2021). This programme of research has shown how it is possible to make use of VR to develop and test theoretical models of the essential nature of visual perception (Glennerster 2016).

Virtual reality also has potential applications in areas of visual cognition beyond perception of the geometric structure of the environment. A critical example of this is the visual information used for social interactions. We are readily able to infer information about others' emotions from their facial expressions and movement. Most research on this topic makes use of stimuli that are artificial, in that they are static, grey-scale, monochrome photographs of actors portraying emotions, which are typically manipulated so as to control for psychophysically important properties such as luminance and contrast, rather than preserving systematic variations of these properties (Gray et al. 2013; Menzel et al. 2018; Webb et al. 2020, 2022). These highly impoverished stimuli are then used in simple forced-choice tasks, such as the detection of the presence of stimuli, or their categorisation as prespecified expressions. While this level of control may be necessary to isolate the relevant properties of the stimuli under investigation, it is important that the methodology is expanded to assess the relevance of such research in understanding how we use visual information in everyday social situations.

This can firstly be achieved by using more natural stimuli, particularly by including dynamic information (Jack and Schyns 2015). Another important extension is to use environments in which virtual actors react to our own behaviours (Geraets et al. 2021) and to use appropriate measures of real-world social understanding. These characteristics – more natural stimuli, reactivity, and the study of real-world behaviours – are all key features of virtual reality, emphasising the importance of the technology in this area of vision science. An important focus for research in this field is to ensure that social agents in VR are believable and accurately convey information about emotions and other social cues.

The focus of the current chapter is on the use of VR in vision research. It should be noted however that a powerful characteristic of VR is that it is a multisensory technology, allowing us to include spatial audio and haptic cues. The provision of multiple cues, and in particular the congruence between cues in a multisensory environment, is an important factor in creating an immersive experience (Servotte et al. 2020). Since VR allows for the independent manipulation of individual cues, it is ideally suited for exploring the role of multisensory information in the perception of our 3D environment.

6 Hardware and Software Characteristics of Virtual Reality

Lanier's fifth definition of VR gives a useful description of what we are trying to achieve, technically, in creating a virtual reality visual display:

> A mirror image of a person's sensory and motor organs, or if you like, an inversion of a person.
>
> (Lanier 2017, page 47)

As the user moves around and interacts with the environment, the goal is to create the ambient optic array that would be experienced in the simulated environment. This reframes the inverse problem not in terms of how the user recreates the 3D structure of the environment, but of how we can use VR to understand their experience and sensory-motor loops. As outlined in the introduction, this is achieved via hardware and software requirements that include (1) creation of the virtual environment, (2) tracking the motion of the user, (3) rendering the images of the environment to be presented, (4) displaying them on the two screens, and ensuring that the displayed images are clearly and comfortably perceived through (5) the use of appropriate headset optics and (6) correct positioning of the display screens and optics relative to the observer. Each of these steps presents technical challenges. While these have all been addressed in current consumer reality to the extent that it now provides a comfortable, immersive, and convincing *feeling* of realism for most users, it is important in vision science to understand how this has been achieved and that differences remain between VR and the real world. In the following section, I highlight issues at each step in this pipeline that are relevant for vision scientists.

6.1 The Environment

Technological advances mean that it is possible to move beyond VR environments composed of simple geometric forms to create complex, naturalistic scenes. This can include, for example, 3D scans of natural objects (Goutcher et al. 2021), landscapes (Liang et al. 2014), and human body movement and facial expressions captured from actors (Nonis et al. 2019). This creates a great deal of scope when designing virtual environments for experiments in vision science, requiring decisions about the level of complexity that is most appropriate for a given research context. It has been argued that, when creating visual displays in general, we should avoid the temptation to increase complexity (Smallman and John 2005). However, in the case of vision science, our goals include understanding the way in which particular aspects of our visual environment influence perception. The potential to create highly realistic, immersive environments opens opportunities for understanding the important characteristics of the ambient optic array, in the same way that digital photographs have done for static and dynamic images (Simoncelli and Olshausen 2001). Decisions about the complexity and realism of the environment are the same in both cases. While simplified artificial stimuli run the risk of excluding important information, the complexity of naturalistic stimuli may make experimental designs intractable (Rust and Movshon 2005).

As environments become more seemingly realistic, there is also the danger that any artefactual differences between the real and virtual environments may go unnoticed (Koenderink 1999). In vision science, there is a risk that our findings may reflect these artefacts. In the case of photographs, for example, the framing and composition of the image may create global statistics that are not representative of mundane, everyday images, while the rectangular pixel lattice will influence local statistical properties (van Hateren and van der Schaaf 1998; Hertzmann 2022). In the case of VR, where we determine the shape, material properties, position, movement, and behaviour of all objects, we increase the scope that any of these properties may diverge from what is characteristic of typical natural scenes, whilst still appearing highly realistic to observers. For example, the rendering of scenes in VR requires models of complex natural phenomena including the movement of, and collisions between, objects and materials, which must be simulated to approximate the real world.

6.2 *Motion Tracking and Latency*

Virtual reality requires us to track the user's position so that the display can be updated in as close to real time as possible. Any latencies in this process will have negative effects on the control of action and the experience of agency, ownership, presence, and immersion (Waltemate et al. 2016), while also contributing to feelings of sickness. While latency has improved in recent VR systems, with a 90 Hz update rate being typical, latencies as low as 7 ms are detectable by observers (Scarfe and Glennerster 2019). As with all aspects of the virtual environment, the deliberate manipulation of these parameters, for example to include high latencies to evaluate its effect on multisensory processing and perceptual experience (van Dam and Stephens 2018), or incorrect positioning of the observer to understand the role of eye-height in the perception of distance (Leyrer et al. 2011; Kim and Interrante 2017), demonstrates how VR provides flexible control of the environment that can be harnessed in experiments.

In addition to ensuring the accurate presentation of the environment, tracking of head and eye movements provides rich behavioural data that allows us to understand how people navigate and interact with virtual environments (Alcañiz et al. 2022; Gulhan et al. 2022).

6.3 Rendering the Images

Early VR used wireframes or simply shaded objects, so that it was obvious to the user that the environment was artificial. Just as our ability to create more complex 3D environments has improved, so has our ability to render these to create images that have a highly realistic appearance (Zibrek et al. 2019). As this level of realism improves, it becomes increasingly difficult to tell natural from artificial environments. While this is of course highly beneficial for applications of VR, when used in vision science there is a risk that our results are influenced by artefacts which may not be noticeable to the participants or experimenters. These will occur in particular due to the technical challenge of accurately rendering lighting (Koenderink 1999).

Some of the complexities here come from the facts that each point on a surface is lit not just by a light source from a single direction, and that once light has hit a matte surface, it will be reflected in all directions, not just towards the observer. This means that light will bounce between surfaces, and objects will cast shadows on one another as they occlude portions of the ambient light array. All of this can be simulated with ray-tracing (Todd et al. 2015; Todd 2020), however this is very computationally intensive and therefore slow to render. Currently, even with high-powered cluster computing, a single frame can take minutes to create rather than the milliseconds available to create stimuli for real-time display. For vision science, it is therefore important either to use physically accurate lighting models or to understand the details and consequences of departures from accuracy for a given experiment design.

6.4 The Display Screen

For each eye, the image is displayed on a screen with a particular spatial resolution, field of view, and dynamic range. The best resolution of the human eye, in the fovea, is around 1 arc min, meaning that we can discriminate between one and two points if they are presented farther apart than this. This means that we can see pixilation if pixels are larger than this separation. Currently pixels in VR headsets are about 3 arc min in size. Screen resolution thus still limits our ability to present very fine-detailed information and to specify the precise location of features through anti-aliasing. This has implications for studying aspects of visual acuity in VR and can create artefactual differences in the statistical properties of scenes in comparison with the natural environment.

Human vision operates over a high dynamic range, sensitive to illumination from as low as 10^{-6} cd m^{-2} to as high as 10^{8} cd m^{-2} (Banterle et al. 2017). This is much greater than the range that is achievable with high dynamic range displays, thus limiting the range of illumination over which vision science can operate, for VR and other displays alike. This limited dynamic range is also a consideration in determining the sense of realism in VR (Vangorp et al. 2014).

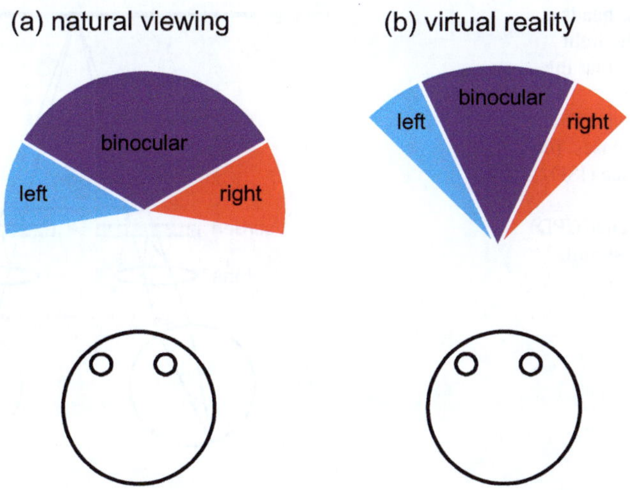

Fig. 3 (**a**) The human binocular field extends to 200°, with 120° of binocular overlap. (**b**) Both the overall field of view and the binocular overlap are smaller than this in VR

The human binocular field of view is approximately 200° horizontally by 100° vertically, with a binocular overlap of 120° (Spector 1990). In current VR, the horizontal and vertical extents of the field of view are around 90° and 36°, respectively, with a 50° horizontal binocular overlap. Thus, while VR provides a very wide field of view in comparison with a traditional display screen, it is still limited in the extent to which it can be used to study peripheral vision (Strasburger et al. 2011). The reduced binocular field of view also means that the contributions of binocular vision, for example to the perception of distance and the observer's movement, may be underestimated when studied using VR (Fig. 3).

6.5 Headset Optics

The display screen in an HMD is positioned just a few millimetres from the viewer. In order to accommodate a screen at this distance, powerful lenses are placed in the headset, creating an image plane that is at some distance from the user (Fig. 4). This image plane appears at a constant, set distance in VR. This distance is determined by the lenses, regardless of the distance to the object in the scene, as specified by binocular convergence and other cues. This creates cue-conflict and difficulties for the coupling of accommodation and convergence responses, which are known to be a source of discomfort in VR and other 3D displays (Shibata et al. 2011). This could theoretically be addressed by the use of multifocal displays, which use multiple overlapping screens at different distances. This technique allows the effective image distance of each point to be varied independently, by distributing it across these multiple screens. This in turn allows the accommodative load to vary with object

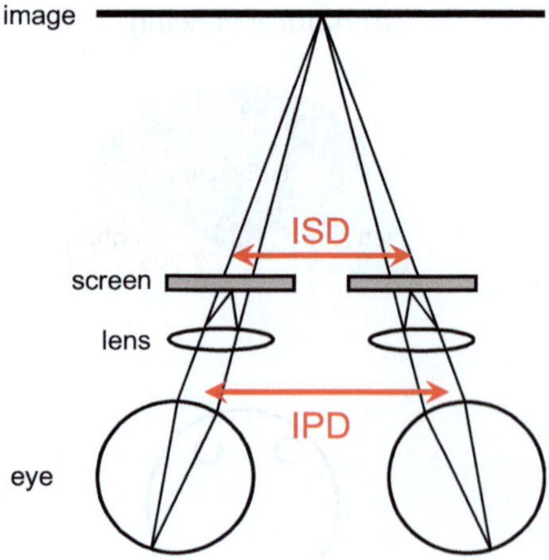

Fig. 4 Within the headset, the optics focus the light from the screen so that the image appears at a comfortable distance. The separation of the lenses, the inter-screen distance (ISD), and the effective interpupillary distance (IPD) used in rendering stimuli should all ideally be matched to the observer's IPD

distance, providing a more natural relationship between convergence and accommodation (Zhong et al. 2021). Multifocal displays are not however currently in use in HMDs.

In natural viewing, objects that are not at the accommodation distance will be blurred. In VR, this optical blurring will not occur, since in this case the amount of blur depends on the distance to the image plane, rather than to the object in the rendered scene. This can be compensated for by including an appropriate blur into the displayed image (Held et al. 2012). Again, by allowing control over the image distance, multifocal displays would remove these incorrect focus cues, introducing natural, optically created defocus to the retinal image.

6.6 *Positioning of the Display Screens and Lenses Relative to the Observer*

When we render the stimuli, we need to specify the optical centres of the two eyes in the environment. Motion tracking allows us to measure the position of the headset. To then determine the positions of the optical centres relative to the headset requires us to specify the interpupillary distance (IPD) – the distance between the user's two eyes (Fig. 4). If there is a mismatch between the observer's IPD and that assumed in rendering and presenting the stimuli, then the binocular disparities in the images will not accurately reflect those that would be experienced when viewing the intended scene (Scarfe and Glennerster 2019). This is an important consideration because IPD varies considerably between individuals. For adults, the overall mean IPD is 63.4 m, with a standard deviation of 3.8 mm (Dodgson 2004). This represents a mean of 63.4 mm for women and 64.7 mm for men. It also increases with age from a mean value of 50 mm at the age of 5, reaching the final adult value at an age of 19 (MacLachlan and Howland 2002). There are a number of important ways in

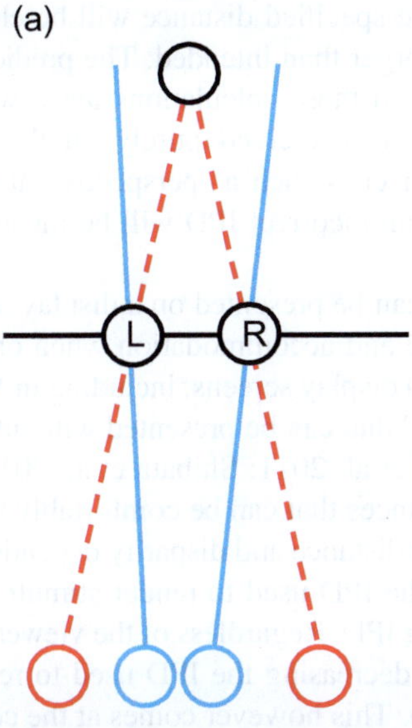

Fig. 5 (**a**) A bird's eye view showing the positioning of a point in the left and right eyes' images, in order to create the appearance of an object (the unfilled circle) at a distance beyond the image plane. The assumed locations of the observer's eyes used to position the points are shown by the red circles. If the observer's eyes (depicted by the blue circles) are closer together than assumed, this can create divergent viewing. (**b**) An incorrectly assumed IPD will also lead to an incorrect distance specified by binocular cues. In this case, the target will appear closer than intended

which a mismatch between the effective IPD in the headset and the user's IPD can negatively affect the viewing experience (Hibbard et al. 2020).

The first problem is the potential for binocular divergence (Fig. 5a). In natural viewing, when we fixate an object in the distance, the directions of the gaze of the two eyes are parallel; as the object moves closer, the eyes will converge, with the angle of convergence increasing with decreasing distance. Divergence of the eyes is therefore never required in order to fixate a target in natural viewing, although in practice this can occur due to imperfect fixation (Darko-Takyi et al. 2016). If the observer's IPD is smaller than that used in creating and rendering the stimuli, divergent viewing will be required to fixate distant stimuli. Our ability to make such divergent eye-movements is limited, leading to a loss of binocular fusion and viewing discomfort (IJsselsteijn et al. 2000; Hoffman et al. 2008; Lambooij et al. 2011; Shibata et al. 2011).

The second problem is that when the assumed and actual IPD are not matched, the perceived distance, shape, and size of objects specified by binocular cues will be incorrect (Fig. 5b). The binocular disparity of a point is determined by the fixation distance, the position of the point relative to fixation, and the viewer's IPD. This means that two viewers with different IPDs will experience different binocular disparities when viewing the same scene. The rendering and display of stimuli with an incorrect IPD will thus present incorrect binocular information to the viewer.

Specifically, if the IPD used is too large, the specified distance will be closer than intended, while if it is too small, it will be larger than intended. The predictions for the misperception of distance and depth from these calculations are a worst-case scenario, however, assuming that distance is perceived purely on the basis of binocular cues. When the influence of other cues such as perspective and motion parallax is taken into account, the effect of an incorrect IPD will be much reduced (Hibbard et al. 2020; Hibbard 2021).

The maximum binocular disparities that can be presented on a display screen are limited by the conflict between convergence and accommodation when objects are presented away from the image plane. On 3D display screens, including in VR, there is a maximum binocular disparity of up to 2° that can be presented without creating discomfort (Hoffman et al. 2008; Lambooij et al. 2011; Shibata et al. 2011). From this limit, we can calculate the range of distances that can be comfortably presented to the viewer. Since the relationship between distance and disparity depends on IPD, this comfortable range will also vary with the IPD used to render stimuli, with the comfortable range decreasing with increasing IPD. Regardless of the viewer's IPD, a larger range of depths can be presented by decreasing the IPD used to render and display the stimuli (Siegel and Nagata 2000). This however comes at the cost of the problems of divergence, misperception of distance, and off-axis viewing discussed elsewhere in this chapter (McIntire et al. 2018).

Even if the images are rendered and positioned appropriately for the observers' IPD, there may still be a misalignment between the positions of the lenses and the eye, creating off-axis viewing. This introduces a number of optical artefacts (Howarth 1999), including prismatic effects, which shift the images away from their intended locations (Peli 1995) and chromatic aberration (Beams et al. 2019). In order to minimise the problems associated with in an incorrect IPD, therefore, it would be necessary to set both the lens separation and the IPD used in software rendering, to match the observer. Although this is not practical in most current research pipelines and practices, it may become possible with future technological development.

7 Conclusions

VR presents many new opportunities for vision sciences. Notably, it allows us to create virtual environments that are immersive and that react to our movements and actions. This more fully reflects our natural environment, and affords a broader range of interactions than is generally possible with a typical display screen. Firstly, it allows us to create wide field-of-view environments that reflect many of the statistical properties of our natural environment. Secondly, it allows the user to actively explore the environment. This is a particularly important characteristic, since our visual inputs are determined not just by the structure of the environment, but also by our own actions within it.

VR also provides many opportunities to extend our theoretical understanding of the ways in which visual processing may be influenced by, or directly linked to, the control of behaviour. In comparison with traditional, forced-choice psychophysical experiments, this introduces a degree of complexity to the data that we gather. The control of visual stimuli, the manner and order in which they are presented, and the responses available to the participant are fundamental to the power of the psychophysical approach. In contrast, the unconstrained, open-world exploration of and interaction with the environment that are possible in VR mean that the data that our experiments generate have the potential to be much less structured. It is therefore important, in designing experiments in VR, to balance the ecological validity that it provides with the need for data that can be interpreted in relation to our theoretical research questions.

VR presents some technical considerations related to the nature of the visual display and the extent to which both software and hardware restrictions allow us to accurately create our intended visual environments. It is important therefore for us to remember that VR is not a substitute for other display technologies and psychophysical techniques. Rather, these different approaches should be seen as complementary. Gibson, for example, argued that theories of vision need to account for our experience of the visual field, or of the visual world giving rise to this visual field (Gibson 1950; Sedgwick 2021). VR is particularly well suited to understanding our interactions with the visual world.

References

Adelson EH, Movshon JA (1982) Phenomenal coherence of moving visual patterns. Nature 300(5892):523–525

Adelson EH, Pentland AP (1996) The perception of shading and reflectance. Perception as Bayesian inference, pp 409–423

Alcañiz M, Chicchi-Giglioli IA, Carrasco-Ribelles LA, Marín-Morales J, Minissi ME, Teruel-García G, Sirera M, Abad L (2022) Eye gaze as a biomarker in the recognition of autism spectrum disorder using virtual reality and machine learning: a proof of concept for diagnosis. Autism Res 15(1):131–145

Backus BT, Banks MS, Van Ee R, Crowell JA (1999) Horizontal and vertical disparity, eye position, and stereoscopic slant perception. Vision Res 39(6):1143–1170

Balboa RM, Tyler CW, Grzywacz NM (2001) Occlusions contribute to scaling in natural images. Vision Res 41(7):955–964

Banterle F, Artusi A, Debattista K, Chalmers A (2017) Advanced high dynamic range imaging. AK Peters/CRC Press

Beams R, Kim AS, Badano A (2019) Transverse chromatic aberration in virtual reality head-mounted displays. Opt Express 27(18):24877–24884

Bhargava A, Lucaites KM, Hartman LS, Solini H, Bertrand JW, Robb AC, Pagano CC, Babu SV (2020) Revisiting affordance perception in contemporary virtual reality. Virtual Reality 24(4): 713–724

Braddick OJ (1974) A short-range process in apparent motion. Vision Res 14:519–527

Braddick O, Campbell FW, Atkinson J (1978) Channels in vision: basic aspects. In: Perception. Springer, Berlin, pp 3–38

Braddick O (1981) Spatial frequency analysis in vision. Nature 291(5810):9–11

Bradshaw MF, Elliott KM, Watt SJ, Hibbard PB, Davies IR, Simpson PJ (2004) Binocular cues and the control of prehension. Spat Vis 17(1-2):95–110

Brainard DH (1997) The psychophysics toolbox. Spat Vis 10(4):433–436

Brindley GS (1960) Physiology of the retina and visual pathway. Edward Arnold, London

Chiao CC, Cronin TW, Osorio D (2000) Color signals in natural scenes: characteristics of reflectance spectra and effects of natural illuminants. J Opt Soc Am A 17(2):218–224

Cleland BG, Dubin MW, Levick WR (1971) Sustained and transient neurones in the cat's retina and lateral geniculate nucleus. J Physiol 217(2):473–496

Cruz-Neira C, Sandin DJ, DeFanti TA, Kenyon RV, Hart JC (1992) The CAVE: audio visual experience automatic virtual environment. Commun ACM 35(6):64–73

Cutting JE, Vishton PM (1995) Perceiving layout and knowing distances: the integration, relative potency, and contextual use of different information about depth. In: Perception of space and motion. Academic Press, pp 69–117

Darko-Takyi C, Khan NE, Nirghin U (2016) A review of the classification of nonstrabismic binocular vision anomalies. Optometry Rep 5(1)

Daugman JG (1985) Uncertainty relation for resolution in space, spatial frequency, and orientation optimized by two-dimensional visual cortical filters. J Opt Soc Am A 2(7):1160–1169

Descartes R (1637/1965) Discourse on method, optics, geometry, and meteorology. Olscamp PJ, Trans. Bobbs-Merrill

Dodgson NA (2004) Variation and extrema of human interpupillary distance. In: Stereoscopic displays and virtual reality systems XI, vol 5291. International Society for Optics and Photonics, pp 36–46

Dong DW, Atick JJ (1995a) Statistics of natural time-varying images. Netw Comput Neural Syst 6(3):345

Dong DW, Atick JJ (1995b) Temporal decorrelation: a theory of lagged and nonlagged responses in the lateral geniculate nucleus. Netw Comput Neural Syst 6(2):159–178

Dosher BA, Lu ZL (1999) Mechanisms of perceptual learning. Vision Res 39(19):3197–3221

Ernst MO, Banks MS (2002) Humans integrate visual and haptic information in a statistically optimal fashion. Nature 415(6870):429–433

Field DJ, Brady N (1997) Visual sensitivity, blur and the sources of variability in the amplitude spectra of natural scenes. Vision Res 37(23):3367–3383

Foulsham T, Walker E, Kingstone A (2011) The where, what and when of gaze allocation in the lab and the natural environment. Vision Res 51(17):1920–1931

Franz VH, Gegenfurtner KR, Buelthoff HH, Fahle M (2000) Grasping visual illusions: no evidence for a dissociation between perception and action. Psychol Sci 11(1):20–25

Geraets CNW, Tuente SK, Lestestuiver BP, van Beilen M, Nijman SA, Marsman JBC, Veling W (2021) Virtual reality facial emotion recognition in social environments: an eye-tracking study. Internet Interv 25:100432

Gibson JJ (1950) The perception of the visual world. Houghton Mifflin

Gibson JJ (1979) The ecological approach to visual perception. Houghton Mifflin Company

Glennerster A (2016) A moving observer in a three-dimensional world. Philos Trans R Soc B Biol Sci 371(1697):20150265

Glennerster A, McKee SP (1999) Bias and sensitivity of stereo judgements in the presence of a slanted reference plane. Vision Res 39(18):3057–3069

Glennerster A, McKee SP, Birch MD (2002) Evidence for surface-based processing of binocular disparity. Curr Biol 12(10):825–828

Glennerster A, Tcheang L, Gilson SJ, Fitzgibbon AW, Parker AJ (2006) Humans ignore motion and stereo cues in favor of a fictional stable world. Curr Biol 16(4):428–432

Goutcher R, Barrington C, Hibbard PB, Graham B (2021) Binocular vision supports the development of scene segmentation capabilities: evidence from a deep learning model. J Vis 21(7): 13–13

Gulhan D, Durant S, Zanker JM (2022) Aesthetic judgments of 3D arts in virtual reality and online settings. Virtual Reality:1–17

Graham NVS (1989) Visual pattern analyzers. Oxford University Press

Gray KL, Adams WJ, Hedger N, Newton KE, Garner M (2013) Faces and awareness: low-level, not emotional factors determine perceptual dominance. Emotion 13(3):537

Hartle B, Wilcox LM (2021) Cue vetoing in depth estimation: physical and virtual stimuli. Vision Res 188:51–64

He ZJ, Ooi TL (2000) Perceiving binocular depth with reference to a common surface. Perception 29(11):1313–1334

Heeger DJ (1992) Normalization of cell responses in cat striate cortex. Vis Neurosci 9(2):181–197

Heeley DW, Buchanan-Smith HM, Cromwell JA, Wright JS (1997) The oblique effect in orientation acuity. Vision Res 37(2):235–242

Held RT, Cooper EA, Banks MS (2012) Blur and disparity are complementary cues to depth. Curr Biol 22(5):426–431

Hertzmann A (2022) The choices hidden in photography. J Vis 22(11)

Hibbard P (2007) A statistical model of binocular disparity. Vis Cogn 15(2):149–165

Hibbard PB (2008) Binocular energy responses to natural images. Vision Res 48(12):1427–1439

Hibbard PB (2021) Estimating the contributions of pictorial, motion and binocular cues to the perception of distance. ECVP. In press

Hibbard PB, Bradshaw MF (2003) Reaching for virtual objects: binocular disparity and the control of prehension. Exp Brain Res 148(2):196–201

Hibbard PB, van Dam LC, Scarfe P (2020) The implications of interpupillary distance variability for virtual reality. In: 2020 international conference on 3D immersion (IC3D), pp 1–7

Hillis JM, Watt SJ, Landy MS, Banks MS (2004) Slant from texture and disparity cues: optimal cue combination. J Vis 4(12):1–1

Hoffman DD, Singh M, Prakash C (2015) The interface theory of perception. Psychon Bull Rev 22(6):1480–1506

Hoffman DM, Girshick AR, Akeley K, Banks MS (2008) Vergence–accommodation conflicts hinder visual performance and cause visual fatigue. J Vis 8(3):33–33

Hornsey RL, Hibbard PB (2021) Contributions of pictorial and binocular cues to the perception of distance in virtual reality. Virtual Reality:1–17

Hornsey RL, Hibbard PB, Scarfe P (2020) Size and shape constancy in consumer virtual reality. Behav Res Methods 52(4):1587

Howarth PA (1999) Oculomotor changes within virtual environments. Appl Ergon 30(1):59–67

Hubel DH, Wiesel TN (1962) Receptive fields, binocular interaction and functional architecture in the cat's visual cortex. J Physiol 160(1):106–154

IJsselsteijn WA, de Ridder H, Vliegen J (2000) Subjective evaluation of stereoscopic images: effects of camera parameters and display duration. IEEE Trans Circuits Syst Video Technol 10(2):225–233

Jack RE, Schyns PG (2015) The human face as a dynamic tool for social communication. Curr Biol 25(14):R621–R634

Johnston EB (1991) Systematic distortions of shape from stereopsis. Vision Res 31(7–8): 1351–1360

Johnston EB, Cumming BG, Parker AJ (1993) Integration of depth modules: stereopsis and texture. Vision Res 33(5–6):813–826

Julesz B (1971) Foundations of cyclopean perception. University of Chicago Press, Chicago

Kanwisher N, Yovel G (2006) The fusiform face area: a cortical region specialized for the perception of faces. Philos Trans R Soc B Biol Sci 361(1476):2109–2128

Keefe BD, Suray PA, Watt SJ (2019) A margin for error in grasping: hand pre-shaping takes into account task-dependent changes in the probability of errors. Exp Brain Res 237(4):1063–1075

Keefe BD, Hibbard PB, Watt SJ (2011) Depth-cue integration in grasp programming: no evidence for a binocular specialism. Neuropsychologia 49(5):1246–1257

Kim J, Interrante V (2017) Dwarf or giant: the influence of interpupillary distance and eye height on size perception in virtual environments. In: 27th international conference on artificial reality and

Telexistence, ICAT 2017 and the 22nd Eurographics symposium on virtual environments, EGVE 2017. Eurographics Association, pp 153–160
Klinghammer M, Schütz I, Blohm G, Fiehler K (2016) Allocentric information is used for memory-guided reaching in depth: a virtual reality study. Vision Res 129:13–24
Koenderink JJ (1984) The concept of local sign. In: van Doorn AJ, van de Grind WA, Koenderink JJ (eds) Limits in perception. VNU Science Press, Utrecht
Koenderink JJ (1999) Virtual psychophysics. Perception 28:669–674
Koenderink JJ (1998) Pictorial relief. Philos Trans R Soc Lond Ser A Math Phys Eng Sci 356(1740):1071–1086
Koenderink JJ (2019) Vision, an optical user interface. Perception 48(7):545–601
Koenderink JJ, van Doorn AJ, Kappers AM, Todd JT (2001) Ambiguity and the 'mental eye' in pictorial relief. Perception 30(4):431–448
Kopiske KK, Bozzacchi C, Volcic R, Domini F (2019) Multiple distance cues do not prevent systematic biases in reach to grasp movements. Psychol Res 83(1):147–158
Lambooij M, IJsselsteijn WA, Heynderickx I (2011) Visual discomfort of 3D TV: assessment methods and modeling. Displays 32(4):209–218
Landy MS, Maloney LT, Johnston EB, Young M (1995) Measurement and modeling of depth cue combination: in defense of weak fusion. Vision Res 35(3):389–412
Lanier J (2017) Dawn of the new everything: a journey through virtual reality. Random House
Lee DN (1980) The optic flow field: the foundation of vision. Philos Trans R Soc Lond B Biol Sci 290(1038):169–179
Leyrer M, Linkenauger SA, Bülthoff HH, Kloos U, Mohler B (2011) The influence of eye height and avatars on egocentric distance estimates in immersive virtual environments. In: Proceedings of the ACM SIGGRAPH symposium on applied perception in graphics and visualization, pp 67–74
Liang Z, Liu D, Zhou M (2014) Research on large scale 3D terrain generation. In: 2014 IEEE 17th international conference on computational science and engineering, pp 1827–1832
Livingstone MS, Hubel DH (1987) Psychophysical evidence for separate channels for the perception of form, color, movement, and depth. J Neurosci 7(11):3416–3468
MacLachlan C, Howland HC (2002) Normal values and standard deviations for pupil diameter and interpupillary distance in subjects aged 1 month to 19 years. Ophthalmic Physiol Opt 22(3): 175–182
Mamassian P, Landy M, Maloney LT (2002) Bayesian modelling of visual perception. Probabilistic Models Brain:13–36
Marr D (1982) Vision. A computational investigation into the human representation and processing of visual information. W H Freeman, New York
Marr D, Nishihara HK (1978) Representation and recognition of the spatial organization of three-dimensional shapes. Proc R Soc Lond Ser B Biol Sci 200(1140):269–294
McIntire JP, Havig PR, Harrington LK, Wright ST, Watamaniuk SNJ, Heft E (2018) Microstereopsis is good, but orthostereopsis is better: precision alignment task performance and viewer discomfort with a stereoscopic 3d display. In: Three-dimensional imaging, visualization, and display
Melmoth DR, Grant S (2006) Advantages of binocular vision for the control of reaching and grasping. Exp Brain Res 171(3):371–388
Menzel C, Redies C, Hayn-Leichsenring GU (2018) Low-level image properties in facial expressions. Acta Psychol (Amst) 188:74–83
Milner D, Goodale M (2006) The visual brain in action. OUP, Oxford
Mon-Williams M, Tresilian JR, Roberts A (2000) Vergence provides veridical depth perception from horizontal retinal image disparities. Exp Brain Res 133(3):407–413
Morgan MJ, Melmoth D, Solomon JA (2013) Linking hypotheses underlying class A and class B methods. Vis Neurosci 30(5-6):197–206
Muryy A, Glennerster A (2021) Route selection in non-Euclidean virtual environments. PLoS One 16(4):e0247818

Nagata S (1991) How to reinforce perception of depth in single two-dimensional pictures. In: Pictorial communication in real and virtual environments. CRC Press, pp 553–573

Newsome WT, Britten KH, Movshon JA (1989) Neuronal correlates of a perceptual decision. Nature 341(6237):52–54

Nonis F, Dagnes N, Marcolin F, Vezzetti E (2019) 3D approaches and challenges in facial expression recognition algorithms – a literature review. Appl Sci 9(18):3904

Olshausen BA, Field DJ (1996) Natural image statistics and efficient coding. Netw Comput Neural Syst 7(2):333

O'Regan JK, Noë A (2001) A sensorimotor account of vision and visual consciousness. Behav Brain Sci 24(5):939–973

Parker AJ, Newsome WT (1998) Sense and the single neuron: probing the physiology of perception. Annu Rev Neurosci 21(1):227–277

Peirce JW (2007) PsychoPy – psychophysics software in python. J Neurosci Methods 162(1-2): 8–13

Peli E (1995) Real vision & virtual reality. Optics Photonics News 6(7):28

Pelli DG (1997) The VideoToolbox software for visual psychophysics: transforming numbers into movies. Spat Vis 10:437–442

Petrov Y, Glennerster A (2004) The role of a local reference in stereoscopic detection of depth relief. Vision Res 44(4):367–376

Pizlo Z (2001) Perception viewed as an inverse problem. Vision Res 41(24):3145–3161

Pizlo Z (2010) 3D shape: its unique place in visual perception. MIT Press

Plumert JM, Kearney JK, Cremer JF, Recker K (2005) Distance perception in real and virtual environments. ACM Trans Appl Percept 2(3):216–233

Rogers B (2021) Optic flow: perceiving and acting in a 3-D world. i-Perception 12(1): 2041669520987257

Rushton SK, Harris JM, Lloyd MR, Wann JP (1998) Guidance of locomotion on foot uses perceived target location rather than optic flow. Curr Biol 8(21):1191–1194

Rust NC, Movshon JA (2005) In praise of artifice. Nat Neurosci 8(12):1647–1650

Scarfe P, Glennerster A (2015) Using high-fidelity virtual reality to study perception in freely moving observers. J Vis 15(9):3–3

Scarfe P, Glennerster A (2019) The science behind virtual reality displays. Annu Rev Vis Sci 5: 529–547

Scarfe P, Glennerster A (2021) Combining cues to judge distance and direction in an immersive virtual reality environment. J Vis 21(4):10–10

Schnapf JL, Kraft TW, Nunn BJ, Baylor DA (1988) Spectral sensitivity of primate photoreceptors. Vis Neurosci 1(3):255–261

Servos P, Goodale MA, Jakobson LS (1992) The role of binocular vision in prehension: a kinematic analysis. Vision Res 32(8):1513–1511

Sedgwick HA (2021) JJ Gibson's "Ground Theory of Space Perception". i-Perception 12(3): 20416695211021111

Servotte JC, Goosse M, Campbell SH, Dardenne N, Pilote B, Simoneau IL, Guillaume M, Bragard I, Ghuysen A (2020) Virtual reality experience: immersion, sense of presence, and cybersickness. Clin Simul Nurs 38:35–43

Shibata T, Kim J, Hoffman DM, Banks MS (2011) The zone of comfort: predicting visual discomfort with stereo displays. J Vis 11(8):11–11

Siegel M, Nagata S (2000) Just enough reality: comfortable 3-d viewing via microstereopsis. IEEE Trans Circuits Syst Video Technol 10(3):387–396

Simoncelli EP, Olshausen BA (2001) Natural image statistics and neural representation. Annu Rev Neurosci 24(1):1193–1216

Simons DJ, Levin DT (1998) Failure to detect changes to people during a real-world interaction. Psychon Bull Rev 5(4):644–649

Smallman HS, John MS (2005) Naïve realism: misplaced faith in realistic displays. Ergon Des 13(3):6–13

Spector RH (1990) Visual fields. In: Walker HK, Hall WD, Hurst JW (eds) Clinical methods: the history, physical, and laboratory examinations. Butterworths, Boston, pp 565–572
Sprague WW, Cooper EA, Tošić I, Banks MS (2015) Stereopsis is adaptive for the natural environment. Sci Adv 1(4):e1400254
Strasburger H, Rentschler I, Jüttner M (2011) Peripheral vision and pattern recognition: a review. J Vis 11(5):13–13
Svarverud E, Gilson S, Glennerster A (2012) A demonstration of 'broken' visual space. PLoS One 7(3):e33782
Tarr MJ, Warren WH (2002) Virtual reality in behavioral neuroscience and beyond. Nat Neurosci 5(11):1089–1092
Todd JT (2020) On the ambient optic array: James Gibson's insights about the phenomenon of chiaroscuro. i-Perception 11(5):2041669520952097
Todd JT, Egan EJ, Kallie CS (2015) The darker-is-deeper heuristic for the perception of 3D shape from shading: is it perceptually or ecologically valid? J Vis 15(15):2–2
Tresilian JR, Mon-Williams M, Kelly BM (1999) Increasing confidence in vergence as a cue to distance. Proc R Soc Lond Ser B Biol Sci 266(1414):39–44
van Dam LC, Stephens JR (2018) Effects of prolonged exposure to feedback delay on the qualitative subjective experience of virtual reality. PLoS One 13(10):e0205145
Van Hateren JH (1993) Spatiotemporal contrast sensitivity of early vision. Vision Res 33(2): 257–267
Vangorp P, Mantiuk RK, Bazyluk B, Myszkowski K, Mantiuk R, Watt SJ, Seidel HP (2014) Depth from HDR: depth induction or increased realism?. In: Proceedings of the ACM symposium on applied perception, pp 71–78
Van Hateren JH, van der Schaaf A (1998) Independent component filters of natural images compared with simple cells in primary visual cortex. Proc Royal Soc Lond Ser B Biol Sci 265(1394):359–366
Vuong J, Fitzgibbon AW, Glennerster A (2019) No single, stable 3D representation can explain pointing biases in a spatial updating task. Sci Rep 9(1):1–13
Wade NJ (2013) Deceiving the brain: pictures and visual perception. Prog Brain Res 204:115–134
Waltemate T, Senna I, Hülsmann F, Rohde M, Kopp S, Ernst M, Botsch M (2016) The impact of latency on perceptual judgments and motor performance in closed-loop interaction in virtual reality. In: Proceedings of the 22nd ACM conference on virtual reality software and technology, pp 27–35
Warren WH, Hannon DJ (1988) Direction of self-motion is perceived from optical flow. Nature 336(6195):162–163
Watt SJ, Akeley K, Ernst MO, Banks MS (2005) Focus cues affect perceived depth. J Vis 5(10):7–7
Watt SJ, Bradshaw MF (2000) Binocular cues are important in controlling the grasp but not the reach in natural prehension movements. Neuropsychologia 38(11):1473–1481
Webb AL, Asher JM, Hibbard PB (2022) Saccadic eye movements are deployed faster for salient facial stimuli, but are relatively indifferent to their emotional content. Vision Res 198:108054
Webb AL, Hibbard PB, O'Gorman R (2020) Contrast normalisation masks natural expression-related differences and artificially enhances the perceived salience of fear expressions. PLoS One 15(6):e0234513
Westheimer G (1972) Visual acuity and spatial modulation thresholds. In: Visual psychophysics. Springer, Berlin, pp 170–187
Wilson HR (1980) A transducer function for threshold and suprathreshold human vision. Biol Cybern 38(3):171–178

Zeki SM (1978) Functional specialisation in the visual cortex of the rhesus monkey. Nature 274(5670):423–428

Zhaoping L, Li Z (2014) Understanding vision: theory, models, and data. Oxford University Press, Oxford

Zhong F, Jindal A, Yöntem Ö, Hanji P, Watt S, Mantiuk R (2021) Reproducing reality with a high-dynamic-range multi-focal stereo display. ACM Trans Graph 40(6):241

Zibrek K, Martin S, McDonnell R (2019) Is photorealism important for perception of expressive virtual humans in virtual reality? ACM Trans Appl Percept 16(3):1–19

VR for Studying the Neuroscience of Emotional Responses

Marta Andreatta, Markus H. Winkler, Peter Collins, Daniel Gromer, Dominik Gall, Paul Pauli, and Matthias Gamer

Contents

Abstract Emotions are frequently considered as the driving force of behavior, and psychopathology is often characterized by aberrant emotional responding. Emotional states are reflected on a cognitive-verbal, physiological-humoral, and motor-behavioral level but to date, human research lacks an experimental protocol for a comprehensive and ecologically valid characterization of such emotional states. Virtual reality (VR) might help to overcome this situation by allowing researchers to study mental processes and behavior in highly controlled but reality-like laboratory settings. In this chapter, we first elucidate the role of presence and immersion as requirements for eliciting emotional states in a virtual environment and discuss different VR methods for emotion induction. We then consider the organization of emotional states on a valence continuum (i.e., from negative to positive) and on this basis discuss the use of VR to study threat processing and avoidance as well as reward processing and approach behavior. Although the potential of VR has not been fully realized in laboratory and clinical settings yet, this technological tool can

M. Andreatta (✉)
Department of Psychology, Educational Sciences, and Child Studies, Erasmus University Rotterdam, Rotterdam, The Netherlands
e-mail: andreatta@essb.eur.nl

M. H. Winkler, P. Collins, D. Gromer, D. Gall, P. Pauli, and M. Gamer (✉)
Department of Psychology, University of Wuerzburg, Wuerzburg, Germany
e-mail: matthias.gamer@uni-wuerzburg.de

Curr Topics Behav Neurosci (2023) 65: 161–188
https://doi.org/10.1007/7854_2022_405
Published Online: 3 January 2023

open up new avenues to better understand the neurobiological mechanisms of emotional responding in healthy and pathological conditions.

Keywords Approach · Avoidance · Emotions · Presence · Virtual reality

1 Introduction

The application of new technology in experimental settings often enables researchers to overcome previous boundaries and thereby opens avenues to novel and unexplored research questions. Virtual reality (VR) is such an ecologically valid and ergonomic new technology which allows researchers to create complex yet highly controlled environments in which humans can be immersed and feel present (Sanchez-Vives and Slater 2005). Like in a real environment, in VR participants can perceive and respond to multisensory exteroceptive stimulation, and their behavior can be monitored allowing for an adjustment of the virtual environment depending on body and head orientation. The possibility and advantages of studying human behavior in such naturalistic but controlled conditions have led to a substantial increase in VR studies in the last decades (Fig. 1).

What makes VR particularly suited for studying human emotions is the unique possibility to simultaneously register subjective, autonomic, neural, and behavioral responses during active exploration of virtual environments. This allows for the comprehensive characterization of emotional states on a cognitive-verbal, physiological-humoral, and motor-behavioral level (Fox 2008). Furthermore, the use of virtual environments facilitates translational work (Fig. 2b), such that certain experimental protocols, which are traditionally used in animals, can now be easily recreated in VR to allow for a more direct comparison between animal and human research (e.g., Biedermann et al. 2017; Madeira et al. 2021).

Traditionally, emotions have been proposed as a specific set of responses (Ekman and Friesen 1971; Ekman 2016). Positive and negative emotions can be considered as two extremes on a continuum, which both motivate behavior and may become dysfunctional in psychopathology (e.g., Boecker and Pauli 2019; Lang 1995; Gray 1990). The influential *motivational priming hypothesis* (Lang 1995) differentiates between a defensive and an appetitive motivational system. The defensive motivational system is activated by threatening situations or by signals of danger eliciting emotions such as anxiety or fear, which in turn prime and facilitate avoidance. By contrast, the appetitive motivational system is activated by joyful situations or signals of reward, which elicit joy or happiness and trigger approach behavior. In this chapter, we focus on both negative and positive emotions as well as their expression in human behavior, subjective, and physiological responses, and we specifically emphasize the advantages of VR for reality-like studies. After providing an overview of the VR requirements for inducing emotional states, we separately

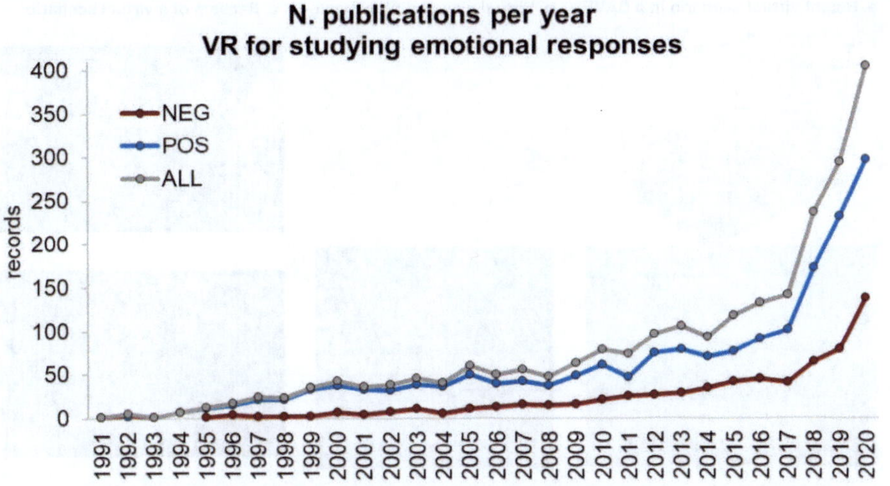

Fig. 1 *Overview of the number of studies applying VR for studying emotional responding*. The search in Web of Science considered articles written in English that were published in the research fields psychology, psychiatry, neuroscience, computer, and behavioral sciences. As keywords, we used ["virtual reality" or VR] and [emotion OR emotions OR anxiety OR fear OR negative OR aversive OR avoidance OR positive OR happy OR joy OR approach], but not [relax OR mindfulness OR therapy OR exposure]. This search returned a total of 2,645 scientific articles (gray line). In order to separately focus on negative and positive emotional responses, we re-run the same search for negative and positive emotions, respectively, excluding either [positive OR happy OR joy OR approach] or [anxiety OR fear OR negative OR aversive OR avoidance]. These additional searches returned 740 scientific articles for negative emotions (NEG, red line) and 1,996 scientific articles for positive emotions (POS, blue line)

focus on negative and positive emotions as well as avoidance and approach behavior. Lastly, we conclude with suggestions for future research.

2 Presence and Immersion

VR simulates real-world experiences by producing sensory-motor contingencies, i.e. the VR system provides sensory information that is consistent with the users' movements in the virtual environment. The most common modality for VR simulation is the visual domain, frequently combined with auditory stimulation. Therefore, VR simulations are mostly based on custom three-dimensional virtual stimuli which are displayed in the first-person view of the user. The complexity of these stimuli comes with the costs of a resource-intensive development process to individualize the virtual environment. In addition to audiovisual content, haptic feedback might provide a promising domain for further sensory-motor contingencies of VR systems (Gall and Latoschik 2018; Gromer et al. 2018). However, to date, technical limitations impede the widespread use of haptic feedback in VR, except for simulated floor

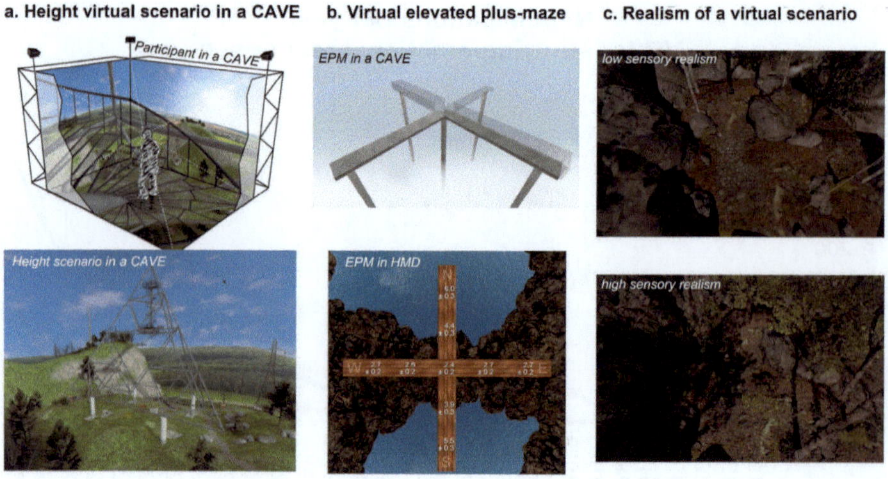

Fig. 2 *Examples of virtual environments for eliciting emotional responses.* (**a**) Participants can be located within a CAVE and by means of a gamepad they can freely move within a virtual environment including, for example, a tower inducing fear of height (Gromer et al. 2018). (**b**) Protocols of animal studies can be easily translated to VR settings as demonstrated with the elevated plus maze (EPM; Biedermann et al. 2017; Madeira et al. 2021). (**c**) The realism of the virtual environments can be manipulated in order to induce a stronger sense of presence (Gromer et al. 2019)

movements that allow for prolonged natural walking in the virtual realm (Nilsson et al. 2018).

Visual-auditory stimulation can be achieved by different technologies, but head-mounted displays (HMD) are most popular. HMDs include two displays, one placed in front of each eye, and a tracking system that tracks the users' head movements. The two screens display stereoscopic images to create an illusion of visual depth. The prevalence of HMDs in the entertainment industry has led to the availability of widely used software tools that facilitate the generation of VR content that can also be used for experimental research in psychology and neuroscience. A downside of HMDs is that they only allow a limited viewing angle and block out all "real" environmental information including the user's own body. Alternatively, augmented reality headsets include visual information of the environment to combine artificial and real-world visual content (for a taxonomy, see Milgram and Kishino 1994).

Cave Automatic Virtual Environment (CAVE) systems use projectors to display stereoscopic images on the walls of a room and track user movements so that participants can see their bodies while moving in a life-size virtual environment (Fig. 2a). Users need to wear special glasses to see the stereoscopic effect and body markers to track their movements within the CAVE. However, CAVE systems are cost-intensive and require customized software to design experiments. Some studies suggest that HMDs and CAVEs produce similar effects, depending on the study's specific requirements (Bouchard and Rizzo 2019).

VR tricks our minds to believe in a world that does not exist. Although VR users are fully aware of the fact that they are in their living room, a laboratory, or a therapist's office when entering an immersive virtual environment, they start to react to the virtual world as if it is real. For example, a height fearful person visiting the terrace of a virtual skyscraper might show the same emotional, physiological, and behavioral reactions as when visiting a real height situation, including fear, sweating, going weak at the knees, and keeping distance from the railing (Gromer et al. 2018, 2019). In short, users react to and behave in virtual environments as if they were really there, i.e. they show indications of a "(suspension of dis-)belief, that they are in a world other than where their real bodies are located" (Slater and Usoh 1993, p. 222). This sense of "being there" in the virtual environment has been termed (spatial) *presence* or *place illusion* (Schubert 2006; Slater 2009; Wirth et al. 2007). Feeling present in a virtual environment has been thought to be a prerequisite for VR to be able to trigger emotional responses (Bouchard and Rizzo 2019; Diemer et al. 2015; Price et al. 2011; Riva et al. 2007). However, this causal relationship has not yet been clearly demonstrated (Diemer et al. 2015). For example, Gromer et al. (2019) manipulated presence by increasing the visual and auditory realism of a virtual height environment (Fig. 2c). Contrary to the hypothesis, increased presence did not result in stronger fear responses in participants with a fear of heights. However, the fear-inducing height situation induced a stronger sense of presence than a control condition, in line with other studies demonstrating that experiencing emotions in VR leads to a stronger sense of presence (Bouchard et al. 2008; Gromer et al. 2019; Pallavicini et al. 2020). These findings highlight the bidirectional relationship between presence and emotional experience and suggest that only a certain level of presence is needed for an effective emotion induction (Bouchard et al. 2012).

The level of presence a user experiences in VR is additionally affected by their *immersion*. Immersion describes the technological characteristics of the system used to deliver the VR experience (Bouchard and Rizzo 2019; Slater and Wilbur 1997). For example, a system that translates both a user's head rotation and position to the virtual experience is thought to be more immersive than a system that only handles a user's head rotation. A meta-analysis showed that the most important aspects of a VR system for achieving a strong sense of presence were frame rate, tracking quality, field of view, and use of stereoscopy (Cummings and Bailenson 2016). However, strong immersion is not a prerequisite for emotional responses. For example, early VR studies with low immersion due to technological constraints still reported substantial emotional responses (e.g., Mühlberger et al. 2001, 2007).

Having discussed VR more generally, the remainder of the chapter will focus on how emotional states can be induced in VR and whether such approach has advantages over traditional research methods in affective science.

3 Emotion Induction in Virtual Reality

Many classic mood induction procedures present stimuli in an explicit manner that participants passively process. Emotional states have been induced, for example, by pictures (e.g., the International Affective Picture System, IAPS, Lang 1995), sounds (e.g., the International Affective Digitalised Sound System, IADS, Bradley and Lang 1999), music (Västfjäll 2001; Kim and André 2008), or videos (Marcusson-Clavertz et al. 2019; Soleymani et al. 2015). A meta-analysis by Westermann et al. (1996) concluded that these methods produce a mood change of approximately one standard deviation and noted a substantial variation in effect sizes across studies. First evidence that VR might be very effective in triggering emotional responses comes from research showing that combining sensory modalities, for instance static images and music (Zhang et al. 2014), audiovisual videos (Fernández-Aguilar et al. 2019), or visual and haptic information, may boost the effects of the manipulation (Westermann et al. 1996; Martin 1990; Ferrer et al. 2015; Yalachkov et al. 2012). Similarly, enhanced emotional responses were observed with pictures moving away or toward participants (Muhlberger et al. 2008), or with animated dynamic emotional facial expressions (Mühlberger et al. 2011; Weyers et al. 2009).

Classical elicitation methods can be implemented in VR by simply presenting traditional emotional stimuli, like pictures, sounds, or videos, using a VR headset (e.g., Orefice et al. 2017; Herrero et al. 2014). The advantage of this procedure compared to non-VR techniques is that the participant is more isolated from the external reality and cannot easily look away from the presented stimulus material or be distracted by exogenous stimuli. The experimenter, therefore, has greater control over the participant's attentional focus. However, potentially more efficacious emotion induction protocols can be implemented in VR by immersing participants into specifically designed virtual environments. For example, Baños et al. (2006) developed a single virtual environment (a park) and combined it with different elements of mood induction procedures to successfully elicit sadness, happiness, anxiety, or relaxation. To induce the different moods, the simulation (lighting conditions and music) was manipulated, and participants had to complete certain tasks in VR (e.g., ordering statements of the Velten technique, associating emotional pictures to different statements). The experimental setting thus included traditional emotion induction methods but presented them within a naturalistic simulated environment.

VR also offers researchers a method of presenting participants with emotion elicitation tasks that would be either dangerous or difficult to standardize outside of VR. For example, in the classic pit room paradigm reported by Slater (2002) participants are instructed to walk into a virtual room with a 10-m precipice at its center and to stand at the edge of the precipice. Although participants are consciously aware that they are not standing at the edge of a cliff, their body reacts as if they are. Thus, the pit room induces reliable changes in heart rate and skin conductance in line with an increase of subjective fear ratings (Gromer et al. 2019; Meehan et al. 2002). It is worth highlighting that affording the user greater freedom over their approach and avoidance behavior may be crucial to support the effective emergence of emotional states. In this example, the participants can already anticipate and experience changes in emotional responses during planning and executing body movements as they approach the pit edge. Collange and Guegan (2020) similarly used the participant's sense of acting in the virtual environment to elicit emotions and emotionally driven behavior. In their scenario, users received help from a virtual benefactor, and when their feelings of gratitude increased, they were also more likely to offer social support in response. In other words, the impact of a virtual environment on emotional responding seems to be in line with real-world experiences, which is further supported by studies reporting similar arousal and valence scores while participants explored a virtual and a physical museum (Marín-Morales et al. 2019). In a well-controlled study by Higuera-Trujillo et al. (2017), physiological (heart rate and skin conductance) and subjective (arousal and presence ratings) responses to a real environment (i.e., during baseline) were compared to the responses elicited by the environment displayed in three different formats: a photograph, a 360° panorama, and a 3D virtual panorama. The VR format elicited the closest approximation of the physiological responses to the physical environment. Both the 360° panorama and the VR formats induced a high sense of presence. The photo format produced the lowest sense of presence and the lowest physiological approximation to the physical environment. Methods for VR emotion induction are also relevant for clinical populations. For example, patients with eating disorders (anorexia and bulimia nervosa) but not healthy controls reported elevated subjective anxiety and showed heart rate and skin conductance responses to both real and virtual food as compared to pictures of food, with no significant difference in anxiety induced by real and virtual food (Gorini et al. 2010).

Taken together, these findings indicate that VR might be advantageous compared to classic procedures in inducing specific emotional states (see Table 1 for illustrative examples of experimental approaches). Not only can classic emotion induction procedures be implemented more effectively, VR scenarios also offer the additional opportunity to induce emotions in a more active fashion that involves users behaving in and interacting with the virtual environment. Furthermore, VR not only allows for inducing emotions and measuring emotional responding with respect to cognitive-verbal and physiological-humoral changes (e.g., the Velten procedure, Kenealy 1986; Konrad et al. 2016), it also offers the unique possibility to track approach and avoidance behavior in a rather naturalistic manner (see, e.g., Gromer et al. 2018; Kinateder et al. 2014, 2015). In the following sections, we will thus separately focus on negative and positive emotional states as well as the accompanying avoidance and approach tendencies. In addition to highlighting VR studies on healthy participants in these domains, we will also address clinical aspects.

Table 1 Examples of emotion induction procedures in virtual reality (VR). Whereas some scenarios were specifically designed for eliciting isolated emotions, others permit the induction of several positive and negative emotional states depending on certain characteristics of the stimulation

Valence	Induced emotion	General VR scenario	Additional features	Reference
Positive	Gratitude	Receiving help during the escape from a threatening situation (fire)	Soundscape of a burning building	Collange and Guegan (2020)
	Sexual arousal and desire	Erotic audiovisual videos presented in VR	–	Milani et al. (2022)
	Relaxation	Virtual forest	Rumble platform for somatosensory stimulation	Valtchanov et al. (2010)
Negative	Fear (innate)	Elevated plus maze	Elements of augmented reality (e.g., wooden maze platform, simulation of wind)	Biedermann et al. (2017)
	Fear (acquired)	Virtual office rooms	Mildly painful electric shocks	Andreatta et al. (2015b)
	Anxiety (innate)	Open field (grassland)	–	Gromer et al. (2021)
	Disgust	Tasting experiment in VR	Real consumption of chocolate with experimental context manipulation	Ammann et al. (2020)
Undefined	Awe	Natural scenes (forest, mountains, earth view from deep space)	Consistent auditory stimulation	Chirico et al. (2018)
Multiple	Sadness, happiness, anxiety, relaxation	Natural park scenery	Varying lighting conditions and music, emotion induction task (e.g., Velten technique)	Baños et al. (2006)
	Joy, anger, boredom, anxiety, sadness	Natural park scenery	Varying lighting and weather conditions, specifically adjusted soundscape	Felnhofer et al. (2015)
	Dimensional emotional responding (valence and arousal)	Virtual museum	VR facsimile of a museum exhibition	Marín-Morales et al. (2019)

Note. This list is not meant to provide an exhaustive overview over all possible scenarios and emotion induction procedures but rather to exemplify the variety of approaches in this field of research. More comprehensive overviews are given by Diniz Bernardo et al. (2021) or Markowitz and Bailenson (in press)

4 Negative Emotions and Avoidance

Well-established taxonomies of emotional qualities differentiate between several negative emotions such as anger, fear, sadness, and disgust (Ekman and Friesen 1971). Due to its enormous relevance for clinical applications, a big amount of VR research on emotional responding has focused on fear. We will also concentrate on this emotion here, but we will differentiate between fear and anxiety since these two aversive emotions have distinct neural and physiological correlates (Davis et al. 2010; Perusini and Fanselow 2015; Tovote et al. 2015, but see Daniel-Watanabe and Fletcher 2022). Fear is defined as a phasic response elicited by imminent and concrete threats, while anxiety is a future-oriented response characterized by sustained apprehension of potential yet not identified threats (Davis et al. 2010; Fanselow 2018). In the following, we will discuss how VR can be used to augment traditional research paradigms to overcome methodological limitations and advance the current knowledge on the neurobehavioral mechanisms of these negative emotional states.

Fear and anxiety are per se adaptive responses. Because of their biological relevance, responses to fear- or anxiety-eliciting stimuli can be easily learned (Fanselow 2018), and classical conditioning is a simple, but well-established laboratory model for these learning processes (Lonsdorf et al. 2017; Pavlov 1927). During a classical conditioning protocol, an initially neutral cue is repeatedly associated with an aversive unconditioned stimulus (US) such as pain. As a consequence, this stimulus becomes a signal predicting the aversive US, then labeled a conditioned stimulus (CS+). Startle potentiation and avoidance behavior are defensive responses elicited by cues associated with an aversive US (Fendt and Fanselow 1999; Hamm and Weike 2005; LeDoux and Pine 2016). These conditioned behaviors (Fendt and Fanselow 1999; Fullana et al. 2016; LeDoux and Pine 2016) and the underlying neural circuits show high similarity across species (Ressler 2020), which suggests that mechanisms of classical conditioning are evolutionarily preserved, and therefore translational work is of great importance (Haaker et al. 2019). Animal conditioning studies typically measure naturalistic behavioral changes as indication of learning, which can be analyzed in response to specific contexts, i.e., the cage which in animal research is easily controllable and changeable (e.g., by selectively placing elements at different locations). In contrast, in human research the typical laboratory setup – with participants sitting in front of a computer screen – does not easily allow researchers to measure naturalistic behavioral responses (but see Roelofs et al. 2010), and neither control nor manipulation of the context can be achieved easily. The flexibility of VR opens up new research avenues by overcoming such physical boundaries.

For example, VR environments can facilitate research on the interaction between contextual information and responses to individual cues (Baas et al. 2004). For instance, seeing a lion in the zoo is surely not as threatening as seeing it in the savanna, because the context informs us about the actual threat. Contextual information is especially important for learning the conditions under which a threat is not dangerous. One type of learning which strongly depends on contextual information is the extinction of conditioned defensive responses (Bouton 2002; Bouton and Moody 2004). Extinction learning is based on the presentation of the CS+ without the US and is a form of inhibitory learning which leads to an additional CS+/no-US memory trace (Quirk and Mueller 2008; Milad and Quirk 2012). The context provides the critical information for disentangling the CS+/US and the CS+/no-US memory traces and enables appropriate responding to the CS+ (i.e., regarding the execution of defensive responses). Numerous human studies have applied VR for studying such context-dependent extinction and confirmed reduction of the physiological and subjective fear response in the context in which the extinction learning occurred (Huff et al. 2011; Simon-Kutscher et al. 2019; Krisch et al. 2020; Baas and Heitland 2015; Heitland et al. 2012; Duits et al. 2016). Importantly, extinguished defensive responses to a CS+ re-occur in a new context, a phenomenon called renewal (Muhlberger et al. 2014), and extinction in multiple virtual contexts may counteract such phenomena and likely prevent relapses after treatment (Shiban et al. 2013; Shiban et al. 2015). Another important observation is that being stressed shortly before virtual classical conditioning seems to annul context- but not cue-dependent aversive learning (Simon-Kutscher et al. 2019). This finding may explain the exaggerated defensive responses to trauma-associated cues in patients with post-traumatic stress disorder (PTSD) which are independent of contextual information (Jovanovic et al. 2012).

Although the acquisition of conditioned fear evoked by discrete stimuli proximal to a threat is a good model for the etiology of fear-related disorders such as specific phobias (Craske et al. 2009; Mineka and Oehlberg 2008), classical context conditioning might be a better model to characterize learned anxiety (Maren et al. 2013; Rudy 2009). In context conditioning, the conditioned contextual stimuli are long-lasting (e.g., several seconds or even minutes) and mostly complex (Maren et al. 2013), thus the exact occurrence of the US remains unpredictable. Consequently, sustained defensive responses, which characterize anxiety, are elicited by contextual stimuli (Davis et al. 2010) that involve activations of the amygdala (Alvarez et al. 2008; Andreatta et al. 2015a) and the bed nucleus of the stria terminalis (BNST, Alvarez et al. 2011).

Applying VR in conditioning studies allows researchers to *systematically manipulate contextual stimuli* without losing experimental control. This type of manipulation is particularly relevant when generalization processes are investigated (Dunsmoor and Paz 2015; Dymond et al. 2015). Generalization of defensive responses is an adaptive mechanism which allows organisms to promptly respond to novel stimuli based on their physical or semantic similarity to a threatening stimulus. By systematically rearranging the elements in virtual environments (i.e., the furniture in a virtual office), we demonstrated that subjective fear and anxiety (assessed by ratings) generalize, whereas physiological responses (assessed by startle responses) generalized for fear, but not for anxiety (Andreatta et al. 2015b, 2017, 2019, 2020a, b; Muhlberger et al. 2014; Andreatta and Pauli 2019; Neueder et al. 2019). Contexts are complex stimuli, which can be perceived as whole (e.g., the office in which one works) or as single elements (e.g., the desk, the chairs, etc., Rudy

2009). The dissociation between physiological and subjective-verbal anxiety responses might be the result of this double context representation. In other words, participants may rely on their global representation when giving the ratings as they are asked to summarize the context in one score (e.g., how anxiogenic it is perceived). In contrast, physiological responses may rely on the elemental representation of the context as these responses promptly occur as the cues appear in the visual field (for a broader discussion, see Andreatta et al. 2020b; Andreatta and Pauli 2021).

Elevated trait anxiety has been linked to a higher risk for anxiety disorders (Sandi and Richter-Levin 2009), and in turn highly anxious individuals share altered learning mechanisms with patient groups. Thus, anxiety patients (Lissek et al. 2014) and highly anxious yet healthy individuals are less able to identify safety than healthy individuals with low levels of anxiety (Sep et al. 2019). Consequently, they feel threatened in numerous situations and defensively respond to cues or contexts which are actually safe. This tendency has been defined as overgeneralization of defensive responses (Lissek et al. 2005). Using virtual offices, we demonstrated that participants with high anxiety sensitivity showed more pronounced generalization of their subjective context conditioned anxiety than low anxiety sensitive individuals (Andreatta et al. 2020b). Considering that high anxiety sensitivity has been specifically linked to a higher risk for panic disorders (McNally 2002), these results suggest that being at risk for panic disorders may lead to a greater generalization of anxiety.

Feeling threatened facilitates avoidance behaviors (Lang 1995). In traditional laboratory research, avoidance behavior is often quantified using simple measures such as response time or response frequency (Krypotos et al. 2018; Pittig et al. 2018). However, in everyday situations avoidance usually includes a chain of movements which allows individuals to avoid or escape from a threat. VR allows the creation of *ecologically valid naturalistic settings* for measuring *avoidance behavior* in several ways; for example, by monitoring how often individuals enter a threatening context (see Fig. 3b, Glotzbach et al. 2012; Grillon et al. 2006), or how much time participants spend in a threatening vs. a control context (Childs et al. 2017). Such studies have revealed that the delay to pick up a virtual object (Betella et al. 2014) or to find an escape platform (Cornwell et al. 2013) is typically prolonged when a threatening cue is presented. Reduced walking distance within a threatening context (Gromer et al. 2018, 2021) or keeping one's distance from an aversive cue (Reichenberger et al. 2017, 2019, 2020) can be used as further indicators of fear-related behavior. Such avoidance behaviors seem to be mediated by the aversiveness of the threat conditions. In studies conducted in natural contexts, the more aversive a threatening context was rated, the more participants avoided it (Glotzbach et al. 2012). Moreover, stronger physiological responses at the end of an extinction phase predicted the amount of avoidance of a threatening cue (Cornwell et al. 2013). These results suggest that avoidance is stronger in those individuals who perceive threats to be more aversive, and this might be even more true for anxious individuals or anxiety patients. The use of VR significantly facilitates such research by providing the possibility to study relatively naturalistic avoidance behavior in well-controlled laboratory conditions.

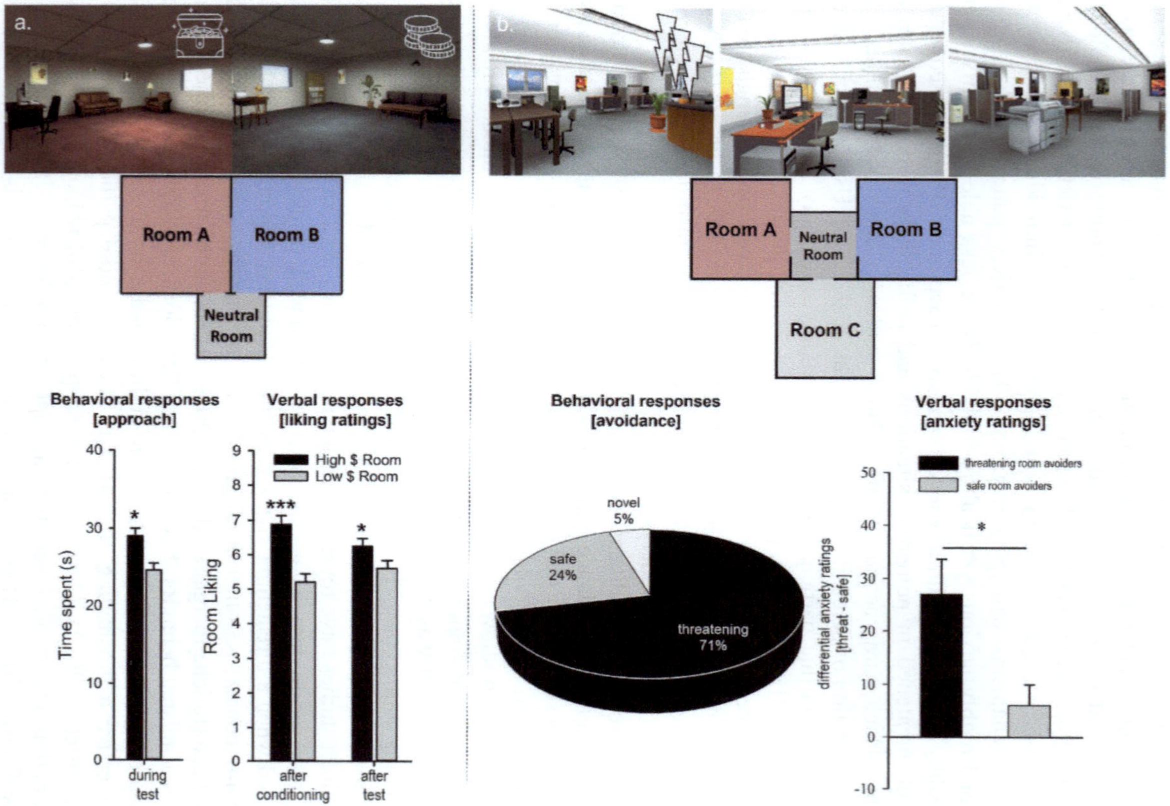

Fig. 3 *Approach and avoidance behavior after classical conditioning in VR*. (**a**) During appetitive conditioning, one virtual living room (High $ room) was associated with higher monetary reward (4.35$), while the other living room (Low $ room) was associated with a low reward (0.60$). During test, no monetary reward was delivered, while participants freely explored both virtual living rooms, which were connected by an open door (upper left panel) for 1 min. Participants spent more time and rated the High $ room more pleasant as compared to the Low $ room (lower left panel, Childs et al. 2017). (**b**) During aversive

Fig. 3 (continued) conditioning, mild painful electric stimulations were unpredictably delivered in one virtual office (threatening context), but never in the other virtual office (safe context). During test, no electric stimulation was delivered, and participants were asked to enter either the threat, the safe or a novel virtual office (upper right panels). The majority of the participants avoided the threatening context, and such behavioral choice was mediated by the ratings such that those participants who rated the threatening context as more anxiogenic were those who avoided the aversive virtual office most (lower right panels; Glotzbach, et al. 2012)

In sum, fear and anxiety are two distinct emotions elicited by imminent threat. By making it possible to create analogous experimental paradigms for humans and animals, VR holds great promise for studying these emotions and the interaction between cognitive and automatic mechanisms as well as the interplay between contextual information and responses linked to discrete cues. A comprehensive understanding of these interactions can greatly contribute to our knowledge on the etiology and maintenance of anxiety disorders.

5 Positive Emotions and Approach

Reward associated stimuli can elicit positive emotions and motivate approach behavior. VR has been used to study motivational responses evoked by naturally rewarding stimuli related to ingestion (Ferrer-Garcia et al. 2015), procreation (Milani et al. 2022), and nurturance (Mochizuki et al. 2017). Research on positive emotional states, like joy (Baños et al. 2012; Felnhofer et al. 2015) and relaxation (Pizzoli et al. 2019; Anderson et al. 2017), has made tremendous progress (Baños et al. 2017) and has recently started to address more complex emotional experiences such as awe (Chirico et al. 2018). This line of research does not only deepen our knowledge of the psychological architecture of appetitive motivation but is also highly relevant for our understanding of clinical disorders related to maladaptive reward processing and the development of novel treatment strategies (Stramba-Badiale et al. 2020; Trost et al. 2021; Baños et al. 2017).

With respect to clinical research questions, substance use disorders are frequently considered a prevalent example of a psychological disorder characterized by aberrant reward processing (Wise and Koob 2014; Robinson and Berridge 1993). Strong craving (i.e., an intense desire or urge to use a drug), compulsive drug-seeking/consumption, and a high propensity to relapse are hallmarks of substance use disorders and represent a fundamental problem for treatment (O'Brien 2005). These core features of the disorder are at least partly situationally specific and frequently evoked by stimuli related to drug-intake, commonly termed *drug cues* (Segawa et al. 2020; Betts et al. 2020; Hone-Blanchet et al. 2014).

Traditionally, human research focused on the effects of *naturalistic* (Betts et al. 2020; Winkler et al. 2011) drug stimuli (e.g., drug-paraphernalia, pictures, videos), in particular on discrete stimuli *proximal* to consumption (e.g., a freshly lit cigarette or a full glass of alcohol, Mucha et al. 2008; Conklin et al. 2008). However, the drug-intake ritual also comprises a multitude of stimuli and events taking place *prior* to or *after* the termination of consumption, with probably different biobehavioral effects (Mucha et al. 2008; Stippekohl et al. 2010; Nees et al. 2012). During the last decade, *distal contextual* stimuli (Conklin et al. 2008; Conklin 2006) have received heightened attention in non-VR research. As drugs are frequently consumed in certain settings (i.e., drug environments, like a bar, a bus station, at home), drug-associated contexts may motivate drug-seeking by themselves (Conklin et al. 2008; Childs and de Wit 2009) or in combination with proximal, discrete cues (Conklin et al. 2019).

VR considerably facilitates this line of research by providing scientists with a logistically convenient tool to create multisensory, three-dimensional drug-intake environments with high ecological validity, strict experimental control, and opportunities for interaction (Segawa et al. 2020; Hone-Blanchet et al. 2014).

Since early seminal VR studies (Bordnick et al. 2005; Jang et al. 2003), the assessment of cue-reactivity in VR has made considerable progress (e.g., Segawa et al. 2020; Hone-Blanchet et al. 2014). This line of research addresses the effects of complex drug-intake situations (e.g., combinations of discrete and contextual stimuli) on drug users (but also see, Paris et al. 2011). Importantly, VR also allows researchers to assess the effects of social stimuli with high experimental control (Segawa et al. 2020). This is particularly relevant for understanding stimulus control in substance use disorders, given that drug consumption frequently takes place in social contexts (de Wit and Sayette 2018). Thus, the presence of other people might promote drug-intake by functioning as a drug-associated stimulus, by altering the effects of a drug, or by enhancing the motivational value of social stimuli (de Wit and Sayette 2018; Segawa et al. 2020).

Overall, there is extensive evidence that virtual drug stimuli evoke subjective craving (Segawa et al. 2020; Hone-Blanchet et al. 2014), as indicated by research on nicotine (Pericot-Valverde et al. 2015), alcohol (Ryan et al. 2010), cannabis (Bordnick et al. 2009), and methamphetamine (Culbertson et al. 2010). Physiological measures of arousal, such as skin conductance or heart rate, are typically used as objective indices of the motivation to consume drugs (Segawa et al. 2020). Future studies could benefit from assessing the effects of drug stimuli on additional response systems as already realized in non-VR research. In particular, physiological measures of motivational valence, like the affect-modulated startle response (Mucha et al. 2008), facial muscle activation (Winkler et al. 2011; Wardle et al. 2018), or neural activity (Stolz et al. 2019; Andreatta et al. 2015a; Versace et al. 2017) might be promising.

Cue- or context-evoked craving and drug-seeking can occur long after the cessation of drug-intake, thereby provoking a risk for relapses (Bedi et al. 2011). Classic cue exposure treatments have been developed to ameliorate cue-reactivity based on the principles of extinction learning (Conklin and Tiffany 2002). During cue exposure, drug stimuli are repeatedly presented without the drug in an attempt to extinguish the association between them (Pericot-Valverde et al. 2014). VR applications have re-ignited research on the therapeutic efficacy of cue exposure treatments as they allow researchers a high degree of control over exposure to drug-related situations (Chen et al. 2018). Despite some mixed results, the findings appear to be promising in general (Segawa et al. 2020; Trahan et al. 2019). Future studies in VR may concentrate on elucidating the mechanisms underlying cue exposure treatments and on ways to further boost therapeutic outcomes (Liu et al. 2020; Chen et al. 2018).

Evidence for the assumption that drug-directed responding is based on associative learning stems from studies which experimentally manipulated the association between the conditioned stimuli and the drug (Winkler et al. 2011). In animals, the conditioned place preference procedure is widely used to investigate the incentive motivational properties of certain stimuli (Napier et al. 2013). This is assessed by measuring the behavioral preference for a physical environment (e.g., amount of time spent at a particular place) previously paired with an appetitive US compared to an environment not paired with the US. In this regard, VR might be a particularly useful tool for adapting the place preference procedure for translational research in humans (see Fig. 3a). Indeed, seminal non-VR studies confirmed a preference for a physical context previously paired with amphetamine or alcohol as indicated by self-reports (Childs and de Wit 2009, 2013) and behavior (Childs and de Wit 2016; Lutz and Childs 2021). VR studies adapted this approach to investigate the effects of music (Molet et al. 2013), food (Astur et al. 2014; van den Akker et al. 2013), monetary (Childs et al. 2017), and gaming point gains (Astur et al. 2016) on conditioned context preferences. Overall, this line of research revealed substantial evidence for a conditioned preference of the US paired environment as indicated by self-report and behavioral measures. Moreover, VR studies revealed a stronger place preference effects in individuals at risk for developing psychological disorders such as eating (Astur et al. 2015) or substance use disorders (Radell et al. 2016).

6 Conclusions

Taken together, VR research has shown great promise for the examination of emotional processing in health and disease. The main advantages of the technology concern (1) the ability to develop and to manipulate scenarios and virtual environments that resemble important characteristics of real emotionally laden conditions, (2) the opportunity to acquire multimodal data during confrontation with threatening and rewarding situations, and (3) the ability to interact with the environment such that realistic approach and avoidance behavior can be expressed and objectively measured.

However, from a methodological perspective it seems important to note that most studies described above have not yet fully realized the potential of VR technology. For instance, VR environments were frequently displayed on a computer screen (e.g., Palmisano and Astur 2020), participants were often passively guided through the environment (e.g., Andreatta et al. 2020a) or they used a joypad for navigation (e.g., Gromer et al. 2019). Moreover, virtual environments were frequently kept simple and emotional stimuli might have appeared to be relatively artificial (e.g., Reichenberger et al. 2017). Fortunately, these issues can be addressed due to the rapid technological advancement that has resulted from the entertainment industry's investments in VR. Devices such as high-quality HMDs are now available even for research groups with a small budget. The same applies to locomotion platforms or omnidirectional treadmills that allow for walking behavior in virtual environments. Finally, powerful software tools based on well-known game engines are increasingly available and allow for developing rich VR scenarios for psychological and neuroscientific research.

Although VR has been used successfully to study emotional processing in the past and will be an important research technology in the future, we are convinced that more systematic research is required to clarify not only the advantages but also the limits of this technology. For example, it is frequently hypothesized that VR allows for generating ecologically valid research environments, but these settings have rarely been compared to traditional laboratory designs or real-world scenarios. Thus, it is currently rather unclear to what degree human behavior and physiological responses in VR match corresponding responses in the real world (see also, Shiban et al. 2016; Wechsler et al. 2019). Research in other domains has provided evidence for qualitative differences in responding between laboratory and field settings (Großekathöfer et al. 2021; Rubo et al. 2020) and it seems important to test whether or under which conditions VR can effectively bride this gap (Rubo and Gamer 2021). Finally, ethical issues have to be considered as VR seems to be able to elicit strong emotional responses in a positive but also negative way (Slater et al. 2020). Thus, although participants may well recognize that they are confronted with a virtual situation, which can be rewarding or threatening, the emotional experience seems real and is accompanied by subjective, physiological, and behavioral changes that resemble emotional states in real life (Gromer et al. 2018).

To sum up, VR has emerged as an important tool for examining emotional responses under relatively unconstrained conditions in the laboratory. It provides crucial evidence on the interplay of neurophysiological, subjective, and behavioral responses and allows researchers to address translational research questions by generating comparable experimental conditions between species. It has therefore the potential to yield integrative knowledge for a more comprehensive understanding of emotion processing in basic science and in clinical and therapeutic settings (see also Ressler 2020).

References

Alvarez RP, Biggs A, Chen G, Pine DS, Grillon C (2008) Contextual fear conditioning in humans: cortical-hippocampal and amygdala contributions. J Neurosci 28:6211–6219. https://doi.org/10.1523/JNEUROSCI.1246-08.2008

Alvarez RP, Chen G, Bodurka J, Kaplan R, Grillon C (2011) Phasic and sustained fear in humans elicits distinct patterns of brain activity. NeuroImage 55:389–400. https://doi.org/10.1016/j.neuroimage.2010.11.057

Ammann J, Hartmann C, Peterhans V, Ropelato S, Siegrist M (2020) The relationship between disgust sensitivity and behaviour: a virtual reality study on food disgust. Food Qual Prefer 80:103833. https://doi.org/10.1016/j.foodqual.2019.103833

Anderson AP, Mayer MD, Fellows AM, Cowan DR, Hegel MT, Buckey JC (2017) Relaxation with immersive natural scenes presented using virtual reality *Aereospace medicine and human performance*, vol 88, pp 520–526. https://doi.org/10.3357/AMHP.4747.2017

Andreatta M, Pauli P (2019) Generalization of appetitive conditioned responses. Psychophysiology 56:e13397. https://doi.org/10.1111/psyp.13397

Andreatta M, Pauli P (2021) Contextual modulation of conditioned responses in humans: a review on virtual reality studies. Clin Psychol Rev 90:102095. https://doi.org/10.1016/j.cpr.2021.102095

Andreatta M, Glotzbach-Schoon E, Mühlberger A, Schulz SM, Wiemer J, Pauli P (2015a) Initial and sustained brain responses to contextual conditioned anxiety in humans. Cortex 63:352–363. https://doi.org/10.1016/j.cortex.2014.09.014

Andreatta M, Leombruni E, Glotzbach-Schoon E, Pauli P, Mühlberger A (2015b) Generalization of contextual fear in humans. Behav Ther 46:583–596. https://doi.org/10.1016/j.beth.2014.12.008

Andreatta M, Neueder D, Glotzbach-Schoon E, Mühlberger A, Pauli P (2017) Effects of context preexposure and delay until anxiety retrieval on generalization of contextual anxiety. Learn Mem 24:43–54. https://doi.org/10.1101/lm.044073.116

Andreatta M, Neueder D, Genheimer H, Schiele MA, Schartner C, Deckert J, Domschke K, Reif A, Wieser MJ, Pauli P (2019) Human BDNF rs6265 polymorphism as a mediator for the generalization of contextual anxiety. J Neurosci Res 97:300–312. https://doi.org/10.1002/jnr.24345

Andreatta M, Genheimer H, Wieser MJ, Pauli P (2020a) Context-dependent generalization of conditioned responses to threat and safety signals. Int J Psychophysiol 155:140–151. https://doi.org/10.1016/j.ijpsycho.2020.06.006

Andreatta M, Neueder D, Herzog K, Genheimer H, Schiele MA, Deckert J, Domschke K, Reif A, Wieser MJ, Pauli P (2020b) Generalization of conditioned contextual anxiety and the modulatory effects of anxiety sensitivity. Neurotherapeutics. https://doi.org/10.1007/s13311-020-00831-8

Astur RS, Carew AW, Deaton BE (2014) Conditioned place preferences in humans using virtual reality. Behav Brain Res 267:173–177. https://doi.org/10.1016/j.bbr.2014.03.018

Astur RS, Palmisano AN, Hudd EC, Carew AW, Deaton BE, Kuhney FS, Niezrecki RN, Santos M (2015) Pavlovian conditioning to food reward as a function of eating disorder risk. Behav Brain Res 291:277–282. https://doi.org/10.1016/j.bbr.2015.05.016

Astur RS, Palmisano AN, Carew AW, Deaton BE, Kuhney FS, Niezrecki RN, Hudd EC, Mendicino KL, Ritter CJ (2016) Conditioned place preferences in humans using secondary reinforcers. Behav Brain Res 297:15–19. https://doi.org/10.1016/j.bbr.2015.09.042

Baas JMP, Heitland I (2015) The impact of cue learning, trait anxiety and genetic variation in the serotonin 1A receptor on contextual fear. Int J Psychophysiol 98:506–514. https://doi.org/10.1016/j.ijpsycho.2014.10.016

Baas JM, Nugent M, Lissek S, Pine DS, Grillon C (2004) Fear conditioning in virtual reality contexts: a new tool for the study of anxiety. Biol Psychiatry 55:1056–1060. https://doi.org/10.1016/j.biopsych.2004.02.024

Baños RM, Liaño V, Botella C, Alcañiz M, Guerrero B, Rey B (2006) Changing induced moods via virtual reality. Berlin

Baños RM, Etchemendy E, Castilla D, García-Palacios A, Quero S, Botella C (2012) Positive mood induction procedures for virtual environments designed for elderly people☆. Interact Comput 24(3):131–138. https://doi.org/10.1016/j.intcom.2012.04.002

Baños RM, Carrillo A, Etchemendy E, Botella C (2017) Positive technologies for understanding and promoting positive emotions. Span J Psychol 20:E50. https://doi.org/10.1017/sjp.2017.42

Bedi G, Preston KL, Epstein DH, Heishman SJ, Marrone GF, Shaham Y, de Wit H (2011) Incubation of Cue-induced cigarette craving during abstinence in human smokers. Biol Psychiatry 69:708–711. https://doi.org/10.1016/j.biopsych.2010.07.014

Bernardo D, Pedro AB, Westwood S, Mograbi DC (2021) Mood induction using virtual reality: a systematic review of recent findings. J Technol Behav Sci 6:3–24. https://doi.org/10.1007/s41347-020-00152-9

Betella A, Zucca R, Cetnarski R, Greco A, Lanatà A, Mazzei D, Tognetti A, Arsiwalla XD, Omedas P, De Rossi D, Verschure PFMJ (2014) Inference of human affective states from psychophysiological measurements extracted under ecologically valid conditions. Front Neurosci 8(286). https://doi.org/10.3389/fnins.2014.00286

Betts JM, Dowd AN, Forney M, Hetelekides E, Tiffany ST (2020) A meta-analysis of Cue reactivity in tobacco cigarette smokers. Nicotine Tob Res 23:249–258. https://doi.org/10.1093/ntr/ntaa147

Biedermann SV, Biedermann DG, Wenzlaff F, Kurjak T, Nouri S, Auer MK, Wiedemann K, Briken P, Haaker J, Lonsdorf TB, Fuss J (2017) An elevated plus-maze in mixed reality for studying human anxiety-related behavior. BMC Biol 15:125. https://doi.org/10.1186/s12915-017-0463-6

Boecker L, Pauli P (2019) Affective startle modulation and psychopathology: implications for appetitive and defensive brain systems. Neurosci Biobehav Rev 103:230–266. https://doi.org/10.1016/j.neubiorev.2019.05.019

Bordnick PS, Graap KM, Copp HL, Brooks J, Ferrer M (2005) Virtual reality Cue reactivity assessment in cigarette smokers. Cyberpsychol Behav 8:487–492. https://doi.org/10.1089/cpb.2005.8.487

Bordnick PS, Copp HL, Traylor A, Graap KM, Carter BL, Walton A, Ferrer M (2009) Reactivity to cannabis cues in virtual reality environments. J Psychoactive Drugs 41:105–112. https://doi.org/10.1080/02791072.2009.10399903

Bouchard S, Rizzo A (2019) Applications of virtual reality in clinical psychology and clinical cognitive neuroscience – an introduction. In: Rizzo A, Bouchard S (eds) Virtual reality for psychological and neurocognitive interventions. Springer, Los Angeles

Bouchard S, St-Jacques J, Robillard G, Renaud P (2008) Anxiety increases the feeling of presence in virtual reality. Presence Teleop Virt 17:376–391. https://doi.org/10.1162/pres.17.4.376

Bouchard S, Robillard G, Larouche S, Loranger C (2012) Description of a treatment manual for in virtuo exposure with specific phobia. In: Eichenberg C (ed) Virtual reality in psychological, medical and pedagogical applications. Sigmund Freud University Vienna, Wien

Bouton ME (2002) Context, ambiguity, and unlearning: sources of relapse after behavioral extinction. Biol Psychiatry 52:976–986. https://doi.org/10.1016/S0006-3223(02)01546-9

Bouton ME, Moody EW (2004) Memory processes in classical conditioning. Neurosci Behav Rev 28:663–674. https://doi.org/10.1016/j.neubiorev.2004.09.001

Bradley MM, Lang PJ (1999) International affective digitized sounds (IADS): stimuli, instruction manual and affective ratings. Gainesville

Chen XJ, Wang DM, Zhou LD, Winkler M, Pauli P, Sui N, Li YH (2018) Mindfulness-based relapse prevention combined with virtual reality cue exposure for methamphetamine use disorder: study protocol for a randomized controlled trial. Contemp Clin Trials 70:99–105. https://doi.org/10.1016/j.cct.2018.04.006

Childs E, de Wit H (2009) Amphetamine-induced place preference in humans. Biol Psychiatry 65: 900–904. https://doi.org/10.1016/j.biopsych.2008.11.016

Childs E, de Wit H (2013) Contextual conditioning enhances the psychostimulant and incentive properties of d-amphetamine in humans. Addict Biol 18:985–992. https://doi.org/10.1111/j.1369-1600.2011.00416.x

Childs E, de Wit H (2016) Alcohol-induced place conditioning in moderate social drinkers. Addiction 111:2157–2165. https://doi.org/10.1111/add.13540

Childs E, Astur RS, de Wit H (2017) Virtual reality conditioned place preference using monetary reward. Behav Brain Res 322(Part A):110–114. https://doi.org/10.1016/j.bbr.2017.01.019

Chirico A, Ferrise F, Cordella L, Gaggioli A (2018) Designing awe in virtual reality: an experimental study. Front Psychol 8. https://doi.org/10.3389/fpsyg.2017.02351

Collange J, Guegan J (2020) Using virtual reality to induce gratitude through virtual social interaction. Comput Hum Behav 113:106473. https://doi.org/10.1016/j.chb.2020.106473

Conklin CA (2006) Environments as cues to smoke: implications for human extinction-based research and treatment. Exp Clin Psychopharmacol 14:12–19. https://doi.org/10.1037/1064-1297.14.1.12

Conklin CA, Tiffany ST (2002) Applying extinction research and theory to cue-exposure addiction treatments. Addiction 97:155–167. https://doi.org/10.1046/j.1360-0443.2002.00014.x

Conklin CA, Robin N, Perkins KA, Salkeld RP, Joseph McClernon F (2008) Proximal versus distal cues to smoke: the effects of environments on smokers' cue-reactivity. Exp Clin Psychopharmacol 16:207–214. https://doi.org/10.1037/1064-1297.16.3.207

Conklin CA, Joseph McClernon F, Vella EJ, Joyce CJ, Salkeld RP, Parzynski CS, Bennett L (2019) Combined smoking cues enhance reactivity and predict immediate subsequent smoking. Nicotine Tob Res 21:241–248. https://doi.org/10.1093/ntr/nty009

Cornwell BR, Overstreet C, Krimsky M, Grillon C (2013) Passive avoidance is linked to impaired fear extinction in humans. Learn Mem 20(3):164–169. https://doi.org/10.1101/lm.028902.112

Craske MG, Rauch SL, Ursano R, Prenoveau J, Pine DS, Zinbarg RE (2009) What is an anxiety disorder? Depress Anxiety 26:1066–1085. https://doi.org/10.1002/da.20633

Culbertson C, Nicolas S, Zaharovits I, London ED, De La Garza R, Brody AL, Newton TF (2010) Methamphetamine craving induced in an online virtual reality environment. Pharmacol Biochem Behav 96:454–460. https://doi.org/10.1016/j.pbb.2010.07.005

Cummings JJ, Bailenson JN (2016) How immersive is enough? A meta-analysis of the effect of immersive technology on user presence. Media Psychol 19:272–309. https://doi.org/10.1080/15213269.2015.1015740

Daniel-Watanabe L, Fletcher PC (2022) Are fear and anxiety truly distinct? Biol Psychiatry Glob Open Sci 2(4):341–349. https://doi.org/10.1016/j.bpsgos.2021.09.006

Davis M, Walker DL, Miles L, Grillon C (2010) Phasic vs sustained fear in rats and humans: role of the extended amygdala in fear vs anxiety. Neuropsychopharmacology 35:105–135. https://doi.org/10.1038/npp.2009.109

de Wit H, Sayette M (2018) Considering the context: social factors in responses to drugs in humans. Psychopharmacology 235:935–945. https://doi.org/10.1007/s00213-018-4854-3

Diemer J, Alpers GW, Peperkorn HM, Shiban Y, Mühlberger A (2015) The impact of perception and presence on emotional reactions: a review of research in virtual reality. Front Psychol 6. https://doi.org/10.3389/fpsyg.2015.00026

Duits P, Cath DC, Heitland I, Baas JMP (2016) High current anxiety symptoms, but not a past anxiety disorder diagnosis, are associated with impaired fear extinction. Front Psychol 7(252). https://doi.org/10.3389/fpsyg.2016.00252

Dunsmoor JE, Paz R (2015) Fear generalization and anxiety: behavioral and neural mechanisms. Biol Psychiatry 78:336–343. https://doi.org/10.1016/j.biopsych.2015.04.010

Dymond S, Dunsmoor JE, Vervliet B, Roche B, Hermans D (2015) Fear generalization in humans: systematic review and implications for anxiety disorder research. Behav Ther 46:561–582. https://doi.org/10.1016/j.beth.2014.10.001

Ekman P (2016) What scientists who study emotion agree about. Perspect Psychol Sci 11:31–34. https://doi.org/10.1177/1745691615596992

Ekman P, Friesen WV (1971) Constants across cultures in the face and emotion. J Pers Soc Psychol 17:124–129. https://doi.org/10.1037/h0030377

Fanselow MS (2018) The role of learning in threat imminence and defensive behaviors. Curr Opin Behav Sci 24:44–49. https://doi.org/10.1016/j.cobeha.2018.03.003

Felnhofer A, Kothgassner OD, Schmidt M, Heinzle A-K, Beutl L, Hlavacs H, Kryspin-Exner I (2015) Is virtual reality emotionally arousing? Investigating five emotion inducing virtual park scenarios. Int J Hum-Comput Stud 82:48–56. https://doi.org/10.1016/j.ijhcs.2015.05.004

Fendt M, Fanselow MS (1999) The neuroanatomical and neurochemical basis of conditioned fear. Neurosci Biobehav Rev 23:743–760

Fernández-Aguilar L, Navarro-Bravo B, Ricarte J, Ros L, Latorre JM (2019) How effective are films in inducing positive and negative emotional states? A meta-analysis. PLoS One 14: e0225040. https://doi.org/10.1371/journal.pone.0225040

Ferrer RA, Grenen EG, Taber JM (2015) Effectiveness of internet-based affect induction procedures: a systematic review and meta-analysis. Emotion 15:752–762. https://doi.org/10.1037/emo0000035

Ferrer-Garcia M, Gutierrez-Maldonado J, Treasure J, Vilalta-Abella F (2015) Craving for food in virtual reality scenarios in non-clinical sample: analysis of its relationship with body mass index and eating disorder symptoms. Eur Eat Disord Rev 23(5):371–378. https://doi.org/10.1002/erv.2375

Fox E (2008) Emotion science: cognitive and neuroscientific approaches to understanding human emotions. Palgrave Macmillan, Basingstoke

Fullana MA, Harrison BJ, Soriano-Mas C, Vervliet B, Cardoner N, Avila-Parcet A, Radua J (2016) Neural signatures of human fear conditioning: an updated and extended meta-analysis of fMRI studies. Mol Psychiatry:1–9. https://doi.org/10.1038/mp.2015.88

Gall D, Latoschik ME (2018) The effect of haptic prediction accuracy on presence. 2018 IEEE conference on virtual reality and 3D user interfaces (VR)

Glotzbach E, Ewald H, Andreatta M, Pauli P, Mühlberger A (2012) Contextual fear conditioning effects predict subsequent avoidance behavior. Cognit Emot 26:1256–1272. https://doi.org/10.1080/02699931.2012.656581

Gorini A, Griez E, Petrova A, Riva G (2010) Assessment of the emotional responses produced by exposure to real food, virtual food and photographs of food in patients affected by eating disorders. Ann General Psychiatry 9:30. https://doi.org/10.1186/1744-859X-9-30

Gray JA (1990) Brain systems that mediate both emotion and cognition. Cognit Emot 4:269–288. https://doi.org/10.1080/02699939008410799

Grillon C, Baas JM, Cornwell B, Johnson L (2006) Context conditioning and behavioral avoidance in a virtual reality environment: effect of predictability. Biol Psychiatry 60:752–759. https://doi.org/10.1016/j.biopsych.2006.03.072

Gromer D, Madeira O, Gast P, Nehfischer M, Jost M, Müller M, Mühlberger A, Pauli P (2018) Height simulation in a virtual reality CAVE system: validity of fear responses and effects of an immersion manipulation. Front Hum Neurosci 12. https://doi.org/10.3389/fnhum.2018.00372

Gromer D, Reinke M, Christner I, Pauli P (2019) Causal interactive links between presence and fear in virtual reality height exposure. Front Psychol 10. https://doi.org/10.3389/fpsyg.2019.00141

Gromer D, Kiser DP, Pauli P (2021) Thigmotaxis in a virtual human open field test. Sci Rep 11(1):6670. https://doi.org/10.1038/s41598-021-85678-5

Großekathöfer JD, Seis C, Gamer M (2021) Reality in a sphere: a direct comparison of social attention in the laboratory and the real world. Behav Res Methods 54:2286. https://doi.org/10.3758/s13428-021-01724-0

Haaker J, Maren S, Andreatta M, Merz CJ, Jan Richter S, Richter H, Drexler SM, Lange MD, Jüngling K, Nees F, Seidenbecher T, Fullana MA, Wotjak CT, Lonsdorf TB (2019) Making translation work: harmonizing cross-species methodology in the behavioural neuroscience of Pavlovian fear conditioning. Neurosci Biobehav Rev 107:329–345. https://doi.org/10.1016/j.neubiorev.2019.09.020

Hamm AO, Weike AI (2005) The neuropsychology of fear learning and fear regulation. Int J Psychophysiol 57:5–14. https://doi.org/10.1016/j.ijpsycho.2005.01.006

Heitland I, Klumpers F, Oosting RS, Evers DJJ, Leon Kenemans J, Baas JMP (2012) Failure to extinguish fear and genetic variability in the human cannabinoid receptor 1. Transl Psychiatry 2:e162. http://www.nature.com/tp/journal/v2/n9/suppinfo/tp201290s1.html

Herrero R, Garcia-Palacios A, Castilla D, Molinari G, Botella G (2014) Virtual reality for the induction of positive emotions in the treatment of fibromyalgia: a pilot study over acceptability, satisfaction, and the effect of virtual reality on mood. Cyberpsychol Behav Soc Netw 17:379–384. https://doi.org/10.1089/cyber.2014.0052

Higuera-Trujillo JL, Maldonado JL-T, Millán CL (2017) Psychological and physiological human responses to simulated and real environments: a comparison between photographs, 360° panoramas, and virtual reality. Appl Ergon 65:398–409. https://doi.org/10.1016/j.apergo.2017.05.006

Hone-Blanchet A, Wensing T, Fecteau S (2014) The use of virtual reality in craving assessment and Cue-exposure therapy in substance use disorders. Front Hum Neurosci 8. https://doi.org/10.3389/fnhum.2014.00844

Huff N, Hernandez JA, Fecteau M, Zielinski D, Brady R, LaBar KS (2011) Revealing context-specific conditioned fear memories with full immersion virtual reality. Front Behav Neurosci 5. https://doi.org/10.3389/fnbeh.2011.00075

Jang HL, Jeonghun K, Kwanguk K, Kim Byoungnyun Y, In K, Byung-Hwan Y, Seok HK, Wiederhold Brenda K, Wiederhold Mark D, Dong-Woo P, Youngsik L, Kim Sun I (2003) Experimental application of virtual reality for nicotine craving through Cue exposure. Cyberpsychol Behav 6(3):275–280. https://doi.org/10.1089/109493103322011560

Jovanovic T, Kazama A, Bachevalier J, Davis M (2012) Impaired safety signal learning may be a biomarker of PTSD. Neuropharmacology 62:695–704. https://doi.org/10.1016/j.neuropharm.2011.02.023

Kenealy PM (1986) The Velten mood induction procedure: a methodological review. Motiv Emot 10:315–335. https://doi.org/10.1007/BF00992107

Kim J, André E (2008) Emotion recognition based on physiological changes in music listening. IEEE Trans Pattern Anal Mach Intell 30:2067–2083. https://doi.org/10.1109/TPAMI.2008.26

Kinateder M, Müller M, Jost M, Mühlberger A, Pauli P (2014) Social influence in a virtual tunnel fire – influence of conflicting information on evacuation behavior. Appl Ergon 45:1649–1659. https://doi.org/10.1016/j.apergo.2014.05.014

Kinateder M, Gromer D, Gast P, Buld S, Müller M, Jost M, Nehfischer M, Mühlberger A, Pauli P (2015) The effect of dangerous goods transporters on hazard perception and evacuation behavior – a virtual reality experiment on tunnel emergencies. Fire Saf J 78:24–30. https://doi.org/10.1016/j.firesaf.2015.07.002

Konrad A, Tucker S, Crane J, Whittaker S (2016) Technology and reflection: mood and memory mechanisms for well-being. Psychol Well-Being 6:5–5. https://doi.org/10.1186/s13612-016-0045-3

Krisch KA, Bandarian-Balooch S, Neumann DL, Zhong J (2020) Eliciting and attenuating reinstatement of fear: effects of an unextinguished CS. Learn Motiv 71:101650. https://doi.org/10.1016/j.lmot.2020.101650

Krypotos A-M, Vervliet B, Engelhard IM (2018) The validity of human avoidance paradigms. Behav Res Ther 111:99–105. https://doi.org/10.1016/j.brat.2018.10.011

Lang PJ (1995) The emotion probe. Studies of motivation and attention. Am Psychol 50:372–385

LeDoux JE, Pine DS (2016) Using neuroscience to help understand fear and anxiety: a two-system framework. Am J Psychiatr 173:1083–1093. https://doi.org/10.1176/appi.ajp.2016.16030353

Lissek S, Powers AS, McClure EB, Phelps EA, Woldehawariat G, Grillon C, Pine DS (2005) Classical fear conditioning in the anxiety disorders: a meta-analysis. Behav Res Ther 43:1391–1424. https://doi.org/10.1016/j.brat.2004.10.007

Lissek S, Kaczkurkin AN, Rabin S, Geraci M, Pine DS, Grillon C (2014) Generalized anxiety disorder is associated with overgeneralization of classically conditioned fear. Biol Psychiatry 75:909–915. https://doi.org/10.1016/j.biopsych.2013.07.025

Liu W, Chen X-J, Wen Y-T, Winkler MH, Paul P, He Y-L, Wang L, Chen H-X, Li Y-H (2020) Memory retrieval-extinction combined with virtual reality reducing drug craving for methamphetamine: study protocol for a randomized controlled trial. Front Psych 11. https://doi.org/10.3389/fpsyt.2020.00322

Lonsdorf TB, Menz MM, Andreatta M, Fullana MA, Golkar A, Haaker J, Heitland I, Hermann A, Kuhn M, Kruse O, Drexler SM, Meulders A, Nees F, Pittig A, Richter J, Römer S, Shiban Y, Schmitz A, Straube B, Vervliet B, Wendt J, Baas JMP, Merz CJ (2017) Don't fear 'fear conditioning': methodological considerations for the design and analysis of studies on human fear acquisition, extinction, and return of fear. Neurosci Biobehav Rev 77:247–285. https://doi.org/10.1016/j.neubiorev.2017.02.026

Lutz JA, Childs E (2021) Alcohol conditioned contexts enhance positive subjective alcohol effects and consumption. Behav Process 187:104340. https://doi.org/10.1016/j.beproc.2021.104340

Madeira O, Gromer D, Latoschik ME, Pauli P (2021) Effects of acrophobic fear and trait anxiety on human behavior in a virtual elevated plus-maze. Front Virtual Real 2. https://doi.org/10.3389/frvir.2021.635048

Marcusson-Clavertz D, Kjell ONE, Persson SD, Cardeña E (2019) Online validation of combined mood induction procedures. PLoS One 14:e0217848. https://doi.org/10.1371/journal.pone.0217848

Maren S, Luan Phan K, Liberzon I (2013) The contextual brain: implications for fear conditioning, extinction and psychopathology. Nat Rev Neurosci 14:417–428. https://doi.org/10.1038/nrn3492

Marín-Morales J, Higuera-Trujillo JL, Greco A, Guixeres J, Llinares C, Gentili C, Scilingo EP, Alcañiz M, Valenza G (2019) Real vs. immersive-virtual emotional experience: analysis of psycho-physiological patterns in a free exploration of an art museum. PLoS One 14:e0223881. https://doi.org/10.1371/journal.pone.0223881

Markowitz DM, Bailenson JN (in press) Virtual reality and emotion: a 5-year systematic review of empirical research (2015-2019). In: Nabi R, Myrick J (eds) Our online emotional selves: the link between digital media and emotional experience. Oxford University Press

Martin M (1990) On the induction of mood. Clin Psychol Rev 10:669–697. https://doi.org/10.1016/0272-7358(90)90075-L

McNally RJ (2002) Anxiety sensitivity and panic disorder. Biol Psychiatry 52:938–946. https://doi.org/10.1016/S0006-3223(02)01475-0

Meehan M, Insko B, Whitton M, Brooks FP (2002) Physiological measures of presence in stressful virtual environments. Proceedings of the 29th annual conference on computer graphics and interactive techniques, San Antonio

Milad MR, Quirk G (2012) Fear extinction as a model for translational neuroscience: ten years of progress. Annu Rev Psychol 63:129–151. https://doi.org/10.1146/annurev.psych.121208.131631

Milani S, Jabs F, Brown NB, Zdaniuk B, Kingstone A, Brotto LA (2022) Virtual reality erotica: exploring general presence, sexual presence, sexual arousal, and sexual desire in women. Arch Sex Behav 51(1):565–576. https://doi.org/10.1007/s10508-021-02111-8

Milgram P, Kishino F (1994) A taxonomy of mixed reality visual displays. IEICE Trans Inf Syst 77:1321–1329

Mineka S, Oehlberg K (2008) The relevance of recent developments in classical conditioning to understanding the etiology and maintenance of anxiety disorders. Acta Psychol 127:567–580

Mochizuki Y, Hsieh R, Agatsuma D, Asano T, Higashida M, Nishikizawa T, Shirai A (2017) Real baby – real family: holdable tangible baby VR. Proceedings of the virtual reality international conference – Laval virtual 2017, Laval, France

Molet M, Billiet G, Bardo MT (2013) Conditioned place preference and aversion for music in a virtual reality environment. Behav Process 92:31–35. https://doi.org/10.1016/j.beproc.2012.10.001

Mucha RF, Pauli P, Weber M, Winkler M (2008) Smoking stimuli from the terminal phase of cigarette consumption may not be cues for smoking in healthy smokers. Psychopharmacology 201:81–95. https://doi.org/10.1007/s00213-008-1249-x

Mühlberger A, Herrmann MJ, Wiedemann G, Ellgring H, Pauli P (2001) Repeated exposure of flight phobics to flights in virtual reality. Behav Res Ther 39(9):1033–1050. https://doi.org/10.1016/S0005-7967(00)00076-0

Mühlberger A, Bülthoff HH, Wiedemann G, Pauli P (2007) Virtual reality for the psychophysiological assessment of phobic fear: responses during virtual tunnel driving. Psychol Assess 19:340–346. https://doi.org/10.1037/1040-3590.19.3.340

Muhlberger A, Neumann R, Wieser MJ, Pauli P (2008) The impact of changes in spatial distance on emotional responses. Emotion 8:192–198

Mühlberger A, Wieser MJ, Gerdes ABM, Frey MCM, Weyers P, Pauli P (2011) Stop looking angry and smile, please: start and stop of the very same facial expression differentially activate threat- and reward related brain networks. Soc Cogn Affect Neurosci 6:321–329. https://doi.org/10.1093/scan/nsq039

Muhlberger A, Andreatta M, Ewald H, Glotzbach-Schoon E, Troger C, Baumann C, Reif A, Deckert J, Pauli P (2014) The BDNF Val66Met polymorphism modulates the generalization of cued fear responses to a novel context. Neuropsychopharmacology 39:1187–1195. https://doi.org/10.1038/npp.2013.320

Napier TC, Herrold AA, de Wit H (2013) Using conditioned place preference to identify relapse prevention medications. Neurosci Biobehav Rev 37:2081–2086. https://doi.org/10.1016/j.neubiorev.2013.05.002

Nees F, Diener C, Smolka MN, Flor H (2012) The role of context in the processing of alcohol-relevant cues. Addict Biol 17:441–451. https://doi.org/10.1111/j.1369-1600.2011.00347.x

Neueder D, Andreatta M, Pauli P (2019) Contextual fear conditioning and fear generalization in individuals with panic attacks. Front Behav Neurosci 13. https://doi.org/10.3389/fnbeh.2019.00152

Nilsson NC, Serafin S, Steinicke F, Nordahl R (2018) Natural walking in virtual reality: a review. Comput Entertain 16.:Article 8:1. https://doi.org/10.1145/3180658

O'Brien CP (2005) Anticraving medications for relapse prevention: a possible new class of psychoactive medications. Am J Psychiatr 162:1423–1431. https://doi.org/10.1176/appi.ajp.162.8.1423

Orefice P-H, Ammi M, Hafez M, Tapus A (2017) Design of an emotion elicitation tool using VR for human-avatar interaction studies. Cham

Pallavicini F, Pepe A, Ferrari A, Garcea G, Zanacchi A, Mantovani F (2020) What is the relationship among positive emotions, sense of presence, and ease of interaction in virtual reality systems? An on-site evaluation of a commercial virtual experience. PRESENCE Virtual Augmented Reality 27:183–201. https://doi.org/10.1162/pres_a_00325

Palmisano AN, Astur RS (2020) Nicotine facilitation of conditioned place preference to food reward in humans. Subst Use Misuse 55:2156–2164. https://doi.org/10.1080/10826084.2020.1795682

Paris MM, Carter BL, Traylor AC, Bordnick PS, Day SX, Armsworth MW, Cinciripini PM (2011) Cue reactivity in virtual reality: the role of context. Addict Behav 36:696–699. https://doi.org/10.1016/j.addbeh.2011.01.029

Pavlov IP (1927) Conditioned reflexes: an investigation of the physiological activity of the cerebral cortex. Oxford University Press, London

Pericot-Valverde I, Secades-Villa R, Gutiérrez-Maldonado J, García-Rodríguez O (2014) Effects of systematic Cue exposure through virtual reality on cigarette craving. Nicotine Tob Res 16:1470–1477. https://doi.org/10.1093/ntr/ntu104

Pericot-Valverde I, Germeroth LJ, Tiffany ST (2015) The use of virtual reality in the production of Cue-specific craving for cigarettes: a meta-analysis. Nicotine Tob Res 18:538–546. https://doi.org/10.1093/ntr/ntv216

Perusini JN, Fanselow MS (2015) Neurobehavioral perspectives on the distinction between fear and anxiety. Learn Mem 22:417–425. https://doi.org/10.1101/lm.039180.115

Pittig A, Treanor M, LeBeau RT, Craske MG (2018) The role of associative fear and avoidance learning in anxiety disorders: gaps and directions for future research. Neurosci Biobehav Rev 88:117–140. https://doi.org/10.1016/j.neubiorev.2018.03.015

Pizzoli SF, Maria KM, Triberti S, Monzani D, Raya MLA, Pravettoni G (2019) User-centered virtual reality for promoting relaxation: an innovative approach. Front Psychol 10. https://doi.org/10.3389/fpsyg.2019.00479

Price M, Mehta N, Tone EB, Anderson PL (2011) Does engagement with exposure yield better outcomes? Components of presence as a predictor of treatment response for virtual reality exposure therapy for social phobia. J Anxiety Disord 25:763–770. https://doi.org/10.1016/j.janxdis.2011.03.004

Quirk GJ, Mueller D (2008) Neural mechanisms of extinction learning and retrieval. Neuropsychopharmacology 33:56–72

Radell ML, Myers CE, Beck KD, Moustafa AA, Allen MT (2016) The personality trait of intolerance to uncertainty affects behavior in a novel computer-based conditioned place preference task. Front Psychol 7. https://doi.org/10.3389/fpsyg.2016.01175

Reichenberger J, Porsch S, Wittmann J, Zimmermann V, Shiban Y (2017) Social fear conditioning paradigm in virtual reality: social vs. electrical aversive conditioning. Front Psychol 8. https://doi.org/10.3389/fpsyg.2017.01979

Reichenberger J, Pfaller M, Forster D, Gerczuk J, Shiban Y, Mühlberger A (2019) Men scare me more: gender differences in social fear conditioning in virtual reality. Front Psychol 10(1617). https://doi.org/10.3389/fpsyg.2019.01617

Reichenberger J, Pfaller M, Mühlberger A (2020) Gaze behavior in social fear conditioning: an eye-tracking study in virtual reality. Front Psychol 11:35. https://doi.org/10.3389/fpsyg.2020.00035

Ressler KJ (2020) Translating across circuits and genetics toward progress in fear- and anxiety-related disorders. Am J Psychiatr 177:214–222. https://doi.org/10.1176/appi.ajp.2020.20010055

Riva G, Mantovani F, Capideville CS, Preziosa A, Morganti F, Villani D, Gaggioli A, Botella C, Alcaniz M (2007) Affective interactions using virtual reality: the link between presence and emotions. Cyberpsychol Behav 10:45–56. https://doi.org/10.1089/cpb.2006.9993

Robinson TE, Berridge KC (1993) The neural basis of drug craving: an incentive-sensitization theory of addiction. Brain Res Rev 18:247–291. https://doi.org/10.1016/0165-0173(93)90013-P

Roelofs K, Hagenaars MA, Stins J (2010) Facing freeze:social threat induces bodily freeze in humans. Psychol Sci 21:1575–1581. https://doi.org/10.1177/0956797610384746

Rubo M, Gamer M (2021) Stronger reactivity to social gaze in virtual reality compared to a classical laboratory environment. Br J Psychol 112:301–314. https://doi.org/10.1111/bjop.12453

Rubo M, Huestegge L, Gamer M (2020) Social anxiety modulates visual exploration in real life – but not in the laboratory. Br J Psychol 111:233–245. https://doi.org/10.1111/bjop.12396

Rudy JW (2009) Context representations, context functions, and the parahippocampal–hippocampal system. Learn Mem 16:573–585. https://doi.org/10.1101/lm.1494409

Ryan JJ, Kreiner DS, Chapman MD, Stark-Wroblewski K (2010) Virtual reality cues for binge drinking in college students. Cyberpsychol Behav Soc Netw 13:159–162. https://doi.org/10.1089/cyber.2009.0211

Sanchez-Vives MV, Slater M (2005) From presence to consciousness through virtual reality. Nat Rev Neurosci 6:332–339. https://doi.org/10.1038/nrn1651

Sandi C, Richter-Levin G (2009) From high anxiety trait to depression: a neurocognitive hypothesis. Trends Neurosci 32:312–320. https://doi.org/10.1016/j.tins.2009.02.004

Schubert TW (2006) The sense of presence in virtual environments. Zeitschrift für Medienpsychologie 15:69–71. https://doi.org/10.1026//1617-6383.15.2.69

Segawa T, Baudry T, Bourla A, Blanc J-V, Peretti C-S, Mouchabac S, Ferreri F (2020) Virtual reality (VR) in assessment and treatment of addictive disorders: a systematic review. Front Neurosci 13. https://doi.org/10.3389/fnins.2019.01409

Sep MSC, Steenmeijer A, Kennis M (2019) The relation between anxious personality traits and fear generalization in healthy subjects: a systematic review and meta-analysis. Neurosci Biobehav Rev 107:320–328. https://doi.org/10.1016/j.neubiorev.2019.09.029

Shiban Y, Pauli P, Mühlberger A (2013) Effect of multiple context exposure on renewal in spider phobia. Behav Res Ther 51:68–74. https://doi.org/10.1016/j.brat.2012.10.007

Shiban Y, Schelhorn I, Pauli P, Mühlberger A (2015) Effect of combined multiple contexts and multiple stimuli exposure in spider phobia: a randomized clinical trial in virtual reality. Behav Res Ther 71:45–53. https://doi.org/10.1016/j.brat.2015.05.014

Shiban Y, Dieme J, Brandl S, Zack R, Mühlberger A, Wüst S (2016) Trier social stress test in vivo and in virtual reality: dissociation of response domains. Int J Psychophysiol 110:47–55. https://doi.org/10.1016/j.ijpsycho.2016.10.008

Simon-Kutscher K, Wanke N, Hiller C, Schwabe L (2019) Fear without context: acute stress modulates the balance of Cue-dependent and contextual fear learning. Psychol Sci 30(8): 1123–1135. https://doi.org/10.1177/0956797619852027

Slater M (2002) Presence and the sixth sense. Presence Teleop Virt 11:435–439. https://doi.org/10.1162/105474602760204327

Slater M (2009) Place illusion and plausibility can lead to realistic behaviour in immersive virtual environments. Philos Trans R Soc B Biol Sci 364:3549–3557. https://doi.org/10.1098/rstb.2009.0138

Slater M, Usoh M (1993) Representations systems, perceptual position, and presence in immersive virtual environments. Presence Teleoper Virtual Environ 2:221–233. https://doi.org/10.1162/pres.1993.2.3.221

Slater M, Wilbur S (1997) A framework for immersive virtual environments (FIVE): speculations on the role of presence in virtual environments. Presence Teleop Virt 6:603–616. https://doi.org/10.1162/pres.1997.6.6.603

Slater M, Gonzalez-Liencres C, Haggard P, Vinkers C, Gregory-Clarke R, Jelley S, Watson Z, Breen G, Schwarz R, Steptoe W, Szostak D, Halan S, Fox D, Silver J (2020) The ethics of realism in virtual and augmented reality. Front Virtual Real 1. https://doi.org/10.3389/frvir.2020.00001

Soleymani M, Pantic M, Pun T (2015) Multimodal emotion recognition in response to videos (extended abstract). 2015 international conference on affective computing and intelligent interaction (ACII), 21–24 Sept 2015

Stippekohl B, Winkler M, Mucha RF, Pauli P, Walter B, Vaitl D, Stark R (2010) Neural responses to BEGIN- and END-stimuli of the smoking ritual in nonsmokers, nondeprived smokers, and deprived smokers. Neuropsychopharmacology 35:1209–1225. https://doi.org/10.1038/npp.2009.227

Stolz C, Endres D, Mueller EM (2019) Threat-conditioned contexts modulate the late positive potential to faces – a mobile EEG/virtual reality study. Psychophysiology 56:e13308. https://doi.org/10.1111/psyp.13308

Stramba-Badiale C, Mancuso V, Cavedoni S, Pedroli E, Cipresso P, Riva G (2020) Transcranial magnetic stimulation meets virtual reality: the potential of integrating brain stimulation with a simulative technology for food addiction. Front Neurosci 14. https://doi.org/10.3389/fnins.2020.00720

Tovote P, Fadok JP, Luthi A (2015) Neuronal circuits for fear and anxiety. Nat Rev Neurosci 16:317–331. https://doi.org/10.1038/nrn3945

Trahan MH, Maynard BR, Smith KS, Farina ASJ, Khoo YM (2019) Virtual reality exposure therapy on alcohol and nicotine: a systematic review. Res Soc Work Pract 29:876–891. https://doi.org/10.1177/1049731518823073

Trost Z, France C, Anam M, Shum C (2021) Virtual reality approaches to pain: toward a state of the science. Pain 162(2)

Valtchanov D, Barton KR, Ellard C (2010) Restorative effects of virtual nature settings. Cyberpsychol Behav Soc Netw 13(5):503–512. https://doi.org/10.1089/cyber.2009.0308

van den Akker K, Jansen A, Frentz F, Havermans RC (2013) Impulsivity makes more susceptible to overeating after contextual appetitive conditioning. Appetite 70:73–80. https://doi.org/10.1016/j.appet.2013.06.092

Västfjäll D (2001) Emotion induction through music: a review of the musical mood induction procedure. Music Sci 5:173–211. https://doi.org/10.1177/10298649020050s107

Versace F, Engelmann JM, Deweese MM, Robinson JD, Green CE, Lam CY, Minnix JA, Karam-Hage MA, Wetter DW, Schembre SM, Cinciripini PM (2017) Beyond Cue reactivity: non-drug-related motivationally relevant stimuli are necessary to understand reactivity to drug-related cues. Nicotine Tob Res 19:663–669. https://doi.org/10.1093/ntr/ntx002

Wardle MC, Lopez-Gamundi P, Flagel SB (2018) Measuring appetitive conditioned responses in humans. Physiol Behav 188:140–150. https://doi.org/10.1016/j.physbeh.2018.02.004

Wechsler TF, Kümpers F, Mühlberger A (2019) Inferiority or even superiority of virtual reality exposure therapy in phobias? – a systematic review and quantitative meta-analysis on randomized controlled trials specifically comparing the efficacy of virtual reality exposure to gold standard in vivo exposure in agoraphobia, specific phobia, and social phobia. Front Psychol 10. https://doi.org/10.3389/fpsyg.2019.01758

Westermann R, Spies K, Stahl G, Hesse FW (1996) Relative effectiveness and validity of mood induction procedures: a meta-analysis. Eur J Soc Psychol 26:557–580. https://doi.org/10.1002/(SICI)1099-0992(199607)26:4<557::AID-EJSP769>3.0.CO;2-4

Weyers P, Mühlberger A, Kund A, Hess U, Pauli P (2009) Modulation of facial reactions to avatar emotional faces by nonconscious competition priming. Psychophysiology 46:328–335. https://doi.org/10.1111/j.1469-8986.2008.00771.x

Winkler MH, Weyers P, Mucha RF, Stippekohl B, Stark R, Pauli P (2011) Conditioned cues for smoking elicit preparatory responses in healthy smokers. Psychopharmacology 213:781–789. https://doi.org/10.1007/s00213-010-2033-2

Wirth W, Hartmann T, Böcking S, Vorderer P, Klimmt C, Schramm H, Saari T, Laarni J, Ravaja N, Gouveia FR, Biocca F, Sacau A, Jäncke L, Baumgartner T, Jäncke P (2007) A process model of the formation of spatial presence experiences. Media Psychol 9:493–525. https://doi.org/10.1080/15213260701283079

Wise RA, Koob GF (2014) The development and maintenance of drug addiction. Neuropsychopharmacology 39:254–262. https://doi.org/10.1038/npp.2013.261

Yalachkov Y, Kaiser J, Naumer MJ (2012) Functional neuroimaging studies in addiction: multisensory drug stimuli and neural cue reactivity. Neurosci Biobehav Rev 36:825–835. https://doi.org/10.1016/j.neubiorev.2011.12.004

Zhang X, Yu HW, Barrett LF (2014) How does this make you feel? A comparison of four affect induction procedures. Front Psychol 5. https://doi.org/10.3389/fpsyg.2014.00689

VR for Cognition and Memory

Nicco Reggente

Contents

Abstract This chapter will provide a review of research into human cognition through the lens of VR-based paradigms for studying memory. Emphasis is placed on why VR increases the ecological validity of memory research and the implications of such enhancements.

Keywords Virtual reality · Cognition · Memory · Cognitive enhancement · Memory enhancement · Cognitive assessments · Memory assessments · Cognitive rehabilitation · Memory rehabilitation · Embodied cognition · Embodied memory · Extended cognition · Extended memory · Environmental enrichment

---

N. Reggente (✉)
Institute for Advanced Consciousness Studies, Santa Monica, CA, USA
e-mail: nicco@advancedconsciousness.org

Curr Topics Behav Neurosci (2023) 65: 189–232
https://doi.org/10.1007/7854_2023_425
Published Online: 14 July 2023

1 Introduction

As the scientific community strives to understand the inner workings and neural correlates supporting cognition, it is of paramount importance that it be studied as naturalistically as possible. Otherwise, findings may not extend to real-world settings. Additionally, if studied outside a holistic perspective, researchers run the risk of deeming sets of findings as incongruent with the scientific corpus when they may indeed be complementary. Readers may find this concept reminiscent of the parable of the blind men and an elephant: since each man only felt one part of the elephant, each was quick to doubt the others when their reports were not congruent. Just as it would be erroneous to say an elephant *is* the way its tusk feels, it would be wrong to conclude about the real-world (RW) nature of cognition based on tests conducted in a sterile laboratory environment. To increase their generalizability to the RW (i.e., veridicality), studies of cognition should be conducted in verisimilar contexts (i.e., contexts appearing as the RW). Such frameworks inherently beget ecological validity.

To study cognition holistically means investigating interconnections between its rich repertoire of functions, including attention, reasoning, language, and memory. Memory is a particularly crucial facet, as it supports and subserves all other aspects of cognition; no cognitive task can be accomplished without memory. We cannot even conceive of the framing and timeline of our own experiences (i.e., autonoetic consciousness) without memory. As such, rigorous investigations of memory can serve as a fulcrum for understanding cognition, akin to equipping one of the blind men with extra limbs.

Fortunately, investigations of memory are arguably most ecologically valid if to-be-remembered information is situated in a spatial context – something virtual environments (VEs) can readily provide. Brain systems supporting memory are responsible for helping from the mental representations used to acquire, code, store, recall, and decode information about the relative locations and attributes of phenomena in everyday and metaphorical spatial environments (Tolman 1948). Such representations of an individual's personal knowledge for guiding behavior have been aptly named "cognitive maps" (O'Keefe and Nadel 1978). Since memory for personally experienced events is always backdropped by a spatio-temporal context (Tulving 1983), VEs serve as a ripe medium with which to present cognitive tasks. Indeed, the environmental customization afforded by VR makes it an ideal tool for studying cognition in an ecologically valid fashion. Through the lens of memory studies, this chapter aims to showcase the ways in which VR has advanced a meaningful and applicable understanding of cognition.

2 Enhancing the Ecological Validity of Memory Research with VR

There exist inherent limitations to generalizing laboratory findings, ripe with deliberate memorization tasks, to the processes normally occurring in people's everyday lives (Parsons 2015). VR offers a powerful means to enhance the ecological validity of memory research by providing realistic VEs in which participants can encode and retrieve information under naturalistic circumstances (Reggente et al. 2018). By utilizing stimuli and paradigms that echo RW demands, VR memory studies stand to more accurately capture the requisite insight to construct holistically grounded neurocognitive models of memory – a necessary first step in developing useful neuropsychological assessments. Given that the Diagnostic and Statistical Manual of Mental Disorders often list "interference with everyday activities" as criteria for diagnosing memory disorders, it follows that ecologically valid neuropsychology should leverage assessments that approximate real-world experiences.

The importance of prioritizing ecological validity goes beyond unveiling the true nature of how healthy and aberrant brains process memories. For example, only ecologically valid memory research can reveal robust and quantifiable signatures of retrieval that can be informative in a criminal justice context. In a recent study, for instance, after committing an RW mock crime, participants were shown laser-scanned, photorealistic models of the crime scene and objects via VR or as 2D images. Detection of concealed recognition was 25% more accurate in the VR condition (Norman et al. 2020), suggesting that VR may enhance stimulus recognition and salience so much that it prevents a criminal's capacity for concealment.

Investigations into validating VR-based tests of memory by comparing outcome distributions with traditional tests have yielded mixed results (Parsons and Rizzo 2008). Such inconsistencies do not necessarily indicate that VR is an unreliable medium for benchmarking traditional memory tests. On the contrary, these outcomes could be an indication that traditional tests are less reliable. Indeed, traditional tests do not typically predict RW behavior (Schultheis et al. 2002), whereas performance on VR tasks correlates with self-reports (Plancher et al. 2012; van der Ham et al. 2010) and observer assessments (Allain et al. 2014) of memory function in daily life. While corroboration with RW reports and studies may indeed be the gold standard when authenticating memory metrics captured by VR, it is also useful to investigate the features of VR that have a relationship with the neural underpinnings of memory.

2.1 Primacy of Space and Context

Both philosophers and psychologists alike postulate that brains have evolved solely to support purposeful and predictable movement (Dennett 1993; Llinas 2002). Glenberg and Hayes (2016) posit that the ontogeny of episodic memory relates to the onset of locomotion during infancy that scales with Hippocampal (Hipp.)

development (which also provides a mechanism for infantile amnesia and age-related episodic memory loss). One source of evidence to support this proposition is in the life cycle of the bluebell tunicate. This filter feeder begins to digest a substantial chunk of its cerebral ganglion once identifying a suitable undersea perch to spend the rest of its existence. This phenomenon suggests that once it has served its purpose as a neural network supporting movement, the cerebral ganglion yields greater utility to the organism as nutrition (Mackie and Burighel 2005). From chemotaxis to cognitive maps, a representation of space is necessary for meaningful movement (Llinas 2002). A neural instantiation of a map that provides spatial bookmarks of an organism's experiences, demarcating the locations of nutrition and enemies within an environment, is a fundamental component of brains (O'Keefe and Nadel 1978).

Indeed, there is a primacy of spatial content in the neural representation of events (Robin et al. 2018). Spatial information is often recalled earliest in the retrieval process (Hebscher et al. 2017), and the degree to which individuals report confidence in their autobiographical memories is predicted by their knowledge of the spatial layout of the setting in which the memory occurred (Rubin et al. 2003). The Method of Loci (a.k.a. Memory Palace) mnemonic has long been appreciated for its ability to increase memory by imagining to-be-remembered information placed at familiar locations. Reggente et al. (2020) used a VR implantation of this technique to suggest that the principal component behind mnemonic efficacy is the explicit binding of the objects to a spatial location and revealed a tight relationship between spatial memory (SM) and free recall of encoded objects. These observations showcase that space and memory are inextricably linked at conceptual and neuronal levels – a notion that has become entrenched in popular culture; the phrase "out of space" is often used when indicating a computer's memory is full.

2.1.1 Context Dependence

If space is the inescapable wallpaper that serves as the backdrop for all experience, then it follows that as our spatial or environmental context changes, so should the neural activity underlying diverse cognitive processes (Lester et al. 2017; Willems and Peelen 2021). Early research on learning and memory focused on ways in which encoding and retrieval contexts impacted memory. For example, in a seminal RW study, underwater and dry-land environments were used to reveal that similarity between encoding and retrieval contexts yielded the best memory – a phenomenon referred to as contextual reinstatement (Godden and Baddeley 1975). VR has provided an opportunity to study such effects with a much wider, rapidly accessible (teleportation), and safer contextual repertoire. Parker et al. (2020) used photorealistic underwater and grassy field VEs to effectively replicate the effects seen by RW Godden and Baddeley (1975). Shin et al. (2021) elaborated on the effect by showing that items deemed to be of high utility within each environment were remembered even more effectively, suggesting that context-dependent memory effects may depend on items being integrated into an active schema.

As VEs become more distinct from one another, so do the neural correlates representing them. This allows for single-context-specific information to be recalled without interference from information encoded in other VEs. During recall, the reinstatement of VE-specific brain activity patterns can act as proxy for the degree to which contextual information was bound to learned information. Essoe et al. (2022) showed that the fidelity of contextual reinstatement during recall (as measured by comparing BOLD activity patterns during recall to those collected when participants were instructed to mentally imagine themselves in the VE) was associated with improved recall performance. Conversely, spatial memory is impacted when Hipp. neural codes are representative of competing environments instead of the current environment (Kyle et al. 2015).

Visual context fidelity also moderates spatial information. If a previously encoded VE is perceptually impoverished by removing landmarks and making the walls and ground uniform, efficiency and accuracy of route retrieval decrease (Rauchs et al. 2008a). Moreover, perceptual elements are not the only modulators of context's impact on memory. Brown et al. (2010) found that successful navigation through virtual mazes that had discrete starting and ending locations but overlapping hallways (i.e., spatial disambiguation) required the retrieval of contextual information relevant to the current navigational episode (i.e., where the subject started the current trial).

Given that internal and environmental cues become bound to learnt information (S. M. Smith 1988), cues presented during both learning and recall facilitate recall – a phenomenon known as contextual support. Meaningfully leveraging contextual support for learning in the RW is difficult since only so much information can be contextually encoded within a single environment and regularly visiting multiple environments can be prohibitive for many learners. Furthermore, change-induced forgetting can also occur when recall fails due to a context change – a phenomenon that is only partially mitigated with mental reinstatement to the learning environment (S. M. Smith 1979). VR's ability to expose users to new contexts positions it as an ideal platform for utilizing contextual support during learning. However, VR researchers must carefully balance the benefit it affords by creating a scaffolding of context for the rich encoding of material with the potential drawback of change-induced forgetting when an individual exits the VE.

2.2 Impact of Immersion and Presence

Immersion is the sense of "being there" in an environment (Steuer 1992). Immersion can be subdivided into engagement and presence (the experience of being in one place while physically situated in another) and captured by questionnaires (Fox et al. 2009; Slater et al. 1994). Immersiveness of an environment is determined by objective characteristics of the VR system (e.g., field of view (FOV), multimodal sensory information, headset type; see Reggente et al. (2018); S. A. Smith (2019)), whereas presence refers to the subjective response to immersion, as if feeling "transported" to the VE. Presence can be increased with interaction, volition, and actions that suspend the disbelief that one is in a digital setting (Dede 2009).

Attempts to mimic RW perception in VR have narrowed in on which facets of sensory experience impact memory. For instance, providing an illusory sense of depth with stereoscopy does not appear to impact memory (S. A. Smith 2019). However, a greater VR FOV does seem to enhance procedure memorization when tested both in a VE and the RW (Ragan et al. 2010). Notably, the impact of FOV on memory is not exclusive to VR – Mania et al. (2003) showed equal memory across groups where subjects either encoded in VR or the RW while wearing custom-made goggles designed to match their FOV with the VR.

Incorporating tactile, olfactory, and/or locomotive cues into what is typically a visually dominant VR experience can help mirror RW experience and bolster memory. For example, 360 VR that allowed subjects to place their hands on RW bicycle handlebars vs. simply watching 2D videos of a motorcycle ride almost doubled recognition scores (Schöne et al. 2019). Triggering RW fans to blow when encountering virtual ones or emitting coffee scents when near virtual pots has improved recall performance (Dinh et al. 1999; Tortell et al. 2007). Given the importance of idiothetic signals in bolstering a sense of presence (Taube et al. 2013), increasing ambulatory movement, even by using tricks like foot pedals in Parkinson's disease studies (Matar et al. 2019), should help to increase immersion (Topalovic et al. 2020). Indeed, using an omnidirectional treadmill and Head Mounted Display (HMD) while learning to navigate a virtual rendition of a campus building can increase RW navigation performance (Hejtmanek et al. 2020) compared to using joystick/mouse/keyboard controls. However, SM performance (specifically judgments of relative direction) in VR did not benefit from the concurrent use of a treadmill and HMD while also increasing motion sickness compared to simply using a joystick and monitor (Huffman and Ekstrom 2019, 2021).

Immersion can also impact neural response profiles. Simple environments that do not foster immersion can permit for memory performance that depends more so on stimulus-response associations (and non-Hipp. processes), whereas highly immersive environments can encourage spatial associations (and thus Hipp. recruitment) (Burgess et al. 2002). For instance, in a simple maze with few distinctive landmarks, an individual may remember to turn left at the first intersection, right at the second intersection, and so on. In contrast, in a complex and highly immersive maze with many distinctive landmarks, an individual might remember that the exit is behind the statue, which is located to the right of the pond, and use this spatial information to form a cognitive map of the maze and navigate to the exit, without relying on a specific series of turns or stimulus-response associations. Indeed, fully immersive 3D VR has been shown to induce a higher sense of presence, enhance success rate of spatial navigation, and increase midline EEG theta during encoding compared to 2D (Slobounov et al. 2015). Increased presence also yields a greater recruitment of regions associated with attention (Kober and Neuper 2012), whereas the cerebellum is more active when immersion is low, suggestive of more reflexive processing and less higher order cognition (Gomez et al. 2014; Hartley et al. 2003). On the other hand, participants recognized more environment-consistent items and showed more confidence in their recognition in quick visits to RW environments compared to VE replicas (Flannery and Walles 2003), suggesting that brief exposures to VEs may not induce immersion necessary to overcome the baseline advantage of RW encoding. Waller et al. (1998) made a similar claim following their observation that short periods of VE training were not more effective than map training when solving a maze in the RW. However, they also found that sufficient exposure to VEs eventually surpassed RW training.

In addition to the impact of exposure duration, somatic cues also influence the efficacy of VR in tasks that involve memory. For instance, memory for dance sequences does not seem to be impacted by whether instruction was delivered via immersive VR or nonimmersive videos (LaFortune and Macuga 2018), suggesting that motor learning may already be maximally attuned to observing bodily movement. Likewise, when tested on map drawing and a computer pointing task, participants that did route-learning in the RW outperformed those that learned the same routes in a VE. However, there was no difference in landmark recognition, estimation of route distance, or object locations (van der Ham et al. 2015), suggesting that VEs reasonably approximate a majority of important SM components – a gap that could be lessened with the insertion of body-based locomotion cues (e.g., with a virtual treadmill).

Finally, context-dependent memory effects have also been linked to immersion in VR-based studies. Participants tasked with learning overlapping sets of words in two foreign languages across two VEs performed better when each language was learned in its own context relative to those who learned both languages in the same context – however, this advantage was only apparent in participants reporting high presence in the VEs (Essoe et al. 2022). The sense of "being there" during an experience directly translates to the recruitment of neuronal populations that have the dual purpose of representing an individual's discrete location in space and encoding memories. Indeed, when participants recall information that was originally encoded in VEs, their performance is associated with the degree to which fMRI patterns of brain activity that represent the VE are reinstated (Essoe et al. 2022). Additionally, this study showed that if there were higher representational similarities between trials when participants imagined themselves in a particular VE and the recall periods for information that was encoded in that VE, participants had improved recall performance. Relatedly, Miller et al. (2013) showed that when retrieving information that was encoded in a VE by driving in a virtual city, the place cells that fired at specific locations during encoding were also activated during recall of the item associated with that location. Such an observation is inherently contingent on the elicitation of a sense of embodiment in the individual.

2.3 *Impact of Embodiment, Enactment, and Extension*

2.3.1 Embodied Cognition

A twenty-first century shift within the cognitive sciences has caused researchers to focus increasing attention on the role of embodied representations as the foundation of cognition. In this view, our cognitive abilities are fundamentally grounded in our capacities for purposeful perception and action. Further, all cognition is inherently situated (Foglia and Wilson 2013). Indeed, all sensory organs are situated in discrete locations on the body and provide a grounding perspective (e.g., sight is only toward the front of the body) that inevitably tags every experience with a sensorimotor component (Varela et al. 1992; Wilson 2002). Given that VR permits for egocentric points of view coupled with volitional movement, it contains the crucial ingredients for the embodiment recipe (Repetto et al. 2016).

In keeping with the idea that many aspects of higher order human cognition, such as long-term memory functions, are fundamentally linked to sensorimotor experience, it has been demonstrated that signals produced by self-motion cues (i.e., idiothetic signals) affect encoding fidelity; subjects encode objects and their position better if they move around a table vs. the table moving (Frances Wang and Simons 1999) – a finding that was extended to improving reaction time (RT) in VR even with only simulated movement (Christou and Bülthoff 1999). Participants who use an HMD and physically walked through a VE made 38% fewer errors than those who physically turned, but moved using a joystick (Ruddle et al. 2011).

Further, active locomotion in VEs, compared to stationary periods, enhances directional sensitivity in the entorhinal and retrosplenial cortices and boosts SM (Nau et al. 2020). Similarly, active navigation enriches SM (Brooks 1999b), source memory (Sauzéon et al. 2016), object recognition (Hahm et al. 2007) compared to passive tours and yields enhanced recall of central and allocentric spatial information as well as binding (Plancher et al. 2012). Even merely watching an experimenter actively search a VE was more effective for learning target locations compared to participants observing the environment from a fixed position (Thomas et al. 2001). Scene recognition, however, does not seem to differ as a function of active vs. passive encoding in VR (Gaunet et al. 2001). The effect of action is present in non-visual modalities as well. Audio-VR provides iconic and spatialized sound cue updates with each step, allowing for an update of heading direction and place in the actual RW environment. Such setups have been used to train blind participants to navigate toward (away from) rewards (predators) in VEs, revealing that ludic (i.e., undirected and playful) exploration yield better shortcut-finding performance than guided tours when tested in an RW analog of the VE they were trained in (Connors et al. 2014). Decision making seems to be the primary component for the acquisitions of topological graph knowledge, whereas idiothetic information is crucial for metric survey knowledge (Chrastil and Warren 2015).

Bréchet et al. (2020) showed that the presence of self-related bodily cues (e.g., virtual hands and legs appearing in the VE when the participant moves their RW counterparts) increases object recognition compared to when these cues are absent. Amazingly, recognition of objects in a scene that was originally encoded without self-related bodily cues was retroactively enhanced by subsequent visits to the same scene where self-related bodily cues were present. It appears the degree to which these virtual renditions of the body have RW fidelity have substantial ramifications. VR manipulation of experienced self-location relative to one's real body (e.g., using the output of a camera placed behind a participant as the input to their HMD) can elicit an "out of body" experience that results in catastrophic EM impairments and decreased Hipp. activity during retrieval (Bergouignan et al. 2014).

When participants see an avatar looking in a particular direction toward an object, they are primed to that perspective and perform better and faster on target displacement tasks when they take that perspective as compared to no avatar conditions (Sulpizio et al. 2016) – an effect theorized to be due to spatial computation being conducted in advance via imagination (Burgess 2006). Deepening this perspective taking, the projection of oneself onto a virtual avatar can also yield consequences on memory. For example, in a study conducted by Ganesh et al. (2012), gamers were asked to rate the extent to which trait words described different aspects of themselves, their long-term avatar, close others, and familiar distant others. A surprise recognition test for the trait words showed that avatar-referent memory was superior to familiar distant other-referent memory. Activity in regions of the brain associated with self-identification (e.g., Inferior Parietal Lobe) was shown on avatar trials (even more than self-trials!) and correlated with an individual's self-reported propensity to incorporate external body enhancements (e.g., prosthetics) into one's bodily identity.

2.3.2 Enacted Cognition

The observation that long-term memory abilities increase with active navigation could also be related to the enactment effect – that is, the finding that participants who perform an action are more likely to recall the event compared to subjects who listen to the action phrase or watch an experimenter do the task (Madan and Singhal 2012). For instance, actively rotating objects vs. passively observing their rotation increases speed of recognition (James et al. 2002). Virtually manipulating body parts vs. watching another do the manipulation increases anatomical memory with greater benefit for individuals with lower baseline spatial abilities (Jang et al. 2017). "Running" in a VE using a joystick evoked a faster understanding of foot-action verbs, but not hand or mouth action verbs (Repetto et al. 2015). Simple interaction with to-be-remembered material also seems to enhance SM for where that information was encountered, but not necessarily the rest of the VE – suggesting a role of proximity to body in the upregulation of encoding during interaction (Reggente et al. 2020).

2.3.3 Extended Cognition

Extended cognition posits that facets of the external environment are *part* of cognitive processes as extensions of mind outside the brain. Draschkow et al. (2021) showed that participants made frequent eye movements (indicative of reliance on the visual scene and not WM) when performing naturalistic tasks in VR like collecting objects and arranging them according to a template. This use of the environment was present even if the number of objects were well within WM range, supporting the idea that there exist a fundamental preference and dependence for memory on external information during goal-directed behavior. The same study revealed that as demands on locomotion increased (i.e., they need to see information that's out of sight), so does reliance on WM. Given the limits of human WM in the context of the expansive workspace within VR, it is not surprising that many VR productivity tools contain easy-to-access visuospatial sketchpads that supplant the need for WM (e.g., sticky notes, whiteboards).

2.4 Impact of Environmental Enrichment

Environmental enrichment was originally observed when mice were given toys and larger, more complex cages, leading to enhanced dendritic arborization and improved learning (van Praag et al. 2000). Relatedly, the use of fantastical and erotic environments increases the efficacy of spatial mnemonic techniques in humans (Bower 1970). Virtual exploration that closely approximates RW factors (e.g., 3D vs. 2D) is a core aspect of enrichment (Clemenson et al. 2015; Freund et al. 2013) and why RW variables like sound in VE can increase Hipp. activity (Andreano et al. 2009). The complexity of structures built in Minecraft scaled with memory improvements (Clemenson et al. 2019), making such mediums far more approachable than similar construction efforts in the RW (Kolarik et al. 2020). Prolonged exposure to enriched virtual environments also appears to confer benefits for RW memory. Video gamers who prefer 3D video games over 2D games performed better on recognition tasks unrelated to the game – a phenomenon that was expanded to naïve gamers who underwent training (Clemenson and Stark 2015).

3 VR Bridges the Gap Between RW and Lab-Based Memories

Given that laboratory and RW stimuli tend to evoke different behavior (Snow et al. 2014) and brain activation profiles (Chen et al. 2017; Chow et al. 2018; Roediger and McDermott 2013), researchers put considerable energy into ensuring they are capturing ecologically valid metrics, especially when memory can be multi-faceted.

For example, the phenomenon of recognition (the ability to recognize previously encountered events, objects, or people) reflects the contribution of both recollection (retrieval of details associated with previous experiences) and familiarity (the feeling that the event was previously experienced) (Yonelinas et al. 2010). VR experiences appear to be retrieved via recollection-based processes similar to those that support autobiographical/recollection memory, whereas retrieval of conventional screen experiences seems more similar to familiarity (Kisker et al. 2021). Elaborating on this point, Schöne et al. (2019) suggest that VR experiences become a part of the autobiographical associative network, whereas conventional presentations remain as episodic events. Support for this comes from the observation that individuals with highly superior autobiographical memory tend to recognize the same number of laboratory events as age and education matched controls (LePort et al. 2012), suggesting that laboratory memory tests are often not ecological tests of recognition else these unique participants would outperform controls.

VR stands to serve as a cost-effective middle-ground that balances the experimental control of a laboratory with ecological validity. A proof of concept for how VR can bridge the gap between RW and lab-based memories comes from two VR-based route-navigation studies (Janzen and Weststeijn 2007; Wegman and Janzen 2011) that showed comparable behavioral and neuroimaging results to an RW study involving the creation of a 3-km outdoor walking course through Philadelphia (Schinazi and Epstein 2010). While RW task paradigms will likely always remain valuable, VR can provide follow-up studies to tease apart nuances. For example, the seminal study that examined static structural correlates in the Hipp. of London taxi drivers that scaled with experience (Maguire et al. 2000) was extended by a study that recorded functional activity in the same population *during* virtual navigation around London, revealing the recruitment of a more distributed network (Woollett et al. 2009). Furthermore, virtual renditions of familiarized RW environments can be used to probe memory for the location of RW objects from precisely positioned frames of reference (Schindler and Bartels 2013) and examine the neural activations supporting recollection of goal location found in familiar RW environments (Retrosplenial and Posterior Hipp.) vs. recently experienced RW environments (Hipp. only; Patai et al. 2019). Being a passenger in a virtual car as it navigates about a VE replica of an RW environment elicits the same SM performance as being an RW passenger (Lloyd et al. 2009). A virtual version of a common false memory test, where participants encountered objects from vendor stands instead of words on a screen replicated some of the findings of the original lab-based study – namely, that younger adults recalled and recognized more correct elements than older adults. However, it also showed that the typically-observed gap between younger and older adults groups in their susceptibility to perceptually and semantically related false recognitions went away – a finding the authors attribute to the use of a naturalistic context in keeping with the idea that VR may provide more "age-fair" and ecological tests of memory (Abichou et al. 2021).

VR can also induce cognitive-impacting situations that may not be commonplace, ethical, or safe in the RW. For example, Martens et al. (2019) showed that despite producing a strong physiological stress response, memory performance was not impacted by asking participants to step off a tall building. The ability to create convincing simulations in a safe and operationalized fashion holds promise in shedding light on how memory works in stressful RW situations (e.g., is there high accuracy in eye witness testimonies of a violent crimes?) while simultaneously supporting investigations and treatment of stress-related disorders.

3.1 Human Analogs of Non-human Research

Feasibility and ethical limitations have prevented human analogs of many non-human research paradigms. For example, the Morris Water Maze (MWM), which places animals in a pool of water and tests their SM based on their ability to find a hidden platform, would be onerous with human subjects. VR's ability to provide precise control over task features has permitted for investigations into the nuances of memory by utilizing the MWM in humans. For example, BOLD activations in the parahippocampus, precuneus, and fusiform (and surprisingly not the Hipp.) were greater during trials where the platform was hidden as opposed to visible (Shipman and Astur 2008). Using a single distal cue (more allocentric) vs. multiple cues (more egocentric) allowed for investigators to reveal how the parietal cortex helps with translating between allocentric coordinates and egocentric directions (Rodriguez 2010).

VR-based MWM investigations have also revealed that higher circulating levels of testosterone (Driscoll et al. 2005), Hipp. volume size (Moffat et al. 2007), and ratios of finger length (2D:4D – an indicator of hormonal ratio in utero; Müller et al. (2018)) are positively correlated with task performance. Pharmacological disruption of the Hipp. with scopolamine allowed for the identification of striatal-based memory systems as a compensatory mechanisms for MWM completion (Antonova et al. 2011). Similarly, it was shown that an intact Hipp. is not necessary for leveraging distal cues to perform well on the MWM (Kolarik et al. 2016).

The Radial Arm Maze (RAM), whose design is a circular central platform that radiates out corridors (arms) that contain rewards not visible from the center, now also has VR variations where the location of the reward will stay in the same place (win-stay; Cyr et al. (2016)) or switch places (win-shift; Demanuele et al. (2015). VR RAMs have been used to differentiate between spatial learners (using landmarks; Hipp. dependent) and response learners (using sequences of turns; caudate dependent) by removing distal landmarks between learning and testing (Bohbot et al. 2004, 2007; Iaria et al. 2003). Similar to the RAM, the hole-board maze (holes instead of arms) has been translated into VR as an evaluation of human place learning that contains facets of working memory (repeat visits to explored locations) and reference memory (prioritized visits to rewarded locations) (Cánovas et al. 2008). This and similar inspired location-based memory task have been validated based on their ability to reproduce commonly observed learning rates and the sexual dimorphism in which men outperform women in spatial memory tasks (Cánovas et al. 2008; Tascón et al. 2017).

Where Leutgeb et al. (2007) created a physical contraption that morphed a square environment to a circular one to investigate changes in rodent behavior, pattern separation, and place-cell remapping, Steemers et al. (2016) created a virtual analog that allowed for two distinct environments to be visually morphed along a continuum – an elegant design that showed changes in reward-finding behavior based on which environment participants thought they were in and also provided support for an attractor dynamic model of pattern completion in the hippocampus, supporting mnemonic processing and decision making.

While visual input seems sufficient for conducting tests of spatial memory in VR without the accompanying RW elements (e.g., swimming in water in the MWM), some task designs may require non-VR elements. For example, in studying associative learning tasks like fear conditioning, there may not be a common visual element that evokes the same response across all participants. In order to utilize the tried-and-true method of instilling fear in rodents, Alvarez et al. (2008) replicated classic fear conditioning experiments by delivering foot shocks to participants as they explored VEs, revealing more Hipp. and Amygdala activity in the VE that was paired with the shock.

A separate analog of non-human research that can be applied to human subjects with the use of VR is the collection of implicit or passively derived metrics. Given that non-humans are incapable of explicit reporting, researchers must rely on behavioral observations to infer cognition and memory which are often remarkably informative and detailed, even sometimes more so than subjective reporting. The ability to extract a plethora of detailed datapoints about a subject's behavior is unprecedented in VEs; real-time position, orienting directions, interactions with objects, and many other metrics can all be used to create derived metrics like time spent in a location, attentional focus, and to eventually yield objective measures of memory.

For instance, much like how freezing behavior exhibited by rodents in particular environments can indicate a contextually based fear memory in the absence of subjective reporting, an individual's behavior in a VE can reveal facets of memory. While technically feasible with RW object tracking, automated classification of learned reward/fear-based behavior through computational ethology is substantially easier and more accurate in VEs and can extract individualized ethograms such as thigmotaxis (tendency to remain close to walls), panic errors, and approach/avoidance/intermittent locomotion (Mobbs et al. 2021). For example, place aversion in open field tests can indicate a fearful memory originally encountered in that place. Computational analyses of VR behavior shed light on an amnesic MTL patient, revealing deficits in spatial precision rather than spatial search strategy, suggesting that an intact Hipp. is not necessary for representing multiple external landmarks during spatial navigation of new environments (Kolarik et al. 2016).

3.2 *Studying Different Types of Memory with VR*

VR affords memory researchers with an unprecedented ability to design paradigms that balance the benefits of RW verisimilitude with the laboratory rigor of isolating variables of interest. Such affordances have significantly advanced the field of memory research by providing an observational lens that can focus in on specific elements without necessarily disrupting the observation of the cohesive whole. For example, researchers have been able to dissociate visual scenes from events, an important tool for examining the binding of/separation of components that make up complex phenomena like EM (Burgess et al. 2001; S. A. Smith 2019). VR has also facilitated the implementation of mnemonic techniques, such as the Method of Loci, which previously relied solely on an individual's imagination of previously visited environments. By utilizing template virtual environments (VEs), VR allows for standardization of the size, detail, and exposure time of environments. This standardization helps control for individual differences in real-world experience (Legge et al. 2012; Reggente et al. 2020).

Additionally, VR exclusively permits for the creation of infinitely large-scale and novel environments, insertion of invisible barriers, event-based rendering of objects, teleportation between locations, impositions of visual route guidance, and interactions with other agents. Such features have bred designs that can disentangle cognitive decision making from other processes of interest (Marsh et al. 2010) or tease apart place-based vs. sequence-based strategies of spatial encoding (Igloi et al. 2015). VR can also present scenes outside of conscious awareness, which has revealed that conscious perception is not mandatory for spatial EM formation (Wuethrich et al. 2018). While VR can allow for visual perspective shifts (e.g., first-person vs. third-person viewing angles), it can also foster perspective taking (e.g., role-playing a skilled scientist) which can help to unbound "trapped intelligence" in poor-performing students by allowing them to step out of their real-world identity and build confidence while embodying another (Dede 2009). The sections that follow describe different types of memory, the ways in which they have been studied traditionally, and how VR paradigms provide a more ecological investigation. Table 1 highlights some particularly potent examples of the ways in which the affordances of VR have uniquely extended the understanding of different types of memory.

3.2.1 Spatial Memory (SM)

SM is defined as the storage and retrieval of information that is needed to remember and plan routes to locations of objects and events. In traditional tests of SM (e.g., finding matching pairs of face-down cards arranged in a grid), the use of egocentric or allocentric processing cannot be distinguished, as the two frames coincide. Furthermore, scenes, objects, and landmarks are encoded differently when discovered during 3D navigation as compared to 2D presentations. While both

Table 1 Highlighted studies of memory that showcase affordances unique to VR

VR affordance	Citation(s)
Rapid teleportation to unique VEs to foster context facilitated learning	Essoe et al. (2022), Parker et al. (2020), Shin et al. (2021)
Creating invisible blockades or targets that impact function but not visual percepts	Shipman and Astur (2008), Kolarik et al. (2016)
Providing participants with interactive perspectives that are not egocentric (i.e., first-person point of view) such as third person or bird's eye points of view	Morganti et al. (2013), Serino and Riva (2015), Bergouignan et al. (2014), Serino et al. (2015), Weniger et al. (2009)
Overlaying toggled route guidance to dissociate the processes responsible for true wayfinding vs. path-following	Hartley et al. (2003)
Maintaining proximal environmental fidelity while altering distal cues to differentiate between spatial and response learning	Bohbot et al. (2004, 2007), Iaria et al. (2003)
Exposing participants to multiple distinct VEs and then computationally blending the visuospatial components of those environments to observe evidence of Hipp. pattern completion and attractor dynamic processes	Steemers et al. (2016)
Create duplicate instantiations of environmental transgressions during route-finding while iterating on other variables (e.g., passive vs. active movement) to isolate differences between multiple navigational pursuits	Janzen and Weststeijn (2007), Wegman and Janzen (2011), Plancher et al. (2012), Gaunet et al. (2001)
Inserting realistic "events" that impact navigational pursuits (e.g., lava blocking a path) as a probe of RW spatial memory and path integration processes	Javadi et al. (2019)
Leveraging VEs as a common, universal template of spatial environments to investigate the mechanisms supporting the Method of Loci – a mnemonic which is typically conducted using mental imagery of familiar locations which vary meaningfully across subjects	Legge et al. (2012), Reggente et al. (2020), Krokos et al. (2019)
Use VEs to impart cognitive-impacting situations mimicking RW situations (e.g., being on a ledge of a tall building, driving a large truck, being chased by a predator) that would be unethical or impractical to induce non-virtually	Martens et al. (2019), Unni et al. (2017), Faul et al. (2020), Mobbs et al. (2007)
Testing the efficacy of procedural memory training in situations where real-life examples are not readily available or practice like surgical procedures or mining expeditions	Siu et al. (2016), Zhang (2017), Wang et al. (2020), Mantovani et al. (2003)
Create VR replicas of RW environments with varying levels of detail to observe the impacts of learning prior to subsequent RW environment exposure	Wallet et al. (2011), Mania et al. (2010), Larrue et al. (2014), Coleman et al. (2019), Brooks (1999b)

(continued)

Table 1 (continued)

VR affordance	Citation(s)
Crafting VEs that become iteratively similar to an individual's feared context like the scene of a victim's road accident or a soldier's battlefield while also substantiating extinguishment over multiple environments	Foa and Kozak (1986), Maples-Keller et al. (2017), Beck et al. (2007), Bohil et al. (2011), Gerardi et al. (2008), Reger and Gahm (2008), Wood et al. (2008), Dunsmoor et al. (2014)

presentations may suffice for recognition, 3D affords participants with the ability to decide upon novel shortcuts or forecast unseen points of view (Burgess et al. 2002).

VR allows for a suite of more ecologically valid measures of SM when tasking participants with navigating from one location to another. During such tasks, critical environmental features can be changed (Dhindsa et al. 2014; Wegman et al. 2014), tests of perspective-based orientation can be conducted (Brown et al. 2014; Dimsdale-Zucker et al. 2018; Gomez et al. 2014; Stokes et al. 2015; N. A. Suthana et al. 2009), and covert tasks can be embedded so as to observe implicit acquisition of object-location memories (Wong et al. 2014). Altering exposure to first-person and/or bird's eye views (e.g., small aerial map overlays) of an environment can also reveal how SM is encoded and retrieved (Serino and Riva 2015). Following navigation, participants can be tasked with pinpointing locations on maps of environments. Such a task requires egocentric-to-allocentric transformation and yields precise quantification of error through measurement of the Euclidean distance between the actual and subject-provided coordinates (Pine et al. 2002; Reggente et al. 2020). VR paradigms have also revealed a sustained neural "memory" of position in VE akin to the classic function of 2D (Hassabis et al. 2009) and 3D (Kim et al. 2017) place cells.

VR has also elucidated the role of sleep on replaying and consolidating SM processes. Perhaps the most salient traditional exemplar of this phenomenon is that the temporal order of place-cell firing in rodents during NREM sleep mirrors that of sequentially visited RW place fields (i.e., a path) before sleep (Davidson et al. 2009). Peigneux et al. (2004) demonstrated in humans that the degree to which hippocampal areas that showed fMRI activation during virtual navigation were also detectable via EEG during subsequent slow wave sleep accounted for individual differences in performance on route retrieval the next day. From a different approach, Rauchs et al. (2008b) made clever use of VEs to demonstrate that post-learning sleep deprivation modulates the neural substrates of both spatial and contextual memories in a way that accounts for individual differences in place-finding efficiency.

Successful navigation to learned locations from different start points, which requires an interplay across memory for environmental maps and routes, is also a straightforward and ecologically valid test of SM. "Optimality of trajectory" (Howett et al. 2019), improvement over runs (Cyr et al. 2016; N. Suthana et al. 2011), time spent on task (Migo et al. 2016), distance traveled (Salgado-Pineda et al. 2016), use of shortcuts (Caglio et al. 2012), and adjusting to events (e.g., lava blocking a learned path (Javadi et al. 2019) or newly precluded landmarks (Wolbers et al.

2004)) are all creative ways in which SM can be objectified in VR. Chapter "VR for Spatial Navigation and Action Science" of this text discusses this topic at length.

3.2.2 Short-Term Memory (STM)

STM is an umbrella term used to describe a capacity for keeping small amounts of information on top of mind, with the ability to regurgitate said information (via recall or recognition) shortly after its initial presentation. The next sections focus on VR research that sheds new light on the mechanisms underlying components of short-term memory such as working memory.

Working Memory (WM)

WM involves the manipulation of information held in STM and can take the form of visuospatial (e.g., remembering the color and location of your drink at a party), verbal (e.g., repeating a phone number in a phonological loop), and/or kinesthetic (e.g., repeating another's hand gestures). WM is usually tested using tools such as the N-back task, where participants are tasked with making a button press when a current presentation of an item (e.g., a word) is the same as one presented N items back in the presentation stream (K. M. Miller et al. 2009) or the List Sorting Test (Tulsky et al. 2014), where participants are tasked with recalling and sorting (e.g., largest to smallest) a list of objects.

In VR, visual-spatial WM has been studied by asking individuals to navigate a previously shown route – a design that permitted researchers to compare epochs of encoding and retrieval and identify that encoding requires more cerebral effort than retrieval (Jaiswal et al. 2010). Similar demarcation of epochs during virtual navigation permitted for the discovery of WM EEG signatures. Weidemann et al. (2009), for instance, observed that frontal theta power was high during navigational efforts and was released when participants released the goal state (and no longer had to utilize their WM). Plancher et al. (2018) extended such work by examining the role of WM in the integration of episodic memories. Participants were tasked with memorizing as much contextual content as they could while navigating a VE and performing a concurrent task that barraged either of two cognitive scaffolds that support WM: the phonological loop (e.g., keeping track of the number of garbage containers colored yellow) or visuospatial sketchpad (memorize a spatial pattern composed by the garbage containers). Results revealed that WM is crucial for EM; blocking the phonological loop impacted the encoding of central elements in the VE and blocking the visuospatial sketchpad interfered with the encoding of temporal context and relational binding.

Researchers have been able to develop more realistic N-back-style tasks like viewing streams of letters on a blackboard in a replica classroom (Coleman et al. 2019) or counting moving fish in a virtual aquarium (Climent et al. 2021). Such developments have equipped researchers with more realistic paradigms for assessing the efficacy of learning aids, especially compared to traditional laboratory tests. For instance, Jang et al. (2021) found that ADHD drugs in children can improve accuracy and RT on an n-back task in a realistic classroom VE – a finding that was accompanied by decreased default mode network activity during high working memory load.

Prospective Memory (PM)

PM is remembering to carry out a previously planned action like taking medication at specific times (time-based) or starting your fitness tracker before a mountain bike ride (event-based). PM is typically tested by giving participants RW time-based (e.g., ask the experimenter for a newspaper after 20 min) and event-based (e.g., change pens after completing seven assignments) instructions before embarking on an intellectually demanding filler task (Groot et al. 2002). In VR, PM can be probed more meticulously and even incidentally by examining if an individual makes stops at instructed spots along virtual routes (Lecouvey et al. 2019) or gathers the right grocery items from virtual stores (Dong et al. 2019). This latter effort elicited significantly more activity in rostral prefrontal cortex than an analogous non-VR version of the task, which corroborates the RW deficits in PM observed in patients with lesions to that area (Volle et al. 2011) and evidences ecological validity of VR-based PM tasks. VR also stands to serve as an encoding environment to upregulate RW PM. Kalpouzos and Eriksson (2013) were able to show that self-efficacy beliefs modulate intentional encoding on delayed real-life intentions by collecting fMRI data while subjects mentally imagined executing tasks (e.g., mail a letter) in VEs that were based on RW environments.

3.2.3 Long-Term Declarative Memory: Semantic Memory

Semantic Memory is the explicit knowledge of general words, concepts, and facts like remembering the name of objects surrounding you or the punch line to a joke. Semantic Memory is typically tested with a picture naming task – a paradigm that has been extended with VR using standardized sets of 3-D objects which carry increased semblance to actual objects compared to pictures (Peeters 2018). Encoding semantics in VR is particularly advantageous for repetitive field training where RW counterparts would be more costly and less readily accessible (e.g., flight simulations for pilots, emergency rooms for health care professionals (Mantovani et al. 2003)). Such utilizations have the additional benefit of increasing motivation, engagement, and retention scores compared to didactic instruction (Chang et al. 2019; Ryan and Poole 2019).

Experiences with virtual routes can provide quantifiable measures of semantic memory (e.g., knowing whether or not you can reach point A from point B without turning). After encoding VEs, tests of semantic memory and survey knowledge revealed a dorsal (survey)/ventral (semantic) dissociation, supporting the putative "where" (dorsal)/"what" (ventral) streams (Aguirre and D'Esposito 1997).

3.2.4 Long-Term Declarative Memory: Episodic

An episode, much like a proper journalistic summary, encapsulates information about the "who," "what," and "where" of an event. As such, EM holds information about temporal-spatial relations among events (Tulving and Thomson 1973); abstract features of an episode are like individual nodes in a network whose temporal linkage forms a collective representation of an event (Eichenbaum and Cohen 2001). An innovative RW test of EM strength has been recognition tasks of stimuli captured from participants' wearable cameras (Chow et al. 2018; Rissman et al. 2016).

The spatio-temporal nature of VR allows for a multitude of experiments conducted in VR to be considered tests of EM (S. A. Smith (2019)). A quintessential study had subjects navigate about a VE and encounter avatars (who) in discrete locations (where) that dispensed objects (what) (Burgess et al. 2001), permitting for discrete probes of EM's constituent components: spatial context (where the object was received), temporal context (which object was received first), which person was involved, and object recognition. Neuroimaging revealed that the Parahipp. was recruited during spatial probes whereas object memory elicited parietal and prefrontal activity. The same task was also utilized in temporal lobectomy patients (Spiers et al. 2001a). The task's ability to precisely target core components of EM revealed that L Temporal patients were impaired on context-dependent memory questions whereas R Temporal patients were impaired on topographical memory and object recognition, suggesting that the R Temporal lobe is more involved in SM, whereas the context-dependent aspects of EM are dependent on the L Temporal lobe. A similar paradigm employed by Buchy et al. (2014) revealed activation of ventrolateral prefrontal cortex during external source memory that scaled with self-reflectiveness.

It should be noted that studies of autobiographical memory in VR may lack some ecological validity because there is currently a richer diversity and temporal separation in actual streams of life events. Such a notion could explain the increase in prefrontal activity during VR-based studies of EM as compared to RW studies (Burgess et al. 2002). Without rigorous design, VEs have higher potential to be self-similar than RW environments which may increase the likelihood of interference and subsequent recruitment of prefrontal regions to disentangle them.

3.2.5 Long-Term Nondeclarative/Procedural Memory

When learning a new skill or routine, memory for the process is typically declarative (e.g., consciously remembering which note comes next when learning to play a new song) before it becomes nondeclarative (i.e., procedural or implicit – as in the case of unconsciously playing the song without thinking about the constituent notes). A classic test of nondeclarative memory is the weather prediction task (Knowlton et al. 1994), where participants use feedback to learn which cues are associated with weather outcomes and improve on the task despite not being able to declare the information that is guiding their behavior.

In VR, procedural memory can be probed by showing and then testing multistep procedures like grasping and moving a series of objects from to specific locations (Ragan et al. 2010). This is particularly useful for testing the efficacy of training in situations where real-life examples are not readily available nor practical (e.g., surgical procedures (Siu et al. 2016) and mining safety (Zhang 2017)). Wang et al. (2020) showed a 12% improvement in learning how to erect scaffolds when using a personalized VR tool vs. conventional methods.

3.2.6 RW Memory Modulators in VR

In addition to creating life-like episodes and providing templates for memory manipulation, it is also possible to bring other components of the RW (e.g., stressful scenarios) into VR, further increasing the ecological validity of research paradigms, while also unveiling the dynamic nature of memory under specific mindsets and settings (e.g., affective states).

3.2.7 Impact of Emotion

Given that the neural architecture supporting SM can become completely remapped once the environment is associated with fear (Moita et al. 2004), the impact of emotion on the encoding of information within VEs can be revelatory. Showing participants task-irrelevant images of negative scenes while they completed an object-location task elicited an increase of Parahipp. activity and faster RTs when later viewing rooms within the VE compared to positive images (Chan et al. 2014). The integration of fearful stimuli into schemas appears to be sensitive to an individual's physical proximity to the stimuli; how close individuals were to an RW bank robbery determined their likelihood of developing PTSD (Frans et al. 2018). VR provides a safe medium with which to dissect this phenomenon. Faul et al. (2020) showed that profiles of neural activity during fear acquisition that occurred within peri-personal space showed recruitment of reactive fear circuits vs. cognitive fear circuits for threats acquired further away. Mobbs et al. (2007) also revealed proximity-based functional differences when participants were far away (ventromedial prefrontal) vs. close to (periaqueductal gray) a virtual predator.

3.2.8 Cognitive Load/Attention

Studying cognitive load and attention in VEs is crucial for understanding how people interact with and respond to complex and dynamic situations, such as driving or navigating unfamiliar environments. Unni et al. (2017) created a realistic driving simulator where participants drove down a virtual highway while monitoring speed signs that appeared every 20 s. Participants were tasked with matching their speed to the sign that appeared N signs back. This putative WM load significantly impacted safety relevant driving behaviors (e.g., variability of brake pedal position and RT). Blondé et al. (2021) had participants passively encode an urban VE while they were randomly probed for reports of mind wandering. Results suggested that attention coupled with a moderate degree of mind wandering creates a hospitable medium for encoding processes by helping to avoid the distraction of inner thoughts while also preventing individuals from being overly focused on the environment. Familiarity with a VE mitigates errors that typically accompany increased cognitive demands (Tascón et al. 2021). Electrodermal, subjective, and behavioral measures of cognitive load as participants carried out simple tasks in RW train stations and their virtual replicas revealed an increased load in novice travelers compared to experts implying that domain expertise can ease situational cognitive load (Armougum et al. 2019). Taken together, these results suggest that simulating RW situations in VR can permit for the development of protocols that detect (and limit) WM load, encourage moderate mind wandering, and develop environmental familiarity to bolster RW safety and performance.

3.2.9 Impact of Volition

Volitional control can improve memory due to the interplay between neural systems related to planning, attention, and item processing (Voss et al. 2011). VR allows researchers to vary volition along component (i.e., stationary or motoric) and degree (e.g., high vs. low-motor control) axes of the interaction being considered (S. A. Smith 2019). This technique has revealed that making itinerary choices or having motor control over a vehicle during virtual navigation enhances SM compared to merely being a passenger (Plancher et al. 2013) – an enhancement more pronounced in older adults (OA; Meade et al. 2019). Further, feature binding is also enhanced for virtual passengers both in cases of itinerary choosing and after experiencing low navigational control (e.g., moving a car along a rail instead of steering), revealing a need for balance between cognitive load and volition when optimizing for memory encoding (Jebara et al. 2014). Likewise, granting participants the ability to place objects along a route resulted in greater recall and SM for those objects compared to having no control over placement (Reggente et al. 2020).

Interestingly, the belief that an avatar was expressing human-based volition (i.e., is controlled by a person and not an algorithm) yielded better factual memory than when participants believed the avatar was a computer program (Okita et al. 2008). Such benefits stand to be increased if human-controlled avatars are attentive – an attribute that promoted emotional security in threatening VEs (Kane et al. 2012).

Finally, virtual navigators seem to activate the anterior Hipp. during wayfinding and the caudate when following routes (Hartley et al. 2003). This outcome can be viewed as a modern extension of early RW work showing the R Hipp. to be associated with knowing places and getting between them, whereas the R caudate was associated with getting there quickly (Maguire et al. 1998).

3.3 VR to RW Transfer

VR to RW transfer is the successful RW recall of VR encoded information. Neural activity patterns expressed during recall remain relatively similar despite encoding in RW vs. equivalent VEs (Spiers and Maguire 2006). When considering the real-world implications of VR-based learning protocols, it would behoove VE designers to carefully balance detail with abstraction and realism with fantasy while also considering other RW components (e.g., bodily information) to increase the likelihood of VR to RW transfer.

High-fidelity VR replicas of RW settings yielded better wayfinding in the RW compared to replicas that removed color and textures (Wallet et al. 2011). Following exposure to a high-detail version of a VE replica of an academic office, participants showed an increase in object recognition compared to when exposed to a low-detail wireframe version (Mourkoussis et al. 2010). Interestingly, the results only held for objects that were consistent with the office (e.g., phone and not a cash register). However, not all detail is created equal – object recognition in the RW was better after encoding low-fidelity flat-shaded (no textures used, but color tone preserved) virtual objects compared to those rendered with radiosity (color blending intended to suggest the presence of a realistic light source) (Mania et al. 2010). The authors suggest that variations from the RW could recruit stronger attentional resources.

Encoding routes in a VE replica of an RW environment with access to full body-based information (treadmill with rotation) promoted better transfer of SM to the actual RW environment over encoding done without body-based information (joysticks or treadmills without rotation) (Larrue et al. 2014). Reciprocally, early studies have shown a reliable transfer of RW spatial knowledge to VE replicas (Ruddle et al. 1997; Wesley Regian and Yadrick 1994; Witmer et al. 1996), emphasizing the ability to create hybrid studies that can be seeded with RW experience and enhanced in VEs. Similarly, differences in SM and wayfinding between OA and YA found in RW environments were also found in their virtual replicas (Taillade et al. 2016). Learning locations in either a familiar or novel RW settings or virtual equivalents show bi-directional RW/VR transfer of SM (Clemenson et al. 2020).

Children with attention disorders who completed 5 weeks of memory training using a "virtual classroom continuous performance task" showed substantial improvement in what the authors describe as a "real-life scenario" of classroom learning, making transfer far more likely (Coleman et al. 2019). Procedural memory supporting an object movement sequence task can also be transferred to the RW (Ragan et al. 2010). Other exemplars of VR/RW transfer are discussed in the section on memory rehabilitation.

4 VR-Based Memory Assessments

Conventional measures of memory typically focus on core content (i.e., the "what") instead of the true binding that happens in actual episodes (i.e., "what," where," and "when"). They also often use verbal materials, which makes the test sensitive to performance in non-memory domains (Helmstaedter et al. 2009), permitting for compensatory strategies which could erroneously reveal normal "memory."

Subjective reports rarely scale with performance on traditional memory tests, warranting criticism that such measures wrongly estimating memory capacities for everyday situations. For example, patients reporting topographical memory deficits have preserved ability in tabletop tests of spatial or geographical knowledge (Habib and Sirigu 1987). Cognitive complaints in amnesiacs typically show little correlation with verbal memory tests used in clinical settings (Chaytor and Schmitter-Edgecombe 2003). Pflueger et al. (2018) designed a VR version of a common verbal memory test and found that the gap between age groups in memory performance was smaller than the traditional assessment, again positioning VR as a more "age-fair" medium. Taken together, clinical cognitive assessments should leverage the affordances of VR paradigms: personalized, objective, reliable, and ecologically valid assessments that are impervious to subjectivity or compensatory strategies and yield behavioral metrics that capture subjective reports (Bohil et al. 2011; Parsons 2015; van Bennekom et al. 2017).

Remembering information and events while completing tasks in a complex environment is one of the crucial abilities needed for the preservation of autonomy in individuals with cognitive impairment (Perneczky et al. 2006). Practicality, however, precludes assessment in standardized RW surroundings (Cushman et al. 2008). "Serious games" have the potential to be effective tools in the prognosis, management, and potentially treatment of cognitive impairments, especially in scenarios that simulate daily activities (Zucchella et al. 2014). VR tests of memory that emphasize feature binding and recall/recognition tests of VE details following virtual executions of everyday life tasks tend to show strong relationships with self-reports (Plancher et al. 2010; Widmann et al. 2012). Likewise, a VR kitchen can capture the RW impairment in Alzheimer's disease (AD); patients performed virtual tasks like preparing a cup of coffee with the same aptitude as in the RW and the objective measures scaled with caregiver reports (Allain et al. 2014).

Virtual supermarkets have been shown as a viable tool for measuring executive function in patients with mild cognitive impairment (MCI) and healthy older adults (OAs; Werner et al. 2009; Zygouris et al. 2015), making it an effective screening tool that can be done at-home over multiple periods – a benefit that increases classification accuracy, sensitivity, and specificity (Zygouris et al. 2017). Tasks in grocery stores can also probe spatial orientation performance that can discriminate AD and frontotemporal dementia groups. Grocery tasks that test prospective memory (PM; e.g., return to pharmacist when hearing your number) and retrospective memory (e.g., securing items from a learned shopping list) are not susceptible to compensatory benefits of executive function, allowing such tools to be specific and reliable assessors (Parsons and Barnett 2017). Similarly, using a fire evacuation task that is sensitive to subtle errors can serve as a tool for both dementia screening (Tarnanas et al. 2013) and MCI prognosis (Tarnanas et al. 2014).

PM memory strength in life-like VR scenarios such as carrying out errands can differentiate OA and mild AD while also ascertaining an individual's ability to be trusted with such tasks in the RW (Knight and Titov 2009; Lecouvey et al. 2019). VR-based memory tests of cognition that are sensitive to population-specific behaviors and deficits are helping with clinical classification (Déjos et al. 2012; La Corte et al. 2019), permitting for early interventions that may dampen disease impact. A seminal VR study showed that an MCI patient had recall deficits when changing viewpoints (King et al. 2002), despite normal recognition of topographical scenes (Spiers et al. 2001b). Intact performance when using egocentric navigation strategies vs. depreciated performance when using allocentric strategies in a VR MWM was also crucial for dissociating another amnesic from controls (Goodrich-Hunsaker et al. 2010) and amnestic MCI (aMCI) from AD (Laczó et al. 2010). A reduction in ability to leverage solutions shown from a bird's eye view perspective when later navigating a first-person VR maze navigation (i.e., allo- to egocentric spatial memory [SM]) is pronounced in AD patients (Morganti et al. 2013) who also show SM and non-verbal EM impairment when navigating to a temporally ordered series of goal locations (Kalová et al. 2005).

VR researchers have also designed diagnostically-sensitive tasks that necessarily recruit the Entorhinal cortex – a region contributing to remembering and navigating to learned places along novel paths and one of the first to exhibit neurodegeneration in AD (Howett et al. 2019). Paired with fMRI, Agosta et al. (2020) postulate that such tasks can identify compensatory mechanisms during performance on memory tasks even if outcome is matched to controls. Skouras et al. (2020) cleverly approached assessment from a neurofeedback angle: individuals were tasked with mentally increasing their velocity in a VE (yoked to downregulation of Hipp. CA1 activity), revealing a reliable functional signature that was characteristic of advanced AD stages evident in high-risk individuals that were cognitively unimpaired. Showing decreased grid-cell like representations during navigation in VEs has also been shown in adults with a genetic risk for AD (Kunz et al. 2015). Similarly, since the neural response of the VR radial arm maze (RAM) is so well known, compensatory mechanisms can be used to predict risk or severity in a variety of psychiatric disorders despite normal task performance (Astur et al. 2005; Migo et al. 2016; Wilkins et al. 2017).

4.1 Profiling Memory-Impaired Populations

Given that both AD and aMCI are characterized by episodic memory impairment, it can be difficult to differentiate the two. Spatial navigation disturbances can provide robust neural signatures of MCI and AD like recruitment of lower-order, task-irrelevant cerebral systems (Drzezga et al. 2005). VE landmark and object recognition tasks show less mistakes while showing more confidence in foils – an impairment that was more pronounced in AD than aMCI (Zakzanis et al. 2009). Serino et al. (2015) showed a deficit in aMCI and AD patients in ability to encode and store allocentric viewpoints, but that AD had difficulty storing allocentric viewpoints and syncing them with viewpoint-dependent representation. Individual differences on allocentric memory assessments following a driving task where patients were either passengers or drivers were capable of dissociating aMCI, AD, and OA (Plancher et al. 2012). Showing SM impairment on the RAM has also been shown to be predictive of which aMCI patients will convert to AD (Lee et al. 2014). Cushman et al. (2008) showed close correlations between RW and virtual navigational deficits that distinguished between controls, MCI, and early AD. AD can also be identified amidst aging, aMCI, and frontotemporal lobar degeneration by poorer performance on temporal memory tests that are void of environmental cues like asking participants to reproduce a virtual route using only body-turns (Bellassen et al. 2012).

VR has also been used to determine the impact of lesions and brain damage on RW memory demands, revealing specific profiles of damage and benchmarking recovery (e.g., impacts on egocentric but not allocentric memory (Weniger et al. 2009) or transfer between the two (Carelli et al. 2011)). Epilepsy patients were shown to recruit the MTL contralateral to their seizure focus during VR object-location tasks (Frings et al. 2008). van der Ham et al. (2010) identified deficits in temporal vs. SM in patients with R parieto-occipital damage when doing navigation tasks in a VE replica of Tübingen. Testing for recall of object locations in RW VEs has also been used to quantify and track memory-impaired Traumatic Brain Injury (TBI) patients (Matheis et al. 2007) and differentiate them from other memory-impaired populations (Arvind Pala et al. 2014) – a metric that can be useful to determine ecological readiness like with normally-appearing athletes following injury (Slobounov et al. 2010). Using a VR version of the MWM, Livingstone and Skelton (2007) also found that TBI patients do worse than controls when force to rely on distal features (i.e., allocentric strategy) vs. using proximal cues (i.e., egocentric strategy).

5 VR-Based Cognitive Rehabilitation and Enhancement

VR tasks have been used to quantify memory enhancement following deep-brain stimulation (Suthana et al. 2012). However, VR can also upregulate encoding through its embedding of to-be-remembered information within spatial, episodic, and embodied contexts. For example, in a virtual navigation study, Kessels et al. (2011) showed that AD patients had higher implicit memory for objects encountered at key decision points. Could VR experiences be explicitly designed to take advantage of such observations with the goal of enhancing memory in patient populations? Can VR be used to repurpose intact functionality to compensate for deficits?

My grandfather's SM was impressive – he could take me through a detailed Google Maps tour of his childhood neighborhood in Italy while completely forgetting that we had shared the same experience mere minutes ago. Can we use VR to drape non-spatial information over the inherently stronger scaffolds that support SM? Indeed, instructions that favor the link between content and its context serve as an effective compensatory strategy for deficient memory processing (Luo and Craik 2008). Empirically driven designs of VR paradigms can provide engaging opportunities to practice tasks, learn compensatory strategies, or intentionally upregulate depreciating brain regions.

5.1 Healthy Aging

Severity of age-related SM decline is correlated with the use of response over spatial strategies (see "Impact of Immersion and Presence" section above) when navigating a virtual maze and is hallmarked by reduced Hipp. gray matter (Konishi and Bohbot 2013), suggesting that the use of spatial encoding could mitigate deviant aging deficits. VR allows older adults to become unbound from the small and often monotonous spaces that their limited mobility confines them to while receiving meaningful, movement-focused mental stimulation in enriched environments that encourages stronger encoding. Benoit et al. (2015) showed that VR is well tolerated by older adults and stimulates autobiographical recollection and conveys scene familiarity – positioning VR as a reminiscence therapy tool (Repetto et al. 2016). Long-term training of OAs also appears viable: 6 months of VR training powerfully increased long-term recall (Optale et al. 2010).

In older adults, active encoding during virtual navigation (Sauzéon et al. 2016) or walking (Tascón et al. 2018) and deciding the itinerary/actively controlling VR navigation (Diersch and Wolbers 2019; Jebara et al. 2014) strengthen distinctive memory traces, enrich source memory, and enhance EM. Training on tasks that mimic daily life yields upticks in visual memory, attention, and cognitive flexibility in older adults (Gamito et al. 2019). Even simple, fun, and engaging tasks like playing Super Mario 64 can increase Hipp. gray volume in both older and younger adults (Kühn et al. 2014; West et al. 2017). A realistic room-scale VR task where OAs with memory complaints needed to memorize a list of grocery objects, engage in conversation mid-task, and then gather the items amongst semantically similar distractors showed marked improvement in auditory recall (Boller et al. 2021).

5.2 Back to Baseline

Where memory for an event is the crux of a disorder (e.g., PTSD), VR can be used to create stimuli that are maximally similar to an individual's feared stimuli (Foa and Kozak 1986; Maples-Keller et al. 2017), as has been shown in cases of road accident victims (Beck et al. 2007) and soldiers (Bohil et al. 2011; Gerardi et al. 2008; Reger and Gahm 2008; Wood et al. 2008) in order to support extinguishment over multiple environments (Dunsmoor et al. 2014).

The use of VR in brain damage rehab can help to train cognitive functioning for everyday activities (Rose et al. 2005). Vascular brain injury patients have shown enhanced SM after VR training (D. Rose et al. 1999). A TBI patient showed memory improvement and increased activation of Hipp. regions following a virtual navigation task that they genuinely enjoyed (Caglio et al. 2012). By using realistic event-based PM tasks, Acquired Brain Injury patients showed significant improvement in both VR and RW PM (Yip and Man 2013).

Amnesic patients were able to learn routes through their hospital after training in a virtual reconstruction of the same environment (Brooks 1999a) and an AD case study showed that relearning cooking activities in VR transfers to real life and remains stable over time (Foloppe et al. 2018). For more information, see Chapter "VR for Spatial Navigation and Action Science" of this volume, D'Cunha et al. (2019), García-Betances et al. (2015), Clay et al. (2020), and Larson et al. (2014).

5.3 *Above Baseline*

The pursuit of cognitive enhancement, especially through extending the breadth of memory, is a perennial modus operandi. Virtual makeovers of the ancient, highly effective mnemonic technique known as the Method of Loci can increase recall by 28% within a single session (Reggente et al. 2020), and its benefits increase as a function of subjective immersion (Krokos et al. 2019). Simply navigating novel VEs has increased motivation during Hipp. dependent measures of memory (Schomaker et al. 2014). Mere hours playing a racing game can yield microstructural changes supporting localized neuroplasticity in the Hipp. of both humans and rats (Sagi et al. 2012). While this chapter has emphasized the importance of context for memory, encoding multiple times in the same environment (e.g., RW classroom) can induce "contextual crutch" and interference (S. M. Smith and Handy 2016) – phenomena that VR can mitigate with multiple distinctive VEs (Perfect and Lindsay 2013; Smith and Handy 2014; Essoe et al. 2022).

High-gamma oscillations during exploration of novel VEs are crucial for successful encoding (Park et al. 2014). Such biomarkers could be utilized to trigger stimuli presentation when individuals are in more impressionable states. Similarly, VR could embed emotionally valent material during encoding to bolster memory strength (Chan et al. 2014). VEs can also help encode information passively. For example, an individual could rapidly encode an environment that has two rooms: one with an elephant in it and then a door to another with a monkey in front of a car that has two balloons on it. By subsequently revealing a "key" (e.g., take the first letter of each object, let each door be an equal sign, and use balloons as exponents), individuals will realize they have passively encoded $E = MC^2$.

6 Outro

This chapter has positioned VR as a potent environmental simulator in which cognition can be measured and modified with unparalleled granularity. Given the intimate relationship between memory and spatial processing, VEs have served as a ripe medium for ecologically valid investigations into memory and the facets of cognition it subserves. If the blind men from the parable in the introduction are researchers looking for the veridical underpinnings of cognition, then VR studies of memory have equipped them with dozens of highly sensitive hands.

The portability of VR in conjunction with other easily accessible data devices (e.g., activity trackers) sets the stage for conducting at-home crowd-sourced studies. Such non-laboratory-based efforts can still be financially incentivized while remaining anonymized with blockchain technology. Proceedings would allow for the creation of large, longitudinal normative databases that permit for investigations of how lifestyle attributes like sleep quality, activity levels, and drug use impact memory. The inherently engaging qualities of VR, coupled with its ability to implicitly quantify and enhance memory, make it a powerful tool in populations spanning from pediatrics to the elderly. As the era of "quantified self" is continually ushered into the collective, the evolution of VR and its incorporation into daily routines will provide pervasive upregulation of encoding processes as well as life-long benchmarks of function that can identify and prevent insidious descents of cognition.

References

Abichou K, La Corte V, Sperduti M, Gaston-Bellegarde A, Nicolas S, Piolino P (2021) The production of false recognition and the associated state of consciousness following encoding in a naturalistic context in aging. Conscious Cogn 90:103097. https://doi.org/10.1016/j.concog.2021.103097

Agosta F, Canu E, Filippi M (2020) Virtual reality and real-time neurofeedback functional MRI: a breakthrough in foreseeing Alzheimer's disease? Brain 143(3):722–726. https://doi.org/10.1093/brain/awaa038

Aguirre GK, D'Esposito M (1997) Environmental knowledge is subserved by separable dorsal/ventral neural areas. J Neurosci 17(7):2512–2518. https://doi.org/10.1523/JNEUROSCI.17-07-02512.1997

Allain P, Foloppe DA, Besnard J, Yamaguchi T, Etcharry-Bouyx F, Gall DL, Nolin P, Richard P (2014) Detecting everyday action deficits in Alzheimer's disease using a nonimmersive virtual reality kitchen. J Int Neuropsychol Soc 20(5):468–477. https://doi.org/10.1017/S1355617714000344

Alvarez RP, Biggs A, Chen G, Pine DS, Grillon C (2008) Contextual fear conditioning in humans: cortical-hippocampal and amygdala contributions. J Neurosci Off J Soc Neurosci 28(24): 6211–6219. https://doi.org/10.1523/JNEUROSCI.1246-08.2008

Andreano J, Liang K, Kong L, Hubbard D, Wiederhold BK, Wiederhold MD (2009) Auditory cues increase the hippocampal response to unimodal virtual reality. Cyberpsychol Behav 12(3): 309–313. https://doi.org/10.1089/cpb.2009.0104

Antonova E, Parslow D, Brammer M, Simmons A, Williams S, Dawson GR, Morris R (2011) Scopolamine disrupts hippocampal activity during allocentric spatial memory in humans: an fMRI study using a virtual reality analogue of the Morris water maze. J Psychopharmacol 25(9): 1256–1265. https://doi.org/10.1177/0269881110379285

Armougum A, Orriols E, Gaston-Bellegarde A, Marle CJ-L, Piolino P (2019) Virtual reality: a new method to investigate cognitive load during navigation. J Environ Psychol 65:101338. https://doi.org/10.1016/j.jenvp.2019.101338

Arvind Pala P, N'Kaoua B, Mazaux J-M, Simion A, Lozes S, Sorita E, Sauzéon H (2014) Everyday-like memory and its cognitive correlates in healthy older adults and in young patients with traumatic brain injury: a pilot study based on virtual reality. Disabil Rehabil Assist Technol 9(6):463–473. https://doi.org/10.3109/17483107.2014.941952

Astur RS, St. Germain SA, Baker EK, Calhoun V, Pearlson GD, Constable RT (2005) FMRI hippocampal activity during a VirtualRadial arm maze. Appl Psychophysiol Biofeedback 30(3): 307–317. https://doi.org/10.1007/s10484-005-6385-z

Beck JG, Palyo SA, Winer EH, Schwagler BE, Ang EJ (2007) Virtual reality exposure therapy for PTSD symptoms after a road accident: an uncontrolled case series. Behav Ther 38(1):39–48. https://doi.org/10.1016/j.beth.2006.04.001

Bellassen V, Iglói K, de Souza LC, Dubois B, Rondi-Reig L (2012) Temporal order memory assessed during spatiotemporal navigation as a behavioral cognitive marker for differential Alzheimer's disease diagnosis. J Neurosci 32(6):1942–1952. https://doi.org/10.1523/JNEUROSCI.4556-11.2012

Benoit M, Guerchouche R, Petit P-D, Chapoulie E, Manera V, Chaurasia G, Drettakis G, Robert P (2015) Is it possible to use highly realistic virtual reality in the elderly? A feasibility study with image-based rendering. Neuropsychiatr Dis Treat 11:557–563. https://doi.org/10.2147/NDT.S73179

Bergouignan L, Nyberg L, Ehrsson HH (2014) Out-of-body–induced hippocampal amnesia. Proc Natl Acad Sci 111(12):4421–4426. https://doi.org/10.1073/pnas.1318801111

Blondé P, Makowski D, Sperduti M, Piolino P (2021) In medio stat virtus: intermediate levels of mind wandering improve episodic memory encoding in a virtual environment. Psychol Res 85(4):1613–1625. https://doi.org/10.1007/s00426-020-01358-5

Bohbot VD, Iaria G, Petrides M (2004) Hippocampal function and spatial memory: evidence from functional neuroimaging in healthy participants and performance of patients with medial temporal lobe resections. Neuropsychology 18(3):418–425. https://doi.org/10.1037/0894-4105.18.3.418

Bohbot VD, Lerch J, Thorndycraft B, Iaria G, Zijdenbos AP (2007) Gray matter differences correlate with spontaneous strategies in a human virtual navigation task. J Neurosci 27(38): 10078–10083. https://doi.org/10.1523/JNEUROSCI.1763-07.2007

Bohil CJ, Alicea B, Biocca FA (2011) Virtual reality in neuroscience research and therapy. Nat Rev Neurosci 12(12):752–762. https://doi.org/10.1038/nrn3122

Boller B, Ouellet É, Belleville S (2021) Using virtual reality to assess and promote transfer of memory training in older adults with memory complaints: a randomized controlled trial. Front Psychol 12:627242. https://doi.org/10.3389/fpsyg.2021.627242

Bower GH (1970) Analysis of a mnemonic device: modern psychology uncovers the powerful components of an ancient system for improving memory. Am Sci 58(5):496–510

Bréchet L, Hausmann SB, Mange R, Herbelin B, Blanke O, Serino A (2020) Subjective feeling of re-experiencing past events using immersive virtual reality prevents a loss of episodic memory. Brain Behav 10(6):e01571. https://doi.org/10.1002/brb3.1571

Brooks BM (1999a) Route learning in a case of amnesia: a preliminary investigation into the efficacy of training in a virtual environment. Neuropsychol Rehabil 9(1):63–76. https://doi.org/10.1080/713755589

Brooks BM (1999b) The specificity of memory enhancement during interaction with a virtual environment. Memory 7(1):65–78. https://doi.org/10.1080/741943713

Brown TI, Ross RS, Keller JB, Hasselmo ME, Stern CE (2010) Which way was I going? Contextual retrieval supports the disambiguation of well learned overlapping navigational routes. J Neurosci 30(21):7414–7422. https://doi.org/10.1523/JNEUROSCI.6021-09.2010

Brown TI, Hasselmo ME, Stern CE (2014) A high-resolution study of hippocampal and medial temporal lobe correlates of spatial context and prospective overlapping route memory. Hippocampus 24(7):819–839. https://doi.org/10.1002/hipo.22273

Buchy L, Hawco C, Bodnar M, Izadi S, Dell'Elce J, Messina K, Lepage M (2014) Functional magnetic resonance imaging study of external source memory and its relation to cognitive insight in non-clinical subjects. Psychiatry Clin Neurosci 68(9):683–691. https://doi.org/10.1111/pcn.12177

Burgess N (2006) Spatial memory: how egocentric and allocentric combine. Trends Cogn Sci 10(12):551–557. https://doi.org/10.1016/j.tics.2006.10.005

Burgess N, Maguire EA, Spiers HJ, O'Keefe J (2001) A temporoparietal and prefrontal network for retrieving the spatial context of lifelike events. Neuroimage 14(2):439–453. https://doi.org/10.1006/nimg.2001.0806

Burgess N, Maguire EA, O'Keefe J (2002) The human hippocampus and spatial and episodic memory. Neuron 35(4):625–641. https://doi.org/10.1016/S0896-6273(02)00830-9

Caglio M, Latini-Corazzini L, D'Agata F, Cauda F, Sacco K, Monteverdi S, Zettin M, Duca S, Geminiani G (2012) Virtual navigation for memory rehabilitation in a traumatic brain injured patient. Neurocase 18(2):123–131. https://doi.org/10.1080/13554794.2011.568499

Cánovas R, Espínola M, Iribarne L, Cimadevilla JM (2008) A new virtual task to evaluate human place learning. Behav Brain Res 190(1):112–118. https://doi.org/10.1016/j.bbr.2008.02.024

Carelli L, Rusconi ML, Scarabelli C, Stampatori C, Mattioli F, Riva G (2011) The transfer from survey (map-like) to route representations into virtual reality mazes: effect of age and cerebral lesion. J Neuroeng Rehabil 8(1):6. https://doi.org/10.1186/1743-0003-8-6

Chan E, Baumann O, Bellgrove MA, Mattingley JB (2014) Negative emotional experiences during navigation enhance Parahippocampal activity during recall of place information. J Cogn Neurosci 26(1):154–164. https://doi.org/10.1162/jocn_a_00468

Chang C-Y, Sung H-Y, Guo J-L, Chang B-Y, Kuo F-R (2019) Effects of spherical video-based virtual reality on nursing students' learning performance in childbirth education training. Interact Learn Environ:1–17. https://doi.org/10.1080/10494820.2019.1661854

Chaytor N, Schmitter-Edgecombe M (2003) The ecological validity of neuropsychological tests: a review of the literature on everyday cognitive skills. Neuropsychol Rev 13(4):181–197. https://doi.org/10.1023/B:NERV.0000009483.91468.fb

Chen H-Y, Gilmore AW, Nelson SM, McDermott KB (2017) Are there multiple kinds of episodic memory? An fMRI investigation comparing autobiographical and recognition memory tasks. J Neurosci 37(10):2764–2775. https://doi.org/10.1523/JNEUROSCI.1534-16.2017

Chow TE, Westphal AJ, Rissman J (2018) Multi-voxel pattern classification differentiates personally experienced event memories from secondhand event knowledge. Neuroimage 176:110–123. https://doi.org/10.1016/j.neuroimage.2018.04.024

Chrastil ER, Warren WH (2015) Active and passive spatial learning in human navigation: acquisition of graph knowledge. J Exp Psychol Learn Mem Cogn 41(4):1162–1178. https://doi.org/10.1037/xlm0000082

Christou C, Bülthoff HH (1999) The perception of spatial layout in a virtual world. https://pure.mpg.de/pubman/faces/ViewItemOverviewPage.jsp?itemId=item_1793661

Clay F, Howett D, FitzGerald J, Fletcher P, Chan D, Price A (2020) Use of immersive virtual reality in the assessment and treatment of Alzheimer's disease: a systematic review. J Alzheimers Dis 75(1):23–43. https://doi.org/10.3233/JAD-191218

Clemenson GD, Stark CEL (2015) Virtual environmental enrichment through video games improves hippocampal-associated memory. J Neurosci 35(49):16116–16125. https://doi.org/10.1523/JNEUROSCI.2580-15.2015

Clemenson GD, Lee SW, Deng W, Barrera VR, Iwamoto KS, Fanselow MS, Gage FH (2015) Enrichment rescues contextual discrimination deficit associated with immediate shock. Hippocampus 25(3):385–392. https://doi.org/10.1002/hipo.22380

Clemenson GD, Henningfield CM, Stark CEL (2019) Improving hippocampal memory through the experience of a rich minecraft environment. Front Behav Neurosci 0. https://doi.org/10.3389/fnbeh.2019.00057

Clemenson GD, Wang L, Mao Z, Stark SM, Stark CEL (2020) Exploring the spatial relationships between real and virtual experiences: what transfers and what doesn't. Front Virtual Real. https://doi.org/10.3389/frvir.2020.572122

Climent G, Rodríguez C, García T, Areces D, Mejías M, Aierbe A, Moreno M, Cueto E, Castellá J, Feli González M (2021) New virtual reality tool (Nesplora aquarium) for assessing attention and working memory in adults: a normative study. Appl Neuropsychol Adult 28(4):403–415. https://doi.org/10.1080/23279095.2019.1646745

Coleman B, Marion S, Rizzo A, Turnbull J, Nolty A (2019) Virtual reality assessment of classroom – related attention: an ecologically relevant approach to evaluating the effectiveness of working memory training. Front Psychol 10:1851. https://doi.org/10.3389/fpsyg.2019.01851

Connors EC, Chrastil ER, Sánchez J, Merabet LB (2014) Virtual environments for the transfer of navigation skills in the blind: a comparison of directed instruction vs. video game based learning approaches. Front Hum Neurosci 8:223. https://doi.org/10.3389/fnhum.2014.00223

Cushman LA, Stein K, Duffy CJ (2008) Detecting navigational deficits in cognitive aging and Alzheimer disease using virtual reality. Neurology 71(12):888–895. https://doi.org/10.1212/01.wnl.0000326262.67613.fe

Cyr M, Wang Z, Tau GZ, Zhao G, Friedl E, Stefan M, Terranova K, Marsh R (2016) Reward-based spatial learning in teens with bulimia nervosa. J Am Acad Child Adolesc Psychiatry 55(11):962–971.e3. https://doi.org/10.1016/j.jaac.2016.07.778

D'Cunha NM, Nguyen D, Naumovski N, McKune AJ, Kellett J, Georgousopoulou EN, Frost J, Isbel S (2019) A mini-review of virtual reality-based interventions to promote well-being for people living with dementia and mild cognitive impairment. Gerontology 65(4):430–440. https://doi.org/10.1159/000500040

Davidson TJ, Kloosterman F, Wilson MA (2009) Hippocampal replay of extended experience. Neuron 63(4):497–507. https://doi.org/10.1016/j.neuron.2009.07.027

Dede C (2009) Immersive interfaces for engagement and learning. Science 323(5910):66–69. https://doi.org/10.1126/science.1167311

Déjos M, Sauzéon H, N'kaoua, B. (2012) Virtual reality for clinical assessment of elderly people: early screening for dementia. Rev Neurol 168(5):404–414. https://doi.org/10.1016/j.neurol.2011.09.005

Demanuele C, Bähner F, Plichta MM, Kirsch P, Tost H, Meyer-Lindenberg A, Durstewitz D (2015) A statistical approach for segregating cognitive task stages from multivariate fMRI BOLD time series. Front Hum Neurosci 9. https://doi.org/10.3389/fnhum.2015.00537

Dennett DC (1993) Consciousness explained. Penguin Adult

Dhindsa K, Drobinin V, King J, Hall GB, Burgess N, Becker S (2014) Examining the role of the temporo-parietal network in memory, imagery, and viewpoint transformations. Front Hum Neurosci 8. https://doi.org/10.3389/fnhum.2014.00709

Diersch N, Wolbers T (2019) The potential of virtual reality for spatial navigation research across the adult lifespan. J Exp Biol 222(Pt Suppl 1):jeb187252. https://doi.org/10.1242/jeb.187252

Dimsdale-Zucker HR, Ritchey M, Ekstrom AD, Yonelinas AP, Ranganath C (2018) CA1 and CA3 differentially support spontaneous retrieval of episodic contexts within human hippocampal subfields. Nat Commun 9(1):1–8. https://doi.org/10.1038/s41467-017-02752-1

Dinh HQ, Walker N, Hodges LF, Song C, Kobayashi A (1999) Evaluating the importance of multi-sensory input on memory and the sense of presence in virtual environments. Proc IEEE Virtual Real:222–228. https://doi.org/10.1109/VR.1999.756955

Dong D, Wong LKF, Luo Z (2019) Assess BA10 activity in slide-based and immersive virtual reality prospective memory task using functional near-infrared spectroscopy (fNIRS). Appl Neuropsychol Adult 26(5):465–471. https://doi.org/10.1080/23279095.2018.1443104

Draschkow D, Kallmayer M, Nobre AC (2021) When natural behavior engages working memory. Curr Biol 31(4):869–874.e5. https://doi.org/10.1016/j.cub.2020.11.013

Driscoll I, Hamilton DA, Yeo RA, Brooks WM, Sutherland RJ (2005) Virtual navigation in humans: the impact of age, sex, and hormones on place learning. Horm Behav 47(3): 326–335. https://doi.org/10.1016/j.yhbeh.2004.11.013

Drzezga A, Grimmer T, Peller M, Wermke M, Siebner H, Rauschecker JP, Schwaiger M, Kurz A (2005) Impaired cross-modal inhibition in Alzheimer disease. PLoS Med 2(10):e288. https://doi.org/10.1371/journal.pmed.0020288

Dunsmoor JE, Ahs F, Zielinski DJ, LaBar KS (2014) Extinction in multiple virtual reality contexts diminishes fear reinstatement in humans. Neurobiol Learn Mem 113:157–164. https://doi.org/10.1016/j.nlm.2014.02.010

Eichenbaum H, Cohen NJ (2001) From conditioning to conscious recollection: memory systems of the brain. Oxford University Press

Essoe JK-Y, Reggente N, Ohno AA, Baek YH, Dell'italia J, Rissman J (2022) Enhancing learning and retention with distinctive virtual reality environments and mental context reinstatement. PsyArXiv. https://doi.org/10.31234/osf.io/vy4b6

Faul L, Stjepanović D, Stivers JM, Stewart GW, Graner JL, Morey RA, LaBar KS (2020) Proximal threats promote enhanced acquisition and persistence of reactive fear-learning circuits. Proc Natl Acad Sci 117(28):16678–16689. https://doi.org/10.1073/pnas.2004258117

Flannery KA, Walles R (2003) How does schema theory apply to real versus virtual memories? Cyberpsychol Behav 6(2):151–159. https://doi.org/10.1089/109493103321640347

Foa EB, Kozak MJ (1986) Emotional processing of fear: exposure to corrective information. Psychol Bull 99(1):20–35. https://doi.org/10.1037/0033-2909.99.1.20

Foglia L, Wilson RA (2013) Embodied cognition. WIREs Cogn Sci 4(3):319–325. https://doi.org/10.1002/wcs.1226

Foloppe DA, Richard P, Yamaguchi T, Etcharry-Bouyx F, Allain P (2018) The potential of virtual reality-based training to enhance the functional autonomy of Alzheimer's disease patients in cooking activities: a single case study. Neuropsychol Rehabil 28(5):709–733. https://doi.org/10.1080/09602011.2015.1094394

Fox J, Bailenson J, Binney J (2009) Virtual experiences, physical behaviors: the effect of presence on imitation of an eating avatar. Presence Teleop Virt 18(4):294–303

Frances Wang R, Simons DJ (1999) Active and passive scene recognition across views. Cognition 70(2):191–210. https://doi.org/10.1016/S0010-0277(99)00012-8

Frans Ö, Åhs J, Bihre E, Åhs F (2018) Distance to threat and risk of acute and posttraumatic stress disorder following bank robbery: a longitudinal study. Psychiatry Res 267:461–466. https://doi.org/10.1016/j.psychres.2018.06.050

Freund J, Brandmaier AM, Lewejohann L, Kirste I, Kritzler M, Krüger A, Sachser N, Lindenberger U, Kempermann G (2013) Emergence of individuality in genetically identical mice. Science 340(6133):756–759. https://doi.org/10.1126/science.1235294

Frings L, Wagner K, Halsband U, Schwarzwald R, Zentner J, Schulze-Bonhage A (2008) Lateralization of hippocampal activation differs between left and right temporal lobe epilepsy patients and correlates with postsurgical verbal learning decrement. Epilepsy Res 78(2):161–170. https://doi.org/10.1016/j.eplepsyres.2007.11.006

Gamito P, Oliveira J, Morais D, Coelho C, Santos N, Alves C, Galamba A, Soeiro M, Yerra M, French H, Talmers L, Gomes T, Brito R (2019) Cognitive stimulation of elderly individuals with instrumental virtual reality-based activities of daily life: pre-post treatment study. Cyberpsychol Behav Soc Netw 22(1):69–75. https://doi.org/10.1089/cyber.2017.0679

Ganesh S, van Schie HT, de Lange PF, Thompson E, Wigboldus DHJ (2012) How the human brain goes virtual: distinct cortical regions of the person-processing network are involved in self-identification with virtual agents. Cereb Cortex 22(7):1577–1585. https://doi.org/10.1093/cercor/bhr227

García-Betances RI, Arredondo Waldmeyer MT, Fico G, Cabrera-Umpiérrez MF (2015) A succinct overview of virtual reality technology use in Alzheimer's disease. Front Aging Neurosci. https://doi.org/10.3389/fnagi.2015.00080

Gaunet F, Vidal M, Kemeny A, Berthoz A (2001) Active, passive and snapshot exploration in a virtual environment: influence on scene memory, reorientation and path memory. Cogn Brain Res 11(3):409–420. https://doi.org/10.1016/S0926-6410(01)00013-1

Gerardi M, Rothbaum BO, Ressler K, Heekin M, Rizzo A (2008) Virtual reality exposure therapy using a virtual Iraq: case report. J Trauma Stress 21(2):209–213. https://doi.org/10.1002/jts.20331

Glenberg AM, Hayes J (2016) Contribution of embodiment to solving the riddle of infantile amnesia. Front Psychol 7:10. https://doi.org/10.3389/fpsyg.2016.00010

Godden DR, Baddeley AD (1975) Context-dependent memory in two natural environments: on land and underwater. Br J Psychol 66(3):325–331. https://doi.org/10.1111/j.2044-8295.1975.tb01468.x

Gomez A, Cerles M, Rousset S, Rémy C, Baciu M (2014) Differential hippocampal and retrosplenial involvement in egocentric-updating, rotation, and allocentric processing during online spatial encoding: an fMRI study. Front Hum Neurosci 8:150. https://doi.org/10.3389/fnhum.2014.00150

Goodrich-Hunsaker NJ, Livingstone SA, Skelton RW, Hopkins RO (2010) Spatial deficits in a virtual water maze in amnesic participants with hippocampal damage. Hippocampus 20(4):481–491. https://doi.org/10.1002/hipo.20651

Groot YCT, Wilson BA, Evans J, Watson P (2002) Prospective memory functioning in people with and without brain injury. J Int Neuropsychol Soc 8(5):645–654. https://doi.org/10.1017/S1355617702801321

Habib M, Sirigu A (1987) Pure topographical disorientation: a definition and anatomical basis. Cortex 23(1):73–85. https://doi.org/10.1016/S0010-9452(87)80020-5

Hahm J, Lee K, Lim S-L, Kim S-Y, Kim H-T, Lee J-H (2007) Effects of active navigation on object recognition in virtual environments. Cyberpsychol Behav 10(2):305–308. https://doi.org/10.1089/cpb.2006.9952

Hartley T, Maguire EA, Spiers HJ, Burgess N (2003) The well-worn route and the path less traveled: distinct neural bases of route following and wayfinding in humans. Neuron 37(5):877–888

Hassabis D, Chu C, Rees G, Weiskopf N, Molyneux PD, Maguire EA (2009) Decoding neuronal ensembles in the human hippocampus. Curr Biol 19(7):546–554

Hebscher M, Levine B, Gilboa A (2017) The precuneus and hippocampus contribute to individual differences in the unfolding of spatial representations during episodic autobiographical memory. Neuropsychologia

Hejtmanek L, Starrett M, Ferrer E, Ekstrom AD (2020) How much of what we learn in virtual reality transfers to real-world navigation? Multisens Res 33(4–5):479–503. https://doi.org/10.1163/22134808-20201445

Helmstaedter C, Wietzke J, Lutz MT (2009) Unique and shared validity of the "Wechsler logical memory test", the "California verbal learning test", and the "verbal learning and memory test" in patients with epilepsy. Epilepsy Res 87(2):203–212. https://doi.org/10.1016/j.eplepsyres.2009.09.002

Howett D, Castegnaro A, Krzywicka K, Hagman J, Marchment D, Henson R, Rio M, King JA, Burgess N, Chan D (2019) Differentiation of mild cognitive impairment using an entorhinal cortex-based test of virtual reality navigation. Brain 142(6):1751–1766. https://doi.org/10.1093/brain/awz116

Huffman DJ, Ekstrom AD (2019) A modality-independent network underlies the retrieval of large-scale spatial environments in the human brain. Neuron 104(3):611–622.e7. https://doi.org/10.1016/j.neuron.2019.08.012

Huffman DJ, Ekstrom AD (2021) An important step toward understanding the role of body-based cues on human spatial memory for large-scale environments. J Cogn Neurosci 33(2):167–179. https://doi.org/10.1162/jocn_a_01653

Iaria G, Petrides M, Dagher A, Pike B, Bohbot VD (2003) Cognitive strategies dependent on the hippocampus and caudate nucleus in human navigation: variability and change with practice. J Neurosci 23(13):5945–5952. https://doi.org/10.1523/JNEUROSCI.23-13-05945.2003

Igloi K, Doeller CF, Paradis A-L, Benchenane K, Berthoz A, Burgess N, Rondi-Reig L (2015) Interaction between hippocampus and cerebellum crus I in sequence-based but not place-based navigation. Cereb Cortex 25(11):4146–4154. https://doi.org/10.1093/cercor/bhu132

Jaiswal N, Ray W, Slobounov S (2010) Encoding of visual-spatial information in working memory requires more cerebral efforts than retrieval: evidence from an EEG and virtual reality study. Brain Res 1347:80–89. https://doi.org/10.1016/j.brainres.2010.05.086

James KH, Humphrey GK, Vilis T, Corrie B, Baddour R, Goodale MA (2002) "Active" and "passive" learning of three-dimensional object structure within an immersive virtual reality environment. Behav Res Methods Instrum Comput 34(3):383–390. https://doi.org/10.3758/BF03195466

Jang S, Vitale JM, Jyung RW, Black JB (2017) Direct manipulation is better than passive viewing for learning anatomy in a three-dimensional virtual reality environment. Comput Educ 106:150–165. https://doi.org/10.1016/j.compedu.2016.12.009

Jang S, Choi J, Oh J, Yeom J, Hong N, Lee N, Kwon JH, Hong J, Kim J, Kim E (2021) Use of virtual reality working memory task and functional near-infrared spectroscopy to assess brain hemodynamic responses to methylphenidate in ADHD children. Front Psych 11:1644. https://doi.org/10.3389/fpsyt.2020.564618

Janzen G, Weststeijn CG (2007) Neural representation of object location and route direction: an event-related fMRI study. Brain Res 1165:116–125. https://doi.org/10.1016/j.brainres.2007.05.074

Javadi A-H, Patai EZ, Marin-Garcia E, Margolis A, Tan H-RM, Kumaran D, Nardini M, Penny W, Duzel E, Dayan P, Spiers HJ (2019) Prefrontal dynamics associated with efficient detours and shortcuts: a combined functional magnetic resonance imaging and magnetoencenphalography study. J Cogn Neurosci 31(8):1227–1247. https://doi.org/10.1162/jocn_a_01414

Jebara N, Orriols E, Zaoui M, Berthoz A, Piolino P (2014) Effects of enactment in episodic memory: a pilot virtual reality study with young and elderly adults. Front Aging Neurosci 6. https://doi.org/10.3389/fnagi.2014.00338

Kalová E, Vlček K, Jarolímová E, Bureš J (2005) Allothetic orientation and sequential ordering of places is impaired in early stages of Alzheimer's disease: corresponding results in real space tests and computer tests. Behav Brain Res 159(2):175–186. https://doi.org/10.1016/j.bbr.2004.10.016

Kalpouzos G, Eriksson J (2013) Memory self-efficacy beliefs modulate brain activity when encoding real-world future intentions. PloS One 8(9):e73850. https://doi.org/10.1371/journal.pone.0073850

Kane HS, McCall C, Collins NL, Blascovich J (2012) Mere presence is not enough: responsive support in a virtual world. J Exp Soc Psychol 48(1):37–44. https://doi.org/10.1016/j.jesp.2011.07.001

Kessels RPC, van Doormaal A, Janzen G (2011) Landmark recognition in Alzheimer's dementia: spared implicit memory for objects relevant for navigation. PloS One 6(4):e18611. https://doi.org/10.1371/journal.pone.0018611

Kim M, Jeffery KJ, Maguire EA (2017) Multivoxel pattern analysis reveals 3D place information in the human hippocampus. J Neurosci 37(16):4270–4279. https://doi.org/10.1523/JNEUROSCI.2703-16.2017

King JA, Burgess N, Hartley T, Vargha-Khadem F, O'Keefe J (2002) Human hippocampus and viewpoint dependence in spatial memory. Hippocampus 12(6):811–820. https://doi.org/10.1002/hipo.10070

Kisker J, Gruber T, Schöne B (2021) Experiences in virtual reality entail different processes of retrieval as opposed to conventional laboratory settings: a study on human memory. Curr Psychol 40(7):3190–3197. https://doi.org/10.1007/s12144-019-00257-2

Knight RG, Titov N (2009) Use of virtual reality tasks to assess prospective memory: applicability and evidence. Brain Impairment 10(1):3–13. https://doi.org/10.1375/brim.10.1.3

Knowlton BJ, Squire LR, Gluck MA (1994) Probabilistic classification learning in amnesia. Learn Mem 1(2):106–120. https://doi.org/10.1101/lm.1.2.106

Kober SE, Neuper C (2012) Using auditory event-related EEG potentials to assess presence in virtual reality. Int J Hum Comput Stud 70(9):577–587. https://doi.org/10.1016/j.ijhcs.2012.03.004

Kolarik BS, Shahlaie K, Hassan A, Borders AA, Kaufman KC, Gurkoff G, Yonelinas AP, Ekstrom AD (2016) Impairments in precision, rather than spatial strategy, characterize performance on the virtual Morris water maze: a case study. Neuropsychologia 80:90–101. https://doi.org/10.1016/j.neuropsychologia.2015.11.013

Kolarik BS, Stark SM, Stark CEL (2020) Enriching hippocampal memory function in older adults through real-world exploration. Front Aging Neurosci. https://doi.org/10.3389/fnagi.2020.00158

Konishi K, Bohbot VD (2013) Spatial navigational strategies correlate with gray matter in the hippocampus of healthy older adults tested in a virtual maze. Front Aging Neurosci. https://doi.org/10.3389/fnagi.2013.00001

Krokos E, Plaisant C, Varshney A (2019) Virtual memory palaces: immersion aids recall. Virtual Reality 23(1):1–15. https://doi.org/10.1007/s10055-018-0346-3

Kühn S, Gleich T, Lorenz RC, Lindenberger U, Gallinat J (2014) Playing super Mario induces structural brain plasticity: gray matter changes resulting from training with a commercial video game. Mol Psychiatry 19(2):265–271. https://doi.org/10.1038/mp.2013.120

Kunz L, Schröder TN, Lee H, Montag C, Lachmann B, Sariyska R, Reuter M, Stirnberg R, Stöcker T, Messing-Floeter PC, Fell J, Doeller CF, Axmacher N (2015) Reduced grid-cell-like representations in adults at genetic risk for Alzheimer's disease. Science 350(6259):430–433. https://doi.org/10.1126/science.aac8128

Kyle CT, Stokes JD, Lieberman JS, Hassan AS, Ekstrom AD (2015) Successful retrieval of competing spatial environments in humans involves hippocampal pattern separation mechanisms. Elife 4:e10499. https://doi.org/10.7554/eLife.10499

La Corte V, Sperduti M, Abichou K, Piolino P (2019) Episodic memory assessment and remediation in normal and pathological aging using virtual reality: a mini review. Front Psychol 10:173. https://doi.org/10.3389/fpsyg.2019.00173

Laczó J, Andel R, Vyhnalek M, Vlcek K, Magerova H, Varjassyova A, Tolar M, Hort J (2010) Human analogue of the Morris water maze for testing subjects at risk of Alzheimer's disease. Neurodegener Dis 7(1–3):148–152. https://doi.org/10.1159/000289226

LaFortune J, Macuga KL (2018) Learning movements from a virtual instructor: effects of spatial orientation, immersion, and expertise. J Exp Psychol Appl 24(4):521–533. https://doi.org/10.1037/xap0000189

Larrue F, Sauzeon H, Wallet G, Foloppe D, Cazalets J-R, Gross C, N'Kaoua B (2014) Influence of body-centered information on the transfer of spatial learning from a virtual to a real environment. J Cogn Psychol 26(8):906–918. https://doi.org/10.1080/20445911.2014.965714

Larson EB, Feigon M, Gagliardo P, Dvorkin AY (2014) Virtual reality and cognitive rehabilitation: a review of current outcome research. NeuroRehabilitation 34(4):759–772. https://doi.org/10.3233/NRE-141078

Lecouvey G, Morand A, Gonneaud J, Piolino P, Orriols E, Pélerin A, Ferreira Da Silva L, de La Sayette V, Eustache F, Desgranges B (2019) An impairment of prospective memory in mild Alzheimer's disease: a ride in a virtual town. Front Psychol 10:241. https://doi.org/10.3389/fpsyg.2019.00241

Lee J-Y, Kho S, Yoo HB, Park S, Choi J-S, Kwon JS, Cha KR, Jung H-Y (2014) Spatial memory impairments in amnestic mild cognitive impairment in a virtual radial arm maze. Neuropsychiatr Dis Treat 10:653–660. https://doi.org/10.2147/NDT.S58185

Legge ELG, Madan CR, Ng ET, Caplan JB (2012) Building a memory palace in minutes: equivalent memory performance using virtual versus conventional environments with the Method of Loci. Acta Psychol (Amst) 141(3):380–390. https://doi.org/10.1016/j.actpsy.2012.09.002

LePort AKR, Mattfeld AT, Dickinson-Anson H, Fallon JH, Stark CEL, Kruggel F, Cahill L, McGaugh JL (2012) Behavioral and neuroanatomical investigation of highly superior autobiographical memory (HSAM). Neurobiol Learn Mem 98(1):78–92. https://doi.org/10.1016/j.nlm.2012.05.002

Lester AW, Moffat SD, Wiener JM, Barnes CA, Wolbers T (2017) The aging navigational system. Neuron 95(5):1019–1035. https://doi.org/10.1016/j.neuron.2017.06.037

Leutgeb JK, Leutgeb S, Moser M-B, Moser EI (2007) Pattern separation in the dentate gyrus and CA3 of the hippocampus. Science 315(5814):961–966. https://doi.org/10.1126/science.1135801

Livingstone SA, Skelton RW (2007) Virtual environment navigation tasks and the assessment of cognitive deficits in individuals with brain injury. Behav Brain Res 185(1):21–31. https://doi.org/10.1016/j.bbr.2007.07.015

Llinas RR (2002) I of the vortex: from neurons to self. MIT Press

Lloyd J, Persaud NV, Powell TE (2009) Equivalence of real-world and virtual-reality route learning: a pilot study. Cyberpsychol Behav 12(4):423–427. https://doi.org/10.1089/cpb.2008.0326

Luo L, Craik FI (2008) Aging and memory: a cognitive approach. Can J Psychiatry 53(6):346–353. https://doi.org/10.1177/070674370805300603

Mackie GO, Burighel P (2005) The nervous system in adult tunicates: current research directions. Can J Zool 83(1):151–183. https://doi.org/10.1139/z04-177

Madan C, Singhal A (2012) Using actions to enhance memory: effects of enactment, gestures, and exercise on human memory. Front Psychol 3:507. https://doi.org/10.3389/fpsyg.2012.00507

Maguire EA, Burgess N, Donnett JG, Frackowiak RS, Frith CD, O'keefe, J. (1998) Knowing where and getting there: a human navigation network. Science 280(5365):921–924

Maguire EA, Gadian DG, Johnsrude IS, Good CD, Ashburner J, Frackowiak RSJ, Frith CD (2000) Navigation-related structural change in the hippocampi of taxi drivers. Proc Natl Acad Sci 97(8):4398–4403. https://doi.org/10.1073/pnas.070039597

Mania K, Troscianko T, Hawkes R, Chalmers A (2003) Fidelity metrics for virtual environment simulations based on spatial memory awareness states. Presence 12(3):296–310. https://doi.org/10.1162/105474603765879549

Mania K, Badariah S, Coxon M, Watten P (2010) Cognitive transfer of spatial awareness states from immersive virtual environments to reality. ACM Trans Appl Percept 7(2):1–14. Article No. 9. https://doi.org/10.1145/1670671.1670673

Mantovani F, Castelnuovo G, Gaggioli A, Riva G (2003) Virtual reality training for health-care professionals. Cyberpsychol Behav 6(4):389–395. https://doi.org/10.1089/109493103322278772

Maples-Keller JL, Yasinski C, Manjin N, Rothbaum BO (2017) Virtual reality-enhanced extinction of phobias and post-traumatic stress. Neurotherapeutics 14(3):554–563. https://doi.org/10.1007/s13311-017-0534-y

Marsh R, Hao X, Xu D, Wang Z, Duan Y, Liu J, Kangarlu A, Martinez D, Garcia F, Tau GZ, Yu S, Packard MG, Peterson BS (2010) A virtual reality-based FMRI study of reward-based spatial learning. Neuropsychologia 48(10):2912–2921. https://doi.org/10.1016/j.neuropsychologia.2010.05.033

Martens MA, Antley A, Freeman D, Slater M, Harrison PJ, Tunbridge EM (2019) It feels real: physiological responses to a stressful virtual reality environment and its impact on working memory. J Psychopharmacol 33(10):1264–1273. https://doi.org/10.1177/0269881119860156

Matar E, Shine JM, Gilat M, Ehgoetz Martens KA, Ward PB, Frank MJ, Moustafa AA, Naismith SL, Lewis SJG (2019) Identifying the neural correlates of doorway freezing in Parkinson's disease. Hum Brain Mapp 40(7):2055–2064. https://doi.org/10.1002/hbm.24506

Matheis RJ, Schultheis MT, Tiersky LA, DeLuca J, Millis SR, Rizzo A (2007) Is learning and memory different in a virtual environment? Clin Neuropsychol 21(1):146–161. https://doi.org/10.1080/13854040601100668

Meade ME, Meade JG, Sauzeon H, Fernandes MA (2019) Active navigation in virtual environments benefits spatial memory in older adults. Brain Sci 9(3) Article 3. https://doi.org/10.3390/brainsci9030047

Migo EM, O'Daly O, Mitterschiffthaler M, Antonova E, Dawson GR, Dourish CT, Craig KJ, Simmons A, Wilcock GK, McCulloch E, Jackson SHD, Kopelman MD, Williams SCR, Morris RG (2016) Investigating virtual reality navigation in amnestic mild cognitive impairment using fMRI. Aging Neuropsychol Cognit 23(2):196–217. https://doi.org/10.1080/13825585.2015.1073218

Miller KM, Price CC, Okun MS, Montijo H, Bowers D (2009) Is the N-back task a valid neuropsychological measure for assessing working memory? Arch Clin Neuropsychol 24(7):711–717. https://doi.org/10.1093/arclin/acp063

Miller JF, Neufang M, Solway A, Brandt A, Trippel M, Mader I, Hefft S, Merkow M, Polyn SM, Jacobs J, Kahana MJ, Schulze-Bonhage A (2013) Neural activity in human hippocampal formation reveals the spatial context of retrieved memories. Science 342(6162):1111–1114. https://doi.org/10.1126/science.1244056

Mobbs D, Petrovic P, Marchant JL, Hassabis D, Weiskopf N, Seymour B, Dolan RJ, Frith CD (2007) When fear is near: threat imminence elicits prefrontal-periaqueductal gray shifts in humans. Science 317(5841):1079–1083. https://doi.org/10.1126/science.1144298

Mobbs D, Wise T, Suthana N, Guzmán N, Kriegeskorte N, Leibo JZ (2021) Promises and challenges of human computational ethology. Neuron 109(14):2224–2238. https://doi.org/10.1016/j.neuron.2021.05.021

Moffat SD, Kennedy KM, Rodrigue KM, Raz N (2007) Extrahippocampal contributions to age differences in human spatial navigation. Cereb Cortex 17(6):1274–1282. https://doi.org/10.1093/cercor/bhl036

Moita MAP, Rosis S, Zhou Y, LeDoux JE, Blair HT (2004) Putting fear in its place: remapping of hippocampal place cells during fear conditioning. J Neurosci Off J Soc Neurosci 24(31):7015–7023. https://doi.org/10.1523/JNEUROSCI.5492-03.2004

Morganti F, Stefanini S, Riva G (2013) From allo- to egocentric spatial ability in early Alzheimer's disease: a study with virtual reality spatial tasks. Cogn Neurosci 4(3–4):171–180. https://doi.org/10.1080/17588928.2013.854762

Mourkoussis N, Rivera FM, Troscianko T, Dixon T, Hawkes R, Mania K (2010) Quantifying fidelity for virtual environment simulations employing memory schema assumptions. ACM Trans Appl Percept 8(1):1–21. Article No. 2. https://doi.org/10.1145/1857893.1857895

Müller N, Campbell S, Nonaka M, Rost TM, Pipa G, Konrad BN, Steiger A, Czisch M, Fernández G, Dresler M, Genzel L (2018) 2D:4D and spatial abilities: from rats to humans. Neurobiol Learn Mem 151:85–87. https://doi.org/10.1016/j.nlm.2018.04.012

Nau M, Navarro Schröder T, Frey M, Doeller CF (2020) Behavior-dependent directional tuning in the human visual-navigation network. Nat Commun 11:3247. https://doi.org/10.1038/s41467-020-17000-2

Norman DG, Wade KA, Williams MA, Watson DG (2020) Caught virtually lying – crime scenes in virtual reality help to expose suspects' concealed recognition. J Appl Res Mem Cogn 9(1):118–127. https://doi.org/10.1016/j.jarmac.2019.12.008

O'Keefe J, Nadel L (1978) The hippocampus as a cognitive map. Oxford University Press

Okita S, Bailenson J, Schwartz D (2008) Mere belief of social action improves complex learning. https://repository.isls.org//handle/1/3144

Optale G, Urgesi C, Busato V, Marin S, Piron L, Priftis K, Gamberini L, Capodieci S, Bordin A (2010) Controlling memory impairment in elderly adults using virtual reality memory training: a randomized controlled pilot study. Neurorehabil Neural Repair 24(4):348–357. https://doi.org/10.1177/1545968309353328

Park J, Lee H, Kim T, Park GY, Lee EM, Baek S, Ku J, Kim IY, Kim SI, Jang DP, Kang JK (2014) Role of low- and high-frequency oscillations in the human hippocampus for encoding environmental novelty during a spatial navigation task. Hippocampus 24(11):1341–1352. https://doi.org/10.1002/hipo.22315

Parker JA, Kaplan AD, Volante WG, Abich J, Sims VK (2020) Exploring the encoding specificity principle and context-dependent recognition in virtual reality. Proc Hum Factors Ergon Soc Annu Meet 64(1):1481–1485. https://doi.org/10.1177/1071181320641353

Parsons TD (2015) Virtual reality for enhanced ecological validity and experimental control in the clinical, affective and social neurosciences. Front Hum Neurosci. https://doi.org/10.3389/fnhum.2015.00660

Parsons TD, Barnett M (2017) Validity of a newly developed measure of memory: feasibility study of the virtual environment grocery store. J Alzheimers Dis 59(4):1227–1235. https://doi.org/10.3233/JAD-170295

Parsons TD, Rizzo AA (2008) Initial validation of a virtual environment for assessment of memory functioning: virtual reality cognitive performance assessment test. Cyberpsychol Behav 11(1):17–25. https://doi.org/10.1089/cpb.2007.9934

Patai EZ, Javadi A-H, Ozubko JD, O'Callaghan A, Ji S, Robin J, Grady C, Winocur G, Rosenbaum RS, Moscovitch M, Spiers HJ (2019) Hippocampal and retrosplenial goal distance coding after long-term consolidation of a real-world environment. Cereb Cortex 29(6):2748–2758. https://doi.org/10.1093/cercor/bhz044

Peeters D (2018) A standardized set of 3-D objects for virtual reality research and applications. Behav Res Methods 50(3):1047–1054. https://doi.org/10.3758/s13428-017-0925-3

Peigneux P, Laureys S, Fuchs S, Collette F, Perrin F, Reggers J, Phillips C, Degueldre C, Del Fiore G, Aerts J, Luxen A, Maquet P (2004) Are spatial memories strengthened in the human hippocampus during slow wave sleep? Neuron 44(3):535–545. https://doi.org/10.1016/j.neuron.2004.10.007

Perfect TJ, Lindsay DS (2013) The SAGE handbook of applied memory. SAGE

Perneczky R, Pohl C, Sorg C, Hartmann J, Tosic N, Grimmer T, Heitele S, Kurz A (2006) Impairment of activities of daily living requiring memory or complex reasoning as part of the MCI syndrome. Int J Geriatr Psychiatry 21(2):158–162. https://doi.org/10.1002/gps.1444

Pflueger MO, Stieglitz R-D, Lemoine P, Leyhe T (2018) Ecologically relevant episodic memory assessment indicates an attenuated age-related memory loss – a virtual reality study. Neuropsychology 32(6):680–689. https://doi.org/10.1037/neu0000454

Pine DS, Grun J, Maguire EA, Burgess N, Zarahn E, Koda V, Fyer A, Szeszko PR, Bilder RM (2002) Neurodevelopmental aspects of spatial navigation: a virtual reality fMRI study. Neuroimage 15(2):396–406. https://doi.org/10.1006/nimg.2001.0988

Plancher G, Gyselinck V, Nicolas S, Piolino P (2010) Age effect on components of episodic memory and feature binding: a virtual reality study. Neuropsychology 24(3):379–390. https://doi.org/10.1037/a0018680

Plancher G, Tirard A, Gyselinck V, Nicolas S, Piolino P (2012) Using virtual reality to characterize episodic memory profiles in amnestic mild cognitive impairment and Alzheimer's disease: influence of active and passive encoding. Neuropsychologia 50(5):592–602. https://doi.org/10.1016/j.neuropsychologia.2011.12.013

Plancher G, Barra J, Orriols E, Piolino P (2013) The influence of action on episodic memory: a virtual reality study. Q J Exp Psychol 66(5):895–909. https://doi.org/10.1080/17470218.2012.722657

Plancher G, Gyselinck V, Piolino P (2018) The integration of realistic episodic memories relies on different working memory processes: evidence from virtual navigation. Front Psychol 9:47. https://doi.org/10.3389/fpsyg.2018.00047

Ragan ED, Sowndararajan A, Kopper R, Bowman DA (2010) The effects of higher levels of immersion on procedure memorization performance and implications for educational virtual environments. Presence Teleop Virt 19(6):527–543. https://doi.org/10.1162/pres_a_00016

Rauchs G, Orban P, Balteau E, Schmidt C, Degueldre C, Luxen A, Maquet P, Peigneux P (2008a) Partially segregated neural networks for spatial and contextual memory in virtual navigation. Hippocampus 18(5):503–518. https://doi.org/10.1002/hipo.20411

Rauchs G, Orban P, Schmidt C, Albouy G, Balteau E, Degueldre C, Schnackers C, Sterpenich V, Tinguely G, Luxen A, Maquet P, Peigneux P (2008b) Sleep modulates the neural substrates of both spatial and contextual memory consolidation. PloS One 3(8):e2949. https://doi.org/10.1371/journal.pone.0002949

Reger GM, Gahm GA (2008) Virtual reality exposure therapy for active duty soldiers. J Clin Psychol 64(8):940–946. https://doi.org/10.1002/jclp.20512

Reggente N, Essoe JK-Y, Aghajan ZM, Tavakoli AV, McGuire JF, Suthana NA, Rissman J (2018) Enhancing the ecological validity of fMRI memory research using virtual reality. Front Neurosci 12. https://doi.org/10.3389/fnins.2018.00408

Reggente N, Essoe JKY, Baek HY, Rissman J (2020) The Method of Loci in virtual reality: explicit binding of objects to spatial contexts enhances subsequent memory recall. J Cogn Enhanc 4(1): 12–30. https://doi.org/10.1007/s41465-019-00141-8

Repetto C, Cipresso P, Riva G (2015) Virtual action and real action have different impacts on comprehension of concrete verbs. Front Psychol 6:176. https://doi.org/10.3389/fpsyg.2015.00176

Repetto C, Serino S, Macedonia M, Riva G (2016) Virtual reality as an embodied tool to enhance episodic memory in elderly. Front Psychol 7:1839. https://doi.org/10.3389/fpsyg.2016.01839

Rissman J, Chow TE, Reggente N, Wagner AD (2016) Decoding fMRI signatures of real-world autobiographical memory retrieval. J Cogn Neurosci 28(4):604–620. https://doi.org/10.1162/jocn_a_00920

Robin J, Buchsbaum BR, Moscovitch M (2018) The primacy of spatial context in the neural representation of events. J Neurosci. https://doi.org/10.1523/JNEUROSCI.1638-17.2018

Rodriguez PF (2010) Human navigation that requires calculating heading vectors recruits parietal cortex in a virtual and visually sparse water maze task in fMRI. Behav Neurosci 124(4): 532–540. https://doi.org/10.1037/a0020231

Roediger HL, McDermott KB (2013) Two types of event memory. Proc Natl Acad Sci 110(52): 20856–20857. https://doi.org/10.1073/pnas.1321373110

Rose DF, Brooks BM, Attree EA, Parslow DM, Leadbetter AG, McNeil JE, Jayawardena S, Greenwood R, Potter J (1999) A preliminary investigation into the use of virtual environments in memory retraining after vascular brain injury: indications for future strategy? Disabil Rehabil 21(12):548–554. https://doi.org/10.1080/096382899297206

Rose FD, Brooks B, M., and Rizzo, A. A. (2005) Virtual reality in brain damage rehabilitation: review. Cyberpsychol Behav 8(3):241–262. https://doi.org/10.1089/cpb.2005.8.241

Rubin DC, Schrauf RW, Greenberg DL (2003) Belief and recollection of autobiographical memories. Mem Cognit 31(6):887–901. https://doi.org/10.3758/bf03196443

Ruddle RA, Payne SJ, Jones DM (1997) Navigating buildings in "desk-top" virtual environments: experimental investigations using extended navigational experience. J Exp Psychol Appl 3(2): 143–159. https://doi.org/10.1037/1076-898X.3.2.143

Ruddle RA, Volkova E, Mohler B, Bülthoff HH (2011) The effect of landmark and body-based sensory information on route knowledge. Mem Cognit 39(4):686–699. https://doi.org/10.3758/s13421-010-0054-z

Ryan E, Poole C (2019) Impact of virtual learning environment on students' satisfaction, engagement, recall, and retention. J Med Imaging Radiat Sci 50(3):408–415. https://doi.org/10.1016/j.jmir.2019.04.005

Sagi Y, Tavor I, Hofstetter S, Tzur-Moryosef S, Blumenfeld-Katzir T, Assaf Y (2012) Learning in the fast lane: new insights into neuroplasticity. Neuron 73(6):1195–1203. https://doi.org/10.1016/j.neuron.2012.01.025

Salgado-Pineda P, Landin-Romero R, Portillo F, Bosque C, Pomes A, Spanlang B, Franquelo JC, Teixido C, Sarró S, Salvador R, Slater M, Pomarol-Clotet E, McKenna PJ (2016) Examining hippocampal function in schizophrenia using a virtual reality spatial navigation task. Schizophr Res 172(1–3):86–93. https://doi.org/10.1016/j.schres.2016.02.033

Sauzéon H, N'Kaoua B, Arvind Pala P, Taillade M, Guitton P (2016) Age and active navigation effects on episodic memory: a virtual reality study. Br J Psychol 107(1):72–94. https://doi.org/10.1111/bjop.12123

Schinazi VR, Epstein RA (2010) Neural correlates of real-world route learning. Neuroimage 53(2): 725–735. https://doi.org/10.1016/j.neuroimage.2010.06.065

Schindler A, Bartels A (2013) Parietal cortex codes for egocentric space beyond the field of view. Curr Biol 23(2):177–182

Schomaker J, van Bronkhorst MLV, Meeter M (2014) Exploring a novel environment improves motivation and promotes recall of words. Front Psychol. https://doi.org/10.3389/fpsyg.2014.00918

Schöne B, Wessels M, Gruber T (2019) Experiences in virtual reality: a window to autobiographical memory. Curr Psychol 38(3):715–719. https://doi.org/10.1007/s12144-017-9648-y

Schultheis MT, Himelstein J, Rizzo AA (2002) Virtual reality and neuropsychology: upgrading the current tools. J Head Trauma Rehabil 17(5):378–394

Serino S, Riva G (2015) How different spatial representations interact in virtual environments: the role of mental frame syncing. Cogn Process 16(2):191–201. https://doi.org/10.1007/s10339-015-0646-4

Serino S, Morganti F, Di Stefano F, Riva G (2015) Detecting early egocentric and allocentric impairments deficits in Alzheimer's disease: an experimental study with virtual reality. Front Aging Neurosci. https://doi.org/10.3389/fnagi.2015.00088

Shin YS, Masís-Obando R, Keshavarzian N, Dáve R, Norman KA (2021) Context-dependent memory effects in two immersive virtual reality environments: on Mars and underwater. Psychon Bull Rev 28(2):574–582. https://doi.org/10.3758/s13423-020-01835-3

Shipman SL, Astur RS (2008) Factors affecting the hippocampal BOLD response during spatial memory. Behav Brain Res 187(2):433–441

Siu K-C, Best BJ, Kim JW, Oleynikov D, Ritter FE (2016) Adaptive virtual reality training to optimize military medical skills acquisition and retention. Mil Med 181(suppl_5):214–220. https://doi.org/10.7205/MILMED-D-15-00164

Skouras S, Torner J, Andersson P, Koush Y, Falcon C, Minguillon C, Fauria K, Alpiste F, Blenow K, Zetterberg H, Gispert JD, Molinuevo JL, The ALFA Study (2020) Earliest amyloid and tau deposition modulate the influence of limbic networks during closed-loop hippocampal downregulation. Brain 143(3):976–992. https://doi.org/10.1093/brain/awaa011

Slater M, Usoh M, Steed A (1994) Depth of presence in virtual environments. Presence Teleop Virt 3(2):130–144. https://doi.org/10.1162/pres.1994.3.2.130

Slobounov SM, Zhang K, Pennell D, Ray W, Johnson B, Sebastianelli W (2010) Functional abnormalities in normally appearing athletes following mild traumatic brain injury: a functional MRI study. Exp Brain Res 202(2):341–354. https://doi.org/10.1007/s00221-009-2141-6

Slobounov SM, Ray W, Johnson B, Slobounov E, Newell KM (2015) Modulation of cortical activity in 2D versus 3D virtual reality environments: an EEG study. Int J Psychophysiol 95(3): 254–260. https://doi.org/10.1016/j.ijpsycho.2014.11.003

Smith SM (1979) Remembering in and out of context. J Exp Psychol Hum Learn Mem 5(5): 460–471. https://doi.org/10.1037/0278-7393.5.5.460

Smith SM (1988) Environmental context – dependent memory. In: Memory in context: context in memory. Wiley, pp 13–34

Smith SA (2019) Virtual reality in episodic memory research: a review. Psychon Bull Rev 26(4): 1213–1237. https://doi.org/10.3758/s13423-019-01605-w

Smith SM, Handy JD (2014) Effects of varied and constant environmental contexts on acquisition and retention. J Exp Psychol Learn Mem Cogn 40(6):1582–1593. https://doi.org/10.1037/xlm0000019

Smith SM, Handy JD (2016) The crutch of context-dependency: effects of contextual support and constancy on acquisition and retention. Memory 24(8):1134–1141. https://doi.org/10.1080/09658211.2015.1071852

Snow JC, Skiba RM, Coleman TL, Berryhill ME (2014) Real-world objects are more memorable than photographs of objects. Front Hum Neurosci 8:837. https://doi.org/10.3389/fnhum.2014.00837

Spiers HJ, Maguire EA (2006) Thoughts, behaviour, and brain dynamics during navigation in the real world. Neuroimage 31(4):1826–1840. https://doi.org/10.1016/j.neuroimage.2006.01.037

Spiers HJ, Burgess N, Maguire EA, Baxendale SA, Hartley T, Thompson PJ, O'Keefe J (2001a) Unilateral temporal lobectomy patients show lateralized topographical and episodic memory deficits in a virtual town. Brain 124(12):2476–2489. https://doi.org/10.1093/brain/124.12.2476

Spiers HJ, Maguire EA, Burgess N (2001b) Hippocampal amnesia. Neurocase 7(5):357–382. https://doi.org/10.1076/neur.7.5.357.16245

Steemers B, Vicente-Grabovetsky A, Barry C, Smulders P, Schröder TN, Burgess N, Doeller CF (2016) Hippocampal attractor dynamics predict memory-based decision making. Curr Biol 26(13):1750–1757. https://doi.org/10.1016/j.cub.2016.04.063

Steuer J (1992) Defining virtual reality: dimensions determining telepresence. J Commun 42(4): 73–93. https://doi.org/10.1111/j.1460-2466.1992.tb00812.x

Stokes J, Kyle C, Ekstrom AD (2015) Complementary roles of human hippocampal subfields in differentiation and integration of spatial context. J Cogn Neurosci 27(3):546–559. https://doi.org/10.1162/jocn_a_00736

Sulpizio V, Committeri G, Lambrey S, Berthoz A, Galati G (2016) Role of the human retrosplenial cortex/parieto-occipital sulcus in perspective priming. Neuroimage 125:108–119. https://doi.org/10.1016/j.neuroimage.2015.10.040

Suthana NA, Ekstrom AD, Moshirvaziri S, Knowlton B, Bookheimer SY (2009) Human hippocampal CA1 involvement during allocentric encoding of spatial information. J Neurosci Off J Soc Neurosci 29(34):10512–10519. https://doi.org/10.1523/JNEUROSCI.0621-09.2009

Suthana N, Ekstrom A, Moshirvaziri S, Knowlton B, Bookheimer S (2011) Dissociations within human hippocampal subregions during encoding and retrieval of spatial information. Hippocampus 21(7):694–701. https://doi.org/10.1002/hipo.20833

Suthana N, Haneef Z, Stern J, Mukamel R, Behnke E, Knowlton B, Fried I (2012) Memory enhancement and deep-brain stimulation of the entorhinal area. N Engl J Med 366(6): 502–510. https://doi.org/10.1056/NEJMoa1107212

Taillade M, N'Kaoua B, Sauzéon H (2016) Age-related differences and cognitive correlates of self-reported and direct navigation performance: the effect of real and virtual test conditions manipulation. Front Psychol. https://doi.org/10.3389/fpsyg.2015.02034

Tarnanas I, Schlee W, Tsolaki M, Müri R, Mosimann U, Nef T (2013) Ecological validity of virtual reality daily living activities screening for early dementia: longitudinal study. JMIR Serious Games 1(1):e2778. https://doi.org/10.2196/games.2778

Tarnanas I, Tsolaki M, Nef T, Müri M, R., and Mosimann, U. P. (2014) Can a novel computerized cognitive screening test provide additional information for early detection of Alzheimer's disease? Alzheimers Dement 10(6):790–798. https://doi.org/10.1016/j.jalz.2014.01.002

Tascón L, García-Moreno LM, Cimadevilla JM (2017) Almeria spatial memory recognition test (ASMRT): gender differences emerged in a new passive spatial task. Neurosci Lett 651:188–191. https://doi.org/10.1016/j.neulet.2017.05.011

Tascón L, Castillo J, León I, Cimadevilla JM (2018) Walking and non-walking space in an equivalent virtual reality task: sexual dimorphism and aging decline of spatial abilities. Behav Brain Res 347:201–208. https://doi.org/10.1016/j.bbr.2018.03.022

Tascón L, Di Cicco C, Piccardi L, Palmiero M, Bocchi A, Cimadevilla JM (2021) Sex differences in spatial memory: comparison of three tasks using the same virtual context. Brain Sci 11(6) Article 6. https://doi.org/10.3390/brainsci11060757

Taube JS, Valerio S, Yoder RM (2013) Is navigation in virtual reality with FMRI really navigation? J Cogn Neurosci 25(7):1008–1019

Thomas KGF, Hsu M, Laurance HE, Nadel L, Jacobs WJ (2001) Place learning in virtual space III: investigation of spatial navigation training procedures and their application to fMRI and clinical neuropsychology. Behav Res Methods Instrum Comput 33(1):21–37. https://doi.org/10.3758/BF03195344

Tolman EC (1948) Cognitive maps in rats and men. Psychol Rev 55(4):189–208. https://doi.org/10.1037/h0061626

Topalovic U, Aghajan ZM, Villaroman D, Hiller S, Christov-Moore L, Wishard TJ, Stangl M, Hasulak NR, Inman CS, Fields TA, Rao VR, Eliashiv D, Fried I, Suthana N (2020) Wireless programmable recording and stimulation of deep brain activity in freely moving humans. Neuron 108(2):322–334.e9. https://doi.org/10.1016/j.neuron.2020.08.021

Tortell R, Luigi DP, Dozois A, Bouchard S, Morie JF, Ilan D (2007) The effects of scent and game play experience on memory of a virtual environment. Virtual Reality 11(1):61–68. https://doi.org/10.1007/s10055-006-0056-0

Tulsky DS, Carlozzi N, Chiaravalloti ND, Beaumont JL, Kisala PA, Mungas D, Conway K, Gershon R (2014) NIH toolbox cognition battery (NIHTB-CB): the list sorting test to measure working memory. J Int Neuropsychol Soc 20(6):599–610. https://doi.org/10.1017/S135561771400040X

Tulving E (1983) Elements of episodic memory. Oxford University Press

Tulving E, Thomson DM (1973) Encoding specificity and retrieval processes in episodic memory. Psychol Rev 80(5):352–373. https://doi.org/10.1037/h0020071

Unni A, Ihme K, Jipp M, Rieger JW (2017) Assessing the driver's current level of working memory load with high density functional near-infrared spectroscopy: a realistic driving simulator study. Front Hum Neurosci 11:167. https://doi.org/10.3389/fnhum.2017.00167

van Bennekom MJ, de Koning PP, Denys D (2017) Virtual reality objectifies the diagnosis of psychiatric disorders: a literature review. Front Psych 8. https://doi.org/10.3389/fpsyt.2017.00163

van der Ham IJM, van Zandvoort MJE, Meilinger T, Bosch SE, Kant N, Postma A (2010) Spatial and temporal aspects of navigation in two neurological patients. Neuroreport 21(10):685–689. https://doi.org/10.1097/WNR.0b013e32833aea78

van der Ham IJM, Faber AME, Venselaar M, van Kreveld MJ, Löffler M (2015) Ecological validity of virtual environments to assess human navigation ability. Front Psychol 6:637. https://doi.org/10.3389/fpsyg.2015.00637

van Praag H, Kempermann G, Gage FH (2000) Neural consequences of enviromental enrichment. Nat Rev Neurosci 1(3):191–198. https://doi.org/10.1038/35044558

Varela FJ, Rosch E, Thompson E (1992) The embodied mind: cognitive science and human experience. MIT Press

Volle E, Gonen-Yaacovi G, de Lacy Costello A, Gilbert SJ, Burgess PW (2011) The role of rostral prefrontal cortex in prospective memory: a voxel-based lesion study. Neuropsychologia 49(8):2185–2198. https://doi.org/10.1016/j.neuropsychologia.2011.02.045

Voss JL, Gonsalves BD, Federmeier KD, Tranel D, Cohen NJ (2011) Hippocampal brain-network coordination during volitional exploratory behavior enhances learning. Nat Neurosci 14(1):115. https://doi.org/10.1038/nn.2693

Waller D, Hunt E, Knapp D (1998) The transfer of spatial knowledge in virtual environment training. Presence Teleop Virt 7(2):129–143. https://doi.org/10.1162/105474698565631

Wallet G, Sauzéon H, Pala PA, Larrue F, Zheng X, N'Kaoua B (2011) Virtual/real transfer of spatial knowledge: benefit from visual fidelity provided in a virtual environment and impact of active navigation. Cyberpsychol Behav Soc Netw 14(7–8):417–423. https://doi.org/10.1089/cyber.2009.0187

Wang P, Wu P, Chi H-L, Li X (2020) Adopting lean thinking in virtual reality-based personalized operation training using value stream mapping. Autom Constr 119:103355. https://doi.org/10.1016/j.autcon.2020.103355

Wegman J, Janzen G (2011) Neural encoding of objects relevant for navigation and resting state correlations with navigational ability. J Cogn Neurosci 23(12):3841–3854. https://doi.org/10.1162/jocn_a_00081

Wegman J, Tyborowska A, Janzen G (2014) Encoding and retrieval of landmark-related spatial cues during navigation: an fMRI study. Hippocampus 24(7):853–868. https://doi.org/10.1002/hipo.22275

Weidemann CT, Mollison MV, Kahana MJ (2009) Electrophysiological correlates of high-level perception during spatial navigation. Psychon Bull Rev 16(2):313–319. https://doi.org/10.3758/PBR.16.2.313

Weniger G, Ruhleder M, Wolf S, Lange C, Irle E (2009) Egocentric memory impaired and allocentric memory intact as assessed by virtual reality in subjects with unilateral parietal cortex lesions. Neuropsychologia 47(1):59–69. https://doi.org/10.1016/j.neuropsychologia.2008.08.018

Werner P, Rabinowitz S, Klinger E, Korczyn AD, Josman N (2009) Use of the virtual action planning supermarket for the diagnosis of mild cognitive impairment. Dement Geriatr Cogn Disord 27(4):301–309. https://doi.org/10.1159/000204915

Wesley Regian J, Yadrick RM (1994) Assessment of configurational knowledge of naturally- and artificially-acquired large-scale space. J Environ Psychol 14(3):211–223. https://doi.org/10.1016/S0272-4944(94)80057-X

West GL, Zendel BR, Konishi K, Benady-Chorney J, Bohbot VD, Peretz I, Belleville S (2017) Playing super Mario 64 increases hippocampal grey matter in older adults. PloS One 12(12): e0187779. https://doi.org/10.1371/journal.pone.0187779

Widmann CN, Beinhoff U, Riepe MW (2012) Everyday memory deficits in very mild Alzheimer's disease. Neurobiol Aging 33(2):297–303. https://doi.org/10.1016/j.neurobiolaging.2010.03.012

Wilkins LK, Girard TA, Herdman KA, Christensen BK, King J, Kiang M, Bohbot VD (2017) Hippocampal activation and memory performance in schizophrenia depend on strategy use in a virtual maze. Psychiatry Res Neuroimaging 268:1–8. https://doi.org/10.1016/j.pscychresns.2017.07.007

Willems RM, Peelen MV (2021) How context changes the neural basis of perception and language. IScience 24(5). https://doi.org/10.1016/j.isci.2021.102392

Wilson M (2002) Six views of embodied cognition. Psychon Bull Rev 9(4):625–636. https://doi.org/10.3758/BF03196322

Witmer BG, Bailey JH, Knerr BW, Parsons KC (1996) Virtual spaces and real world places: transfer of route knowledge. Int J Hum Comput Stud 45(4):413–428. https://doi.org/10.1006/ijhc.1996.0060

Wolbers T, Weiller C, Büchel C (2004) Neural foundations of emerging route knowledge in complex spatial environments. Cogn Brain Res 21(3):401–411. https://doi.org/10.1016/j.cogbrainres.2004.06.013

Wong CW, Olafsson V, Plank M, Snider J, Halgren E, Poizner H, Liu TT (2014) Resting-state fMRI activity predicts unsupervised learning and memory in an immersive virtual reality environment. PloS One 9(10):e109622. https://doi.org/10.1371/journal.pone.0109622

Wood DP, Murphy JA, Center KB, Russ C, McLay RN, Reeves D, Pyne J, Shilling R, Hagan J, Wiederhold BK (2008) Combat related post traumatic stress disorder: a multiple case report using virtual reality graded exposure therapy with physiological monitoring. Stud Health Technol Inform 132:556–561

Woollett K, Spiers HJ, Maguire EA (2009) Talent in the taxi: a model system for exploring expertise. Philos Trans R Soc B Biol Sci 364(1522):1407–1416. https://doi.org/10.1098/rstb.2008.0288

Wuethrich S, Hannula DE, Mast FW, Henke K (2018) Subliminal encoding and flexible retrieval of objects in scenes. Hippocampus 28(9):633–643. https://doi.org/10.1002/hipo.22957

Yip BCB, Man DWK (2013) Virtual reality-based prospective memory training program for people with acquired brain injury. NeuroRehabilitation 32(1):103–115. https://doi.org/10.3233/NRE-130827

Yonelinas AP, Aly M, Wang W-C, Koen JD (2010) Recollection and familiarity: examining controversial assumptions and new directions. Hippocampus 20(11):1178–1194. https://doi.org/10.1002/hipo.20864

Zakzanis KK, Quintin G, Graham SJ, Mraz R (2009) Age and dementia related differences in spatial navigation within an immersive virtual environment. Med Sci Monit 15(4):CR140–CR150

Zhang H (2017) Head-mounted display-based intuitive virtual reality training system for the mining industry. Int J Min Sci Technol 27(4):717–722. https://doi.org/10.1016/j.ijmst.2017.05.005

Zucchella C, Sinforiani E, Tassorelli C, Cavallini E, Tost-Pardell D, Grau S, Pazzi S, Puricelli S, Bernini S, Bottiroli S, Vecchi T, Sandrini G, Nappi G (2014) Serious games for screening pre-dementia conditions: from virtuality to reality? A pilot project. Funct Neurol 29(3):153–158

Zygouris S, Giakoumis D, Votis K, Doumpoulakis S, Ntovas K, Segkouli S, Karagiannidis C, Tzovaras D, Tsolaki M (2015) Can a virtual reality cognitive training application fulfill a dual role? Using the virtual supermarket cognitive training application as a screening tool for mild cognitive impairment. J Alzheimers Dis 44(4):1333–1347. https://doi.org/10.3233/JAD-141260

Zygouris S, Ntovas K, Giakoumis D, Votis K, Doumpoulakis S, Segkouli S, Karagiannidis C, Tzovaras D, Tsolaki M (2017) A preliminary study on the feasibility of using a virtual reality cognitive training application for remote detection of mild cognitive impairment. J Alzheimers Dis 56(2):619–627. https://doi.org/10.3233/JAD-160518

Virtual Reality for Awe and Imagination

Alice Chirico and Andrea Gaggioli

Contents

Abstract Empirical research has explored the potential of the emotion of awe to shape creativity, while theoretical work has sought to understand the link between this emotion and transformation in terms of imagining new possible worlds. This branch of study relies on the transformative potential of virtual reality (VR) to examine and invite cognitive and emotional components of transformative experiences (TEs) within the interdisciplinary model of Transformative Experience Design (TED) and the Appraisal-Tendency Framework (ATF). TED suggests using the epistemic and emotional affordances of interactive technologies, such as VR, to invite TEs. The ATF can provide insight into the nature of these affordances and their relationship. This line of research draws on empirical evidence of the

A. Chirico (✉)
Department of Psychology, Research Center in Communication Psychology, Universitá Cattolica del Sacro Cuore, Milan, Italy
e-mail: Alice.chirico@unicatt.it

A. Gaggioli
Department of Psychology, Research Center in Communication Psychology, Universitá Cattolica del Sacro Cuore, Milan, Italy

Applied Technology for Neuropsychology Lab (ATNP-Lab), Italian Auxologico Institute of Milan, Milan, Italy

Curr Topics Behav Neurosci (2023) 65: 233–254
https://doi.org/10.1007/7854_2023_417
Published Online: 18 February 2023

awe-creativity link to broaden the discourse and consider the potential impact of this emotion on core beliefs about the world. The combination of VR with these theoretical and design-oriented approaches may enable a new generation of potentially transformative experiences that remind people that they can aspire to more and inspire them to work toward imagining and creating a new possible world.

Keywords Awe · Beliefs · Creativity · Imagination · Possibility · Transformation · Virtual reality

1 Introduction

The challenge addressed in this chapter is three-fold. First, we introduce the sophisticated and unique nature of the complex transformative emotion of awe within the broader field of emotion science. Drawing from these conditions, we propose adopting an approach informed by the Transformative Experience Design (TED) model (Gaggioli 2016) to outline the unique potential of Virtual Reality for studying and inviting transformative experiences at any scale, even in the lab. TED posits that new technologies have epistemic and emotional affordances that can elicit TEs. Next, we deepen the discourse by providing a novel operationalization of the concept of "epistemic" and "emotional" affordances by drawing from the Appraisal-Tendency framework (ATF) that allows for testable assumptions on how emotions relate to specific cognitive processes, motivational aspects, and behaviors. Specifically, we rely on empirical evidence of the awe-creativity link and broaden the focus to outline how this complex emotion could impact not just creative thinking but even individuals' basic beliefs of the world, thus paving the way for transformation. Finally, we suggest concrete examples of ways in which the combination of awe and VR can facilitate the emergence of transformative experiences (TEs) by supporting individuals' imagination and experience of a new world through the systematic violation and expansion of their basic beliefs of the world, that is, Primals (Clifton 2020).

2 An Overview on Transformation and Transformative Experiences

In this chapter, we examine how virtual reality (VR) can support transformative experiences (TEs). A transformative experience is as challenging to define exhaustively as it is difficult to *invite*. However, a recent cross-disciplinary and cross-domain working definition of TEs has been proposed: "Transformative experiences can be defined as brief experiences, perceived as extraordinary and unique, entailing durable and/or irreversible outcomes, which contribute to changing individuals' self-conception, worldviews, and view of others, as well as their own personality and identity by involving an epistemic expansion (as new forms of knowledge of the self, others, and the world) and a heightened emotional complexity (emotional variability, high intensity, mixed emotions), as the two core phenomenological features. They are usually remembered vividly. These experiences emerge suddenly, either apparently spontaneously or they can be invited by specific elicitors/facilitating conditions encompassing both state (related to contingencies) and trait elements (related to more stable conditions of the experiencer). Elicitors, usually, are perceived as novel stimuli, able to challenge an individual's mental schema, thus also resulting in disruption" (Chirico et al. 2022; p. 14).

Four main cross-disciplinary and cross-domain assets emerge in regard to transformation: (1) emotional complexity; (2) epistemic expansion; (3) facilitating conditions/elicitors, and (4) specific effects on the recipient (after-effects). These four assets can be applied also to describe the most recent cross-disciplinary framework on TEs, that is, the Transformative Experience Design (TED) model (Gaggioli 2016). Compared to other models on TEs, this framework posits that TEs – which hold the transformative potential to change people's worldview – can be better "designed" in virtual reality, due to its unprecedented capability to induce complex emotions and alter cognitive schema in an ecological but controlled way. Indeed, TEs would emerge by combining and manipulating specific (1) emotional and (2) epistemic affordances embedded in new technologies, especially VR. Specifically, emotional affordances can be perceived as cues that elicit complex emotions (e.g., awe, admiration, elevation, the sublime) that would be infrequent in real-world contexts and have been extensively studied in recent years (e.g., see Piff et al. 2015; Chirico 2020; Gordon et al. 2017; Onu 2016). Epistemic affordances are cognitive hints that promote knowledge restructuring. These cues can be meaningful structured narratives, dilemmas, or counterfactual situations that stimulate self-reflection and/or bewilderment, potentially leading to a restructuring of beliefs. In contrast to emotional affordances, we are just beginning to study the cognitive side of a TE, in which one of the main pillars concerns individuals' worldviews of core beliefs of the world (Chirico et al. 2022). Please, see Fig. 1 for a pictorial representation of the rational of this chapter.

2.1 *The Emotional Side of Transformation: Awe as the Acme of Emotion Science*

Some decades ago, the scientific study of emotions might have seemed an unusual and unconventional topic. However, it is now frequently required in various fields, such as education (e.g., Pekrun and Linnenbrink-Garcia 2014), marketing (e.g., Khatoon and Rehman 2021), and psychotherapy (see Rottenberg and Gross 2007). Emotions can be seen as pervasive, but fleeting phenomena that organize individuals' response systems in relation to specific stimuli from both the inner and outer

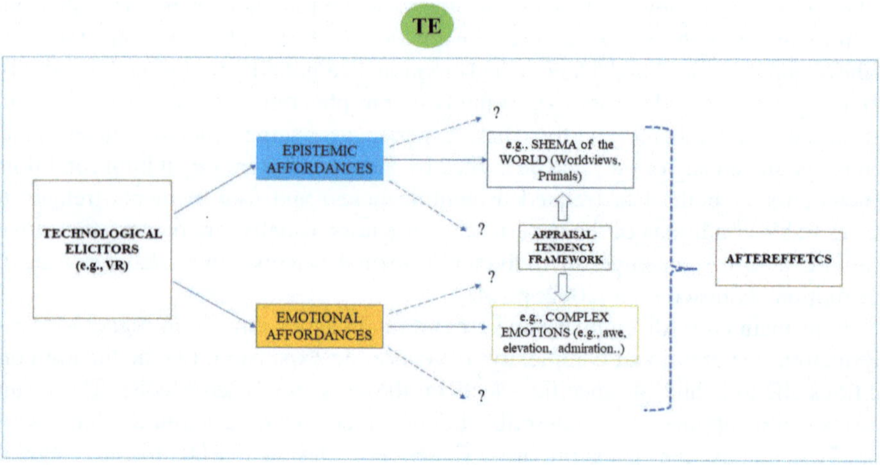

Fig. 1 Pictorial representation of the rational of this chapter

world (Lewis 2000; Fischer and Van Kleef 2010). Recently, the social value of emotions has evolved alongside their functional value (Keltner et al. 2006a; Lench et al. 2015) as key drivers of survival (Plutchik 1982). However, the science of emotions has never stopped evolving since the first attempts in the nineteenth century (James 1884), and it continues to reveal a more sophisticated understanding of how emotions are structured and organized (Barrett 2017) and how they can impact behavioral, physiological, experiential, and cognitive systems (Cacioppo et al. 2000; Larsen et al. 2008; Lench 2018; Moors et al. 2013; Tong 2015).

Defining emotions is complex, and the study of awe is no exception. Initially, researchers tried to outline the structure of awe by including it among the so-called basic emotions. According to Paul Ekman (1992), basic emotions are fleeting, partially automatic phenomena that are characterized by universal and distinctive features at the expressive (e.g., facial expressions, vocalizations) and physiological (specific autonomous nervous system activation patterns) levels.

Building on earlier studies, scholars have attempted to define awe's evolutionary function (Chirico and Yaden 2018; Keltner and Haidt 2003), building upon the initial theory of basic emotions. As a result, two main proposals have emerged regarding awe's evolutionary value. Each of these theories attempts to explain the kind of survival problem(s) that awe addresses. First, Keltner and Haidt (2003) framed awe's functional value in terms of maintaining social hierarchy, suggesting that it arises in response to a special type of vastness, i.e. power from people (leaders). Therefore, the prototypical elicitors of awe should be considered social, and subsequently, the awe response generalized to phenomena in nature. The other proposal reversed this order. Drawing from the Prospect-Refuge theory (Appleton 1996), it is proposed that the evolutionary function of awe can be traced back to an earlier period when our ancestors sought safe shelters in places characterized by elevated positions offering a vast view of their surroundings. These are also the typical elicitors of awe.

Recently, two main features have been identified as underlying almost all of the elicitors of awe. First, awe-inducing stimuli should be vast both conceptually and perceptually (Chirico et al. 2021; Clewis et al. n.d.; Graziosi and Yaden 2019; Yaden et al. 2018). For example, not only magnificent panoramic views can be suitable inductors of awe (perceptual vastness), but also the sudden understanding of a complex theory (e.g., the Theory of Relativity, Einstein 1922) can embed a strong conceptual vastness resulting in awe. This vastness should not be considered as acting for its own sake, but rather, the vaster a stimulus is perceived, the more it is associated with a need to accommodate current mental schema to the new incoming information. Additionally, according to the more recent socio-functionalist approach, on which Keltner and Haidt's first conceptualization relied, awe should not be considered an armored emotion category with fixed boundaries, but rather, it would be better defined as an "emotion family" (Ekman and Cordaro 2011; Keltner and Haidt 2003; Keltner et al. 2006b).

These theoretical advancements have led other scholars to move away from the initial view of awe as a basic emotion toward a new conceptualization of awe as a non-basic emotion (Ekman and Wallace 1975; Oatley and Johnson-Laird 1987) stemming from basic emotional sub-components, such as joy, wonder, fear, and dread (Arcangeli et al. 2020; Chirico and Gaggioli 2018; Clewis 2021; Clewis et al. 2021; Pearsall 2007). Awe should be considered an inherently complex emotion that unfolds over time and is able to stimulate "increasingly sophisticated changes" (Chirico and Gaggioli 2021) ranging from transient physiological fluctuations to long-term self-transcendence or personal transformation. Within this continuum of changes, it has become evident that awe strongly impacts neurological and physiological activity (both central and peripheral), as well as behavior and cognition (Chirico 2020; Yaden et al. 2018; Chirico et al. 2018a, b; Valdesolo and Graham 2014; Taylor and Uchida 2019). Nevertheless, many open questions remain, particularly regarding the relationship between awe and cognition. Before exploring this link further, it is useful to situate this line of research within the broader field of research into the relationship between cognition and emotion.

2.2 *The Epistemic Side of Transformation*

2.2.1 The Link Between Emotion and Cognition: A Précis

If lay psychology still tends to view emotion and cognition as opposing phenomena, scholars from a range of disciplines have recently begun to view them as two sides of the same coin or as inseparably interrelated (Pessoa 2008; Colombo et al. 2020; Leventhal and Scherer 1987). This delay in understanding may be due to the recent debate concerning the link between emotions and cognition, which only began about three decades ago with the publication of the first issue of the Cognition and Emotion journal (Rothermund and Koole 2018). A milestone in this evolution was the seminal article by Lerner and Keltner (2000), which attempted to overcome the traditional valence-affect model of affect used in risk perception studies to explain why emotions of the same valence but driven by different appraisals (e.g., anger and fear) have different effects on cognitive processes. This work can be seen as the beginning of a new, integrated approach to the study of emotions and cognition, known as the "Appraisal-tendency Framework" (ATF) (Lerner and Keltner 2000; Lerner and Tiedens 2006).

2.2.2 Awe and Cognition

According to Moors et al. (2013), appraisal is the "process that detects and assesses the significance of the environment for well-being" (p. 120). This implies an exchange between the person appraising and the event they are appraising (Lazarus 1991). For example, anger is more likely to arise when someone else is deemed responsible for a negative event, but there is still an appraisal of control over the situation (e.g., Lazarus 1991; Ortony et al. 1988).

The ATF posits that emotions not only create specific cognitive properties, such as specific appraisal patterns, but also motivational ones, such as action tendencies (Frijda 1986). For instance, anger can trigger the desire to hurt a specific target (e.g., Roseman 1994). This motivational component may not result in specific actions, but it prepares the individual to act in response to a given situation.

The ATF approach emphasizes both the adaptive function of emotions and their ability to uniquely impact other cognitive processes, from decision-making (Kozlowski et al. 2017) to moral judgment (Horberg et al. 2011) and other high-level cognitive processes (Zhang et al. 2017). The relationship between an emotion and cognitive processes is shaped by specific patterns of appraisal that prioritize concerns semantically linked to the emotion's appraisals (Oveis et al. 2010; Smith and Ellsworth 1985). For example, individuals with a high tendency toward fear tend to make more pessimistic risk judgments (Lerner and Keltner 2001). Furthermore, exposure to fearful images of faces (compared to neutral images) increases the likelihood of risk-averse decisions in a gambling task framed as a gain situation (Habib et al. 2015). This result can depend on the main appraisal-tendency of fear, which is concern about uncertainty and loss of control over the eliciting event (Lerner and Keltner 2001).

Let us consider the emotion of awe. Awe is elicited by stimuli that are so vast that they require the individuals to adjust their mental schema in order to make sense of them. Therefore, people who tend to experience awe – either spontaneously or through induction in the lab – should display a more open mindset, more exploratory behaviors, and increased curiosity, as research has shown (Rudd et al. 2018; Silvia et al. 2015). Consistent with awe's core appraisal themes of vastness and the need for accommodation, it has been found that awe "expands" the width and depth of an individual's mental frames (Gocłowska et al. 2021). At the trait level, awe is positively correlated with extraversion and openness to experience personality traits

(Shiota et al. 2006; Silvia et al. 2015). Additionally, awe can encourage people to seek out new information outside of their usual domains of interest (Chirico et al. 2018b), or to deepen their understanding of existing ones (Rudd et al. 2018) in order to gain a renewed comprehension of the surrounding world and of themselves (Pearsall 2007). Awe can even inspire individuals to learn by re-conceptualizing familiar topics, objects, and phenomena (Valdesolo et al. 2017). This may be due to the uncertainty inherent in awe, which can engender a variety of explanation-seeking tendencies and behaviors. Awe arises from the awareness that something is missing and that more is needed, as our expectations may have been violated (Chirico 2020) and need to be updated. It may not be a coincidence that individuals more prone to experiencing awe also tend to exhibit a more flexible way of thinking and are better able to deal with uncertainty (Shiota et al. 2007). However, when awe is induced in the lab, individuals also struggle to deal with uncertainty (Valdesolo and Graham 2014).

Awe appears to push individuals beyond the boundaries of their self by diminishing the importance usually attributed to the self (e.g., Bai et al. 2017). Indeed, one of the key consequences of this emotion is the generation of a sense of a smaller self (Piff et al. 2015). Awe also unveils what individuals do not know (e.g., McPhetres 2019) and can push them toward the exploration of this (Valdesolo et al. 2017), which can be both thrilling and terrifying (Luo et al. 2021). This unique profile of awe opens up a new area of study focused on higher-order cognitive processes that may also tap into creativity and, in turn, the imagination of new worlds. In the following section, we delve deeper into these aspects, particularly in light of the current controversy surrounding contradictory findings and the unique role that virtual reality can play in advancing this field.

Awe: The Case for Creativity

What does awe hold for the study of creativity? At the cognitive level, awe holds unique potential to foster new and useful ideas and to support flexible thinking, where even distant ideas are linked in a meaningful way, according to the definitions of creative ideation (Plucker et al. 2006; Runco et al. 2001) and creative thinking (Guilford 1982; Torrance 1972). Importantly, awe, as a composite emotion, can have a mixed or double valence (Chirico et al. 2017; Chaudhury et al. 2021). Keltner and Haidt (2003) suggested that different nuances of awe exist (e.g., threat, beauty, ability, virtue, supernatural causality), which can give this emotion a varied valence. Therefore, awe can be considered an ambivalent emotion – especially threat-based awe (Chaudhury et al. 2021), also conceivable as a family of emotions in which both positive and negative affect can occur at the same time (Fong 2006). This is thought to have a strong impact on creativity, as being primed with an ambiguous emotion that has the co-presence of opposite valences can increase sensitivity to unusual associations among challenging ideas or concepts, leading to enhanced cognitive flexibility (i.e., the ability to discover new links among ideas; Guilford 1967; Torrance 1969).

Overall, it should be noted that, based on this evidence, a suitable awe-induction procedure, such as virtual reality, should be suggested to reproduce this sophisticated and nuanced emotion in the lab. In this regard, some preliminary experimental evidence has shown the specific impact of awe elicited by virtual reality on specific components of the creative thinking process compared to a neutral scenario in VR. Specifically, it was found that awe enhanced three out of four dimensions of creative thinking (compared to a neutral video), namely elaboration, flexibility, and fluency.

There are several effective methods for eliciting awe that have been used to study its relationship with creativity. For example, some studies have compared awe and amusement both elicited using autobiographical recall. They found that those recalling awe experiences improved on convergent thinking tasks measuring the extent to which participants were willing to consider unconventional associations (Zhang et al. 2017). Interestingly, the same study also reported that those who recalled awe experiences showed improved originality,[1] but not fluency or flexibility. Furthermore, in this cross-cultural study involving individuals from the USA, Iran, and Malaysia (Zhang et al. 2017), both divergent (i.e., creating something new) and convergent (i.e., applying restructured knowledge to an existing problem) forms of creative thinking (Cropley 2006) and self-reported creative personality were strongly and uniquely related to dispositional awe (as opposed to dispositional amusement).

However, slightly different results were found by Chirico et al. (2018a, b), who used an immersive 360° video of tall trees in a green forest to elicit awe. This video was validated along with three other videos in a pilot study. Thirty-six participants watched four video contents: (1) amusing, (2) awe-inspiring featuring a grand vista on the mountains, (3) awe-inspiring depicting tall trees in a forest, and (4) neutral (hens wandering on grass; see Chirico et al. 2017; Chaudhury et al. 2021 for further details). Participants were asked to rate the extent to which they experienced different emotional states using single items (1 = not at all, 7 = extremely) for anger, disgust, fear, pride, sadness, joy, amusement, and awe.

Similarly, in Chirico et al. (2018a, b), the neutral state was conveyed using the validated neutral video. Each participant watched each video in a counterbalanced order. In all participants, creative thinking was measured after viewing either the awe-eliciting or neutral immersive video using the subtest 4 and 5 of the Torrance test of creative thinking (TTCT; Torrance et al. 1989). Dimensions of originality, fluidity, flexibility, and elaboration were computed by an independent rater and were found to significantly increase after exposure to awe (vs. neutral content), with originality being the only variable not affected by awe.

The studies by Zhang et al. and Chirico et al. differed in several ways. First, Chirico et al. (2018a, b) used a basic form of virtual reality (VR) to elicit awe (i.e., 360° videos), while Zhang et al. (2017) opted for more traditional methods of awe induction such as autobiographical recall ("Please try to recall an event in your life when you saw a particular panoramic view for the first time"; Zhang et al. 2017, p. 7) or flat screen videos. Secondly, Chirico et al. (2018a, b) compared awe-eliciting

[1] Originality was measured by an experimenter who rated each idea or suggestion on a scale from 1 (not original at all) to 5 (very original) (Zhang et al. 2017, p. 10).

stimuli with neutral ones, while Zhang et al. chose to also compare awe's creative potential with amusement based on their "different" appraisals (Piff et al. 2015).

In a recent study, Zhang et al. (2021) expanded upon previous research by Chirico et al. (2018a, b) and Zhang et al. (2017) to investigate the relationship between awe and different aspects of creativity in different cultures (USA, Iranian, and Malaysian). The authors used the HEXACO Personality Inventory (Ashton and Lee 2007) to measure creative personality, the Duncker's Candle Problem (Duncker 1945) to measure convergent creativity, and a diary method to measure everyday creativity. They found a positive between awe and creative personality as well as with convergent thinking, which was not mediated by amusement. The authors also found evidence that curiosity plays a key role by mediating the relationship between trait awe and daily creative activities. However, the study did not establish a causal relationship between awe and creativity. The authors suggested further research, specifically longitudinal studies, could shed light on this relationship. We would also suggest exploring the use of VR-based awe-induction techniques, which have been shown to be effective in eliciting intense awe in the laboratory.

Other scholars had also identified the need for new methods in this field (Silvia et al. 2015), and also prior to Chirico et al. (2017) there had been early successful proposals and implementations in the experimental induction of awe (e.g., Gallagher et al. 2015). Following up this line, in their study, Chirico et al. (2017) disentangled the role of media (360° vs. 2D flat screen videos) and content (awe-inspiring vs. neutral) in eliciting intense awe in the laboratory, as measured by physiological signals and self-reports. The presentation of awe-eliciting content through a more immersive and realistic medium (360° videos in VR) resulted in higher intensity awe, as indicated by arousal measures (heart rate variability indexes) and self-reported items on awe and its sub-dimensions of vastness and need for accommodation. This study informed the design of Chirico et al. (2018a, b), in which the role of specific media content combinations was further explored in relation to creative thinking. In this study, awe was elicited using 360° videos to maximize the likelihood of detecting an effect on creativity measures after a single 2-min exposure. However, there are still many unanswered questions about the optimal elicitation of awe using VR experiences. For example, the role of interactivity in VR awe-inspiring experiences remains to be explored. Chirico et al. (2018a) compared the effect sizes of awe elicited by immersive and immersive and interactive VR and found higher effect sizes in the immersive VR format (360° videos) compared to immersive and interactive VR scenarios that could be navigated.

Virtual reality (VR) has increasingly been used as a powerful tool for simulating complex experiences in a controlled way, making it an ideal candidate for eliciting intense instances of awe in the laboratory (Parsons 2015). VR also has the potential to improve the study of the effects of awe on creativity and other related cognitive abilities, and how they relate to creative thinking (Chirico et al. 2018a, b). In the following section, we will discuss the role of VR in studying the relationship between awe and creativity, and critically examine its potential and limitations.

3 Virtual Reality for Studying the Awe-Creativity Link

While interest in the potential of VR to enhance creativity (Gong and Georgiev 2020) and elicit intense and ecologically valid instances of emotions and affect (Parsons 2015; Parsons et al. 2017) continues to grow, it is important to consider guidelines for designing VR environments that elicit suitably intense experiences of awe in order to detect a subsequent effect on creativity. One recent approach to eliciting emotions in VR responds to the need for increasing the ecological validity of complex phenomena even in the laboratory. This VR-based approach suggests drawing from emotion appraisal themes (which are considered to be core elements impacting the overall emotional process in other contexts) to design emotion-eliciting scenarios that can induce highly intense target emotions (Chirico et al. 2018a; Triberti et al. 2017). For instance, an effective VR awe-eliciting scenario should include the appraisal dimension of vastness (in terms of height or width) as well as the dimension of the need for accommodation (which can be difficult to operationalize, but could be conceptualized in terms of surprise: something unexpected happening to a user). See Fig. 2 for an overview on how VR can be used to elicit awe in the lab.

The ecological validity of experimentally induced emotions and affect can be enhanced by the unique potential of VR to convey a sense of *presence* (Baños et al. 2004; Diemer et al. 2015; Felnhofer et al. 2015), i.e. the sensation of being in another physical or imaginary place (Riva et al. 2014, 2015; Riva and Waterworth 2003, 2014; Waterworth and Riva 2014; Waterworth et al. 2010, 2015). Highly arousing emotions are typically associated with a high sense of presence (Riva et al. 2007). The sense of presence can be supported by certain technological components of VR, such as immersion – defined as the first-person experience of a virtual reality environment in a condition of sensorial isolation from the real world, along with the technological sophistication of the VR system (Coelho et al. 2006) (Diemer et al. 2015) – and interactivity – defined as the extent to which the form and content of a medium can be manipulated by users (Blackmon and Mellers 1996). Immersion alone is associated with higher emotional intensity (Baños et al. 2008), while the role of interactivity in eliciting emotions depends on the specific emotion being targeted.

For instance, in the case of awe, interactivity does not seem to result in more intense affect, possibly because awe is a primarily contemplative emotion (Darbor et al. 2016; Keltner and Piff 2020). In fact, the most basic form of VR – 360° videos – is as effective as interactive VR formats in eliciting intense awe in the laboratory (Chirico et al. 2018a). Therefore, the challenge remains how to elicit ecologically valid instances of this emotion in the lab. To achieve this, it is important to consider the underlying cognitive determinants of each emotion. In the case of awe, the appraisal of vastness can be effectively represented within a VR scenario by creating the illusion of being immersed in sweeping natural landscapes or standing in front of exceptionally tall natural stimuli. In this case, the immersion provided by VR plays a key role in supporting the experience of vastness – a core appraisal of awe. Additionally, being briefly exposed to spacious natural images has also been shown

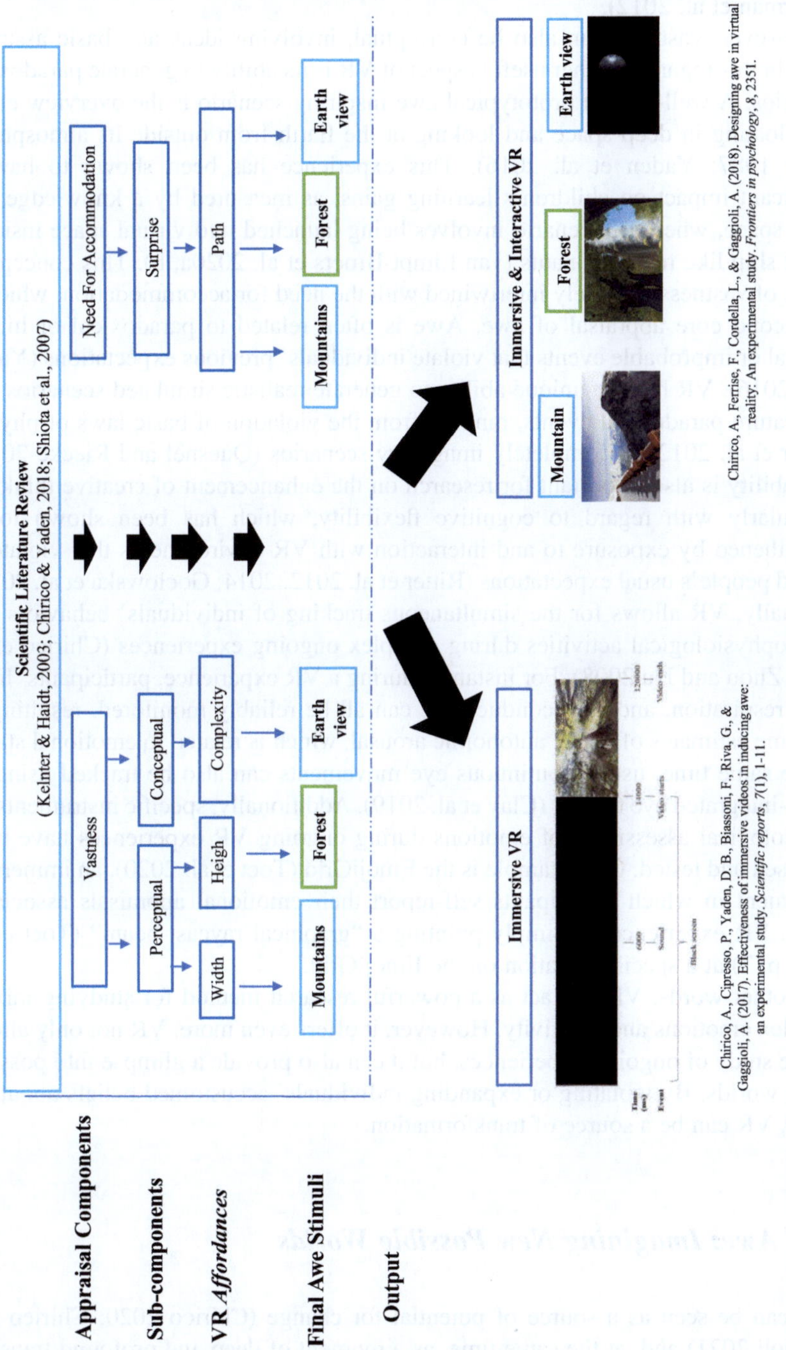

Fig. 2 Guidelines for designing awe in virtual reality

to enhance creative thinking abilities (e.g., fluency and originality) in young people (Liberman et al. 2012).

However, vastness can also be conceptual, involving ideas and basic assumptions. In this regard, another useful aspect of VR is its ability to generate paradoxical scenarios. A well-known prototypical awe-inspiring scenario is the overview effect (i.e., floating in deep space and looking at the Earth from outside its atmosphere; White 1987; Yaden et al. 2016). This experience has been shown to have a significant impact on children's learning gains, as measured by a knowledge test about space, when the scenario involves being launched into virtual space inside a rocket ship like real astronauts (van Limpt-Broers et al. 2020a, b). This conceptual aspect of vastness is closely intertwined with the need for accommodation, which is the second core appraisal of awe. Awe is often related to paradoxical or highly unusual or improbable events that violate individuals' previous expectations (Yaden et al. 2016). VR has the unique ability to generate realistic simulated scenarios that can feature paradoxical events, ranging from the violation of basic laws of physics (Ritter et al. 2012) to completely imaginary scenarios (Quesnel and Riecke 2018). This ability is also important for research on the enhancement of creative thinking, particularly with regard to cognitive flexibility, which has been shown to be strengthened by exposure to and interaction with VR environments that violate or exceed people's usual expectations (Ritter et al. 2012, 2014; Gocłowska et al. 2021).

Finally, VR allows for the simultaneous tracking of individuals' behaviors and psychophysiological activities during complex ongoing experiences (Chirico et al. 2016; Zhou and Hu 2008). For instance, during a VR experience, participants' heart rates, respiration, and skin conductance can all be reliably monitored, resulting in real-time estimates of users' autonomic arousal, which is related to emotional states. At the same time, users' continuous eye movements can also be tracked using an HMD-integrated eye tracker (Clay et al. 2019). Additionally, specific instruments for the ecological assessment of emotions during ongoing VR experiences have been proposed and tested. One example is the EmojiGrid (Toet et al. 2020), an immersive instrument in which participants self-report their emotional appraisals associated with a VR experience by simply pointing a "graphical raycast beam" (Toet et al. 2020; p. 7) at a specific location on the EmojiGrid.

In other words, VR can act as a powerful research method for studying mixed/complex emotions and creativity. However, it offers even more. VR not only allows for the study of ongoing experiences, but it can also provide a glimpse into possible future worlds. By violating or expanding individuals' accustomed beliefs about the world, VR can be a source of transformation.

3.1 Awe: Imagining New Possible Worlds

Awe can be seen as a source of potential for change (Chirico 2020; Chirico and Gaggioli 2021) and, at the same time, as a moment of deep and profound transformation (Gaggioli 2016; Gaggioli et al. 2016). This transformation is characterized by uncertainty – a common correlate of awe elicited in the laboratory (Valdesolo and Graham 2014) – stemming from the sense of "being suspended" in the "not-yet" space of the need for accommodation (Chirico 2020), which can lead to disruptions in the cognitive system and a restructuring of it (e.g., Valdesolo et al. 2017).

At the dispositional level, awe-prone people tend to report less need for cognitive closure (Shiota et al. 2007), which is considered an index of personal discomfort with uncertainty and inconsistency (Webster and Kruglanski 1994). Conversely, philosophers have suggested that moments of awe are associated with uncertainty (James 1902/1987; Kierkegaard 1843/1983), and Valdesolo and Graham not only showed that this occurs after even a single exposure to awe-inspiring videos (e.g., BBC's Plant Earth documentary) (vs. control video vs. amusement-inducing video), but also that this awe-inspiring condition was able to influence agency detection (as measured by the extent to which participants endorsed the idea that the universe was controlled by supernatural forces or by God) [study 1], by decreasing their tolerance for uncertainty [study 2], as measured by the Ambiguity subscale of the Need for Closure Scale (Webster and Kruglanski 1994). This result was replicated in a more "secular" domain: after being exposed to awe-inducing videos (vs. control video vs. amusing video), participants were more prone to interpret even random patterns (12-digit numbers randomly displayed on an iPad) as if they had been generated intentionally by other human beings [study 3].

Secondly, this state of uncertainty that comes along with awe can motivate individuals to seek new explanations for environmental occurrences. In this regard, both religion and science can serve this function, since they provide suitable explanatory frameworks in terms of the causality of daily occurrences (Valdesolo et al. 2016). For instance, after experiencing awe, people who strongly believed in God reported a lower explanatory power for science vs. religion. Nontheistic individuals, on the other hand, did not show such a clear pattern, despite reporting a significant preference for an orderly version of the evolution theory vs. a random one.

According to Valdesolo et al. (2017), awe can act as a key emotional antecedent in science learning by promoting the need for accommodation. Specifically, awe may emerge when encountering major violations of expectations about the world, others, and ourselves. This experience can also generate uncertainty and confusion, but instead of avoiding it, individuals may be motivated to search for new information. The overview effect, when viewing the Earth from space, is a good example of this. The view of the Earth from an unusual perspective can drastically reframe our understanding of our place in the world and the universe, making us feel small in comparison to the vastness of the universe. Gocłowska et al. (2021) found that a positive awe experience was more likely to be associated with stimuli that exceeded previous expectations (i.e., perceived as vast and unique) than with stimuli that disconfirmed expectations (i.e., perceived as inconsistent and uncertain).

Awe can expand our frames of reference by helping us overcome familiar schemas of the world, ourselves, and others. This has been theoretically referred to as the "need for accommodation" dimension of awe. However, the "need for" should not be confused with "accommodation" as an outcome. The former refers to an internal imbalance arising from novel information, which could potentially lead to either the assimilation of this knowledge or the accommodation of an individual's schema according to this data. In other words, awe can lead to new possibilities, ideas, connections, and insights. However, whether awe leads to the realization of these possibilities depends on several factors, including the role played by uncertainty, a key emotional correlate of awe. As the intensity of the emotional experience of awe increases, uncertainty may also increase, making accommodation a real challenge.

Awe can be a destabilizing experience that can change individuals' beliefs about the world. A recent construct called "Primals" has been proposed to describe individuals' inner assumptions about the world (Clifton 2020; Clifton et al. 2019, 2022). Primals can be hierarchically organized (Stahlmann et al. 2020) into at least three levels: one primary Primal (Good), three secondary Primals (Safe, Enticing, and Alive), and several tertiary Primals. Given the potential impact of Primals on individuals' health outcomes and well-being (Clifton and Kim 2020), it is important to understand whether and to what extent these inner assumptions can be changed by specific experiences. It has been suggested that these basic beliefs about the world may be changed through transformative experiences, such as awe (Carel and Kidd 2020; Paul 2014; Riva et al. 2016). Awe can be considered as both a transformative experience and a key emotional dimension of it, according to the TED model (Gaggioli 2016).

If VR has been proposed as the best way to induce awe experiences in the lab, it is still unclear how these virtually induced experiences can lead to significant changes in our existing cognitive schemas of the world. In the following paragraph, the TED model will be used to provide a preliminary proposal for how technological, artistic, and psychological elements can be combined to achieve this goal (Gaggioli 2016).

3.2 Virtual Reality for Inviting TEs by Depicting a New Possible World: A Proposal of Applications

Even short inductions of complex self-transcendent emotions like elevation and admiration via video clips can impact individuals' assumptions of others and the world as benevolent (Van Cappellen et al. 2013), and this result holds at the dispositional level for awe and love (Chirico et al. 2022). Since VR has been shown to be a powerful tool for inducing higher intensity emotions even in the lab (Chirico et al. 2017), it is likely that the link between complex emotions and core assumptions about the world would be strengthened in this case, facilitating the emergence of TEs. As previously mentioned, VR can support users' imagination and experience of new worlds that are different from the ones they are accustomed to. Specifically, VR can facilitate transformation by acting on two key stages of the change process (DiClemente and Prochaska 1998; Prochaska and DiClemente 1982). The first stage involves imagining a new world, and the second involves experiencing this new world in which a person can not only be exposed to new scenarios but can also actively participate in and contribute to creating them. Even basic forms of VR can be used to expose participants to a new world. For example, a person who is accustomed to viewing the world as unintentional (vs. intentional) – that is, a place where events happen by chance rather than according to a higher-order design – may have a negative impact on their sense of meaning in life (Steger 2009). In other words, this specific inner assumption can potentially impact individuals' lives in a pervasive way.

A potential goal of using VR is to transform assumptions that have a detrimental effect. This can be done by repeatedly immersing a person in awe-inspiring virtual reality environments (VREs) that convey a sense of a higher power (Keltner and Haidt 2003; Van Cappellen and Saroglou 2012), nurturing the perception of being part of a world that is imbued with meaning, coherence, and order. Additionally, an individual who perceives the world as ugly and uncontrollable may be placed in a procedurally generated virtual simulation that changes according to their emotional states (Badia et al. 2018; Yannakakis and Togelius 2011) or their interactions with the VRE, framed within a story that emphasizes the dimension of personal control.

To make the idea of helping an individual change their mindset more concrete, the first step could be to identify real-life events that are typically associated with a lack of control over the world. These events can then be rated according to their level of uncontrollability. The experimenter, clinician, or professional can then help the individual to reframe these events by answering the question "what should have happened instead, if I had believed that the world was characterized by order and sense?" This question can serve as a guiding principle for creating simple scenarios, even using 360° videos, in which the participant can experience a world that is imbued with meaning and purpose, according to their own definition of those concepts. This can be seen as a form of counterfactual thinking, allowing the individual to practically experience alternative possible worlds that could have occurred but did not (Byrne 2016), with the goal of preparing them for similar events that may happen in the future.

The design of a VR scenario should be based on the target emotion that the participant should experience. For example, if the goal is to elicit awe, the VR environment should be designed to convey a sense of vastness and a need for accommodation, which are awe's key appraisal themes. Similarly, other emotions can be incorporated into the design of VR environments. By integrating various emotional and cognitive affordances, participants can be exposed to a range of counterfactual scenarios, which can be linked by a cover story developed by the participants themselves. Furthermore, emotional VR scenarios can provide a first-person view of potential future situations, which has been shown to increase participant involvement and impact behavioral intentions (Baek 2020). In addition to enhancing the sense of presence, a well-crafted narrative can be essential to eliciting emotions and tapping into the epistemic dimension of a VR experience (Gorini et al. 2011). Stories can enhance individuals' engagement by making the experience personally relevant and important (Ellard et al. 2012).

Finally, participant progress can be assessed through their behavior, psychophysiological parameters, and self-report measures using instruments like the Primal World Beliefs Inventory (PI-99, PI-18, PI-6; Clifton et al. 2019; Clifton and Yaden 2021).

4 Conclusions

In this chapter, we explored the potential of VR for studying complex psychological phenomena, such as awe and personal transformation. We also proposed a new research question about the relationship between complex emotions and individuals' core beliefs about the world. Our goal was to understand the mechanisms underlying these phenomena and derive design guidelines for inviting personally transformative experiences. This analysis was guided by a broader framework known as the TED model (Gaggioli 2016).

Although VR has been heralded as a promising medium for eliciting transformative experiences (Gaggioli et al. 2016; Kitson and Riecke 2018; Kitson et al. 2019), this research area is still in its early stages. The TED framework offers comprehensive guidelines for eliciting transformative experiences by focusing on the emotional and epistemic affordances of technology-mediated experiences. However, these two high-level constructs still need to be further elaborated and operationalized. In this chapter, we focused on awe as a prototypical example of an emotional affordance associated with transformative experiences and proposed specific cognitive variables (worldviews) to fill the "epistemic affordance" dimension, as defined by the Primals model (Clifton et al. 2019).

While much progress has been made in understanding the emotional aspects of transformation, more work needs to be done at the cognitive level. VR has proven to be a valuable medium for studying complex emotional experiences in the lab, including their relationship with sophisticated cognitive processes. For example, research on awe has demonstrated that VR can be an effective technique for inducing awe in the laboratory, allowing researchers to study the effects of this emotion on related cognitive processes, such as creativity. VR-induced awe has been shown to enhance dimensions of creativity such as fluidity, flexibility, and elaboration. However, the dimension of originality was not affected by this emotion, and conventional emotion-induction techniques (such as videos) have been more effective at stimulating original thoughts.

These findings suggest that the medium itself can influence the relationship between complex emotional states and cognitive processes. In the case of awe, VR interactivity and immersion may not always result in better outcomes compared to a simply immersive scenario. More research is needed to design, develop, and test VR scenarios with varying levels of technological sophistication and their ability to elicit specific complex emotions, such as admiration, elevation, nostalgia, or the sublime, compared to a control condition.

The "Appraisal-tendency Framework" (ATF) suggests that appraisals of emotions play a key role in determining their impact on cognitive processes. In the context of transformation, we propose that research should focus on the role of core beliefs about the world, known as Primals. These constructs have been well-defined, and several instruments are available for their cross-cultural assessment (Clifton et al. 2019; Clifton and Yaden 2021). However, the relationship between complex emotions and core beliefs remains unclear and warrants further investigation, including at the correlational level. It is possible that certain emotions are more closely tied to specific Primals, and that some emotions have a greater impact on certain beliefs than others. Once these questions have been addressed, it will be possible to design transformative interventions using VR simulations that consider appraisals of complex emotions and their associated Primals. These interventions can be applied in both clinical and non-clinical settings.

References

Appleton J (1996) The experience of landscape. Wiley, Chichester, England

Arcangeli M, Sperduti M, Jacquot A, Piolino P, Dokic J (2020) Awe and the experience of the sublime: a complex relationship. Front Psychol 11:1340

Ashton MC, Lee K (2007) Empirical, theoretical, and practical advantages of the HEXACO model of personality structure. Personal Soc Psychol Rev 11:150–166. https://doi.org/10.1177/1088868306294907

Badia SB, Quintero LV, Cameirao MS, Chirico A, Triberti S, Cipresso P, Gaggioli A (2018) Towards emotionally-adaptive virtual reality for mental health applications. IEEE J Biomed Health Inform:1. https://doi.org/10.1109/JBHI.2018.2878846

Baek E (2020) Glimpse into the future experience: when is virtual reality sampling beneficial? In: Advances in consumer research. Association for Consumer Research, pp 1032–1037

Bai Y, Maruskin LA, Chen S, Gordon AM, Stellar JE, McNeil GD et al (2017) Awe, the diminished self, and collective engagement: universals and cultural variations in the small self. J Pers Soc Psychol 113(2):185

Baños R, Botella C, Liaño V, Guerrero B, Rey B, Alcañiz M (2004) Sense of presence in emotional virtual environments. Proc Presence:156–159

Baños RM, Botella C, Rubió I, Quero S, Garcia-Palacios A, Alcañiz M (2008) Presence and emotions in virtual environments: the influence of stereoscopy. Cyberpsychol Behav 11(1):1–8

Barrett LF (2017) How emotions are made: the secret life of the brain. Houghton Mifflin Harcourt

Blackmon, T. T., & Mellers, B. A. (1996). The effects of pictorial realism, delay of visual feedback, and observer interactivity on the subjective sense of presence

Byrne RM (2016) Counterfactual thought. Annu Rev Psychol 67:135–157

Cacioppo JT, Berntson GG, Larsen JT, Poehlmann KM, Ito TA (2000) The psychophysiology of emotion. Handbook of emotions 2:173–191

Carel H, Kidd IJ (2020) Expanding transformative experience. Eur J Philos 28(1):199–213

Chaudhury SH, Garg N, Jiang Z (2021) The curious case of threat-awe: a theoretical and empirical reconceptualization. Emotion

Chirico A (2020) Awe the Palgrave encyclopedia of the possible. Springer, Cham, pp 1–9

Chirico A, Gaggioli A (2018) Awe: "More than a feeling". Humanist Psychol 46(3):274–280. https://doi.org/10.1037/hum0000098

Chirico A, Gaggioli A (2021) The potential role of awe for depression: reassembling the puzzle. Front Psychol 12:1191

Chirico A, Yaden DB (2018) Awe: a self-transcendent and sometimes transformative emotion. In: L. H. (ed) The function of emotions. Springer, Cham

Chirico A, Yaden D, Riva G, Gaggioli A (2016) The potential of virtual reality for the investigation of awe. Front Psychol 7(1766). https://doi.org/10.3389/fpsyg.2016.01766

Chirico A, Cipresso P, Yaden DB, Biassoni F, Riva G, Gaggioli A (2017) Effectiveness of immersive videos in inducing awe: an experimental study. Sci Rep 7(1):1–11

Chirico A, Ferrise F, Cordella L, Gaggioli A (2018a) Designing awe in virtual reality: an experimental study. Front Psychol 8(2351). https://doi.org/10.3389/fpsyg.2017.02351

Chirico A, Glaveanu VP, Cipresso P, Riva G, Gaggioli A (2018b) Awe enhances creative thinking: an experimental study. Creat Res J 30(2):123–131

Chirico A, Clewis RR, David BY, Gaggioli A (2021) Nature versus art as elicitors of the sublime: a virtual reality study. PLoS One

Chirico A, Pizzolante M, Villani D (2022) Self-transcendent dispositions and spirituality: the mediating role of believing in a benevolent world. J Spiritual Ment Health:1–24. https://doi.org/10.1080/19349637.2022.2079041

Chirico A, Pizzolante M, Kitson A, Gianotti E, Riecke BE, Gaggioli A (2022) Defining transformative experiences: a conceptual analysis. Front Psychol 13:790300. https://doi.org/10.3389/fpsyg.2022.790300

Clay V, König P, Koenig S (2019) Eye tracking in virtual reality. J Eye Mov Res 12(1)

Clewis RR (2021) Why the sublime is aesthetic awe. J Aesthet Art Critic 79(3):301–314

Clewis RR, Yaden DB, Chirico A (2021) Intersections between awe and the sublime: an empirical study. Emp Stud Arts

Clewis RR, Yaden DB, Chirico A (n.d.) Intersections between awe and the sublime: a preliminary empirical study. Emp Stud Arts:0276237421994694. https://doi.org/10.1177/0276237421994694

Clifton JD (2020) Testing if primal world beliefs reflect experiences – or at least some experiences identified ad hoc. Front Psychol 11:1145

Clifton JD, Kim ES (2020) Healthy in a crummy world: implications of primal world beliefs for health psychology. Med Hypotheses 135:109463

Clifton JD, Yaden DB (2021) Brief measures of the four highest-order primal world beliefs. Psychol Assess 33(12):1267

Clifton, J. D., Baker, J. D., Park, C. L., Yaden, D. B., Clifton, A. B., Terni, P., . . . Schwartz, H. A. (2019). Primal world beliefs. Psychol Assess, 31(1), 82

Clifton AB, Stahlmann AG, Hofmann J, Chirico A, Cadwallader R, Clifton JD (2022) Improving scale equivalence by increasing access to scale-specific information. Perspect Psychol Sci 17456916221119396

Coelho C, Tichon J, Hine TJ, Wallis G, Riva G (2006) Media presence and inner presence: the sense of presence in virtual reality technologies. From communication to presence: cognition, emotions and culture towards the ultimate communicative experience, pp 25–45

Colombo, D., Fernández-Álvarez, J., Suso-Ribera, C., Cipresso, P., Valev, H., Leufkens, T., . . . Botella, C. (2020). The need for change: understanding emotion regulation antecedents and consequences using ecological momentary assessment. Emotion, 20(1), 30

Cropley A (2006) In praise of convergent thinking. Creat Res J 18(3):391–404

Darbor KE, Lench HC, Davis WE, Hicks JA (2016) Experiencing versus contemplating: language use during descriptions of awe and wonder. Cognit Emot 30(6):1188–1196

DiClemente CC, Prochaska JO (1998) Toward a comprehensive, transtheoretical model of change: stages of change and addictive behaviors. Plenum Press

Diemer J, Alpers GW, Peperkorn HM, Shiban Y, Mühlberger A (2015) The impact of perception and presence on emotional reactions: a review of research in virtual reality. Front Psychol 6

Duncker K (1945) On problem-solving. (L. S. Lees, Trans.). Psychol Monogr 58(5):i–113. https://doi.org/10.1037/h0093599

Einstein A (1922) The general theory of relativity the meaning of relativity. Springer, pp 54–75

Ekman P (1992) An argument for basic emotions. Cognit Emot 6(3–4):169–200

Ekman P, Cordaro D (2011) What is meant by calling emotions basic. Emot Rev 3(4):364–370
Ekman P, Wallace V (1975) Unmasking the face: a guide to recognizing emotions from facial clues. Prentice-Hall, Englewood Cliffs
Ellard KK, Farchione TJ, Barlow DH (2012) Relative effectiveness of emotion induction procedures and the role of personal relevance in a clinical sample: a comparison of film, images, and music. J Psychopathol Behav Assess 34(2):232–243
Felnhofer A, Kothgassner OD, Schmidt M, Heinzle A-K, Beutl L, Hlavacs H, Kryspin-Exner I (2015) Is virtual reality emotionally arousing? Investigating five emotion inducing virtual park scenarios. Int J Human Comp Stud
Fischer AH, Van Kleef GA (2010) Where have all the people gone? A plea for including social interaction in emotion research. Emot Rev 2(3):208–211
Fong CT (2006) The effects of emotional ambivalence on creativity. Acad Manage J 49(5):1016–1030
Frijda NH (1986) The emotions. Cambridge University Press
Gaggioli A (2016) Transformative experience design. In: Gaggioli A, Ferscha A, Riva G, Dunne S, Viaud-Delmon I (eds) Human computer confluence. Transforming human experience through symbiotic technologies. De Gruyter Open, Berlin, pp 96–121
Gaggioli A, Chirico A, Triberti S, Riva G (2016) Transformative interactions: designing positive technologies to Foster self-transcendence and meaning. Annu Rev CyberTherapy Telemed 169
Gallagher S, Janz B, Reinerman L, Trempler J, Bockelman P (2015) A neurophenomenology of awe and wonder: towards a non-reductionist cognitive science. Springer
Gocłowska MA, Elliot AJ, van Elk M, Bulska D, Thorstenson CA, Baas M (2021) Awe arises in reaction to exceeded rather than disconfirmed expectancies. Emotion
Gong Z, Georgiev GV (2020) Literature review: existing methods using VR to enhance creativity. Paper presented at the proceedings of the sixth international conference on design creativity (ICDC 2020)
Gordon AM, Stellar JE, Anderson CL, McNeil GD, Loew D, Keltner D (2017) The dark side of the sublime: distinguishing a threat-based variant of awe. J Pers Soc Psychol 113(2):310
Gorini A, Capideville CS, De Leo G, Mantovani F, Riva G (2011) The role of immersion and narrative in mediated presence: the virtual hospital experience. Cyberpsychol Behav Soc Netw 14(3):99–105
Graziosi M, Yaden D (2019) Interpersonal awe: exploring the social domain of awe elicitors. J Posit Psychol:1–9
Guilford JP (1967) Creativity: yesterday, today and tomorrow. J Creat Behav 1(1):3–14
Guilford J (1982) Is some creative thinking irrational? J Creat Behav
Habib M, Cassotti M, Moutier S, Houdé O, Borst G (2015) Fear and anger have opposite effects on risk seeking in the gain frame. Front Psychol 6:253
Horberg EJ, Oveis C, Keltner D (2011) Emotions as moral amplifiers: an appraisal tendency approach to the influences of distinct emotions upon moral judgment. Emot Rev 3(3):237–244
James W (1884) What is an emotion? Mind 9(34):188–205
James W (1987) The varieties of religious experience: a study in human nature. Library of America, New York, NY. (Original work published 1902)
Keltner D, Haidt J (2003) Approaching awe, a moral, spiritual, and aesthetic emotion. Cognit Emot 17(2):297–314
Keltner D, Piff PK (2020) Self-transcendent awe as a moral grounding of wisdom. Psychol Inq 31(2):160–163
Keltner D, Haidt J, Shiota MN (2006a) Social functionalism and the evolution of emotions. Evol Soc Psychol 115:142
Keltner D, Haidt J, Shiota MN (2006b) Social functionalism and the evolution of emotions
Khatoon S, Rehman V (2021) Negative emotions in consumer brand relationship: a review and future research agenda. Int J Consum Stud
Kierkegaard S (1983) Fear and trembling (H. V. Hong & E. H. Hong, Trans.). Princeton University Press, Princeton, NJ. (Original work published 1843)

Kitson A, Riecke BE (2018, March) Can lucid dreaming research guide self-transcendent experience design in virtual reality? In: In 2018 IEEE workshop on augmented and virtual realities for good (VAR4Good). IEEE, pp 1–4
Kitson A, Buie E, Stepanova ER, Chirico A, Riecke BE, Gaggioli A (2019, May) Transformative experience design: designing with interactive technologies to support transformative experiences. In Extended Abstracts of the 2019 CHI conference on human factors in computing systems (pp 1–5)
Kozlowski D, Hutchinson M, Hurley J, Rowley J, Sutherland J (2017) The role of emotion in clinical decision making: an integrative literature review. BMC Med Educ 17(1):1–13
Larsen JT, Berntson GG, Poehlmann KM, Ito TA, Cacioppo JT (2008) The psychophysiology of emotion. In: Handbook of emotions, vol 3, pp 180–195
Lazarus RS (1991) Emotion and adaptation. Oxford University Press, Oxford, England
Lench HC (ed) (2018) The function of emotions. Springer
Lench HC, Bench SW, Darbor KE, Moore M (2015) A functionalist manifesto: goal-related emotions from an evolutionary perspective. Emot Rev 7(1):90–98
Lerner JS, Keltner D (2000) Beyond valence: toward a model of emotion-specific influences on judgement and choice. Cognit Emot 14(4):473–493
Lerner JS, Keltner D (2001) Fear, anger, and risk. J Pers Soc Psychol 81(1):146
Lerner JS, Tiedens LZ (2006) Portrait of the angry decision maker: how appraisal tendencies shape anger's influence on cognition. J Behav Decis Mak 19(2):115–137
Leventhal H, Scherer K (1987) The relationship of emotion to cognition: a functional approach to a semantic controversy. Cognit Emot 1(1):3–28
Lewis M (2000) The emergence of human emotions. Handbook of emotions 2:265–280
Liberman N, Polack O, Hameiri B, Blumenfeld M (2012) Priming of spatial distance enhances children's creative performance. J Exp Child Psychol 111(4):663–670
Luo L, Mao J, Gao W, Yuan J (2021) Psychological research of awe: definition, functions, and application in psychotherapy. Stress and Brain 1(1):59–75
McPhetres J (2019) Oh, the things you don't know: awe promotes awareness of knowledge gaps and science interest. Cogn Emot 33(8):1599–1615
Moors A, Ellsworth PC, Scherer KR, Frijda NH (2013) Appraisal theories of emotion: state of the art and future development. Emot Rev 5(2):119–124
Oatley K, Johnson-Laird PN (1987) Towards a cognitive theory of emotions. Cognit Emot 1(1): 29–50
Onu D, Kessler T, Smith JR (2016) Admiration: a conceptual review. Emot Rev 8(3):218–230
Ortony A, Clore GL, Collins A (1988) The cognitive structure of emotions cambridge. Cambridge University Press, UK, p 9
Oveis C, Horberg EJ, Keltner D (2010) Compassion, pride, and social intuitions of self-other similarity. J Pers Soc Psychol 98(4):618
Parsons TD (2015) Virtual reality for enhanced ecological validity and experimental control in the clinical, affective and social neurosciences. Front Hum Neurosci 9
Parsons TD, Gaggioli A, Riva G (2017) Virtual reality for research in social neuroscience. Brain Sci 7(4):42
Paul LA (2014) Transformative experience. Oxford University Press, Oxford
Pearsall P (2007) Awe: the delights and dangers of our eleventh emotion. Health Communications, Inc
Pekrun R, Linnenbrink-Garcia L (2014) Introduction to emotions in education. Routledge
Pessoa L (2008) On the relationship between emotion and cognition. Nat Rev Neurosci 9(2): 148–158
Piff PK, Dietze P, Feinberg M, Stancato DM, Keltner D (2015) Awe, the small self, and prosocial behavior. J Pers Soc Psychol 108(6):883
Plucker JA, Runco MA, Lim W (2006) Predicting ideational behavior from divergent thinking and discretionary time on task. Creat Res J 18(1):55–63
Plutchik R (1982) A psychoevolutionary theory of emotions. Sage Publications

Prochaska JO, DiClemente CC (1982) Transtheoretical therapy: toward a more integrative model of change. Psychother Theory Res Pract 19(3):276
Quesnel D, Riecke BE (2018) Are you awed yet? How virtual reality gives us awe and goose bumps. Front Psychol 9:2158
Ritter SM, Damian RI, Simonton DK, van Baaren RB, Strick M, Derks J, Dijksterhuis A (2012) Diversifying experiences enhance cognitive flexibility. J Exp Soc Psychol 48(4):961–964
Ritter SM, Kühn S, Müller BC, Van Baaren RB, Brass M, Dijksterhuis A (2014) The creative brain: Corepresenting schema violations enhances TPJ activity and boosts cognitive flexibility. Creat Res J 26(2):144–150
Riva G, Waterworth JA (2003) Presence and the self: a cognitive neuroscience approach. Presence Connect 3(3)
Riva G, Waterworth JA (2014) Being present in a virtual world. In: Grinshaw M (ed) The oxford handbook of virtuality. Oxford University Press, Oxford, pp 205–221
Riva, G., Mantovani, F., Capideville, C. S., Preziosa, A., Morganti, F., Villani, D., . . . Alcañiz, M. (2007). Affective interactions using virtual reality: the link between presence and emotions. Cyberpsychol Behav, 10(1), 45–56
Riva G, Waterworth J, Murray D (2014) Interacting with presence: HCI and the sense of presence in computer-mediated environments. Walter de Gruyter GmbH & Co KG
Riva G, Mantovani F, Waterworth EL, Waterworth JA (2015) Intention, action, self and other: an evolutionary model of presence immersed in media. Springer, pp 73–99
Riva G, Baños RM, Botella C, Mantovani F, Gaggioli A (2016) Transforming experience: the potential of augmented reality and virtual reality for enhancing personal and clinical change. Frontiers Psychiatry 7(164). https://doi.org/10.3389/fpsyt.2016.00164
Roseman (1994) Emotions and emotion families in the emotion system. 171–175. https://doi.org/10.7282/T31N836X
Rothermund K, Koole SL (2018) Three decades of cognition & emotion: a brief review of past highlights and future prospects. Cognit Emot 32(1):1–12
Rottenberg, J., & Gross, J. J. (2007). Emotion and emotion regulation: a map for psychotherapy researchers
Rudd M, Hildebrand C, Vohs KD (2018) Inspired to create: awe enhances openness to learning and the desire for experiential creation. J Mark Res 55(5):766–781
Runco MA, Plucker JA, Lim W (2001) Development and psychometric integrity of a measure of ideational behavior. Creat Res J 13(3–4):393–400
Shiota MN, Keltner D, John OP (2006) Positive emotion dispositions differentially associated with Big Five personality and attachment style. J Posit Psychol 1(2):61–71
Shiota MN, Keltner D, Mossman A (2007) The nature of awe: elicitors, appraisals, and effects on self-concept. Cognit Emot 21(5):944–963
Silvia PJ, Fayn K, Nusbaum EC, Beaty RE (2015) Openness to experience and awe in response to nature and music: personality and profound aesthetic experiences. Psychol Aesthet Creat Arts 9(4):376–384
Smith CA, Ellsworth PC (1985) Patterns of cognitive appraisal in emotion. J Pers Soc Psychol 48(4):813
Stahlmann AG, Hofmann J, Ruch W, Heintz S, Clifton JD (2020) The higher-order structure of primal world beliefs in German-speaking countries: adaptation and initial validation of the German Primals Inventory (PI-66-G). Personal Individ Differ 163:110054
Steger, M. F. (2009). Meaning in life
Taylor PM, Uchida Y (2019) Awe or horror: differentiating two emotional responses to schema incongruence. Cognit Emot 33(8):1548–1561
Toet A, Heijn F, Brouwer AM, Mioch T, van Erp JB (2020) An immersive self-report tool for the affective appraisal of 360 VR videos. Frontiers Virtual Reality 14
Tong EM (2015) Differentiation of 13 positive emotions by appraisals. Cognit Emot 29(3):484–503
Torrance EP (1969) Creativity. What research says to the teacher, series. National Education Association, Washington, DC

Torrance EP (1972) Predictive validity of the torrance tests of creative thinking. J Creat Behav
Torrance E, Sprini G, Tomasello S (1989) Test di pensiero creativo:(Torrance tests of creative thinking), Bensenville, IL. OrganizzazioniSpeciali (16) (PDF) awe enhances creative thinking: an experimental study
Triberti S, Chirico A, La Rocca G, Riva G (2017) Developing emotional design: emotions as cognitive processes and their role in the design of interactive technologies. Front Psychol 8:1773
Valdesolo P, Graham J (2014) Awe, uncertainty, and agency detection. Psychol Sci 25(3):844–844
Valdesolo P, Park J, Gottlieb S (2016) Awe and scientific explanation. Emotion 16(7):937
Valdesolo P, Shtulman A, Baron AS (2017) Science is awe-some: the emotional antecedents of science learning. Emot Rev 9(3):215–221
Van Cappellen P, Saroglou V (2012) Awe activates religious and spiritual feelings and behavioral intentions. Psychol Relig Spiritual 4(3):223
Van Cappellen P, Saroglou V, Iweins C, Piovesana M, Fredrickson BL (2013) Self- transcendent positive emotions increase spirituality through basic world assumptions. Cognit Emot 27(8): 1378–1394. https://doi.org/10.1080/02699931.2013.787395
van Limpt-Broers H, Louwerse MM, Postma M (2020a) Awe yields learning: a virtual reality study. Paper presented at the CogSci
van Limpt-Broers HAT, Postma M, Louwerse MM (2020b) Creating ambassadors of planet earth: the overview effect in K12 education. Front Psychol 11:2548
Waterworth J, Riva G (2014) Feeling present in the physical world and in computer-mediated environments. Palgrave Macmillan, Basingstoke
Waterworth JA, Waterworth EL, Mantovani F, Riva G (2010) On feeling (the) present. J Conscious Stud 17(1–2):167–188
Waterworth JA, Waterworth EL, Riva G, Mantovani F (2015) Presence: form, content and consciousness. In: Lombard M, Biocca F, Freeman J, IJsselsteijn W, Schaevitz RJ (eds) Immersed in media: telepresence theory, measurement & technology. Springer, Berlin, pp 35–58
Webster DM, Kruglanski AW (1994) Individual differences in need for cognitive closure. J Pers Soc Psychol 67(6):1049
White F (1987) The overview effect: space exploration and human evolution. Houghton Mifflin, Boston
Yaden DB, Iwry J, Slack KJ, Eiechstaedt JC, Zhao Y, Vaillant GE, Newberg AB (2016) The overview effect: awe and self-transcendent experience in space flight. Psychol Conscious Theory Res Pract 3(1):1
Yaden DB, Kaufman SB, Hyde E, Chirico A, Gaggioli A, Zhang JW, Keltner D (2018) The development of the AWE experience scale (AWE-S): a multifactorial measure for a complex emotion. J Posit Psychol:1–15
Yannakakis GN, Togelius J (2011) Experience-driven procedural content generation. IEEE Trans Affect Comput 2(3):147–161
Zhang JW, Anderson C, Razavi P, Mello Z, Shaban-Azad H, Monroy M, Keltner D (2017) Trait and state based experience of awe promotes creativity: eScholarship. University of California
Zhang JW, Howell RT, Razavi P, Shaban-Azad H, Chai WJ, Ramis T et al (2021) Awe is associated with creative personality, convergent creativity, and everyday creativity. Psychol Aesthet. Creativity, and the arts
Zhou H, Hu H (2008) Human motion tracking for rehabilitation – a survey. Biomed Signal Process Control 3(1):1–18

Using Extended Reality to Study the Experience of Presence

Keisuke Suzuki, Alberto Mariola, David J. Schwartzman, and Anil K. Seth

Contents

K. Suzuki (✉)
Center for Human Nature, Artificial Intelligence and Neuroscience (CHAIN), Hokkaido University, Sapporo, Hokkaido, Japan

Sackler Centre for Consciousness Science, School of Engineering and Informatics, University of Sussex, Brighton, UK
e-mail: ksk@chain.hokudai.ac.jp

A. Mariola
Sackler Centre for Consciousness Science, School of Engineering and Informatics, University of Sussex, Brighton, UK

School of Engineering and Informatics, University of Sussex, Brighton, UK

Sussex Neuroscience, School of Life Sciences, University of Sussex, Brighton, UK

D. J. Schwartzman
Sackler Centre for Consciousness Science, School of Engineering and Informatics, University of Sussex, Brighton, UK

School of Engineering and Informatics, University of Sussex, Brighton, UK

A. K. Seth
Sackler Centre for Consciousness Science, School of Engineering and Informatics, University of Sussex, Brighton, UK

School of Engineering and Informatics, University of Sussex, Brighton, UK

Program on Brain, Mind, and Consciousness, Canadian Institute for Advanced Research, Toronto, ON, Canada

Curr Topics Behav Neurosci (2023) 65: 255–286
https://doi.org/10.1007/7854_2022_401
Published Online: 3 January 2023

Abstract Extended reality (XR), encompassing various forms of virtual reality (VR) and augmented reality (AR), has become a powerful experimental tool in consciousness research due to its capability to create holistic and immersive experiences of oneself and surrounding environments through simulation. One hallmark of a successful XR experience is when it elicits a strong sense of presence, which can be thought of as a subjective sense of reality of the self and the world. Although XR research has shed light on many factors that may influence presence (or its absence) in XR environments, there remains much to be discovered about the detailed and diverse phenomenology of presence, and the neurocognitive mechanisms that underlie it. In this chapter, we analyse the concept of presence and relate it to the way in which humans may generate and maintain a stable sense of reality during both natural perception and virtual experiences. We start by reviewing the concept of presence as developed in XR research, covering both factors that may influence presence and potential ways of measuring presence. We then discuss the phenomenological characteristics of presence in human consciousness, drawing on clinical examples where presence is disturbed. Next, we describe two experiments using XR that investigated the effects of sensorimotor contingency and affordances on a specific form of presence related to the sense of objects as really existing in the world, referred to as 'objecthood'. We then go beyond perceptual presence to discuss the concept of 'conviction about reality', which corresponds to people's beliefs about the reality status of their perceptual experiences. We finish by exploring how the novel XR method of 'Substitutional Reality' can allow experimental investigation of these topics, opening new experimental directions for studying presence beyond the 'as-if' experience of fully simulated environments.

Keywords Experience of presence · Extended reality · Reality monitoring · Sense of presence · Substitutional reality · Virtual reality

1 Introduction

Extended reality (XR) has for many years offered substantial promise to the scientific study of consciousness. XR is a general term encompassing virtual reality (VR), augmented reality (AR), as well newer methods such as substitutional reality (SR), which is described later in this chapter. Recent advances in XR technologies related to Head-Mounted Displays (HMD) and motion capture systems have provided powerful new tools enabling experimenters to manipulate specific aspects of the experienced world, or self, within highly realistic virtual environments (Bohil et al. 2011; Foreman 2010; Wilson and Soranzo 2015). Experiments capitalising on XR in these ways have addressed topics including spatial navigation (Ekstrom et al. 2003; Hartley et al. 2003; Shelton and Hedley 2004; Voermans et al. 2004), multisensory bodily perception (Ehrsson 2007; Lenggenhager et al. 2007; Slater et al. 2010; Suzuki et al. 2019a), and social neuroscience (de Borst and de Gelder 2015; Parsons 2015; Parsons et al. 2017; Pertaub et al. 2002). In the context of consciousness research, XR appears to offer the unique advantage of creating holistic, immersive conscious experiences of oneself and the environment within a simulated reality, while still permitting the manipulation of sensory input in a highly controlled manner – and crucially, enabling embodied interactions to be studied and manipulated (Bohil et al. 2011; Parsons et al. 2020).

What does it mean to feel presence within a virtual environment? Imagine that you put on a HMD and find yourself suddenly immersed in a highly realistic virtual environment. You are standing on a ledge of a very tall skyscraper, and looking down, you see the streets far below. You remain aware that this experience is not actually real: you know that the visual and auditory information your senses are receiving arise from the technology you are wearing. However, you nevertheless find yourself responding to these sensory signals 'as if' they were real: you are reluctant to move closer to the edge, you feel a sense of vertigo, your heart rate increases, and so on. Cognitively, you know that you are not in any danger, but you still experience the danger in some sense as being real. This contrast, verging on a contradiction, is at the heart of the concept of presence within XR environments.

In general, the feeling of presence has been defined as the subjective sense of reality of the world and of the self within the world (Metzinger 2003a; Sanchez-Vives and Slater 2005). More specifically within the context of XR, presence has been operationalised as the sense of being present in a virtual environment, rather than the place where one's body is actually located (Sanchez-Vives and Slater 2005; Lombard and Ditton 1997; Felton and Jackson 2021). Thanks to the growing body of research using XR, we now have considerable knowledge regarding the technological and psychological factors that modulate the sense of presence within XR, which we will discuss below (Bailenson and Yee 2008; Bowman and McMahan 2007; Slater et al. 1995).

Beyond XR research, there has been a growing interest more broadly within psychology and neuroscience to understand how the subjective experience of presence arises – and, more generally how perceptions and beliefs regarding our sense of 'reality' about the self and the world arise (Metzinger 2003a; Sanchez-Vives and Slater 2005; Seth et al. 2012). The topic of presence has also attracted the attention of clinicians investigating and treating mental disorders such as depersonalisation/derealisation disorder (DPDR), in which patients report a loss of their sense of presence in the world. Further research into conditions such as these promises to shed light on the cognitive and neural mechanisms of presence more generally.

In this chapter, we will analyse the concept of presence in terms of the general question of how humans generate and maintain a stable sense of reality during both natural perception and virtual experience. We will distinguish several aspects of 'presence' as it is generally construed. First is the notion of presence as the subjective *impression of being present* within a virtual environment. Second, there is the characteristic of normal perceptual phenomenology that the contents of perception seem to 'really exist' – we call this the *perceptual presence*.[1] Finally, there are higher-level beliefs that the contents of perceptual experience not only *seem* real but actually *are* real. Normally these two aspects go together, but they can come apart – for example, in lucid dreams. Crucially, XR experiences and technological design have tended to emphasise the first aspect of presence while neglecting the latter two. Indeed, most participants in XR, though they have the subjective impression of being present within a virtual environment, neither experience their virtual environment as being fully real, nor believe it to be real. Broadening the analysis of presence to these additional dimensions will add value to XR development and may catalyse insightful interactions between XR and perceptual presence in normal perceptual experience, as well as in a number of relevant clinical conditions.

2 Presence

2.1 Presence in XR

In this section, we first expand the above analysis of presence, alongside various methods to measure it in XR, and its relationship to conscious experience. We will also address similar concepts that have been discussed within the fields of neuropsychology and psychiatry in the context of specific mental disorders, such as DPDR.

We begin by outlining some of the many different definitions of presence that have been proposed within the XR literature (Felton and Jackson 2021; Lee 2004; Lombard and Ditton 1997), all of which arguably share the central idea that presence refers to the feeling of being physically present within a virtual environment, irrespective of one's actual physical reality (Witmer and Singer 1998). Alternative or complementary definitions, again within the XR context, consider presence as a loss of awareness of the media through which perceived objects or environments are conveyed (e.g. losing awareness of a HMD as mediating an XR experience). This conception of presence is somewhat coextensive with the philosophical notion of 'transparency' (Lee 2004; Lombard and Ditton 1997; Metzinger 2003b, 2014). Perhaps the most precise general definition of presence emerging from this background comes from Felton and Jackson (2021), who define the feeling of presence as 'the extent to which something (environment, person, object, or any other stimulus) appears to exist in the same physical world as the observer'. Notably, this definition could apply equally to normal (non-XR-mediated) experiences, as well as to XR.

[1] Note that perceptual contents seem real does not mean that they are, in some objective sense, real. Colours, for example, often seem to really exist in the world, even though they do not exist in any mind-independent way.

Although, as we will discover, many factors affecting presence are related to the perceptual domain (e.g. visual immersion, field of view, or synchronisation of multisensory information), many have argued that the experience of presence should not be viewed simply as an attribute of some other, more specific, perceptual content. There is an ongoing discussion about the nature of presence in conscious experience, i.e. 'what the experience of presence is like'. For example, Seth (2009) proposed that presence is a 'structural property' of consciousness – one that can apply to perceptual experiences in general – rather than to a specific instantiation of perceptual contents (e.g. a coffee cup on a table). Others have gone further, suggesting that presence within XR should be viewed as a distinct state of consciousness (Felton and Jackson 2021), based on similarities in the shifts of presence that may occur within XR to other distinct states of consciousness such as dream states (Biocca 2003), psychotic hallucinations (Bentall 1990), or during drug-induced mental states (Klüver 1942). These views emphasise presence as having its own distinctive signature in perceptual awareness, that is, presence cannot be reduced to other perceptual attributes, and has its own distinctive perceptual phenomenology. An alternative view suggests that the concept of presence refers to a type of 'metacognitive feeling' (Dokic and Martin 2017). This perspective proposes that the sense of presence is a cognitive rather than a perceptual experience, which is constructed based on judgements about reality. However, proponents of this view also emphasise that presence has an affective component, which they describe as a 'feeling', which again contrasts with the idea of presence as a primarily world-related perceptual attribute of some specific content (such as, for example, the experienced spatial dimensionality of a perceived object; Dokic and Martin 2017).

XR research has largely focused on uncovering the factors that are necessary for presence (as defined above) to emerge, and designing reliable measures that assess the degree to which presence is experienced. This line of research has been motivated, at least in part, by a need to assess the effectiveness of XR technologies for virtual training, entertainment, and applications in psychotherapy. Overall, the factors that have been shown to affect the feeling of presence within XR can be grouped into three categories: sensory factors; coupling of sensation and action (sensorimotor coupling); and embodiment.

Many of the *sensory factors* that affect the sense of presence in XR relate to basic properties of the media technologies used to create virtual environments. For example, the degree to which a HMD occludes a person's vision from the physical world, providing a field of view within the HMD that is roughly equivalent to the human visual field (e.g. 180°), and using a minimum graphics frame-rate of $\sim$15 Hz have all been shown to increase the feeling of presence within XR (Axelsson et al. 2001; Duh et al. 2002; Lin et al. 2002; Slater et al. 1996; Witmer and Singer 1998). These types of differences in the properties of the media hardware are also tightly related to the concept of immersiveness (i.e. the degree to which sensory information appears to originate in an external world rather than from the technological device itself; Weech et al. 2019).

The addition of multisensory feedback, for example, haptic or auditory feedback, within virtual environments has also been shown to increase reported feelings of presence (Bailenson and Yee 2008; Cooper et al. 2015; Freeman and Lessiter 2001). While general haptic feedback has yet to be fully achieved due to the non-trivial technical challenges associated with this form of sensory feedback (although see here https://haptx.com/), it can be cheaply simulated with the so-called static haptics or tactile feedback. A static haptic is a simple version of a physical object located in the real world that corresponds to a virtual counterpart, such as a plank of wood on the floor corresponding to the edge of a simulated building. There are many methods of producing tactile feedback in XR. For example, in a previous experiment from our lab that investigated the sense of agency induced by virtual hand movements within XR (Suzuki et al. 2019b), we simulated the tactile sensation associated with a virtual button press by attaching a vibrating pad to the participant's fingertips, which vibrated whenever the participant pressed a virtual button – here, the tactile feedback from the pad created a sense of touching a physical object. The experience of presence can also be enhanced by combining multiple multisensory cues (Cooper et al. 2015; Dinh et al. 1999), possibly due to multiple sources of sensory information confirming a particular perceptual interpretation of a virtual environment. Altogether, these parameters provide a repertoire of presence-affecting factors which researchers can use to increase or otherwise modulate the felt presence of a virtual environment within XR experiments.

As well as considering the fidelity, transparency, and multimodality of sensory data, presence may also depend in substantial ways on the incorporation of action – in the form of *sensorimotor couplings* or contingencies within XR environments (Flach and Holden 1998; Grabarczyk and Pokropski 2016; Zahorik and Jenison 1998). The central idea here is that presence is shaped by the range of actions it is possible to perform within a virtual environment – and by their sensory consequences (Sanchez-Vives and Slater 2005). This view inherits from the concept of affordances (Dalgarno and Lee 2010; Gibson 1979), which we will discuss in more depth in the next section. Within this view, the degree of interactivity and sensorimotor coupling that XR technology can provide is critical in creating a sense of presence (Lallart et al. 2009; Slater and Usoh 1993; Steuer 1995). One key aspect of this view is that sensorimotor coupling is fundamental in eliciting a sense of *agency*, which can be thought of as the experience of controlling one's own actions, more generally, of being the cause of things that happen, and which is considered to be a key part of the experience of the 'self' (Gallagher 2000; Haggard 2017).

A further dimension that has been shown to affect presence within XR can broadly be termed *embodiment*. Using virtual avatars within immersive VR environments allows users to experience the sensation of ownership over a virtual body, which in turn leads to the feeling of being 'embodied' in such a virtual avatar (Grabarczyk and Pokropski 2016; Schultze 2010; Slater et al. 2009a). For instance, people report an increased sense of presence when their avatar is a complete virtual body, compared to a more simple representation such as a 3D cursor (Slater and Usoh 1993). Embodying a virtual avatar has been shown to produce particularly interesting effects within social contexts: research has found that participants experienced a greater sense of interactivity and immersion when the appearance of the avatar reflected their ideal self rather than their actual appearance (Jin 2009; Kafai et al. 2010). Finally, the experience of 'embodying' a virtual avatar within XR in a broader context has been used to investigate the experience of the bodily self (Blanke and Metzinger 2009; Lenggenhager et al. 2007; Slater et al. 2010). These studies have extended the rubber hand illusion (RHI), a cornerstone of the scientific literature of embodiment, in which people (to varying degrees) report experiencing illusionary ownership of a rubber hand when synchronous visuo-tactile stimulation is applied to the visible rubber hand and to the participant's own occluded hand (Botvinick and Cohen 1998; Roseboom and Lush 2022).

Although research into how these factors contribute to the feeling of presence within XR has been considerable, one underappreciated factor, particularly relevant to embodiment, is the potential role for demand characteristics (Orne 1962) and – more generally – participant expectations in shaping aspects of any XR experience, including reported presence. Demand characteristics refer to the long-standing concern in psychology that characteristics of experimental design may lead people to 'know the correct answer' that the experimenter is 'looking for' – and that this knowledge, or expectation, whether implicit or explicit, may influence or explain their responses. It is difficult to think of a situation more prone to embedding demand characteristics and participant expectations than XR environments, which are usually designed explicitly to bring about immersive and highly specific experiences. Importantly, demand characteristics may exert their effects not only through participants merely making responses according to what they think the experimenter wants ('behavioural compliance'), but also by shaping or generating subjective experiences to fit experimental demands (Kirsch and Council 1989; Olson et al. 2020).

For example, a recent study found that people who are more capable of producing experiences consistent with task demands in a hypnotic context, may also be able to respond to demand characteristics in general, outside of that context, by changing their actual experience (Dienes et al. 2020). In a series of follow-up studies this group investigated the effects of individual differences in suggestibility (trait phenomenological control) on the RHI (Lush et al. 2020; Lush and Seth 2022; Roseboom and Lush 2022; see also Ehrsson et al. 2022). Together, these studies revealed that individual differences in suggestibility confounded the subjective measurements of ownership in the RHI. The authors concluded that it is not possible to exclude the possibility that the experience of embodiment in the RHI is due to implicit imaginative suggestion effects – to phenomenological control. These findings are not a concern only for the RHI, but likely extend to any experimental setting involving measuring subjective experiences that does not adequately control for the effects of demand characteristics. Future studies investigating the sense of presence in XR should carefully consider the implicit demand characteristics inbuilt into the design of their experiments and develop suitable conditions to control for these factors. Such an approach will allow the field to develop a better understanding of the mechanisms and processes underlying empirical, phenomenological, behavioural, and physiological observations in XR, while also broadening our understanding of the role top-down expectations play in the construction of perceptual experiences in both real and virtual worlds.

2.2 Measuring Presence

Adding complexity to the presence research landscape is the lack of consensus about how best to measure it, either within XR contexts or more generally. Questionnaires are the most commonly used method, due to their controlled and structured nature and simplicity of administration (Van Baren 2004). The use of questionnaires does not require special equipment and can be performed easily before or after the experimental sessions (Witmer and Singer 1998). However, questionnaire-based methods for assessing presence have been shown to be unstable, in that prior information can change the results (Freeman et al. 1999), and as mentioned above, they are also susceptible to the effects of demand characteristics (Van Baren 2004). A more subtle worry is that reports of presence may only arise in XR when participants are probed to provide them (Slater 2004), suggesting that researchers need to move away from the current heavy reliance on questionnaires in order to make progress in this area.

Besides questionnaires, researchers have sought to identify behavioural signatures of presence in XR. Such behavioural signatures revolve around the intuition that if participants within a XR experience behave as if they are in an equivalent real environment, then they are likely to be experiencing presence within that virtual environment. For example, if a participant ducks in response to a looming stimulus or shows postural sway in response to moving visual field, then this may be used as a sign of presence (Freeman et al. 2000; Held and Durlach 1992; Sanchez-Vives and Slater 2005). An extension of this approach is to use physiological measures, based on a similar assumption that if a participant's normal physiological response to a particular situation is replicated in a virtual environment, then this is a sign that they are experiencing a high degree of presence. So far, expected alterations in heart rate (Meehan et al. 2002), skin conductance (Slater et al. 2009b), and electroencephalography (Terkildsen and Makransky 2019) have all been demonstrated in response to fear inducing situations within virtual environments.

Although physiological measures like these may be less subject to explicit bias than direct subjective reports, it is important to note that they are not immune from the effects of demand characteristics or implicit imaginative suggestion. For example, imaginative suggestion has been repeatedly demonstrated for more than half a century to affect skin conductance responses including responses to the RHI (Barber and Coules 1959; Kekecs et al. 2016; Lush et al. 2020). Furthermore, there are concerns about the degree to which behavioural and physiological measures actually reflect the subjective sense of presence, as objective and subjective measures of presence have been shown to diverge in some cases (Bailey et al. 2009; Freeman et al. 2000; IJsselsteijn et al. 2002; Riva et al. 2003; Wiederhold et al. 1998).

Given the current state of art regarding measuring presence, there is a need to continue to develop new behavioural measures that correlate with presence, while carefully avoiding the demand characteristics mentioned above. For example, behavioural measures such as proprioceptive drift (Botvinick and Cohen 1998) or intentional binding (Haggard et al. 2002) have been considered as behavioural signatures of (respectively) body ownership and agency, which could relate to presence in the context of embodiment and sensorimotor coupling. However, great care needs to be taken to measure participant expectancies for such measures in order to avoid confounds due to demand characteristics (Lush et al. 2021). Outside the context of embodiment, there is a need to develop behavioural measures that correlate with presence in simple perceptual tasks. In Sect. 3, we introduce one possibility along these lines, using a version of continuous flash suppression.

Another possible avenue towards developing measures that remain unconfounded by demand characteristics is to use a version of XR called substitutional reality (SR), which we will describe in detail in Sect. 3. For now, it suffices to say that the defining feature of an SR experience is that participants are unable to distinguish it from the corresponding real-world situation. In such cases, one primary demand characteristic confound is removed since participants should not be able to strategically behave in a particular way in the virtual situation. However other, more general demand characteristics related to expected performance in the experiment as a whole may remain.

Studies attempting to identify the neural correlates of presence remain rare, in part due to the difficulties of bringing a virtual reality environment into an fMRI scanner (Clemente et al. 2014; Jäncke et al. 2009; Sjölie et al. 2014). Among the few studies that have been conducted, the dorsolateral prefrontal cortex (DLPFC) has emerged as one of the relevant brain areas in shaping the reported degree of presence and immersiveness (Baumgartner et al. 2008; Jäncke et al. 2009). It is interesting to relate these findings to neuroimaging studies of DPDR, which have shown both reduced activation in neural regions typically implicated in the generation of affective responses towards salient stimuli (e.g. insula and amygdala) along with increased activation (e.g. hyperactivation) of prefrontal regions (Jay et al. 2014; Phillips and Sierra 2003; Sierra and Berrios 1998). Furthermore, Dresler et al. (2012) using fMRI found increased activation in right DLPFC and in frontopolar areas in lucid dreams compared to REM sleep, while Filevich et al. (2015) showed greater grey matter volume in the frontopolar cortex for people reporting higher dream lucidity compared to the low lucidity group. Together, these neuroimaging results – while sparse and largely indirect – nonetheless offer some support for metacognitive accounts of presence (e.g. Dokic and Martin 2017), while also highlighting the delicate balance between monitoring mechanisms, interoceptive and emotional processes in orchestrating the experience of presence (see also Sects. 2.3 and 4.1).

2.3 *Disorders of Presence in Clinical Conditions*

In our everyday lives our sense of reality is so pervasive that it tends to be taken for granted. Only when it is disturbed pathologically do we even appreciate that it existed in the first place (Jaspers 1973, 1997). For example, in DPDR people report feeling that their sense of reality is attenuated or diminished. DPDR is a type of dissociative disorder in which the patient reports persistent or recurrent feelings of being detached (dissociated) from their body or mental processes, usually accompanied by a feeling of being an 'outside observer' of their life (depersonalisation), or of being detached from their surroundings (derealisation), both of which are related to the experience of presence (Phillips et al. 2001; Shorvon 1946; Sierra et al. 2005; Sierra and David 2011) Interestingly, people with DPDR usually have normal or near-normal cognitive and perceptual capabilities, suggesting that DPDR may occur due to a separation between the experience of specific perceptual content and a more general (structural) experience of presence.

Another commonly reported clinical feature of DPDR in conjunction with a loss of presence is emotional numbness, characterised by attenuated emotional colouring of subjective experiences (Sierra and David 2011). The observation that patients with DPDR display attenuated emotional responses supports theories of presence that suggest that it is a (possibly metacognitive) affective experience, or a *feeling,* that rejects the idea of presence as being a purely world-directed perceptual experience (Dokic and Martin 2017).

Although the neurological causes of DPDR are not fully understood, it has been suggested that aberrant integration of autonomic signals from the body, which are thought to be crucial in generating normative emotional responses, may underlie some features of DPDR (Lemche et al. 2008; Sierra et al. 2002). Supportive of this theory are findings that show that DPDR seems to be associated with a reduction in autonomic responses to aversive stimuli (Sierra et al. 2002). In a related line of work, we developed a computational model of DPDR, in which we proposed that the lack of presence in DPDR was due to abnormal inference of (the causes of) interoceptive signals (Seth et al. 2012). Within this interoceptive inference framework, a sense of presence arises when interoceptive signals are successfully explained by top-down predictions supported by hierarchical generative models in the brain. According to our model, the loss of presence associated with DPDR is associated with imprecise predictions relating to interoceptive signals, which cannot explain away the prediction errors. Experimentally, we found that visual feedback of an individual's own cardiac signals can induce illusory experiences of body ownership (Suzuki et al. 2013), which supports the notion that interoceptive signals may have a role in the generation of the bodily sense of presence (Suzuki et al. 2013; see also Aspell et al. 2013). Although, as discussed earlier, the results of these studies may have been confounded by demand characteristics, as this experiment did not control for expectations or individual differences in suggestibility (Lush et al. 2020).

In this section, we introduced the concept of presence as the subjective sense of reality of the world and of the self. We have seen how XR technologies can enhance experienced presence through combinations of sensory factors, sensorimotor coupling, and embodiment. We also emphasised the issue of demand characteristics when measuring any type of subjective experience, an issue that we suggest potentially confounds existing measures of presence. Shifting focus to the phenomenology of presence, we have seen how the experience of presence is pervasive in our everyday experience and how it can be disturbed in certain clinical conditions, notably in DPDR. Clinical cases such as this raise the question: What causes us to experience certain specific perceptual contents as veridical and realistic rather than fake or unreal? In the next section, we discuss the specific sense that external objects seem to really 'exist' out there in the world, a concept we refer to as *perceptual presence*.

3 Perceptual Presence

In this section, we describe XR experiments that connect presence to the sensorimotor theory of consciousness through manipulations of the perception of real and virtual objects.

3.1 Perceptual Presence and 'Mastery' of Sensorimotor Contingencies

In addition to our sense of presence in the world and within the context of VR, we also experience a distinct sense of presence associated with objects in the real world. For example, when we see a coffee cup on a table, we perceive this object as an existing real entity embedded in the world. This perceptual experience that there is a really-existing object out there in the world can be called *perceptual objecthood* – or, for simplicity here, just objecthood. To flesh out this notion, consider again the coffee cup. When we observe a coffee cup in real life, we experience it as having a back (and sides) even though the back (and perhaps sides) are not immediately available within our vision. We experience the cup as having a three-dimensional volumetric extension in the world. The same feeling does not arise from a coffee cup as represented in a photograph or a painting. This notion of objecthood is one way of bringing specificity to the more general notion of 'perceptual presence'.

The property of perceptual presence has motivated the sensorimotor theory of consciousness (Noë 2004; O'regan and Noë 2001), which understands perceptual phenomenology to be shaped by 'mastery' of the sensorimotor contingencies governing how sensory signals respond to actions. This theory views perceptual experience as being instantiated through a closed loop between action and perception, a proposal which inherits from James J. Gibson's concept of *affordance*, which refers to the opportunities an object presents for action (Gibson 1979). According to Noë, the perceptual presence of a real object is given by the learnt 'know-how' or 'mastery' of the sensorimotor contingency that governs how the sensory information is altered by a particular action (Noë 2004; O'regan and Noë 2001). To continue the coffee cup metaphor, I see a coffee cup as really existing in the world because my brain 'knows about' the sensory consequences of moving my eyes or rotating the cup. In this sense, I perceive that the cup has a back even though I cannot directly see it[2] (Noë 2004; O'regan and Noë 2001).

While influential, the sensorimotor theory of consciousness expressed by O'Regan and Noë (2001) has been criticised for lacking a clear implementation in neural circuitry. More recently there have been attempts to incorporate sensorimotor contingencies into a *predictive processing* account of perception (Seth 2014). In the predictive processing framework, perception is viewed as probabilistic inference of the 'hidden causes' of sensory information, by neuronally encoding a generative model that makes predictions about the likely causes of sensory information (Clark 2013; Friston 2010; Rao and Ballard 1999). The Predictive Processing Theory of SensoriMotor Contingencies (PPSMC) proposes that perceptual presence arises when the brain encodes a rich repertoire of predictions using a generative model that can make predictions about how sensory signals would change given specific actions (Seth 2014, 2015).

These theoretical developments have been accompanied by a growing body of empirical work investigating the effects of action, or the opportunity an object presents for action, on visual perception (Bekkering and Neggers 2002; Chan et al. 2013; Lindemann and Bekkering 2009). One of the most promising avenues of research for investigating how actions may affect perceptual presence focuses on how the brain responds to real objects versus images of objects (Gomez et al. 2018; Marini et al. 2019; Snow et al. 2014; Snow and Culham 2021). One such study investigated how a person's attention changes when viewing either real objects or images of the same objects (Gomez et al. 2018). In this experiment, participants were asked to identify the orientation of a target object (a spoon) within a field of distractor objects as quickly as possible. The results revealed that compared with both 2-D and 3-D images, real objects yielded slower response times overall and elicited greater flanker interference effects. Interestingly, when the real spoon was placed outside of the participant's reach, or when a transparent screen was placed between the spoon and the participant the differences in reaction time and interference effects were comparable with that of 2-D images. These results demonstrate that real objects exert a more powerful influence on attention and motor responses compared to representations of objects and suggest that this effect may be due to the affordances that real objects provide for physical interaction. However, returning to the core predictions of the sensorimotor theory of consciousness, few empirical studies have directly addressed if 'mastery' of sensorimotor contingencies contributes to perceptual presence. Fortunately, XR technologies offer new opportunities to

---

[2] One might argue that it is possible to experience objecthood even for a cup in a painting or a video. Consider, for example, that in the painting or video, there is a person trying to grasp the cup. In this example, we might have some sense that the cup is indeed a three-dimensional object. Critically, though, this is with respect to the person in the painting, not with respect to us as external observers. Arguably, the notion of objecthood that pertains in this case is more indirect, and cognitive, rather than the direct and immediate objecthood that is part of our real-world everyday perceptual experience.

investigate this question using naturalistic sensorimotor interactions with virtual objects.

3.2 *Using Binocular Suppression to Measure Perceptual Presence*

An early study investigating the link between action and perception used a binocular rivalry paradigm, in which two dynamic visual stimuli compete for perceptual awareness (Maruya et al. 2007). This study found that when the motion of one stimulus (but not the other) was contingent on a participant's voluntary actions, dominance durations for that stimulus were longer, and suppression times were shorter, compared to stimuli that moved independently from the participant's actions. This finding suggests that disruption of veridical sensorimotor contingencies can affect the formation of a visual percept even outside of awareness. However, the simple visual stimuli used (random dot stereogram), and the trained stereotypical movements used in this study did not address how naturalistic sensorimotor interactions with real-world objects shape subjective visual experience.

To address these issues, we developed a novel experimental setup combining AR and VR technologies that allowed real-time naturalistic sensorimotor interactions with novel virtual 3D objects. Our setup allowed us to investigate the dependence of visual experience on the dynamic causal coupling between actions and their sensory consequences within a real-world setting (Suzuki et al. 2019b). We manipulated the specific sensorimotor contingencies associated with virtual objects within a well-studied visual paradigm called continuous flash suppression (CFS; Jiang et al. 2007; Stein et al. 2011; Tsuchiya and Koch 2005). CFS is a more controllable version of binocular rivalry in which perceptual awareness of a target stimulus presented to one eye is suppressed by a series of rapidly changing, high contrast, Mondrian patterns (see Fig. 1b) presented to the other eye, and the time it takes the target to 'breakthrough' into awareness is measured. Often (as in our paradigm) the relative contrasts ramp up (for the target) and down (for the Mondrian) to ensure breakthrough occurs within a reasonable time frame. Differences in the time it takes a stimulus feature to breakthrough into conscious awareness in CFS have been interpreted by some authors as indicating the amount of unconscious processing associated with a specific stimulus feature (Salomon et al. 2013; Stein and Sterzer 2014). Although earlier studies focused on examining the effects of low-level visual features, such as image contrast, on breakthrough times, later studies found that even high-level visual features can modulate unconscious visual processing within the CFS paradigm. For example, upright faces have been shown to breakthrough into conscious awareness faster than inverted faces (Jiang et al. 2007). In addition, the CFS paradigm has been used to study how multisensory information affects visual awareness by manipulating the congruency between somatosensory and visual information (Salomon et al. 2013, 2015).

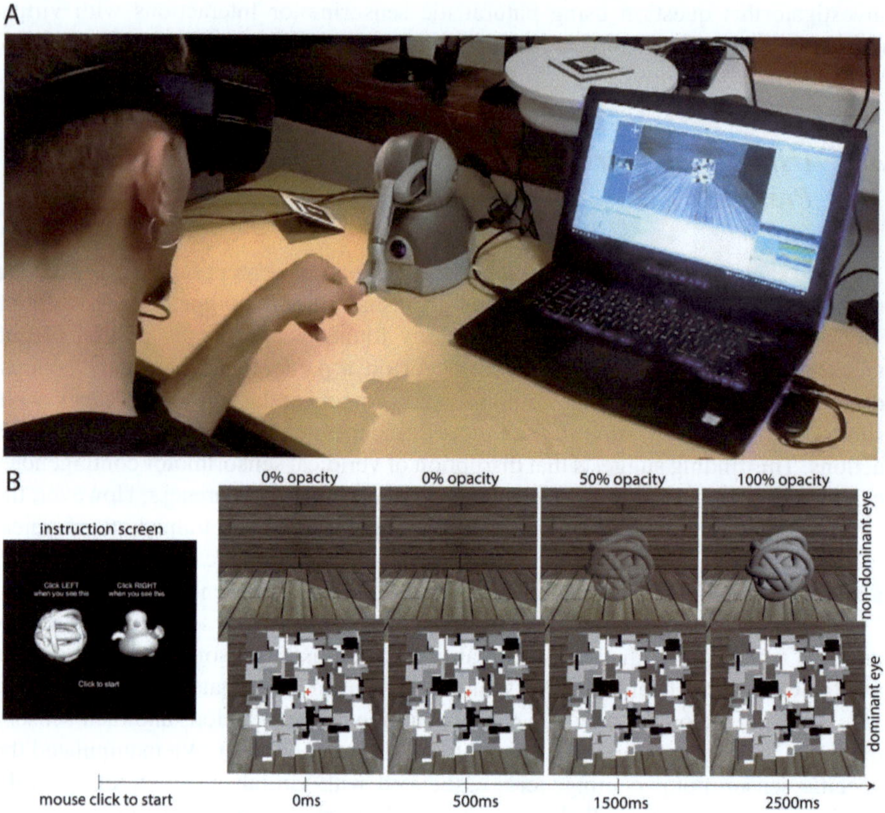

Fig. 1 An illustration of the continuous flash suppression (CFS) experiment used to investigate the effects of sensorimotor contingencies of virtual 3D objects on breakthrough times (adapted from Suzuki et al. 2019a, Crown Copyright © 2019 Published by Elsevier B.V. All rights reserved.). (**a**) Experimental setup, participants viewed a virtual 3D object and dynamic Mondrian through a HMD. They manipulated a virtual object using a motion-tracking stylus with their left hand. (**b**) Single trial structure of the experiment. The object's opacity gradually increased over time and was presented to the non-dominant eye (top) and the Mondrian mask was presented to the dominant eye (bottom)

We used this setup to test the influence of sensorimotor coupling on visual awareness of a virtual object whose visibility was suppressed by a CFS mask (Fig. 1). To exclude the influence of inter-subject differences in sensorimotor knowledge for familiar objects, we used a series of novel 3D virtual objects. For each trial, participants were asked to rotate a motion-tracking device held in one hand, which was reflected by the movements of the 3D object rendered to the non-dominant eye, whilst a CFS mask was presented to the dominant eye inside a head-mounted display. We found that participants' responses were faster when the virtual object was directly coupled to their ongoing actions, compared to when the object's movements were replayed from previous trials, or when the object did not move at all in response to their actions.[3]

While this study did not measure perceptual presence directly, the results support the idea that object perception is enhanced when naturalistic, ecologically valid sensorimotor contingencies are followed. More specifically, the acquisition of the sensorimotor contingencies associated with the virtual object during the live condition led to faster breakthrough times, compared to the replay or static conditions. Breakthrough time under CFS, in this paradigm, could therefore be considered as a potential behavioural proxy for perceptual presence.

Extending and complementary to the results of this study, Korisky and colleagues developed an experimental setup that enabled CFS with real objects (Korisky et al. 2019). Their setup consisted of AR glasses that displayed a CFS suppression mask to one eye, while the other eye viewed an object located in the physical world. Using this setup, they were able to use real objects and two-dimensional pictures of the same objects and measure the time taken for these two classes of stimuli to breakthrough into conscious awareness. It should be noted that, unlike in our experiments, the sensorimotor contingencies of the objects were not manipulated as the objects remained stationary. However, another difference was that the objects used were all familiar, permitting the assumption that the participants had already acquired mastery of the relevant sensorimotor contingencies. Interestingly, they found that real objects took less time to breakthrough into conscious awareness than two-dimensional images of the same objects. Since the object was only presented to one eye, there was no binocular disparity between the 3D real object and the 2D photograph, so the difference in reaction times could not be explained by stereoscopy. What then could account for the difference in breakthrough time between the real object and the two-dimensional photograph?

Korisky and Mudrik examined this question using an elaborate follow-up experiment (Korisky and Mudrik 2021) in which they created 3D replicas of the original familiar objects using a 3D printer. They then cut the 3D replica into pieces, and randomly combined them to create a 3D scrambled object. They then took photographs of the 3D object and of the scrambled counterpart. In vision science, scrambling techniques like these are often used to change the content of visual stimuli, while keeping the low-level statistical properties the same. Their 3D scrambled object provided a neat control condition, which shared the low-level visual features of the object, but removed the high-level features that define the object as a member of a certain category (Fig. 2). Using these four conditions (Intact 3D, Intact 2D, Scrambled 3D, and Scrambled 2D) they compared the time taken for each to reach conscious awareness in the same CFS paradigm as they previously used. They found that familiar 3D objects broke through into conscious awareness faster than their 2D pictorial representations. Critically, they found that the faster processing

[3]Note that we controlled for the speed of the participants' rotational movements of the virtual objects between live and replay conditions, meaning that any differences in breakthrough times were most likely due to the presence or absence of normal sensorimotor contingencies.

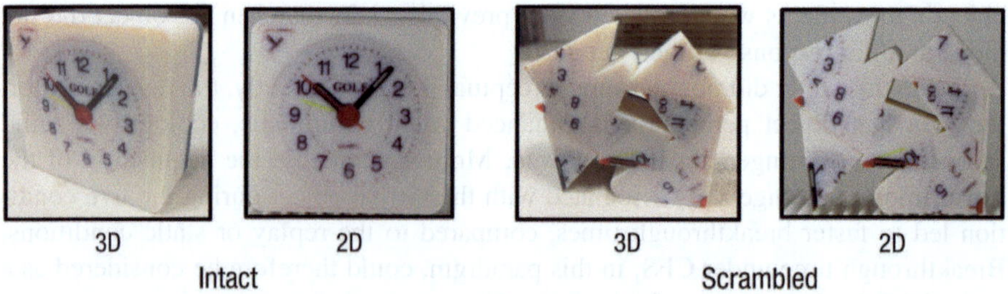

Fig. 2 Example stimuli used in the experiment used in real-world CFS experiments. 3D-printed familiar object and its 2D photograph (Left) and scrambled 3D object and its 2D photograph (Right) (adapted from Korisky and Mudrik 2021, Copyright 2020 Mudrik, Liad; Korisky, Uri)

times of 3D objects than the 2D pictures were not found for scrambled, unfamiliar objects. These results are also consistent with the idea that valid sensorimotor contingencies enhance object perception, with breakthrough times providing a potential behavioural proxy for perceptual presence. The key difference from Suzuki et al., was that, here, these contingencies were already present (since familiar objects were used) rather than learned during the experiment.

Taken together, the results of these CFS experiments provide convergent evidence that objects with learned or familiar sensorimotor contingencies are associated with shorter breakthrough times. What justifies the notion that breakthrough time in this paradigm could be a behavioural surrogate for perceptual presence? A common interpretation of breakthrough time in CFS studies is that it reflects differential unconscious processing (Gayet et al. 2014). Thus, one might interpret the results of these studies as reflecting unconscious processing of sensorimotor contingencies and affordances, both of which have been proposed as factors that influence perceptual presence (Flach and Holden 1998; Grabarczyk and Pokropski 2016; Zahorik and Jenison 1998). In the case of affordances, such an interpretation is in line with findings of enhanced attentional capture by 3D object flankers compared with photographs of flankers (Gomez et al. 2018), as well as with the shorter suppression times reported for useful tools compared with useless tools (Weller et al. 2019). However, while the perceived ability to act on an object (affordance) is clearly closely related to the perceptual impression that the object really exists in the world (perceptual presence), these perceptual properties may not be equivalent to each other, and we again note that neither study explicitly investigated perceptual presence.

In this section, we have discussed the characteristic of normal perceptual phenomenology that the contents of perception seem to 'really exist' – a quality which for objects we call 'perceptual presence' or, equivalently, 'objecthood'. We examined how this aspect of perception may be constructed from the sensorimotor contingencies and affordances associated with perceptual content. Describing the results of two empirical studies that leveraged combinations of XR technologies with a binocular suppression paradigm, we have shown how the validity of sensorimotor contingencies affects object perception. The use of such perceptual tasks combined with modern XR immersive technologies allows for the manipulation of sensory input in a highly controlled manner, providing many advantages compared to classical experimental settings, which can be exploited in future studies to investigate not only presence, but many aspects of conscious experience.

4 When Presence Is Not Enough: Beyond Virtual Reality

While previous sections highlight the capacity of VR and AR to manipulate different aspects of presence, there are certain features of human sense of reality that this method cannot directly address.

4.1 Layers of Veridicality

Whenever a person behaves within a virtual environment, they may sometimes be in a state of suspension of disbelief, in which their judgments about the artificiality of the virtual world are temporarily suspended, in order for them to properly engage with it. Such situations may be characterised by their behaviour and subjective reports being 'as if' the world that the HMD presents to them corresponds to, or is, the 'real' world. People may *feel* that they are immersed and present in the virtual world, but it is highly unlikely that they would ever *believe* that what they are experiencing was actually real. This dissociation between feeling and believing highlights that the concept of 'sense of reality' cannot be reduced to, or described along a single dimension, but necessarily spans domains and layers.

Recently, in the context of his predictive processing model of sensorimotor contingencies (PPSMC), Seth (2014) proposed to differentiate perceptual contents by means of four different dimensions of reality/veridicality, as reproduced in Table 1.

In this formulation, perceptual reality refers to the vividness of the content, whereas veridicality is divided into three subtypes: (1) *subjective*, namely the property of being phenomenologically experienced as part of the real world (here equivalent to objecthood/perceptual presence); (2) *doxastic*, the property of being cognitively considered as part of the real world (e.g. 'I believe that this coffee cup I am perceiving really exists'); and (3) *objective*, which obtains when the contents of perception lawfully correspond to an aspect of the real world (e.g. I perceive this coffee cup as being real, and indeed it is real).

This classification system can help orient us to alterations in experienced reality, whether in VR or otherwise. For instance, the experience of derealised patients can be considered as a deficit at the level of subjective veridicality (and possibly at the vividness level of perceptual reality too), while the doxastic component is retained. Patients typically feel alienated from their bodies and experience the world 'as if' it is flat and unreal (Radovic and Radovic 2002), while their ability to distinguish

Table 1 A schematic subdivision of different kinds of perceptual content and their corresponding veridicality status (adapted from Seth 2014 under CC BY 4.0 licence)

	Perceptual reality	Subjective veridicality	Doxastic veridicality	Objective veridicality
Normal perception	✔	✔	✔	✔
Dreaming (non-lucid)	✔	✔	✔	✗
Hallucinations (with delusions)	✔	✔	✔	✗
Dreaming (lucid)	✔	✔	✗	✗
Hallucinations (without delusions)	✔	✔/✗	✗	✗
Synesthesia (projector)	✔	✗	✗	✗
Afterimages (e.g. retinal)	✔	✗	✗	✗
Synesthesia (associator)	✔/✗	✗	✗	✗
Imagery/associative recall	✗	✗	✗	✗

between what is real and what is not remains intact (Guralnik et al. 2000). More precisely, since people that suffer from derealisation tend to be also depersonalised, it is necessary to characterise their condition by considering the loss of a global sense of presence which – as already mentioned in previous sections – has been hypothesised to be related to aberrant interoceptive prediction errors (Seth et al. 2012).

Similarly, the experience of VR can be conceptualised at the opposite side of the spectrum: people tend to react as if virtual contents are real, while at the same time being fully aware that they are not. Thus, VR allows us to create and maintain both the perceptual reality and (perhaps) subjective veridicality component via sensorimotor coupling, but fails to reach the doxastic level (see Fortier 2018, who criticises the idea that VR and DPDR correspond to completely symmetric experiences).

At first glance, the outlined classification might appear superfluous and unnecessary for pragmatic applications, but it is important to highlight that its usefulness goes beyond VR (and DPDR). As illustrated in Table 1, further examples of dissociation between 'levels of veridicality' can be found in typical forms of dreaming and lucid dreaming. While normal dreams are characterised by a total state of immersion for the dreamer, lucid dreams are distinguished by the presence and maintenance of a level of insight over the experience: a form of metacognitive monitoring that allows the dreamer to recognise that they are dreaming, retaining doxastic insight.

Other clinical examples in which we observe alterations in experienced reality include phenomena such as hallucinations and delusions. Hallucinations are defined as percepts that occur in the absence of a (typically) corresponding external stimulus (Tracy and Shergill 2013) and they constitute one of the most common symptoms of conditions such as schizophrenia and Parkinson's disease. Hallucinations tend to occur in all sensory modalities with a clear prevalence for audition (70% of schizophrenia patients; Hugdahl et al. 2008) and vision (27% of schizophrenia patients;

Waters et al. 2014) and similar experiences have been reported in 'neurotypical' populations as well as being a consequence of certain pharmacological manipulations (e.g. psychedelics like LSD or psilocybin; Müller et al. 2017; Schmid et al. 2015). Delusions are defined by the Diagnostic and Statistical Manual of Mental Disorders (DSM-5 – American Psychiatric Association, 2013) as false beliefs – not accounted by the person's intelligence, cultural or religious background – that are firmly maintained in the face of overwhelming contrary evidence (Bortolotti et al. 2012; Kiran and Chaudhury 2009).

In spite of their heterogeneous aetiology, hallucinations have often been described in a rather unitary way by assuming that whoever experiences them is unable to distinguish them from 'ordinary reality'. In this context, Martin Fortier (2018) argues that this might not always be the case. In the chapter: 'Sense of Reality, Metacognition and Culture in Schizophrenic and Drug-Induced Hallucinations', he points out that hallucinating patients show at least two differential patterns that the clinical literature has described as 'single bookkeeping' and 'double bookkeeping'. The former refers to the phenomenon by which the sense of reality of hallucinations is equated with ordinary perception and, when accompanied by delusions, patients act coherently with their beliefs. The latter refers to the observation that for some patients, their hallucinatory reality does not feel 'real' in the same way as normal perception and, when accompanied by delusions, these aberrant beliefs are not accompanied by coherent behaviour (Bleuler 1911, 1950; Bortolotti and Broome 2012), As Ratcliffe (2017) write:

> [...] this is consistent with the observation that delusions and hallucinations often involve a kind of double-bookkeeping, where the patient speaks and acts in ways that are in some respects consistent with believing or perceiving that p, but also speaks and acts in other ways that distinguish her attitude toward p from her ordinary perceptions and beliefs (Ratcliffe 2017)

In a similar manner, different drugs seem to produce hallucinations that are experienced with different levels of veridicality (see Table 1). While psychedelics can be related to the phenomenon of double-bookkeeping, the effects of anticholinergic drugs might exemplify single-bookkeeping. Psychedelic compounds like LSD, psilocybin, or N,N-Dimethyltryptamine (DMT) that typically act as agonists to serotonin receptors (mostly 5-HT2A receptors; González-Maeso et al. 2007; Rolland et al. 2014) appear to generate vivid and colourful visual hallucinations that nonetheless are not equated with the same level of reality as normal perception (e.g. psychonauts remain able to distinguish them from ordinary perception, while considering them real). Other hallucinogenic plants like toé (brugmansia), belladonna, or synthetic compounds such as Ditran are known for their powerful anticholinergic activity that causes delirium-like symptoms (Ashton 2002). People administered with these substances show a complete lack of insight and reduced reality monitoring skills such that they experience their hallucinations as being completely indistinguishable from ordinary perception and typically take everything they see at face value.

Fortier (2018) argues that this clinical and pharmacological evidence pose a threat for classical metacognitive monitoring models like the source monitoring model

(Johnson and Raye 1981; Simons et al. 2017) and the online reality monitoring model (ORM, Dokic and Martin 2017). By considering hallucinations simply as a unitary metacognitive tagging error (either offline or online) of an internal source that is mistakenly considered to be external, the two models might have difficulties in accommodating the different levels of reality reported for hallucinatory contents. Future work in this direction might usefully consider these phenomena in light of newer metacognitive monitoring models such as Lau's 'perceptual reality monitoring' model (Lau 2019; but see also Dijkstra et al. 2022, Gershman 2019).

Considering Fortier's argument in light of Table 1, it is possible to see that what Seth (2014) broadly defined years before as 'hallucinations without delusions' possesses the same features as serotonergic hallucinatory experiences (and double-bookkeeping): while the level of perceptual reality (vividness) and subjective veridicality makes the person feel that these constructs are real, the fact that they do not reach the doxastic level is reflected by preserved insight into the unreal nature of these experiences of patients and/or psychonauts (similarly with lucid dreams). Conversely, 'hallucinations with delusions' reflect anticholinergic hallucinations (and single-bookkeeping) since they maintain all the components listed above: the individual does not only experience them as real, but the lack of insight makes them fully convinced about their reality, thus leading to states of delirium (similarly to non-lucid dreaming).

This discussion makes clear that while most of the literature involved in the debate around sense of reality has focused on the concept of perceptual presence, of equal importance is the different (albeit related) concept of conviction about reality (CR), namely its higher order/metacognitive counterpart (or *doxastic veridicality* as in Table 1). To capture CR, technologies like VR appear to be still too limited at present. It is therefore necessary to examine methods which do not rely on digital environments per se, but rather leverage novel video production/recording capacities and/or *augment* them.

4.2 *Substitutional Reality: A Promising Naturalistic XR Framework*

The substitutional reality system (SR) is a novel experimental platform developed by Suzuki et al. (2012). The system is composed of an HMD coupled with a frontal (or detached) live camera. SR allows switching between the live feed of the camera and pre-recorded 360° videos (see Fig. 3, panel A). The goal of the system is to manipulate participants' CR by allowing them to believe that what they are experiencing is real and is happening live in that moment, even though it can be a pre-recorded, and perhaps manipulated video.

In the original description of SR (Suzuki et al. 2012), the authors reported that participants ($n = 21$) failed to recognise that what they were experiencing corresponded to a pre-recorded scene in a variety of different conditions other than

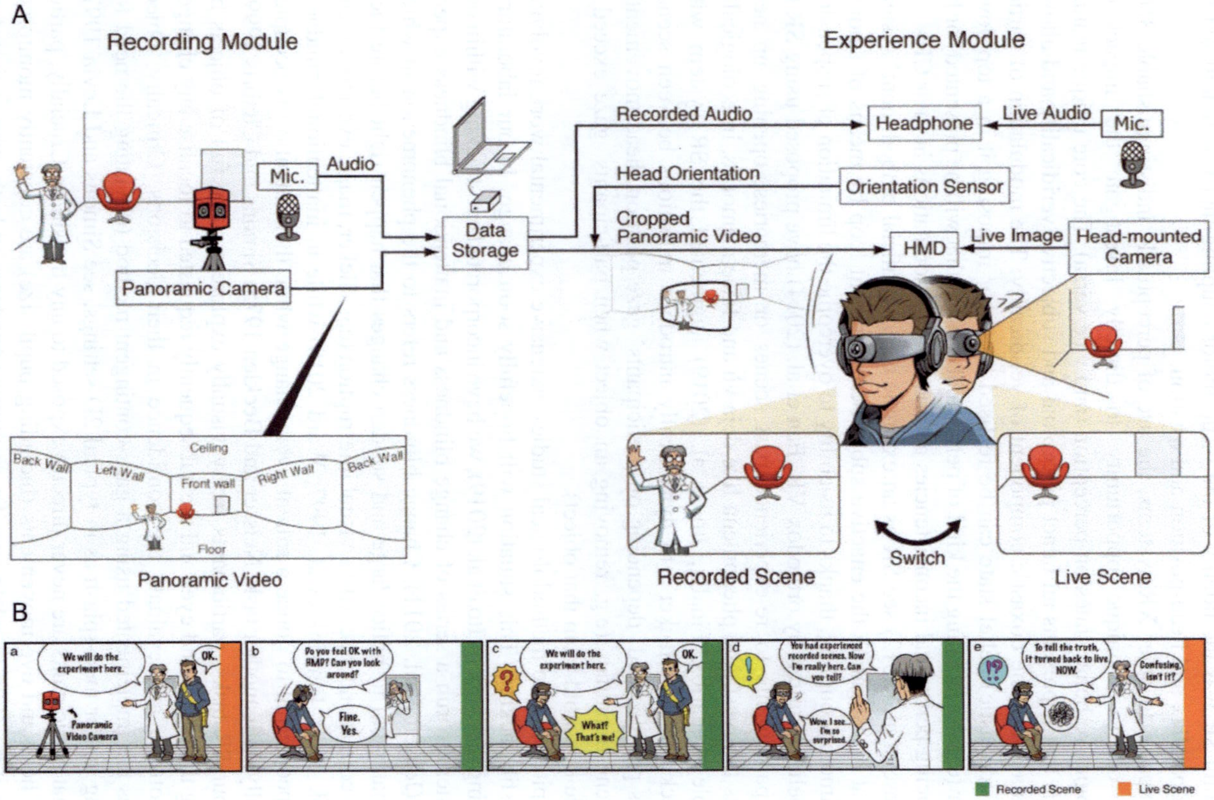

Fig. 3 (**a**) A schematic illustration of the substitutional reality system. (**b**) Potential experimental scenario using the SR system (adapted from Suzuki et al. 2012, under CC BY 4.0 licence)

one in which participants viewed the recording of themselves entering the room before the start of the experiment (the 'Doppelgänger' condition). Additional experiments confirmed that the major factor contributing to the maintenance of a stable CR within SR was the fidelity of the visuo-motor coupling that was modulated by motion parallax and head-position and rotation.

Compared to other XR systems, the use of perceptually indistinguishable 360° videos offers a unique opportunity to directly investigate the impact of metacognitive belief states on perception (and vice versa). The core unique feature of SR is that it presents an intact perceptual and subjective veridicality and allows manipulation of the doxastic component of veridicality. The modulation of participants' (doxastic) belief state can be realised by either imposing it in a top-down manner (e.g. by labelling the kind of perceptual content that will be presented) or by introducing perceptual inconsistencies and observing modulations in their CR.

Froese et al. (2012) see SR as an example of an 'artificial embodiment' system aimed at investigating the enactive subject-environment loop by means of smooth sensorimotor coupling, thanks to its ability to overcome the limitation of suspension of disbelief required by orthodox VR. Fan et al. (2014) have proposed using SR to allow participants to re-experience past scenes or memories, opening up new avenues to investigate phenomena like deja vù and false memories, in ecologically plausible scenarios. Finally, Ito et al. (2019) augmented the SR system with eye-tracking ('EyeHacker') to dynamically introduce transitions between scenes (live vs pre-recorded) depending on participants' gaze position, head-movements, and scene dynamics (e.g. removing an object when participants' gaze exceed a distance threshold from that object).

In spite of these methodological studies, extensive experimental work involving SR is still lacking. This situation will hopefully soon change. In our laboratory, extending the work of Ito et al. (2019), we have incorporated eye-tracking within our SR system to run a series of change blindness and inattentional blindness experiments (Jensen et al. 2011). Change blindness refers to the phenomenon in which participants fail to notice large and sudden changes that happen right before their eyes in correspondence with a visual interruption (e.g. flicker, mask, eye-movement, Rensink 2000; Rensink et al. 1997; Ward 2018) while in inattentional blindness tasks, people fail to notice a salient event/change while their attention is occupied (typically with another task, Neisser and Becklen 1975; Simons and Chabris 1999).

In our SR tasks, participants freely visually explore a room full of objects by moving their head and eyes and are subsequently requested to localise any changes they noticed, and evaluate the confidence in their decisions. Critically, object changes are implemented using a gaze-contingent method (obviating the need for masking and/or mud-splash as in typical 2D settings, see Simons and Levin 1997) such that participants are never directly exposed to any transition. Crucially, participants' head and eye movements (including pupil size) are constantly monitored, making it possible to relate specific oculomotor patterns to both change detection performance and metacognitive assessments. More generally, such an experimental setting provides the opportunity to study the limits of awareness and attention in a more naturalistic and ecological manner – and critically one that involves the ability to manipulate doxastic veridicality.

Another fascinating potential application for SR is to investigate specific memory distortions. Indeed, Suzuki et al. (2012) noted that the switching between 'live' and pre-recorded scenes might resemble deja vù phenomena, and confabulations. Here, we explore the latter in a little more detail.

Confabulations are typically defined as false or erroneous memories that arise either spontaneously or in response to a memory challenge, demonstrating that the way in which people build up their sense of reality can also be impaired along the temporal dimension (Schnider 2003). While discussing different classification systems of confabulations is outside the scope of this chapter, Schnider (2008); Schnider et al. (2017) proposed four main categories: simple intrusions in memory tests; momentary confabulations provoked in response to questions; behaviourally spontaneous confabulations reflecting confusion of reality, and fully fantastical thoughts that can be found in advanced dementia and psychosis. According to Schnider et al. (2017), one of the crucial symptoms of patients with confabulations (e.g. with Korsakoff Syndrome and/or damages to the posterior orbitofrontal cortex) is the inability to properly locate memories and thoughts in the correct temporal context: the tendency of mistaking memories of the past as part of their current 'now' and erroneously acting according to them.

Directly eliciting confabulations using SR, whether in patients or in healthy controls, is an intriguing possibility, albeit one that is fraught with important ethical complications. Nevertheless, the ability of SR to introduce time shifts and unexpected changes in participants' perceptual experience within an immersive environment allows researchers to test scenarios that are completely impossible to investigate via classical 2D memory paradigms. For instance, we could imagine letting healthy participants explore an environment that they believe to be real and sequentially introducing a series of changes (obfuscating sharp transitions – see above). This would open up a series of questions: will participants notice these changes? If so, will they break their stable or sense of reality or will participants explain them away in order to maintain it? In the context of a continuous live-recorded loop, how would they react?

At present, using panoramic videos and live feed in SR, instead of carefully crafted virtual worlds, comes at the expense of limiting participants' interaction with the environment (currently, to head and eye movements), and the use of an HMD still provides a large degree of separation between reality frames as in VR. However, technological advances may alleviate some of these limitations. Regarding richness of interaction, the SR core principles might be exportable into an augmented reality context, in which the inclusion of virtual objects within the live feed could enhance participants' sense of presence and immersion within the environment. More ambitiously, recording footage using multiple light-field cameras might allow participants to move freely within an SR scene, which would dramatically increase the plausibility and richness of the substituted reality by improving the degree of interactivity and sensorimotor coupling that SR provides.

5 Conclusions

In this chapter, we have described some of the opportunities for consciousness science provided by XR technologies (especially VR). First we introduced the theoretical concept of presence, the factors that affect the sense of presence in XR, a selection of relevant neuroimaging studies, and outlined the need to develop better measures of presence that adequately control for the effects of demand characteristics. We then discussed clinical cases in which a normal sense of presence is disturbed, such as DPDR. We went on to describe a series of experiments that leveraged the unique advantages of XR that go beyond what is possible in classical experimental settings to investigate how sensorimotor coupling affects breakthrough times within a CFS paradigm.

Finally, we expanded on the notion of a 'sense of reality' by highlighting the existence of cases in which perceptual veridicality and subjective veridicality are not sufficient to capture the complexity of human experience, highlighting the need to also consider people's metacognitive state and beliefs – their conviction or reality, or state of doxastic belief. Examples that illustrate this position include the typical dissociation between experiences of XR and DPDR, different kinds of dream states (lucid vs non-lucid) and hallucination types (with and without insight).

We concluded by presenting an experimental paradigm (substitutional reality, SR) aimed specifically at manipulating doxastic veridicality (conviction about reality). Leveraging relatively simple technologies, such as panoramic recordings and concurrent live feed, SR allows researchers to study the relationship between doxastic states and the detection of perceptual inconsistencies, while potentially enabling the translation of classical attention and memory paradigms into a more ecological context.

The experience of 'what is real' is a complex and multifaceted phenomenon. Understanding how its various components are related and constructed requires (1) a multidisciplinary approach that combines different experimental paradigms, (2) solid theoretical and computational foundations, and (3) state-of-the-art XR technologies. Advances in XR technologies provide a suite of unique tools that can be used to answer these questions, complementing and extending other psychological and neuroscientific approaches. Perhaps above all, researchers should avoid resorting to XR simply because it is available and in some sense 'immersive', but should carefully consider which aspect of perceptual presence and the experience of 'what is real' that they wish to investigate.

Acknowledgements The authors are grateful to the Dr. Mortimer and Theresa Sackler Foundation. AKS is also grateful to the Canadian Institute for Advanced Research (CIFAR) Program on Brain, Mind, and Consciousness, and to the European Research Council (Advanced Investigator Grant CONSCIOUS, 101019254) for support. AM is also grateful to Sussex Neuroscience and the Sussex Neuroscience 4-Year PhD Programme for generous support.

References

Ashton H (2002) Delirium and hallucinations. Adv Conscious Res 36:181–204

Aspell JE, Heydrich L, Marillier G, Lavanchy T, Herbelin B, Blanke O (2013) Turning body and self inside out: visualized heartbeats alter bodily self-consciousness and tactile perception. Psychol Sci 24(12):2445–2453

Axelsson A-S, Abelin Å, Heldal I, Schroeder R, Wideström J (2001) Cubes in the cube: a comparison of a puzzle-solving task in a virtual and a real environment. Cyberpsychol Behav 4(2):279–286

Bailenson JN, Yee N (2008) Virtual interpersonal touch: haptic interaction and copresence in collaborative virtual environments. Multimed Tools Appl 37(1):5–14

Bailey R, Wise K, Bolls P (2009) How avatar customizability affects children's arousal and subjective presence during junk food–sponsored online video games. Cyberpsychol Behav 12(3):277–283

Barber TX, Coules J (1959) Electrical skin conductance and galvanic skin response during "hypnosis.". Int J Clin Exp Hypn 7(2):79–92

Baumgartner T, Speck D, Wettstein D, Masnari O, Beeli G, Jäncke L (2008) Feeling present in arousing virtual reality worlds: prefrontal brain regions differentially orchestrate presence experience in adults and children. Front Hum Neurosci 2:8. https://doi.org/10.3389/neuro.09.008.2008

Bekkering H, Neggers SF (2002) Visual search is modulated by action intentions. Psychol Sci 13(4):370–374

Bentall RP (1990) The illusion of reality: a review and integration of psychological research on hallucinations. Psychol Bull 107(1):82

Biocca F (2003) Can we resolve the book, the physical reality, and the dream state problems? From the two-pole to a three-pole model of shifts in presence. EU Future and Emerging Technologies, Presence Initiative Meeting, 12, 13–69

Blanke O, Metzinger T (2009) Full-body illusions and minimal phenomenal selfhood. Trends Cogn Sci 13(1):7–13

Bleuler E (1911) Dementia praecox oder Gruppe der Schizophrenien, vol 4. Deuticke

Bleuler E (1950) Dementia praecox or the group of schizophrenias

Bohil CJ, Alicea B, Biocca FA (2011) Virtual reality in neuroscience research and therapy. Nat Rev Neurosci 12(12):752–762

Bortolotti L, Broome MR (2012) Affective dimensions of the phenomenon of double bookkeeping in delusions. Emot Rev 4(2):187–191. https://doi.org/10.1177/1754073911430115

Bortolotti L, Cox R, Barnier A (2012) Can we recreate delusions in the laboratory? Philos Psychol 25(1):109–131. https://doi.org/10.1080/09515089.2011.569909

Botvinick M, Cohen J (1998) Rubber hands 'feel' touch that eyes see. Nature 391(6669):756–756

Bowman DA, McMahan RP (2007) Virtual reality: how much immersion is enough? Computer 40(7):36–43

Chan D, Peterson MA, Barense MD, Pratt J (2013) How action influences object perception. Front Psychol 4:462

Clark A (2013) Whatever next? Predictive brains, situated agents, and the future of cognitive science. Behav Brain Sci 36(3):181–204. https://doi.org/10.1017/S0140525X12000477

Clemente M, Rey B, Rodríguez-Pujadas A, Barros-Loscertales A, Baños RM, Botella C, Alcañiz M, Ávila C (2014) An fMRI study to analyze neural correlates of presence during virtual reality experiences. Interact Comput 26(3):269–284. https://doi.org/10.1093/iwc/iwt037

Cooper N, Milella F, Cant I, Pinto C, White M, Meyer G (2015) The effects of multisensory cues on the sense of presence and task performance in a virtual reality environment

Dalgarno B, Lee MJ (2010) What are the learning affordances of 3-D virtual environments? Br J Educ Technol 41(1):10–32

de Borst AW, de Gelder B (2015) Is it the real deal? Perception of virtual characters versus humans: an affective cognitive neuroscience perspective. Front Psychol 6:576

Dienes, Z., Palfi, B., & Lush, P. (2020). Controlling phenomenology by being unaware of intentions
Dijkstra N, Kok P, Fleming SM (2022) Perceptual reality monitoring: neural mechanisms dissociating imagination from reality. Neurosci Biobehav Rev 135:104557. https://doi.org/10.1016/j.neubiorev.2022.104557
Dinh HQ, Walker N, Hodges LF, Song C, Kobayashi A (1999) Evaluating the importance of multisensory input on memory and the sense of presence in virtual environments. Proceedings IEEE virtual reality (cat. No. 99CB36316), pp 222–228
Dokic J, Martin J-R (2017) Felt reality and the opacity of perception. Topoi 36(2):299–309. https://doi.org/10.1007/s11245-015-9327-2
Dresler M, Wehrle R, Spoormaker VI, Koch SP, Holsboer F, Steiger A et al (2012) Neural correlates of dream lucidity obtained from contrasting lucid versus non-lucid REM sleep: a combined EEG/fMRI case study. Sleep 35(7):1017–1020
Duh HB-L, Lin JJ, Kenyon RV, Parker DE, Furness TA (2002) Effects of characteristics of image quality in an immersive environment. Presence Teleop Virt Environ 11(3):324–332
Ehrsson HH (2007) The experimental induction of out-of-body experiences. Science 317(5841): 1048–1048
Ehrsson HH, Fotopoulou A, Radziun D, Longo MR, Tsakiris M (2022) No specific relationship between hypnotic suggestibility and the rubber hand illusion. Nat Commun 13(1):564. https://doi.org/10.1038/s41467-022-28177-z
Ekstrom AD, Kahana MJ, Caplan JB, Fields TA, Isham EA, Newman EL, Fried I (2003) Cellular networks underlying human spatial navigation. Nature 425(6954):184–188
Fan K, Sugiura Y, Minamizawa K, Wakisaka S, Inami M, Fujii N (2014) Ubiquitous substitutional reality: re-experiencing the past in immersion 1
Felton WM, Jackson RE (2021) Presence: a review. Int J Hum Comput Interact 38(1):1–18. https://doi.org/10.1080/10447318.2021.1921368
Filevich E, Dresler M, Brick TR, Kühn S (2015) Metacognitive mechanisms underlying lucid dreaming. J Neurosci 35(3):1082–1088. https://doi.org/10.1523/JNEUROSCI.3342-14.2015. PMID: 25609624; PMCID: PMC6605529
Flach JM, Holden JG (1998) The reality of experience: Gibson's way. Presence 7(1):90–95
Foreman N (2010) Virtual reality in psychology. Themes Sci Technol Educ 2(1–2):225–252
Fortier M (2018) Sense of reality, metacognition, and culture in schizophrenic and drug-induced hallucinations, vol 1. Oxford University Press. https://doi.org/10.1093/oso/9780198789710.003.0016
Freeman J, Lessiter J (2001) Here, there and everywhere: the effects of multichannel audio on presence
Freeman J, Avons SE, Pearson DE, IJsselsteijn, W. A. (1999) Effects of sensory information and prior experience on direct subjective ratings of presence. Presence Teleop VEs 8(1):1–13. https://doi.org/10.1162/105474699566017
Freeman J, Avons SE, Meddis R, Pearson DE, IJsselsteijn WA (2000) Using behavioral realism to estimate presence: a study of the utility of postural responses to motion stimuli. Presence Teleop Virt Environ 9(2):149–164. https://doi.org/10.1162/105474600566691
Friston K (2010) The free-energy principle: a unified brain theory? Nat Rev Neurosci 11(2): 127–138. https://doi.org/10.1038/nrn2787
Froese T, Suzuki K, Ogai Y, Ikegami T (2012) Using human–computer interfaces to investigate 'Mind-As-It-Could-Be' from the first-person perspective. Cogn Comput 4(3):365–382. https://doi.org/10.1007/s12559-012-9153-4
Gallagher S (2000) Philosophical conceptions of the self: implications for cognitive science. Trends Cogn Sci 4(1):14–21
Gayet S, Van der Stigchel S, Paffen CL (2014) Breaking continuous flash suppression: competing for consciousness on the pre-semantic battlefield. Front Psychol 5:460
Gershman SJ (2019) The generative adversarial brain. Front Artif Intell 2. https://doi.org/10.3389/frai.2019.00018

Gibson JJ (1979) The theory of affordances. The ecological approach to visual perception. Houghton Mifflin, Boston
Gomez MA, Skiba RM, Snow JC (2018) Graspable objects grab attention more than images do. Psychol Sci 29(2):206–218
González-Maeso J, Weisstaub NV, Zhou M, Chan P, Ivic L, Ang R, Lira A, Bradley-Moore M, Ge Y, Zhou Q, Sealfon SC, Gingrich JA (2007) Hallucinogens recruit specific cortical 5-HT2A receptor-mediated Signaling pathways to affect behavior. Neuron 53(3):439–452. https://doi.org/10.1016/j.neuron.2007.01.008
Grabarczyk P, Pokropski M (2016) Perception of affordances and experience of presence in virtual reality. Avant 7(2):25–44
Guralnik O, Schmeidler J, Simeon D (2000) Feeling unreal: cognitive processes in depersonalization. Am J Psychiatr 157(1):103–109. https://doi.org/10.1176/ajp.157.1.103
Haggard P (2017) Sense of agency in the human brain. Nat Rev Neurosci 18(4):196–207
Haggard P, Clark S, Kalogeras J (2002) Voluntary action and conscious awareness. Nat Neurosci 5(4):382–385
Hartley T, Maguire EA, Spiers HJ, Burgess N (2003) The well-worn route and the path less traveled: distinct neural bases of route following and wayfinding in humans. Neuron 37(5): 877–888
Held RM, Durlach NI (1992) Telepresence. Presence Teleop Virt Environ 1(1):109–112
Hugdahl K, Løberg E-M, Specht K, Steen VM, van Wageningen H, Jørgensen HA (2008) Auditory hallucinations in Schizophrenia: the role of cognitive, brain structural and genetic disturbances in the left temporal lobe. Front Hum Neurosci 1. https://doi.org/10.3389/neuro.09.006.2007
IJsselsteijn WA, Bouwhuis DG, Freeman J, De Ridder H (2002) P-15: presence as an experiential metric for 3-D display evaluation. SID Symposium Digest of Technical Papers, 33(1), 252–255
Ito D, Wakisaka S, Izumihara A, Yamaguchi T, Hiyama A, Inami M (2019) EyeHacker: gaze-based automatic reality manipulation. ACM SIGGRAPH 2019 emerging technologies on - SIGGRAPH '19, 1–2. https://doi.org/10.1145/3305367.3327988
Jäncke L, Cheetham M, Baumgartner T (2009) Virtual reality and the role of the prefrontal cortex in adults and children. Front Neurosci 3(1):52–59. https://doi.org/10.3389/neuro.01.006.2009
Jaspers K (1973) Allgemeine psychopathologie. Springer
Jaspers K (1997) General psychopathology, vol 2. JHU Press
Jay E-L, Sierra M, Van den Eynde F, Rothwell JC, David AS (2014) Testing a neurobiological model of depersonalization disorder using repetitive transcranial magnetic stimulation. Brain Stimul 7(2):252–259. https://doi.org/10.1016/j.brs.2013.12.002
Jensen MS, Yao R, Street WN, Simons DJ (2011) Change blindness and inattentional blindness. Wiley Interdiscip Rev Cogn Sci 2(5):529–546. https://doi.org/10.1002/wcs.130
Jiang Y, Costello P, He S (2007) Processing of invisible stimuli: advantage of upright faces and recognizable words in overcoming interocular suppression. Psychol Sci 18(4):349–355
Jin S-AA (2009) Avatars mirroring the actual self versus projecting the ideal self: the effects of self-priming on interactivity and immersion in an exergame, Wii Fit. Cyberpsychol Behav 12(6): 761–765
Johnson MK, Raye CL (1981) Reality monitoring. Psychol Rev 88(1):67–85. https://doi.org/10.1037/0033-295X.88.1.67
Kafai YB, Fields DA, Cook MS (2010) Your second selves: player-designed avatars. Games Cult 5(1):23–42
Kekecs Z, Szekely A, Varga K (2016) Alterations in electrodermal activity and cardiac parasympathetic tone during hypnosis. Psychophysiology 53(2):268–277
Kiran C, Chaudhury S (2009) Understanding delusions. Ind Psychiatry J 18(1):3. https://doi.org/10.4103/0972-6748.57851
Kirsch I, Council JR (1989) Response expectancy as a determinant of hypnotic behavior
Klüver H (1942) Mechanisms of hallucinations

Korisky U, Mudrik L (2021) Dimensions of perception: 3D real-life objects are more readily detected than their 2D images. Psychol Sci 32(10):1636–1648. https://doi.org/10.1177/09567976211010718

Korisky U, Hirschhorn R, Mudrik L (2019) "Real-life" continuous flash suppression (CFS)-CFS with real-world objects using augmented reality goggles. Behav Res Methods 51(6):2827–2839

Lallart E, Lallart X, Jouvent R (2009) Agency, the sense of presence, and schizophrenia. Cyberpsychol Behav 12(2):139–145

Lau H (2019) Consciousness, metacognition, & perceptual reality monitoring. https://doi.org/10.31234/osf.io/ckbyf

Lee KM (2004) Presence, explicated. Commun Theory 14(1):27–50

Lemche E, Anilkumar A, Giampietro VP, Brammer MJ, Surguladze SA, Lawrence NS, Gasston D, Chitnis X, Williams SC, Sierra M (2008) Cerebral and autonomic responses to emotional facial expressions in depersonalisation disorder. Br J Psychiatry 193(3):222–228

Lenggenhager B, Tadi T, Metzinger T, Blanke O (2007) Video ergo sum: manipulating bodily self-consciousness. Science 317(5841):1096–1099

Lin J-W, Duh HB-L, Parker DE, Abi-Rached H, Furness TA (2002) Effects of field of view on presence, enjoyment, memory, and simulator sickness in a virtual environment. Proceedings IEEE Virtual Reality 2002:164–171

Lindemann O, Bekkering H (2009) Object manipulation and motion perception: evidence of an influence of action planning on visual processing. J Exp Psychol Hum Percept Perform 35(4):1062

Lombard M, Ditton T (1997) At the heart of it all: the concept of presence. J Comput-Mediat Commun 3(2):JCMC321

Lush P, Seth AK (2022) Reply to: no specific relationship between hypnotic suggestibility and the rubber hand illusion. Nat Commun 13(1):563. https://doi.org/10.1038/s41467-022-28178-y

Lush P, Vazire S, Holcombe A (2020) Demand characteristics confound the rubber hand illusion. Collabra: Psychology 6(1)

Lush P, Seth AK, Dienes Z (2021) Hypothesis awareness confounds asynchronous control conditions in indirect measures of the rubber hand illusion. R Soc Open Sci 8(11):210911. https://doi.org/10.1098/rsos.210911

Marini F, Breeding KA, Snow JC (2019) Distinct visuo-motor brain dynamics for real-world objects versus planar images. NeuroImage 195:232–242

Maruya K, Yang E, Blake R (2007) Voluntary action influences visual competition. Psychol Sci 18(12):1090–1098

Meehan M, Insko B, Whitton M, Brooks FP Jr (2002) Physiological measures of presence in stressful virtual environments. Acm Transactions on Graphics (Tog) 21(3):645–652

Metzinger T (2003a) Being no one: the self-model theory of subjectivity. MIT Press, pp xii, 699

Metzinger T (2003b) Phenomenal transparency and cognitive self-reference. Phenomenol Cogn Sci 2(4):353–393

Metzinger T (2014) How does the brain encode epistemic reliability? Perceptual presence, phenomenal transparency, and counterfactual richness. Cogn Neurosci 5(2):122–124

Müller F, Lenz C, Dolder P, Lang U, Schmidt A, Liechti M, Borgwardt S (2017) Increased thalamic resting-state connectivity as a core driver of LSD-induced hallucinations. Acta Psychiatr Scand 136(6):648–657. https://doi.org/10.1111/acps.12818

Neisser U, Becklen R (1975) Selective looking: attending to visually specified events. Cogn Psychol 7(4):480–494

Noë A (2004) Action in perception. MIT Press

O'regan JK, Noë A (2001) A sensorimotor account of vision and visual consciousness. Behav Brain Sci 24(5):939–973

Olson JA, Suissa-Rocheleau L, Lifshitz M, Raz A, Veissiere SP (2020) Tripping on nothing: placebo psychedelics and contextual factors. Psychopharmacology 237(5):1371–1382

Orne MT (1962) On the social psychology of the psychological experiment: with particular reference to demand characteristics and their implications. Am Psychol 17(11):776

Parsons TD (2015) Virtual reality for enhanced ecological validity and experimental control in the clinical, affective and social neurosciences. Front Hum Neurosci 9:660
Parsons TD, Gaggioli A, Riva G (2017) Virtual reality for research in social neuroscience. Brain Sci 7(4):42
Parsons TD, Gaggioli A, Riva G (2020) Extended reality for the clinical, affective, and social neurosciences. Brain Sci 10(12):E922. https://doi.org/10.3390/brainsci10120922
Pertaub D-P, Slater M, Barker C (2002) An experiment on public speaking anxiety in response to three different types of virtual audience. Presence 11(1):68–78
Phillips ML, Sierra M (2003) Depersonalization disorder: a functional neuroanatomical perspective. Stress 6(3):157–165. https://doi.org/10.1080/1025389031000138538
Phillips ML, Medford N, Senior C, Bullmore ET, Suckling J, Brammer MJ, Andrew C, Sierra M, Williams SC, David AS (2001) Depersonalization disorder: thinking without feeling. Psychiatry Res Neuroimaging 108(3):145–160
Radovic F, Radovic S (2002) Feelings of unreality: a conceptual and phenomenological analysis of the language of depersonalization. Philos Psychiatry Psychol 9(3):271–279. https://doi.org/10.1353/ppp.2003.0048
Rao RPN, Ballard DH (1999) Predictive coding in the visual cortex: a functional interpretation of some extra-classical receptive-field effects. Nat Neurosci 2(1):79. https://doi.org/10.1038/4580
Ratcliffe M (2017) Real hallucinations: psychiatric illness, intentionality, and the interpersonal world. MIT Press
Rensink RA (2000) Seeing, sensing, and scrutinizing. Vis Res 40(10):1469–1487. https://doi.org/10.1016/S0042-6989(00)00003-1
Rensink RA, O'regan JK, Clark JJ (1997) To see or not to see: the need for attention to perceive changes in scenes. Psychol Sci 8(5):368–373
Riva G, Davide F, IJsselsteijn WA (2003) Measuring presence: subjective, behavioral and physiological methods. In: Being there: concepts, effects and measurement of user presence in synthetic environments, 5, pp 110–118
Rolland B, Jardri R, Amad A, Thomas P, Cottencin O, Bordet R (2014) Pharmacology of hallucinations: several mechanisms for one single symptom? [research article]. Biomed Res Int. https://doi.org/10.1155/2014/307106
Roseboom W, Lush P (2022) Serious problems with interpreting rubber hand "illusion" experiments. Collabra: Psychology 8(1):32274. https://doi.org/10.1525/collabra.32274
Salomon R, Lim M, Herbelin B, Hesselmann G, Blanke O (2013) Posing for awareness: proprioception modulates access to visual consciousness in a continuous flash suppression task. J Vis 13(7):2–2
Salomon R, Kaliuzhna M, Herbelin B, Blanke O (2015) Balancing awareness: vestibular signals modulate visual consciousness in the absence of awareness. Conscious Cogn 36:289–297
Sanchez-Vives MV, Slater M (2005) From presence to consciousness through virtual reality. Nat Rev Neurosci 6(4):332. https://doi.org/10.1038/nrn1651
Schmid Y, Enzler F, Gasser P, Grouzmann E, Preller KH, Vollenweider FX, Brenneisen R, Müller F, Borgwardt S, Liechti ME (2015) Acute effects of lysergic acid diethylamide in healthy subjects. Biol Psychiatry 78(8):544–553. https://doi.org/10.1016/j.biopsych.2014.11.015
Schnider A (2003) Spontaneous confabulation and the adaptation of thought to ongoing reality. Nat Rev Neurosci 4(8):662. https://doi.org/10.1038/nrn1179
Schnider A (2008) The confabulating mind: how the brain creates reality. Oxford University Press
Schnider A, Nahum L, Ptak R (2017) What does extinction have to do with confabulation? Cortex 87:5–15
Schultze U (2010) Embodiment and presence in virtual worlds: a review. J Inf Technol 25(4):434–449
Seth A (2009) Explanatory correlates of consciousness: theoretical and computational challenges. Cogn Comput 1(1):50–63

Seth AK (2014) A predictive processing theory of sensorimotor contingencies: explaining the puzzle of perceptual presence and its absence in synesthesia. Cogn Neurosci 5(2):97–118. https://doi.org/10.1080/17588928.2013.877880

Seth AK (2015) Presence, objecthood, and the phenomenology of predictive perception. Cogn Neurosci 6(2–3):111–117. https://doi.org/10.1080/17588928.2015.1026888

Seth AK, Suzuki K, Critchley HD (2012) An interoceptive predictive coding model of conscious presence. Front Psychol 2. https://doi.org/10.3389/fpsyg.2011.00395

Shelton BE, Hedley NR (2004) Exploring a cognitive basis for learning spatial relationships with augmented reality. Technol Instr Cogn Learn 1(4):323

Shorvon HJ (1946) The depersonalization syndrome. SAGE Publications

Sierra M, Berrios GE (1998) Depersonalization: neurobiological perspectives. Biol Psychiatry 44(9):898–908. https://doi.org/10.1016/s0006-3223(98)00015-8

Sierra M, David AS (2011) Depersonalization: a selective impairment of self-awareness. Conscious Cogn 20(1):99–108

Sierra M, Senior C, Dalton J, McDonough M, Bond A, Phillips ML, O'Dwyer AM, David AS (2002) Autonomic response in depersonalization disorder. Arch Gen Psychiatry 59(9):833–838

Sierra M, Baker D, Medford N, David AS (2005) Unpacking the depersonalization syndrome: an exploratory factor analysis on the Cambridge Depersonalization Scale. Psychol Med 35(10): 1523–1532

Simons DJ, Chabris CF (1999) Gorillas in our midst: sustained inattentional blindness for dynamic events. Perception 28(9):1059–1074

Simons DJ, Levin DT (1997) Change blindness. Trends Cogn Sci 1(7):261–267

Simons JS, Garrison JR, Johnson MK (2017) Brain mechanisms of reality monitoring. Trends Cogn Sci 21(6):462–473. https://doi.org/10.1016/j.tics.2017.03.012

Sjölie D, Kalpouzos G, Eriksson J (2014) Neural correlates of disrupted presence: strange disruptions in a naturalistic virtual environment. In: Proceedings of the 24th international conference on artificial reality and telexistence and the 19th Eurographics symposium on virtual environments, 21–28

Slater M (2004) How colorful was your day? Why questionnaires cannot assess presence in virtual environments. Presence 13(4):484–493

Slater M, Usoh M (1993) Representations systems, perceptual position, and presence in immersive virtual environments. Presence Teleop Virt Environ 2(3):221–233

Slater M, Usoh M, Steed A (1995) Taking steps: the influence of a walking technique on presence in virtual reality. ACM Trans Comput Hum Interact (TOCHI) 2(3):201–219

Slater M, Linakis V, Usoh M, Kooper R (1996) Immersion, presence and performance in virtual environments: an experiment with tri-dimensional chess. Proceedings of the ACM symposium on virtual reality software and technology, pp 163–172

Slater M, Lotto B, Arnold MM, Sanchez-Vives MV (2009a) How we experience immersive virtual environments: the concept of presence and its measurement. Anuario de Psicologia 40(2): 193–210

Slater M, Pérez Marcos D, Ehrsson H, Sanchez-Vives MV (2009b) Inducing illusory ownership of a virtual body. Front Neurosci 3:29

Slater M, Spanlang B, Sanchez-Vives MV, Blanke O (2010) First person experience of body transfer in virtual reality. PLoS One 5(5):e10564

Snow JC, Culham JC (2021) The treachery of images: how realism influences brain and behavior. Trends Cogn Sci. https://doi.org/10.1016/j.tics.2021.02.008

Snow JC, Skiba RM, Coleman TL, Berryhill ME (2014) Real-world objects are more memorable than photographs of objects. Front Hum Neurosci 8:837

Stein T, Sterzer P (2014) Unconscious processing under interocular suppression: getting the right measure. Front Psychol 5:387

Stein T, Hebart MN, Sterzer P (2011) Breaking continuous flash suppression: a new measure of unconscious processing during interocular suppression? Front Hum Neurosci 5:167

Steuer J (1995) Defining virtual reality: dimensions determining presence. Commun Age Virtual Reality:33–56
Suzuki K, Wakisaka S, Fujii N (2012) Substitutional reality system: a novel experimental platform for experiencing alternative reality. Sci Rep 2(1). https://doi.org/10.1038/srep00459
Suzuki K, Garfinkel SN, Critchley HD, Seth AK (2013) Multisensory integration across exteroceptive and interoceptive domains modulates self-experience in the rubber-hand illusion. Neuropsychologia 51(13):2909–2917
Suzuki K, Lush P, Seth AK, Roseboom W (2019a) Intentional binding without intentional action. Psychol Sci 30(6):842–853. https://doi.org/10.1177/0956797619842191
Suzuki K, Schwartzman DJ, Augusto R, Seth AK (2019b) Sensorimotor contingency modulates breakthrough of virtual 3D objects during a breaking continuous flash suppression paradigm. Cognition 187:95–107
Terkildsen T, Makransky G (2019) Measuring presence in video games: an investigation of the potential use of physiological measures as indicators of presence. Int J Hum Comput Stud 126: 64–80. https://doi.org/10.1016/j.ijhcs.2019.02.006
Tracy DK, Shergill SS (2013) Mechanisms underlying auditory hallucinations—understanding perception without stimulus. Brain Sci 3(2):642–669. https://doi.org/10.3390/brainsci3020642
Tsuchiya N, Koch C (2005) Continuous flash suppression reduces negative afterimages. Nat Neurosci 8(8):1096–1101
Van Baren J (2004) Measuring presence: a guide to current measurement approaches. Deliverable of the OmniPres Project IST-2001-39237
Voermans NC, Petersson KM, Daudey L, Weber B, Van Spaendonck KP, Kremer HP, Fernández G (2004) Interaction between the human hippocampus and the caudate nucleus during route recognition. Neuron 43(3):427–435
Ward EJ (2018) Downgraded phenomenology: how conscious overflow lost its richness. Philos Trans R Soc Lond B Biol Sci 373(1755):20170355. https://doi.org/10.1098/rstb.2017.0355
Waters F, Collerton D, Ffytche DH, Jardri R, Pins D, Dudley R, Blom JD, Mosimann UP, Eperjesi F, Ford S, Larøi F (2014) Visual hallucinations in the psychosis spectrum and comparative information from neurodegenerative disorders and eye disease. Schizophr Bull 40(Suppl 4):S233–S245. https://doi.org/10.1093/schbul/sbu036
Weech S, Kenny S, Barnett-Cowan M (2019) Presence and cybersickness in virtual reality are negatively related: a review. Front Psychol 10:158
Weller PD, Rabovsky M, Abdel Rahman R (2019) Semantic knowledge enhances conscious awareness of visual objects. J Cogn Neurosci 31(8):1216–1226
Wiederhold BK, Davis R, Wiederhold MD (1998) The effects of immersiveness on physiology. Stud Health Technol Inform 58:52–60
Wilson CJ, Soranzo A (2015) The use of virtual reality in psychology: a case study in visual perception. Comput Math Methods Med 2015:151702. https://doi.org/10.1155/2015/151702. (Review)
Witmer BG, Singer MJ (1998) Measuring presence in virtual environments: a presence questionnaire. Presence 7(3):225–240
Zahorik P, Jenison RL (1998) Presence as being-in-the-world. Presence 7(1):78–89

Part III
Applications of VR

Virtual Reality for Learning

David Checa and Andres Bustillo

Contents

Abstract The consumer age of the Personal Computer and mobile devices has opened up a new world of opportunities for innovative teaching methodologies, many based on serious games and virtual worlds. Similar levels of market penetration are expected for the use of Immersive Virtual Reality (iVR) over upcoming decades, once all the core technologies for game engines and head-mounted displays are available on the market at affordable prices. In this chapter, a general overview of the state of the art of iVR learning experiences is presented. Firstly, the advantages of iVR over traditional learning are described – advantages that must be considered when defining iVR experiences for the optimization of student learning and satisfaction. Secondly, the relationship between learning theories and iVR experiences is briefly summarized; an area where constructivist theories appear to be the most commonly used theory in iVR experiences. Thirdly, some examples of the success of iVR applications at different learning levels, from primary school to higher education, are summarized. Fourthly, the key factors for the successful design and use of an iVR experience in education are identified, from the predesign stage to the final evaluation – with special attention given to the different possibilities of each type of

D. Checa and A. Bustillo (✉)
University of Burgos, Burgos, Spain
e-mail: dcheca@ubu.es; abustillo@ubu.es

Curr Topics Behav Neurosci (2023) 65: 289–308
https://doi.org/10.1007/7854_2022_404
Published Online: 3 January 2023

HMD for different kinds of educational experiences. Finally, the main limitations of iVR for learning today and the future trends of this technology for teaching are also identified and discussed.

Keywords Active learning · Educational games · Educational technology · Games for learning · Head-mounted displays · Immersive virtual reality

1 Introduction

Learning related to abstract thinking is one of the most complex human activities. Acquiring new knowledge depends on many factors and is not always achieved in the same way. There is consensus over two of the most effective strategies to improve student learning rates: (1) make the new concepts easy and exciting to grasp; and (2) integrate the concepts into previously acquired student knowledge. These strategies imply a natural and engaging learning process. However, this motivational point of view is in direct contradiction with reality: learning material is not always easy to understand or easy to connect with daily life. Most of the time, it is difficult to reformulate what should be taught in an engaging way. Moreover, core competencies – such as mathematical and computational reasoning, basic science knowledge, computer literacy, critical thinking, writing, and problem-solving abilities must all be developed before learners are able to make meaningful contributions in the discipline of their choice.

In parallel with the challenges to K-12 learning, continuous refreshment of knowledge and skills in many different professional contests is also required. Courses designed to meet this need may be difficult to reconcile with traditional learning methodologies, as they should be autonomous and flexible, in terms of schedule and content. However, both in formal school settings and in life-long learning contexts, the challenges that educators face can be mitigated by merging Virtual Reality Environments (VREs) with traditional teaching methodologies (Buttussi and Chittaro 2017). Indeed, recent pedagogical strategies have sought to attract the attention of students through active visual presentations of new concepts, so that boredom will not lead to demotivation, placing the learner at the center of the learning process, and allowing both hands-on learning strategies and multimedia tools to become central in the classroom.

In line with these new pedagogical strategies, the hands-on approaches of immersive Virtual Reality (iVR) support active learning in many different disciplines. Although the state of the art of iVR technology still faces certain limitations, it is expected that in the near future, it will also provide photorealistic 3D environments that will hardly be distinguishable from reality (Checa and Bustillo 2020a). Although it might be thought that these advantages could also be associated with educational games played on a 2D screen, iVR environments add fundamental new advantages to the learning process. First, they provide the learner with a strong sense of presence. Presence is, within the context of Virtual Reality, defined as the illusion of "being there": the perceptual illusion of being in the virtual world (rendered by the iVR system), to the extent that the virtuality of the environment goes unnoticed and becomes felt as the dominant reality (Barfield et al. 1995). Presence in iVR environments facilitates learning by engaging students with the teaching material (Makransky and Lilleholt 2018). Interactivity and immersion, two characteristics that underlie immersive virtual reality, have repeatedly been associated with presence (Makransky and Petersen 2019). The terms "presence" and "immersion" are often used interchangeably, though they should not be confused. Immersion refers to the objective sensory fidelity of an iVR system (Slater 2003), although presence, as it is defined above, is the subjective psychological response of a user to an iVR system experience.

Another key advantage of an iVR environment that forms part of a learning process is the way in which knowledge and concepts are implemented through bodily activities: a concept that is defined as embodied learning. It is based on the idea that cognitive processes are closely related with bodily interaction in the world (Wilson 2002). These theories contend that human cognition – including abstract thinking – is firmly grounded in multisensory processes and bodily experiences of the world (Walsh and Simpson 2013) (Corcoran 2018). iVR has the potential to foster embodied forms of teaching and learning, due to its highly immersive and modern interfaces that can use the human body as an input (Jensen and Konradsen 2018). However, insufficient attention has been lent to research into the role of the senses and the body in VR educational experiences within the classroom (Kavanagh et al. 2017).

2 State of the Art of VR-Learning

In this chapter, a general overview of the state of the art of iVR in learning contexts will be provided. Firstly, the advantages of iVR vs. traditional learning, which must be considered when defining iVR experiences, will be described. Secondly, the relationship between learning theories and iVR experiences will be briefly reviewed. Thirdly, some examples of the success of iVR applications at different learning levels will be summarized. Fourthly, the key facts needed for the successful design and use of a Virtual-Reality Learning Environment (VRLE) in education and the different possibilities of each type of HMD in education will be presented. Finally, the main limitations of iVR in current learning contexts will also be identified.

2.1 *Advantages of Virtual Reality vs. Traditional Learning*

Virtual reality represents the next step in the advancement of new technologies for the classroom (as was previously the case with videos, computers, and so forth). Yet iVR is costly and not necessarily effective. So, what does it bring to the table compared to traditional education? These key advantages of immersive virtual reality for learning improvement have been identified (Checa and Bustillo 2020a):

- *Motivation and engagement*: Maintaining student interest and motivation is a challenge for any teacher. A motivated learner will be more engaged and more persistent when trying to understand the learning material, as well as more resilient to potential obstacles when understanding it (Parong and Mayer 2018). Most investigations measuring motivation and engagement have concluded that the use of immersive virtual reality leads to increased interest and engagement when compared to conventional learning environments or other media such as videos (Checa and Bustillo 2020a; Radianti et al. 2020; Makransky et al. 2019a).
- *Interaction*: Immersive virtual reality, even in its less immersive forms, offers more hands-on, body-based engagement than conventional educational methodologies. Traditional education has always been language-based, conceptual and abstract – which distances it from practical learning. On the other hand, iVR supports 'doing' rather than only observing, which leads to a constructivist approach toward learning where students can interact with virtual objects and even collaborate with other students. In doing so, students can experiment, investigate, and obtain instant feedback in a personalized experience that can improve learning (Roussou and Slater 2017). VR experiences bring space and action into the learning processes, which requires a commitment to embodied pedagogy. It is important to highlight that many of these applications are single-user, since the aspect of social interaction, through the inclusion of virtual human partners to simulate interpersonal encounters, is still in its early stages of development (Bombari et al. 2015).
- *The impossible becomes possible*: iVR applications can be used to transform abstract concepts into concrete perceptions and experiences. Processes that occur at both microscopic and macroscopic levels and that are not easily observable in real life can be examined in detail. iVR can also be used for complex technical, dangerous, and even expensive experiences (Kwon 2019). Finally, iVR can create a window to another place and time, allowing students to discover far-away or inaccessible places and sites that no longer exist today (Checa and Bustillo 2020b).
- *Soft-skills training*: Besides knowledge acquisition, soft-skills training is central to the needs of our society. For instance, iVR can be used for empathy training, enabling students to empathize with others and to broaden their range of perspectives and experiences beyond their normal spheres of interaction. iVR has been used to build empathy toward homelessness (Herrera et al. 2018) and to counter racism (Roswell et al. 2020). Likewise, virtual reality can be used to produce virtual body ownership (a virtual body coincident in space with the real body and seen as the first person, a perspective that can generate the illusion that it is the person's body (Slater 2017)) that can lead to implicit changes. For example, if a person embodies a body of different race, their implicit bias against people of that race may decrease (Banakou et al. 2016).

2.2 Learning Theories and iVR

A prerequisite for an effective iVR educational application is its pedagogical approach and the learning theory motivating its design (Mikropoulos and Natsis 2011), though unfortunately, it is common to find iVR applications that do not appear grounded in any theory of learning, user experience, or design logic (Fowler 2015). Learning theories can be categorized according to how learners assimilate, process, and retain the information they have learned (Pritchard 2017). For instance, according to constructivist views, learning does not rely on passive acceptance of knowledge, but rather, requires learner participation by giving students responsibility for the learning process in order to improve understanding. The characteristics of iVR and the axioms of constructivist learning are entirely compatible according to Winn (1993). Indeed, the constructivist learning paradigm is central to the use of iVR as a pedagogical tool, as iVR can provide students with exploratory learning environments that enable learning through experimentation (Fosnot and Perry 1996) and can support a learning process that offers flexibility for repetition, is self-directed, and discovery driven. To this end, iVR-based educational applications should provide learning in and from context and must present an authentic problem, similar to the problems found in the real world, in an engaging and interesting way (Jen Chen 2009).

Other theoretically significant learning paradigms include behaviorism, cognitivism, and connectivism (Table 1). The main assumption of behaviorism is that appropriate instructional stimuli will lead to the desired learning outcomes,

Table 1 Relations between the learning theories and iVR learning applications

Learning frameworks	Student role	Type of learning	Relation to learning in iVR
Constructivism	Active	Social hands-on contextual	Students actively participate in their own learning process Student-centered, self-directed, experimental, and discovery-driven learning process Importance of interaction and experience
Behaviorism	Passive – reactive	Task-based	Stimulation of learner's attention and engagement Learners learn by doing – trial and error – iVR presents a safe and engaging space for practical training
Cognitivism	Reactive	Reasoning problem-solving	Internalizing knowledge construction Organizing new knowledge as 'related' to existing knowledge
Connectivism	Proactive	Connection Collective	Enabling internal and external knowledge networks in order to incorporate and to interpret new knowledge or constructing new meaning to add to existing knowledge iVR as a socially constructed process where learners interact in pursuit of a shared goal

emphasizing practice and performance (Skinner 1989). The application of this viewpoint plays out in virtual reality through the capacity to learn through trial and error in a safe space, in addition to its capability to stimulate participation, and motivation for learning in an extrinsic way.

Cognitivism focuses on understanding mental processes. Learning is thought to flourish in environments that foster discovery and assimilation of knowledge (Shuell 1986). Cognitive strategies seek to address the learner's prior knowledge, motivation and reflection on the act of learning. iVR-based learning can reinforce cognitivist learning design by constructing a learning experience based on reasoning and problem-solving that replicates mental models and organizes new knowledge as "related" to existing knowledge (Dede 2009).

Finally, connectivism suggests that people do not stop their learning after completing formal education. By making use of new technological tools, they continue learning and acquiring knowledge outside the traditional channels of education (Michelle 2011). Connectivism also approaches learning as a socially constructed process in which learners interact to work toward a shared goal. This proactive and collaborative perspective is enhanced by the qualities of the iVR that can facilitate collective learning environments for interacting in the pursuit of acquiring new knowledge or constructing new meaning to existing knowledge.

Of course, there are many varieties of learning theories, in addition to the main paradigms listed here, but these main theories provide a pedagogical framework and an effective procedural basis for the development and use of iVR for learning. iVR learning experiences have the potential to achieve learning objectives across cognitive processes and knowledge dimensions. Most iVR learning experiences found in the literature combine two or more instructional strategies, depending on the designer's pedagogical goals and the extent or complexity of the design experience. Thus, it could be argued that iVR learning is based on a fusion of principles from multiple pedagogical perspectives. Regardless of the learning theories and the underlying paradigm that the iVR researcher may choose, it is crucial to ground the development of iVR applications for higher education in existing learning theories, because they offer guidelines on motivations, learning processes, and learning outcomes for the learners.

2.3 *Examples of Success of iVR Applications in Learning*

The studies in which the effectiveness of iVR has been compared with conventional educational methods have consistently yielded results that favor iVR with regard to motivational outcomes, including enjoyment and motivation, but their results for learning outcomes have been varied. (di Natale et al. 2020; Checa and Bustillo 2020a; Radianti et al. 2020). Few examples are found that target elementary school students – the most likely reasons being related to health and safety. Individuals aged 13 or who overuse HMDs may be mentioned (Freina and Ott 2015). Certain companies advise children under 12 (Sony PlayStation VR) or 13 years old

(Samsung Gear VR, Google Daydream, Oculus) not to use their HMDs, while others such as HTC advise young children not to use their products without specifying an age limit. Notably, studies in the literature support the use of this technology among children without noteworthy effects on visuomotor functions (Tychsen and Foeller 2020), such as postural instability or maladaptation of the vestibulo-ocular reflex.

On the other hand, most of these studies endorse virtual reality as a powerful tool for improving learning in elementary school. For instance, Liu et al. (2020) examined the effects of iVR-based science lessons on several learning-related dimensions of primary school students, including cognitive engagement, behavioral engagement, emotional engagement, and social engagement. Their results revealed that the experimental group obtained significantly higher academic achievement and engagement scores than the control group. Other works support these conclusions for teaching mathematics (Stranger-Johannessen 2018), climate change (Petersen et al. 2020), water-safety skills (Araiza-Alba et al. 2021a), music (Innocenti et al. 2019), and problem-solving skills (Araiza-Alba et al. 2021b).

There are also some studies that support the value of iVR as an educational tool in middle school. Isabwe et al. (2018) pointed to the usefulness of iVR solutions for experiential learning of abstract concepts such as chemistry with students aged between 14 and 16. In another example, Albus et al. (Albus et al. 2021) used textual annotations that labeled important components represented in a picture. This study reported that annotations helped learners to improve learning outcomes for recall, but not for questions of comprehension or transfer. It also highlighted the need to take into account prior knowledge and intrinsic motivation during learning. However, a lack of intrinsic motivation can, according to the same study, be overcome through additional support elements, such as annotations.

In a related area, Checa et al. (Checa and Bustillo 2020b) evaluated the possibilities and limitations of iVR environments for teaching topics related to cultural heritage. In this research, teenagers learned about medieval through topics ranging from historical knowledge to urban layout. In this experience, they could freely walk through a virtual urban reconstruction of a fifteenth century city in Spain. The design of the teaching experience was for students to learn historical facts, to increase visually acquired knowledge of urban structures, and to recall the spatial positioning of the main buildings and services of the city. The design of this iVR learning experience was based on constructivist learning theory. Its validation was performed with a total of 100 students, comparing the iVR experience (Fig. 1) against a conventional teaching procedure such as viewing the same content on video followed by a brief class discussion supported through traditional multimedia devices. Learning was assessed through questions of historical knowledge and urban layout. The results showed significantly higher satisfaction levels among students in the group that performed the iVR experience. Likewise, students in the VR group performed better in the retention of visually acquired knowledge and were able to better define the location of the main buildings and places in the city. On the other hand, historical facts and urban layout seemed to be better conveyed through conventional teaching procedures.

Fig. 1 Middle school students playing a seated 6DoF iVR experience

At the high-school level, a study of iVR applications for learning English reported that students participating in the iVR group learned more words than students participating in the control group (Repetto et al. 2021). Additionally, this study also examined the feasibility and effectiveness of using 360° immersive videos as homework, adopting non-immersive videos as a control standard that showed the same visual content and that had been enriched with an auditory description of the environment. Students with more positive attitudes toward technology were found to have watched the videos more than those with less positive attitudes, indicating that teachers interested in including VR technologies in their learning activities should pay attention to the acceptance of that technology among students.

Another study centered on history (Calvert and Abadia 2020) drew a parallel link between enhanced learning outcomes and enhanced perceptual experience. The effects of using a virtual reality learning environment with an immersive linear narrative on affective and cognitive factors among high school and college students were investigated and A virtual reality experience was developed to explain one of the most important military campaigns between Australian and Japanese soldiers in World War II. Key moments of the campaign were described and linked in a series of 12 scenarios, giving a first-hand view of the equipment in use, the soldiers involved, the victories, and the defeats. In this study, the impact of this experience on affective and cognitive factors among high school and college students was compared in a 6DOF vs. 3DOF virtual reality learning environment. Analysis of variables such as participant engagement, presence, empathy, and knowledge acquisition showed that students who participated in the iVR learning environment (6DOF) scored significantly higher at identifying and recalling information than those who used the 360° video (3DOF).

Post-secondary education is the most common target group in VRLE research. University students constitute a convenient target group, because they are easier to recruit and parental consent is not required (Alhalabi 2016). Examples of studies revealing positive learning outcomes can be found across disciplines such as chemistry (Miller et al. 2021), history (Checa and Bustillo 2020b), science (Pande et al. 2021; Allcoat and von Mühlenen 2018), and engineering (Alhalabi 2016).

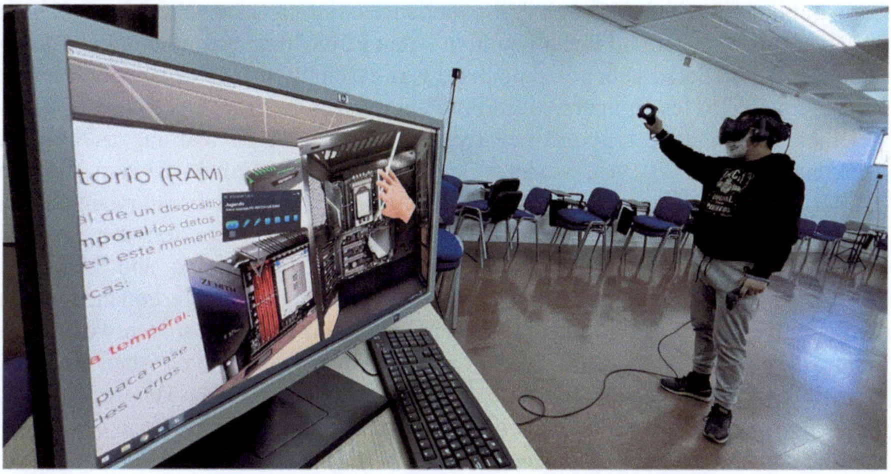

Fig. 2 Undergraduates with high-quality experiences tethered to a workstation

Additionally, the learning of concepts associated with computer hardware assembly was examined (Checa et al. 2021) with the aim of validating a custom-designed iVR model – the operational learning model, which was informed by the idea that the contexts, activities, and social interactions in the iVR learning environment promote the construction of new knowledge (Zhou et al. 2018). It was investigated whether an immersive iVR experience under development was more effective than other learning methodologies adapted to online learning, such as teaching conventional online classes or using the same experience, but on a desktop PC (Fig. 2). The iVR game was designed to improve the acquisition of basic computer concepts. It presented a hands-on learning environment that supported practice-oriented content learning rather than memorization of facts. The results revealed a significantly higher level of satisfaction among students in the iVR experience group vs. the groups using non-immersive approaches.

This result is a common finding in iVR studies when compared to conventional teaching methodologies. However, it is especially interesting against the backdrop of the COVID-19 crisis and student isolation, when student health and well-being should receive special support. In the same study (Checa et al. 2021), students also found it significantly easier to interact in iVR than in other less immersive mediums, such as a desktop PC. This effect has previously been reported in some general studies and points to the major role that usability plays in student satisfaction (Chen et al. 2013). In this study, iVR was found to promote "Remembering," defined as the ability to recognize (identify) and to recall (recall) – which was related in these experiments to visual recognition. Remembering was clearly better acquired in the iVR environments compared to traditional approaches.

Finally, those studies that found mixed results or that found no advantages or positive learning outcomes within a VRLE must in all fairness be highlighted. Some of the research presented above revealed that not all types of knowledge will be learned more effectively through iVR experiences. For instance, Checa et al. (Checa and Bustillo 2020b) found lower retention – that is, recalling essential information – of concepts associated with computer hardware assembly using operational learning models relative to less immersive learning approaches.

Other studies have reported negative effects of iVR on learning as well. In (Makransky et al. 2019b), students learned from a simulation via a desktop (PC) or 3DOF head-mounted display (VR) in a virtual biology laboratory simulation of mammalian transient protein expression. The authors reported that learners experienced higher presence in the VR scenario, but learned less and felt that it delivered a significantly higher cognitive load. This outcome can be couched within the cognitive theory of multimedia learning, which predicts that students will learn more with a well-designed slide than from an iVR application, even when reporting lower levels of interest and motivation (Parong and Mayer 2018). The theory states that adding iVR to a lesson might create extraneous processing demands, thereby limiting the student's ability to engage with the material in an optimal and effective way. Extraneous processing might accentuate the challenge of selecting, organizing, and integrating relevant information for the student (Mayer 2005). The results of this study and others like it (Makransky et al. 2021) suggest that it is not appropriate to take a solely technology-focused approach and expect that adapting learning material to immersive VR will automatically lead to improved learning outcomes. According to this research, if the goal is to promote learning, converting science lab simulations from a computer-based medium into an immersive VR medium may not be the best approach.

A similar pattern has been observed in many other experiments in which virtual reality garnered strong user preferences over more traditional methods of content delivery, but yielded minimal impact on mean performance levels in pre- and post-tests of knowledge relating to the control condition. For instance, Madden et al. (2018) used iVR to teach moon-phase activity and compared participant learning outcomes when using either the iVR simulation or traditional methods. Their results showed that iVR is as good at teaching moon phases as traditional methods, despite the strong preference of participants for iVR rather than traditional methods. However, the authors emphasized that the inexperience of the participants with iVR controllers and the technical limitations (frame rates, low fields of view and simulator sickness) distracted users from the concepts being taught as much as hands-on activities and were the main limitations on providing a truly immersive experience. Additionally, Moro et al. (2017) assessed whether learning structural anatomy with either VR or AR was as effective as tablet-based applications, and whether those modes enhanced student learning, engagement, and performance. No significant differences were found between mean assessment scores in any of the studies. This example only underscores the need for further research on the role that design elements and learning contents play in the effectiveness of iVR experiences for learning.

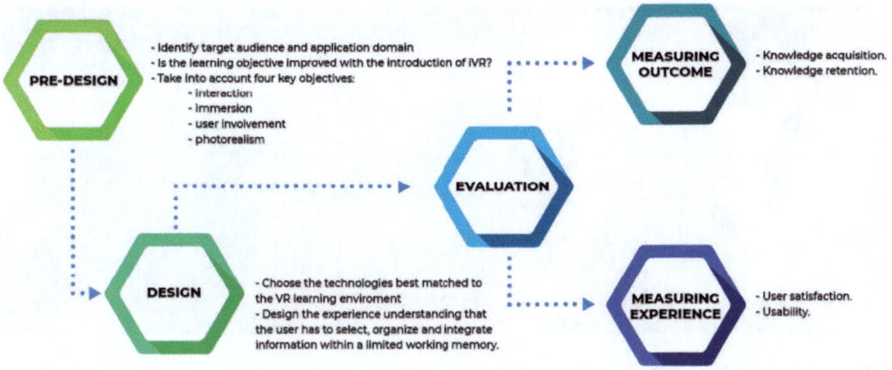

Fig. 3 Flow chart for the design and implementation of a VRIE adapted from (Checa and Bustillo 2020a)

2.4 *Developing iVR for Learning*

Three stages are followed for the development of an educational application in iVR: pre-design, design, and evaluation (Checa and Bustillo 2020a). Figure 3 summarizes the flow chart for the design and implementation of a VRLE. In the initial pre-design stage, the target audience and the application domain are established. The learning objective must be one that can plausibly be enhanced by the introduction of iVR technologies, based on well-established learning theories, so that the user will not become lost in amusement within the VRLE. The four key objectives for a VRLE must also be taken into account: interaction, immersion, user involvement and, to a lesser extent, photorealism (Roussos et al. 1999), balanced by the target audience and the scope of the application.

In the second stage, the design stage, some questions have to be answered, such as: which are the best technologies to be used to develop the VR environment? Which is the best game design for a certain application? The learner has to select, to organize, and to integrate information within the limitations of attention spans and working memory, so a VRLE should be designed to support these limitations. For example, interactivity should be clearly designed with an effective ramp for beginners in the technology and with a game structure that offers genuine play, rather than quiz-style questions and answers. Bodily movements are also integrated in the learning experience. Finally, to keep up student motivation and engagement, the iVR application should include game-based learning elements that support the motivational needs of competence, autonomy, and relatedness.

Finally, the third stage consists of the evaluation of the VRLE when improvements are expected in the short term. Reviews outline that, up to now, most of the evaluations of iVR applications are run with small samples, without replication, using mostly qualitative methods for data analysis, with neither a well-defined evaluation method, nor comparisons with other learning strategies (Checa and

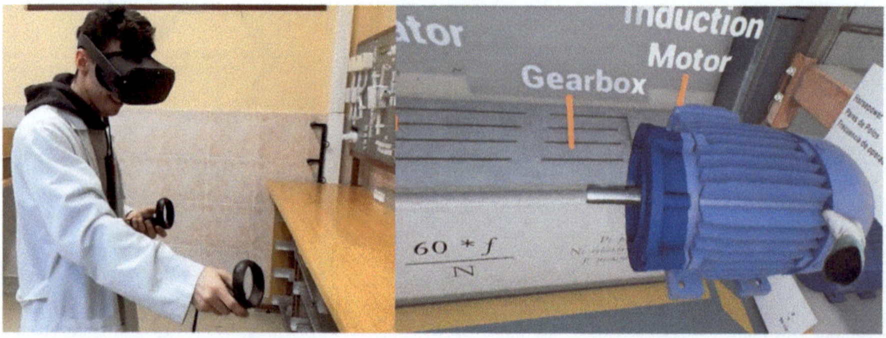

Fig. 4 High school student with stand-alone 6DoF system interacting with an iVR simulation centered on condition-based maintenance of induction motors (the right panel depicts what the student sees)

Bustillo 2020a; Egenfeldt-Nielsen 2006; Mikropoulos and Natsis 2011). Four different evaluation procedures are commonly used: questionnaires, user interviews, data recording, and direct user observation. However, most studies use only one of them, while a combination of two of them, especially questionnaires and indicators extracted from data recording, could increase the validity of the research results, especially if standardized questionnaires were created (Petri et al. 2017).

In Fig. 4, high-school students are shown using standalone 6DoF HMDs to engage in a naturalistic, interactive iVR experience – which, in this case, supports learning basic knowledge and practical skills related to condition-based maintenance of induction motors (Checa et al. 2022). The testing of games developed with end-users before a research study is important, since the usability of iVR experiences is directly connected with user satisfaction with the experience and performance.

2.5 Limitations to the Application of Virtual Reality in Learning

Despite the growing popularity of iVR and its many advantages for education, as with any other technology, it is not without its drawbacks. First, HMD prices are still relatively expensive, considering that sufficient devices must be purchased to satisfy the needs of a group. Budget limitations have other consequences for the development of a VRLE. VR experiences tend to be short in duration, because their development is costly in terms of both time and money. Short exposure times can limit the learning rate (Ritterfeld et al. 2004). While short viewing times were expected in the past – due in part to the immaturity of HMD technology and the incipient awareness of VR simulation sickness (Bustillo et al. 2015) many such problems appear to have been resolved with the new generation of HMDs.

Participants have been found to exhibit fewer signs of discomfort in high-fidelity virtual environments if the VR provided multiple sensory information (Chang et al. 2020).

Secondly, the use of the technology itself precipitates extraneous processing during learning, potentially leading to negative impacts on learning outcomes (Allcoat and von Mühlenen 2018) – an effect known as the novelty effect. VR is a relatively new technology in classrooms and may for many learners be their first opportunity to wear an HMD. The unfamiliar experiences associated with wearing the headset, learning how to use the controllers in a specific way, and the novelty of the VR interface could all increase extraneous cognitive load. New users also tend to feel lost in virtual environments. Slow and progressive familiarization, visual clues, and guidance incorporated in the software should all be used to help the user and to overcome these limitations.

Thirdly, from an application development perspective, iVR presents a number of challenges. For example, in VR applications, users have a lot of space to control outcomes and a wide range of possible interactions with the environment. It means that interaction accuracy can be lower in general, especially for more novice users (e.g. grab or drop actions will not have the same accuracy than a real-world training activity in motor assembly). A balance between immersion, freedom, and comfort must be sought in the design of an iVR experience. Audio is another key aspect and must be spatialized to keep players immersed. There are also other open issues, such as font size for readability of text when it is included in an experience (Baceviciute et al. 2020).

3 Future Trends in VR-Learning

Although iVR is a very promising technology to be applied in educational contexts, it still needs further development in some areas. Firstly, although many companies manufacture HMDs, there is no standard for application development that ensures the interoperability of either hardware or software. The development of educational VR applications, therefore, becomes a challenge when more than one vendor is supported, and the technology is still at an experimental stage. Nevertheless, the future is promising, and some companies, such as Oculus/Meta, have already adopted standards like OpenXR (The Khronos Group 2021), which facilitate the process of development and the implementation of VR learning applications in school curricula.

Secondly, more complex and realistic iVR environments are desirable, allowing for an increased sense of presence and making the experience more believable and engaging. Similarly, as stated before, longer VR experiences have to be developed. Although learning times are a key factor for learning rates (Ritterfeld et al. 2004), the duration of learning experiences tends to be short, due to the limitations on time and money within the multidisciplinary research teams with specific skill sets that are needed to develop most educational applications. Future interventions will have to use appropriate instructional activities and material designs, to minimize the cognitive load and to consider a longer time of exposure or repeated VR sessions to promote relevant learning.

Thirdly, another important milestone for VRLE applications is real-time social collaboration in immersive virtual worlds. Collaborative educational interventions have previously been shown to be more efficient than individual task solving (Johnson and Johnson 1999). Some references to social collaboration with VR have been reported (Šašinka et al. 2018), but more research is needed on some of the drawbacks, such as communication in iVR with other users. The limitations of current technology mean, for example, that user avatars show no facial expressions, making communication much more difficult. New HMD features are constantly being introduced, to overcome these limitations, with such functions as eye-tracking and cameras that capture and transfer lip movement to the avatar. Eyetracking is one of the most promising technologies for mass implementation in HMDs. It has advantages for the user when interacting with elements, and it also facilitates the incorporation of technologies such as foveated rendering (Albert et al. 2017). However, its use has raised some doubts over the privacy of our future digital footprint within these environments.

Fourthly, the use of large sets of heterogeneous acquired data can play a major role in both addressing the diverse learning needs of students and improving current educational practices with VRLEs. The use of learning analytics, from the learners' point of view, will enable students to improve their learning, by identifying pathways to help achieve learning objectives. Educators will be able to improve the quality of teaching based on real-time data that reflects student performance, participation, and engagement with the subject matter. Besides, iVR developers will be able to improve applications based on the analysis of the most commonly used elements – feedback from students on the interventions and comments from teachers.

Finally, for the adoption of VRLEs in academic curricula at all levels, these applications must be evaluated in terms of technical feasibility (software engineering standpoint), and learning outcomes (pedagogical standpoint) (Radianti et al. 2020). Thus, future research needs to include evidence-based, ongoing, and unobtrusive assessments, embedded within VRLEs. Shute et al. (Shute et al. 2008) used the term "stealth assessment." As the learner interacts within the VRLE, stealth assessment serves to analyze user actions, to estimate user proficiency, and to create a model that is continuously updated. This information allows the VRLE to provide relevant feedback to the learner in real time and/or to adapt the VRLE to the needs of learners (Shute et al. 2017).

4 Conclusions

One central advantage of iVR over traditional learning is its potential to elicit higher levels of student motivation and engagement. Furthermore, the interactive and immersive experiences that are afforded within virtual environments have some potential to boost learning through diverse mechanisms. Firstly, student interaction with new technologies engages the student in an immediate way in the first person and can boost student motivation. Secondly, iVR has the potential to foster embodied forms of learning, due to its highly immersive and advanced interfaces that can use the human body as input. In this way, students perceive iVR experiences in education as exciting and challenging opportunities once a minimum of expertise in the iVR interface has been gained. Thirdly, the hands-on affordances of iVR experiences have been shown to boost learning. iVR has proven itself a useful tool for promoting different types of learning, such as "comprehension," defined as the ability to interpret, exemplify, classify, infer, compare and explain – and "remembering," which is defined as the ability to recognize/identify and recall. Visual recognition is clearly better acquired in iVR than in traditional learning environments.

These advantages of iVR experiences require further empirical validation before iVR gains widespread acceptance as a reliable pedagogical method. It is an achievable aim, although the importance of grounding iVR experiences within established learning theories is required, and robust evaluation methods must be used. Additionally, iVR educational experiences should follow a multi-stage design process considering learning objectives and technology possibilities. An exciting world can be created in the short-term future: iVR serious games with high user interactivity, well-designed ramps for iVR beginners, a game structure that offers genuine play, and reliable integrated evaluation methods.

Acknowledgments This work was partially supported by the ACIS project (Reference Number INVESTUN/21/BU/0002) of the *Consejeria de Empleo of the Junta de Castilla y León* (Spain) and the Erasmus+ RISKREAL Project (Reference Number 2020-1-ES01-KA204-081847) of the European Commission.

References

Albert R, Patney A, Luebke D, Kim J (2017) Latency requirements for foveated rendering in virtual reality. ACM Trans Appl Percept 14(4). https://doi.org/10.1145/3127589

Albus P, Vogt A, Seufert T (2021) Signaling in virtual reality influences learning outcome and cognitive load. Comput Educ 166. https://doi.org/10.1016/j.compedu.2021.104154

Alhalabi WS (2016) Virtual reality systems enhance students' achievements in engineering education. Behav Inform Technol. https://doi.org/10.1080/0144929X.2016.1212931

Allcoat D, von Mühlenen A (2018) Learning in virtual reality: effects on performance, emotion and engagement. Res Learn Technol 26. https://doi.org/10.25304/rlt.v26.2140

Araiza-Alba P, Keane T, Chen WS, Kaufman J (2021a) Immersive virtual reality as a tool to learn problem-solving skills. Comput Educ 164. https://doi.org/10.1016/j.compedu.2020.104121

Araiza-Alba P, Keane T, Matthews B, Simpson K, Strugnell G, Chen WS, Kaufman J (2021b) The potential of 360-degree virtual reality videos to teach water-safety skills to children. Comput Educ 163. https://doi.org/10.1016/j.compedu.2020.104096

Baceviciute S, Mottelson A, Terkildsen T, Makransky G (2020) Investigating representation of text and audio in educational VR using learning outcomes and EEG. In: Conference on human factors in computing systems – proceedings. https://doi.org/10.1145/3313831.3376872

Banakou D, Hanumanthu PD, Slater M (2016) Virtual embodiment of white people in a black virtual body leads to a sustained reduction in their implicit racial bias. Front Hum Neurosci 10 (NOV2016). https://doi.org/10.3389/fnhum.2016.00601

Barfield W, Zeltzer D, Sheridan TB, Slater M (1995) Presence and performance within virtual environments. In: Virtual environments and advanced Interface design. Oxford University Press, Oxford, pp 473–541

Bombari D, Mast MS, Canadas E, Bachmann M (2015) Studying social interactions through immersive virtual environment technology: virtues, pitfalls, and future challenges. Front Psychol. https://doi.org/10.3389/fpsyg.2015.00869

Buttussi F, Chittaro L (2017) Effects of different types of virtual reality display on presence and learning in a safety training scenario. IEEE Trans Vis Comput Graph. https://doi.org/10.1109/TVCG.2017.2653117

Bustillo A, Alaguero M, Miguel I, Saiz JM, Iglesias LS (2015) A flexible platform for the creation of 3D semi-immersive environments to teach cultural heritage. Digit Appl Archaeol Cult Herit 2 (4):248–259. https://doi.org/10.1016/j.daach.2015.11.002

Calvert J, Abadia R (2020) Impact of immersing university and high school students in educational linear narratives using virtual reality technology. Comput Educ 159. https://doi.org/10.1016/j.compedu.2020.104005

Chang, Eunhee, Hyun Taek Kim, and Byounghyun Yoo. (2020). Virtual reality sickness: a review of causes and measurements. Int J Hum-Comput Int. 1658–1682. doi:https://doi.org/10.1080/10447318.2020.1778351

Checa D, Bustillo A (2020a) A review of immersive virtual reality serious games to enhance learning and training. Multimed Tools Appl 79(9–10):5501–5527. https://doi.org/10.1007/s11042-019-08348-9

Checa D, Bustillo A (2020b) Advantages and limits of virtual reality in learning processes: Briviesca in the fifteenth century. Virtual Reality 24(1):151–161. https://doi.org/10.1007/s10055-019-00389-7

Checa D, Miguel-Alonso I, Bustillo A (2021) Immersive virtual-reality computer-assembly serious game to enhance autonomous learning. Virtual Reality. https://doi.org/10.1007/s10055-021-00607-1

Checa D, Saucedo-Dorantes JJ, Osornio-Rios RA, Antonino-Daviu JA, Bustillo A (2022) Citation: Checa, D virtual reality training application for the condition-based maintenance of induction motors. https://doi.org/10.3390/app12010414

Chen S, Pan Z, Zhang M, Shen H (2013) A case study of user immersion-based systematic design for serious heritage games. Multimed Tools Appl. https://doi.org/10.1007/s11042-011-0864-4

Corcoran RP (2018) An embodied cognition approach to enhancing Reading achievement in New York City public schools: promising evidence. Teach Teach Educ 71. https://doi.org/10.1016/j.tate.2017.11.010

Dede C (2009) Immersive interfaces for engagement and learning. Science. https://doi.org/10.1126/science.1167311

Egenfeldt-Nielsen S (2006) Overview of research on the educational use of video games. Digital Kompetanse. https://doi.org/10.1353/dia.2006.0003

Fosnot CT, Perry RS (1996) Constructivism: a psychological theory of learning. In: Constructivism: theory, perspectives, and practices. Teachers College Press, New York

Fowler C (2015) Virtual reality and learning: where is the pedagogy? Br J Educ Technol. https://doi.org/10.1111/bjet.12135

Freina L, Ott M (2015) A literature review on immersive virtual reality in education: state of the art and perspectives. In: ELearning & software for education. National Defence University, pp 133–141

Herrera F, Bailenson J, Weisz E, Ogle E, Zak J (2018) Building long-term empathy: a large-scale comparison of traditional and virtual reality perspective-taking. PLoS One 13(10). https://doi.org/10.1371/journal.pone.0204494

Innocenti ED, Geronazzo M, Vescovi D, Nordahl R, Serafin S, Ludovico LA, Avanzini F (2019) Mobile virtual reality for musical genre learning in primary education. Comput Educ Pergamon 139:102–117. https://doi.org/10.1016/J.COMPEDU.2019.04.010

Isabwe GMN, Moxnes M, Ristesund M, Woodgate D (2018) Children's interactions within a virtual reality environment for learning chemistry. In: Advances in intelligent systems and computing. https://doi.org/10.1007/978-3-319-60018-5_22

Jen Chen C (2009) Theoretical bases for using virtual reality in education. *Themes Sci Technol Educ*. no. Special Issue

Jensen L, Konradsen F (2018) A review of the use of virtual reality head-mounted displays in education and training. Educ Inf Technol 23(4). https://doi.org/10.1007/s10639-017-9676-0

Johnson DW, Johnson RT (1999) Making cooperative learning work. Theory Pract 38(2). https://doi.org/10.1080/00405849909543834

Kavanagh S, Luxton-Reilly A, Wuensche BC, Plimmer B (2017) A systematic review of virtual reality in education. Themes Sci Technol Educ 10(2):85–119

Kwon C (2019) Verification of the possibility and effectiveness of experiential learning using HMD-based immersive VR technologies. Virtual Reality 23(1):101–118. https://doi.org/10.1007/s10055-018-0364-1

Liu R, Wang L, Lei J, Wang Q, Ren Y (2020) Effects of an immersive virtual reality-based classroom on students' learning performance in science lessons. Br J Educ Technol 51(6). https://doi.org/10.1111/bjet.13028

Madden JH, Won AS, Schuldt JP, Kim B, Pandita S, Sun Y, Stone TJ, Holmes NG (2018) Virtual reality as a teaching tool for moon phases and beyond. In: Physics education research conference proceedings, Vol. 2018. American Association of Physics Teachers. https://doi.org/10.1119/perc.2018.pr.Madden

Makransky G, Borre-Gude S, Mayer RE (2019a) Motivational and cognitive benefits of training in immersive virtual reality based on multiple assessments. J Comput Assist Learn 35(6). https://doi.org/10.1111/jcal.12375

Makransky G, Terkildsen TS, Mayer RE (2019b) Adding immersive virtual reality to a science lab simulation causes more presence but less learning. Learn Instruct 60 (April). Pergamon:225–236. https://doi.org/10.1016/J.LEARNINSTRUC.2017.12.007

Makransky G, Andreasen NK, Baceviciute S, Mayer RE (2021) Immersive virtual reality increases liking but not learning with a science simulation and generative learning strategies promote learning in immersive virtual reality. J Educ Psychol 113(4). https://doi.org/10.1037/edu0000473

Makransky G, Lilleholt L (2018) A structural equation modeling investigation of the emotional value of immersive virtual reality in education. Educ Technol Res Dev 66(5). https://doi.org/10.1007/s11423-018-9581-2

Makransky G, Petersen GB (2019) Investigating the process of learning with desktop virtual reality: a structural equation modeling approach. Comput Educ 134. https://doi.org/10.1016/j.compedu.2019.02.002

Mayer RE (2005) Cognitive theory of multimedia learning. In: The Cambridge handbook of multimedia learning. Cambridge University Press, pp 31–48. https://doi.org/10.1017/CBO9780511816819.004

Michelle KD (2011) Connectivism: learning theory and pedagogical practice for networked information landscapes. In: Reference services review 39(4). Emerald Group Publishing Limited, pp 675–85. https://doi.org/10.1108/00907321111186686

Mikropoulos TA, Natsis A (2011) Educational virtual environments: a ten-year review of empirical research (1999–2009). Comput Educ 56:769–780. https://doi.org/10.1016/j.compedu.2010.10.020

Miller MD, Castillo G, Medoff N, Hardy A (2021) Immersive VR for organic chemistry: impacts on performance and grades for first-generation and continuing-generation university students. Innov High Educ. https://doi.org/10.1007/s10755-021-09551-z

Moro C, Štromberga Z, Raikos A, Stirling A (2017) The effectiveness of virtual and augmented reality in health sciences and medical anatomy. Anat Sci Educ 10(6). https://doi.org/10.1002/ase.1696

Natale AF d, Repetto C, Riva G, Villani D (2020) Immersive virtual reality in K-12 and higher education: a 10-year systematic review of empirical research. Br J Educ Technol. https://doi.org/10.1111/bjet.13030

Pande P, Thit A, Sørensen AE, Mojsoska B, Moeller ME, Jepsen PM (2021) Long-term effectiveness of immersive vr simulations in undergraduate science learning: lessons from a media-comparison study. Res Learn Technol 29. https://doi.org/10.25304/rlt.v29.2482

Parong J, Mayer RE (2018) Learning science in immersive virtual reality. J Educ Psychol 110(6). https://doi.org/10.1037/edu0000241

Petersen GB, Klingenberg S, Mayer RE, Makransky G (2020) The virtual field trip: investigating how to optimize immersive virtual learning in climate change education. Br J Educ Technol 51(6). https://doi.org/10.1111/bjet.12991

Petri G, Gresse C, von Wangenheim. (2017) How games for computing education are evaluated? A systematic literature review. Comput Educ. https://doi.org/10.1016/j.compedu.2017.01.004

Pritchard A (2017) Ways of learning: learning theories for the classroom. Routledge

Radianti J, Majchrzak TA, Fromm J, Wohlgenannt I (2020) A systematic review of immersive virtual reality applications for higher education: design elements, lessons learned, and research agenda. Comput Educ 147. https://doi.org/10.1016/j.compedu.2019.103778

Repetto C, Flavia A, di Natale D, Villani ST, Germagnoli S, Riva G (2021) The use of immersive 360° videos for foreign language learning: a study on usage and efficacy among high-school students. Interact Learn Environ. https://doi.org/10.1080/10494820.2020.1863234

Ritterfeld U, Weber R, Fernandes S, Vorderer P (2004) Think science!: entertainment education in interactive theaters. Comput Entertain. https://doi.org/10.1145/973801.973819

Roswell RO, Cogburn CD, Tocco J, Martinez J, Bangeranye C, Bailenson JN, Wright M, Mieres JH, Smith L (2020) Cultivating empathy through virtual reality: advancing conversations about racism, inequity, and climate in medicine. Acad Med. https://doi.org/10.1097/ACM.0000000000003615

Roussos M, Johnson A, Moher T, Leigh J, Vasilakis C, Barnes C (1999) Learning and building together in an immersive virtual world. Presence Teleop Virt. https://doi.org/10.1162/105474699566215

Roussou M, Slater M (2017) Comparison of the effect of interactive versus passive virtual reality learning activities in evoking and sustaining conceptual change. IEEE Trans Emerg Top Comput. https://doi.org/10.1109/TETC.2017.2737983

Šašinka Č, Stachoň Z, Sedlák M, Chmelík J, Herman L, Kubíček P, Šašinková A et al (2018) Collaborative immersive virtual environments for education in geography. ISPRS Int J Geo Inf 8(1). https://doi.org/10.3390/ijgi8010003

Shuell TJ (1986) Cognitive conceptions of learning. Rev Educ Res 56(4). https://doi.org/10.3102/00346543056004411

Shute VJ, Hansen EG, Almond RG (2008) You Can't fatten a hog by weighing it-or can you? Evaluating an assessment for learning system called ACED. Int J Artif Intell Educ 18(4)

Shute VJ, Ke F, Wang L (2017) Assessment and adaptation in games. In: Instructional techniques to facilitate learning and motivation of serious games. https://doi.org/10.1007/978-3-319-39298-1_4

Skinner BF (1989) The origins of cognitive thought. Am Psychol 44(1). https://doi.org/10.1037/0003-066X.44.1.13

Slater M (2003) A note on presence terminology. Presence Connect 3 (January)

Slater M (2017) Implicit learning through embodiment in immersive virtual reality. In: Liu D, Dede C, Huang R, Richards J (eds) Virtual, augmented, and mixed realities in education. Smart computing and intelligence. Springer, Singapore. https://doi.org/10.1007/978-981-10-5490-7_2

Stranger-Johannessen E (2018) Exploring math achievement through gamified virtual reality. In: Lecture notes in computer science (including subseries lecture notes in artificial intelligence and lecture notes in bioinformatics), Vol. 11082. LNCS. https://doi.org/10.1007/978-3-319-98572-5_57

The Khronos Group (2021) OpenXR Overview – The Khronos Group Inc. Khronos. https://www.khronos.org/openxr/

Tychsen L, Foeller P (2020) Effects of immersive virtual reality headset viewing on young children: visuomotor function, postural stability, and motion sickness. Am J Ophthalmol 209. https://doi.org/10.1016/j.ajo.2019.07.020

Walsh M, Simpson A (2013) Touching, tapping thinking examining the dynamic materiality of touch pad devices for literacy learning. Aust J Lang Lit 36(3)

Wilson M (2002) Six views of embodied cognition. Psychon Bull Rev. https://doi.org/10.3758/BF03196322

Winn W (1993) A conceptual basis for educational applications of virtual reality. University of Washington, Human Interface Technology Laboratory, Washington Technology Centre, Seattle,Washington, Technical Publication R-93-9

Zhou Y, Ji S, Tao X, Wang Z (2018) Promoting knowledge construction: a model for using virtual reality interaction to enhance learning. Proc Comput Sci 130:239–246. https://doi.org/10.1016/j.procs.2018.04.035

VR for Pain Relief

Marta Matamala-Gomez, Tony Donegan, and Justyna Świdrak

Contents

Abstract The present chapter explores how immersive virtual reality (VR) systems can be used for pain research and treatment. Pain is a universal, yet entirely subjective and multifaceted unpleasant experience. One of the earliest VR studies

---

M. Matamala-Gomez (✉)
Department of Cognition, Development and Educational Psychology, University of Barcelona, Barcelona, Spain

Cognition and Brain Plasticity Group, Barcelona, Spain
e-mail: marta.matamala@ub.edu

T. Donegan
Cortical Networks and Virtual Environments in Neuroscience Lab, IDIBAPS, Barcelona, Spain

Experimental Virtual Environments for Neuroscience and Technology, University of Barcelona, Barcelona, Spain

J. Świdrak
Cortical Networks and Virtual Environments in Neuroscience Lab, IDIBAPS, Barcelona, Spain

Experimental Virtual Environments for Neuroscience and Technology, University of Barcelona, Barcelona, Spain

Institute of Psychology, Polish Academy of Sciences, Warsaw, Poland

Curr Topics Behav Neurosci (2023) 65: 309–336
https://doi.org/10.1007/7854_2022_402
Published Online: 3 January 2023

on pain highlighted the role of attention in pain modulation. However, the role of body representation in pain modulation has also been described as a crucial factor. Through virtual reality systems, it is possible to modulate both attention to pain and body representation. In this chapter, first we define how immersive VR can be used to create the illusion of being present in immersive VR environments and argue why VR can be an effective tool for distracting patients from acute pain. However, distraction seems to be less useful in chronic pain treatment. Chronic pain can be highly disabling and can significantly impact not only the sufferer's quality of life, but also their perceptions of the bodily self. Close neural connections between the body matrix and pain open a chance for influencing pain through bodily illusions. This chapter explores approaches to inducing body ownership illusions in VR and discusses how they have been applied in pain research. The present chapter also covers a set of practical indications and methodological caveats of immersive VR and solutions for overcoming them. Finally, we outline several promising future research directions and highlight several yet unexplored areas.

Keywords Body representation · Embodiment · Pain relief · Virtual reality

1 Introduction

In 1965, Ronald Melzack and Pat Wall transformed the current views on pain with the introduction of gate control theory (Melzack and Wall 1965), which argued that the brain plays a crucial role in pain perception, as it filters, selects, and modulates the inputs arriving at or produced by the body (Melzack 1999a, b). The theory postulated that pain perception is not a simple sensory stimulus-response model, as our previous Cartesian understanding suggested, but a complex process in which the brain contributes not only to the ultimate perception of pain, but also to the nature of pain itself.

The primary purpose of pain is protection. It motivates the organism to stop what it is doing and seek help. From an experiential point of view then, pain is necessarily unpleasant (Merskey 2002) and demands attention (Eccleston and Crombez 1999). Further, pain reduces cortical processing capacity (Derbyshire et al. 1998), induces slowed decision-making (Crombez et al. 1996), and increases cognitive error rates (Buckelew et al. 1986). In addition, it modifies immune activity (Watkins and Maier 2000), the activity of the hypothalamus–pituitary–adrenal axes, and alters the sympathetic nervous system (Melzack 1999a, b), as well as reducing reproductive system function (Rivier 1995), and activating visuo-motor systems (Price 1999). From a neurobiological point of view, the following brain areas were previously thought to be pain specific: the primary (S1) and secondary (S2) somatosensory cortices, the insula, and the anterior cingulate cortex (ACC) (Iannetti and Mouraux 2010). This extensive network is known as the "pain matrix" for mediating pain experience itself (Ploghaus et al. 1999), although activation of these areas has subsequently been shown in other non-painful but strong sensory stimuli (Salomons et al. 2016).

In this regard, the concept of a pain matrix is often used to understand the neural mechanisms of pain in health and disease. For instance, it is known that chronic pain conditions, where pain persists for longer than 3 months, are associated with changes in the structure of the brain as a result of central plasticity (May 2008). Indeed, there is evidence demonstrating an altered localized brain chemistry and functional reorganization of cortical networks in patients suffering from chronic back pain (Flor 2003). Neuroplastic changes related to function, chemical profile, or structure during chronic pain conditions have been described for both the peripheral and central nervous systems in terms of reorganization of receptors and ion channels, changes in neurotransmitters at the peripheral level, and functional changes in cortical representational fields at the central level (May 2008; Yang and Chang 2019) Other examples of neuroplastic change due to chronic pain include spinal cord sensitization or disinhibition and the interaction of the nervous system with the immune system and higher cognitive functions (May 2008).

Accordingly, a large number of studies suggest that cortical reorganization takes place at a functional level in patients suffering from chronic pain conditions, such as in amputee patients with phantom limb pain (Flor et al. 1995), and in patients with chronic back pain (Flor et al. 1997). Moreover, such functional cortical reorganization has also been shown in patients with complex regional pain syndrome, in which a shrinkage of the representational field in the brain of the affected arm was found and was highly correlated with pain intensity (Maihöfner et al. 2003; Pleger et al. 2004) although this finding is controversial, with later studies showing enlarged representation of the healthy hand on the contralateral hemisphere (Di Pietro et al. 2015) and no interhemispheric differences (Mancini et al. 2019). In this regard, the neuromatrix pain theory, which was introduced by Melzack (1996), recognizes the influence of both ascending and descending inputs to the conscious experience of pain, as well as the important contributions of memory and past experiences to pain perception (Melzack 2001).

The neuromatrix pain theory has paved the way for the introduction of new nonpharmacologic methods for pain relief based on behavioral interventions. A crucial concept of this approach is that pain relies on a body image that is held by the brain like a "virtual body" (Moseley 2003). Pain perception may therefore induce some changes in the "virtual body." In patients suffering from phantom limb pain, for example, they may perceive a telescoped effect, which is the feeling that the proximal portion of the amputated limb is missing or has shrunk with the more distal portion floating near, attached to, or "within" the stump (Flor et al. 2006; Giummarra et al. 2007). Indeed, we know that in the case of chronic pain conditions, both the nociceptive system and the "virtual body" may experience profound changes that increase the sensitivity to noxious or non-noxious stimuli, and as a consequence of this may disrupt the integrity of motor output (Moseley 2003). In this case, the "virtual body" in the brain is continuously updated by the inputs from the sensory system.

Since the 1990s, new behavioral therapies for pain relief have attempted to modulate such a distorted or corrupted internal "virtual body" by inducing body ownership illusions through mirrors (Ramachandran and Altschuler 2009), fake bodies (Hegedüs et al. 2014), and virtual bodies (Matamala-Gomez et al. 2019b, 2021b). Growing interest in new technologies such as VR systems is paving the way for the development of new clinical interventions to manage pain via non-pharmacological means (Carter et al. 2014). This chapter aims to show how VR and specifically virtual body ownership illusions can be applied for pain relief interventions. It also reviews the most recent literature and outlines several promising research lines which have become possible with current and future technological advancements.

2 VR and Pain

VR was introduced to the field of pain at the beginning of the twenty-first century. One of the first applications for pain relief in a clinical context was proposed by Hoffman and colleagues (Hoffman et al. 2000a, b, c), who found that adolescent and adult burns patients perceived less pain when playing a video game whilst having their dressings changed – a notoriously painful procedure. Later, an fMRI (functional-magnetic resonance imaging) study conducted by the same group found that VR exposure while experiencing a painful heat stimulus significantly reduced pain in five brain regions of interest related to pain: the anterior cingulate cortex, primary and secondary somatosensory cortex, insula, and thalamus (Hoffman et al. 2004a, b). Further, some years later, another fMRI study showed that when using VR systems for pain relief, the reduction of pain was comparable to the analgesic effect of a moderate dose of hydromorphone pain medication (Hoffman et al. 2007). Moreover, the effectiveness of VR systems for pain relief has been demonstrated in patients with mild and severe pain states (Hoffman et al. 2000a, b, c, 2011, 2014).

Whereas the first VR applications were non-immersive (i.e., they were presented on a computer screen), new VR applications are continuing to be developed for immersive VR systems. Immersive VR is a computer-generated 360°, 3D environment that is rendered through a head-tracked, head-mounted display (HMD) or in a cave automatic virtual environment (CAVE) (Slater 2018). Immersive VR systems allow users to be transported to an artificial world and even give them a full virtual body with which to explore it. Further, both the virtual world and body can be fully controlled and modified by the experimenter or clinician with the intention of achieving specific clinical outcomes, making VR systems an optimal tool for modulating the internal "virtual body," the sensory system, as well as the cognitive functions of patients suffering from pain conditions.

Nowadays, the dramatically lower costs for buying a VR system, with standalone headset systems available for less than €300, opens up the possibility for using VR for assessment and treatment of different pain conditions in hospital environments or even in patients' homes (Matamala-Gomez et al. 2021a; Scuteri et al. 2020; Slater 2014; Stasolla et al. 2021). It has been demonstrated that VR systems can effectively reduce pain perception in both (1) healthy subjects experiencing painful stimuli and (2) in patients with clinical pain, via two main mechanisms: (1) the distractive potential of VR and (2) the capability of modulating internal body representation (Donegan et al. 2020; Matamala-Gomez et al. 2019b). In addition, some evidence demonstrates that the positive pain-relieving effects of VR may also be mediated through a reduction in anxiety and through the user experiencing positive emotions such as a sense of fun when experiencing it (Triberti et al. 2014).

2.1 Evidence for Using Immersive VR for Pain Distraction

Distraction has been considered an effective cognitive behavioral procedure for pain relief, as a time-honored psychological pain intervention (Blount et al. 2003; Dahlquist 1999a, b). Some distractive strategies for pain relief are deep breathing, listening to soothing music, watching a video, or playing a video game (Austin and Siddall 2021; Bondin and Dingli 2021; Krupić et al. 2021). It is argued that distractive strategies consume some of our attentional resources leaving less cognitive capacity available for processing pain (McCaul and Malott 1984). The success of these techniques for pain relief has led to the use of VR systems to maximize distraction (Wiederhold et al. 2014a, b).

Today, there is a growing interest in using VR systems as a distractive intervention for pain reduction (Botella et al. 2008; Gorman 2006; Hoffman et al. 2007, 2011; Donegan et al. 2020; Matamala-Gomez et al. 2019a). Through VR, it is possible to draw attention away from the patients' mental processing, decreasing their pain perception (Wiederhold et al. 2014a, b). It is generally thought that the amount of attention directed to the VR intervention is inversely proportional to the available remaining attentional resources that can process incoming nociceptive signals. VR interventions for pain relief have been found to be effective in reducing reported pain in patients undergoing burn wound care, chemotherapy, or dental procedures (Bani Mohammad and Ahmad 2019; Hoffman et al. 2001, 2000a, b, c; Schneider et al. 2003; Schneider and Workman 1999; Tanja-Dijkstra et al. 2014; Wiederhold and Wiederhold 2012; Wiederhold et al. 2014a, b). Hence, distraction strategies using VR are considered a cognitive target with specific therapeutic utility in acute and procedural pain contexts (Trost et al. 2021).

Importantly, we know that the distracting effects of VR interventions are able to change how the brain integrates pain, not just the perception of painful stimuli (Birckhead et al. 2021). To verify this, Birckhead and colleagues have proposed to evaluate the effectiveness of three forms of VR for patients with chronic lower back pain in a forthcoming three-arm clinical trial. In detail, the authors aim to compare the effects of (1) a skills-based VR, a program incorporating principles of cognitive behavioral therapy, mindful meditation, and physiological biofeedback therapy using embedded biometric sensors; (2) a distraction-based VR program using 360°

immersive videos designed to distract users from pain; and (3) a sham VR program in which the patients were allocated to a non-immersive program using two-dimensional videos within a VR headset on pain perception.

It has also been shown that VR interventions facilitate positive affective states, which also contribute to the reduction of pain perception and bolster the diversionary power of the intervention (Sharar et al. 2016). To date, results from experimental research support VR as a tool for distraction and affecting modulation (Indovina et al. 2018; Malloy and Milling 2010 for reviews). These studies provide evidence that VR distraction is an effective intervention for reducing experimental pain in healthy subjects, as well as the pain associated with burn injury care, and other medical procedures such as wound care, and physical therapy. Moreover, VR seemed to decrease cancer-related symptoms in different settings, including during chemotherapy (Chirico et al. 2016). The studies from these reviews showed a clear pattern that immersive VR technology is more likely than non-immersive VR technology to generate relief from pain.

Further, the sense of presence induced during a VR experience and the possibility to interact with the virtual environment correlate inversely with the perception of external pain induced during the experimental session (Gutierrez-Maldonado et al. 2011; Gutiérrez-Martínez et al. 2011; Hoffman et al. 2004b; Riva et al. 2007; Wender et al. 2009). Interestingly, a recent study conducted by Hoffman reported that interacting with virtual objects through embodied virtual hands increased the sense of presence within the virtual environment – an effect that was accompanied by diminished attentional resources available to perform an attention-demanding task in healthy subjects, as well as a decrease in perceived pain induced by brief thermal stimuli during the experimental session (Hoffman 2021). The results from this study implicate an attentional mechanism for how VR reduces pain and help us to understand how VR influences pain perception. Further, the results from this study indicate that the immersiveness of a VR system can be increased substantially (e.g., through avatars) with little or no increase in VR side effects – unlike opioids, which show a dose–response increase in side effects (e.g., increased nausea and constipation) with higher doses. Further, opioid side effects linger for hours after a medical procedure has been completed. Such results pave the way to include immersive VR as a behavioral non-pharmacological intervention for treating pain conditions.

In relation to the above-mentioned study by Hoffman and colleagues, the effects of virtual embodiment on pain relief are more related to brain plasticity changes to the internal body representation than on distractive mechanisms (Matamala-Gomez et al. 2021b). Specifically, these changes rely on the predictive coding hypothesis, which argues that the brain maintains an internal model of the body and the space around it (i.e., the body matrix) which allows the brain to create predictions about the upcoming sensory stimuli arriving at the body and to optimally interact with the dynamic environment around the body (Barrett 2017; Riva et al. 2019). Then, top-down and bottom-up multisensory processes converge into the body matrix and redefine the place of the self, inside the body, consequently modulating the internal body representation as we interact with the surrounding environment (Apps and Tsakiris 2014; Holmes and Spence 2004; Serino et al. 2018). More specifically, some authors argue that the brain creates an embodied simulation of the body to effectively control and regulate the body in the world, which includes predicting people's actions, concepts, and emotions (Riva et al. 2019). Along these lines, VR experiences attempt to replicate the sensory consequences of the individual's actions, providing them with the same scene or body representation that they can see in the real world. To achieve this, the VR system, like the brain, maintains a model (simulation) of the body and the space around it (Riva et al. 2019). Hence, the effectiveness of virtual body ownership illusions relies on its capability of simulating a body representation within a virtual environment while allowing the possibility to modulate the bodily experience by designing targeted virtual bodies and environments (De Oliveira et al. 2016).

2.2 *Evidence for Using Virtual Embodiment for Pain Relief*

Embodiment is a concept that has been defined in various ways. From the philosophical perspective it is a part of the general discussion on how one defines and experiences oneself (Blanke and Metzinger 2009), which means how the cognitive system utilizes the environment and the body as external informational structures that complement internal representations (Barsalou 2010). For cognitive neuroscience and psychology, it is concerned with the question of how the brain represents the body (Berlucchi and Aglioti 1997; Graziano and Botvinick 2002). It denotes the sense of having a body, and the body can be considered to be both the subject and object of medical science and practice (Gallagher 2001). Moreover, it can be also defined as the sense of being inside, having, and controlling a body, especially when referring to the sense of embodiment toward a virtual avatar (Kilteni et al. 2012). Indeed, in VR, embodiment is frequently associated with a sense of body ownership (Lopez et al. 2008), the concept of self-location (Arzy et al. 2006a, b), and a sense of agency (Newport et al. 2010). Hence, having a sense of embodiment toward a virtual body refers to the feeling of our self as being inside a body, a body that moves according to our intentions, and that interacts with the surrounding virtual environment.

A sense of self-location is described as one of the key components of inducing a sense of embodiment toward a fake body (Arzy et al. 2006a, b; Blanke and Metzinger 2009; Lenggenhager et al. 2007). However, some studies suggest that feeling embodied within a fake body does not require a sense of body ownership. In fact, it appears that these two phenomena can be dissociated, as shown in the study conducted by de Preester and Tsakiris, in which participants could feel a sense of embodiment toward a tool, that is the feeling that the tool is part of one's own body (de Preester and Tsakiris 2009) without reporting feelings of ownership (De Vignemont 2011). These studies support the idea of a pre-existing body-model (representation) that allows for the incorporation of objects into the current body representation model. This body-model is also a basis for the distinction between body extensions (e.g., in the case of tool-use) and incorporation (e.g., in

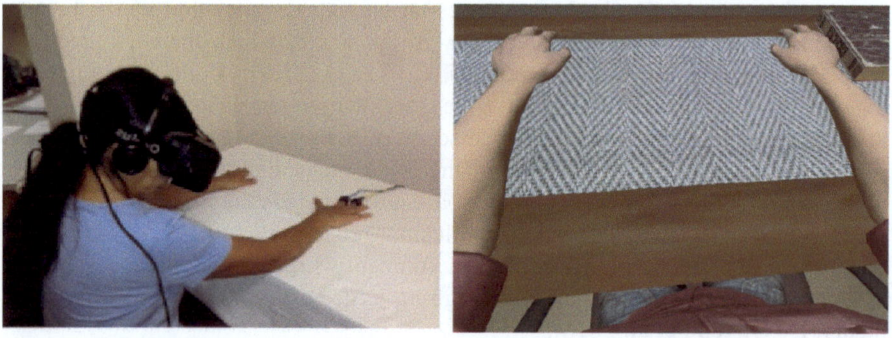

Fig. 1 Sense of embodiment in virtual reality

the case of successful prosthesis use). More specifically, it was demonstrated that in the case of incorporation, changes in the sense of body ownership involve a reorganization of the body-model, whereas extension of the body with tools does not involve changes in the sense of body ownership.

Another crucial factor for inducing embodiment within a fake body is experiencing a sense of agency, which has been demonstrated to give coherence to the internal body representation (Tsakiris et al. 2006). Further, it is shown that a lack of agency can inhibit the sense of embodiment, as in the case, for example, of inducing the illusion that the participant's finger was being stretched to twice its normal length until it snapped and the tip came off, and moving it independently without participant's control (Newport and Preston 2010).

In summary, through VR it is possible to induce the three subcomponents of embodiment – body ownership, a sense of self-location, and a sense of agency – so that VR can be used to modulate a user's internal body representation. In turn, a users' changed body representation can influence his or her behavioral, cognitive, and physiological responses as the person engages with virtual surroundings (Moseley et al. 2012). In this way, VR can be an effective tool for inducing virtual body ownership illusions through which participants can feel fully embodied in the virtual body, the appearance of which can be determined by the experimenter (Kilteni et al. 2012) (Fig. 1). For instance, some studies suggest, that assuming the role in VR of a person of a different race can impact empathy and perspective taking (Thériault et al. 2021; Banakou et al. 2020). Given these studies, it may prove fruitful to explore the impact of empathy through the use of virtual embodiment on pain modulation.

Body Perception and Pain

In recent years, interest has grown in investigating how observing the body while being in pain may impact pain perception (Martini 2016). Cross-modal interactions between the vision of the body and somatosensory responses have been widely investigated (Macaluso and Maravita 2010; Medina and Coslett 2010; Serino and Haggard 2010; Wesslein et al. 2014). It was reported that in healthy subjects observing video clips of other hands receiving painful stimulation, while receiving a painful laser stimulation on their real hands, the neural processing of the early nociceptive-related responses was modulated by decreasing neural activity in the pain matrix sensory node, compared to the observation of a non-human control condition (Valeriani et al. 2008). Later, another study conducted by Longo and colleagues reported that seeing one's own painful body part can itself exert an analgesic effect (Longo et al. 2009). In three separate experiments with healthy subjects, the authors showed that while subjects were looking at their own painfully stimulated hand, they felt less pain compared to when they were looking at a box or even at somebody else's hand. Further, they observed a reduction of the late N2/P2 components of the laser-evoked potentials, which showed an analgesic effect related to the vision of the painful body part. The authors argued that this effect was driven by the visually induced activation of the inhibitory GABAergic interneurons in the brain somatosensory areas (SI, SII). Moreover, a different electroencephalogram (EEG) study with healthy subjects showed that seeing the body, compared to seeing a neutral object, increased beta oscillatory activity bilaterally in sensorimotor brain areas, indicating cortical inhibitory activity toward nociceptive stimuli processing (Mancini et al. 2013). Another neuroimaging study found that images of a body part subjected to painful stimulation increased the functional coupling between brain areas of the "pain matrix" and body representation brain areas within the posterior parietal cortex and the occipito-temporal areas (Longo et al. 2012). Interestingly, it has been shown that the analgesic effect while observing a body under painful conditions is site-specific, meaning that there is only pain relief to the site of the body that is homologous to the one being subjected to pain while the participant watches (Diers et al. 2013).

Intriguingly, visual modification of a body part has been found to shape pain perception. For instance, it has been shown in both healthy subjects (Mancini et al. 2011; Romano and Maravita 2014) and clinical populations (Diers et al. 2013; Moseley et al. 2008; Preston and Newport 2011; Ramachandran et al. 2009; Stanton et al. 2018) that by changing the size of the observed painful limb there is a modulation of pain perception in the viewer proportional to the enlargement or reduction of the size of the seen body part. Interestingly, resizing illusions has led to contrasting results in clinical populations with chronic pain, as well as healthy individuals in some cases (for a review, see Martini 2016).

A possible explanation for these contradictory results could reside in the altered neural representation of the body in chronic pain patients (Tsay et al. 2015). It is known that consequences of experiencing body representation alterations include changes in the perception of the size of the painful limb, as has been shown in patients with complex regional pain syndrome (Lewis et al. 2007), hand osteoarthritis (Themelis and Newport 2018), and painful phantom limb (Ramachandran and Hirstein 1998). Interestingly, altered body perceptions analogous to those reported in pathological conditions can be induced in healthy individuals using controlled experimental paradigms (Matamala-Gomez et al. 2020a; Moseley et al. 2012). All experimental manipulations that induce body illusions toward a fake body rely on exposing participants to altered multisensory stimulation through synchronous visuo-tactile or visuo-motor congruent stimulation. For instance, in the study conducted by Petkova and Ehrsson (2008), the authors applied synchronous visuo-tactile stimulation over a mannequin fake body and to the participants real body to induce the sense of ownership over the fake mannequin body. In the same way, Slater et al. (2008) induced a sense of ownership toward a virtual body using synchronous visuo-tactile stimulation. Further, others induced a sense of ownership toward fake virtual bodies using synchronous visuo-motor correlations (Kokkinara et al. 2015). This research supports the overall view that self-body representations in the brain are built dynamically through multisensory integration processes and through our existing knowledge about the human body (Maselli and Slater 2013; Tajadura-Jiménez et al. 2012).

These findings are particularly exciting for VR researchers in that they suggest the possibility of inducing full body ownership illusions through VR by applying synchronous visuo-tactile or visuo-motor congruent stimulations. Moreover, through VR, it is possible to observe the virtual body from a first-person perspective – which could represent a crucial factor to feeling fully embodied in a virtual avatar (Maselli and Slater 2013). Hence, immersive VR systems are a promising tool for both representing the possible body distortions due to the chronic pain condition and for normalizing the altered representation of the painful part of the body reducing pain perception. Then, one may postulate that virtual body ownership illusions can be used for both the assessment and treatment of chronic pain conditions.

Virtual Embodiment for Pain Relief

In line with the ideas outlined above, a large number of studies have attempted over the past decade to leverage the power of virtual embodiment by inducing virtual body ownership illusions in healthy and clinical populations (see Matamala-Gomez et al. 2019b, 2021b; Donegan et al. 2020; Martini 2016). The benefits of virtual body ownership illusions in clinical populations can be explained by recourse to the predictive coding hypothesis, which argues that the brain maintains an internal representation of the body and the space around it. These representations allow the brain to predict upcoming sensory stimuli arriving at the body and to optimally interact with the dynamic environment around the body (Barrett 2017; Riva et al. 2019). Then, top-down and bottom-up multisensory signals arriving to the body redefine the place of the self, inside the body, consequently modulating internal body representation as we interact with the surrounding environment, whether real or virtual (Holmes and Spence 2004; Ribu et al. 2013).

The effectiveness of virtual embodiment stems from its capacity to simulate a virtual body representation that is experienced as genuinely present in the virtual environment, while also allowing this virtual body to be felt as one's own even if its appearance differs from the real body (De Oliveira et al. 2016; Slater and Sanchez-Vives 2016). VR offers the possibility of inducing a sense of body ownership toward fake virtual bodies that are present in the virtual environment through the use of multisensory stimulation (e.g., visuo-tactile or visuo-motor), such that the researcher can modify the morphological characteristics of the virtual body, designing targeted virtual bodies depending on the aim of the intervention. Some evidence

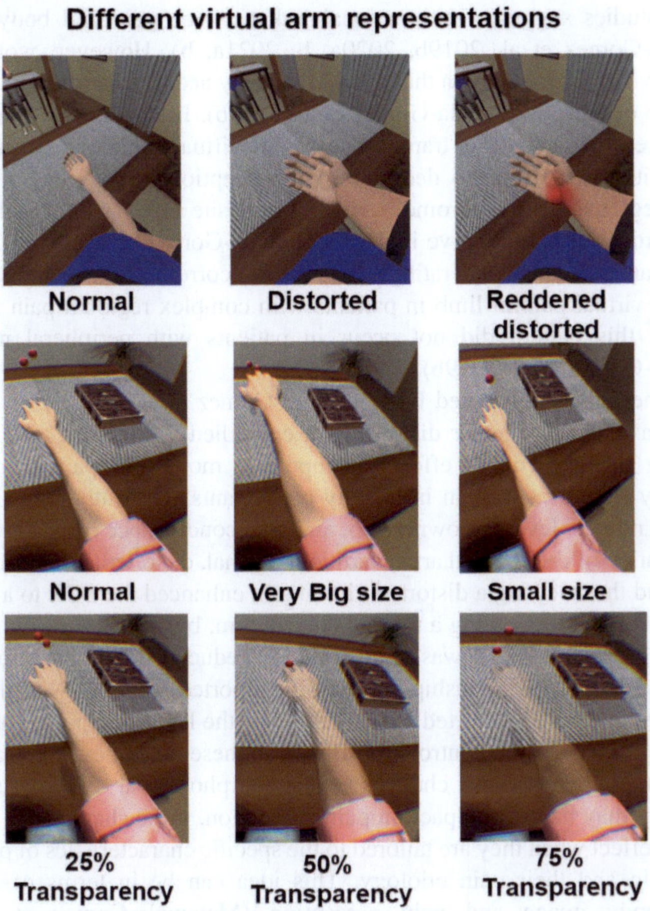

Fig. 2 Virtual body distortions used for upper limb pain relief. Pictures were retrieved from Matamala-Gomez et al. (2019b, 2020a, b)

demonstrates that by changing the morphological characteristics of the virtual body, it is possible to modulate aspects of participants' behavior (Slater and Sanchez-Vives 2016). For instance, it has been shown that a sense of ownership over virtual bodies with different skin colors can be instilled in VR users, decreasing implicit racial bias (Banakou et al. 2016). Similarly, inducing ownership over a child's virtual body has allowed researchers to assess changes in object size estimation and behavioral attitudes (Banakou et al. 2013). Likewise, it has been shown that virtual body ownership illusions can be used in order to prompt functional brain alterations in patients with chronic pain and to decrease pain perception by changing the morphological characteristics of the virtual body (Matamala-Gomez et al. 2019b, 2021b) (Fig. 2).

These studies suggest a bidirectional link between pain and body perception (Matamala-Gomez et al. 2019b, 2020a, b, 2021a, b). However, work from the authors' own group has shown that the best strategy needs to be tuned to the etiology of a source of pain (Matamala-Gomez et al. 2019b). In detail, it was demonstrated that by increasing the level of transparency of the virtual homolog of a painful part of the body, it was possible to decrease pain perception in patients suffering from complex regional pain syndrome – but the opposite effect was found in patients suffering from peripheral nerve injury (Matamala-Gomez et al. 2019b). Further, in this study, an increase in pain ratings was found to correspond with an increase in the size of the virtual painful limb in patients with complex regional pain syndrome – and again, this pattern did not occur in patients with peripheral nerve injury (Matamala-Gomez et al. 2019b).

In another study conducted by Matamala-Gomez et al. (2020a, b), the authors aimed to investigate whether distorting an embodied virtual arm in virtual reality (simulating the telescoping effect in amputees) modulated pain perception and anticipatory responses to pain in healthy participants. The authors evaluated pain/discomfort ratings, levels of ownership, and skin conductance response (SCR) after the observation of each virtual arm condition (normal, distorted, reddened-distorted). It was found that viewing a distorted virtual arm enhanced the SCR to a threatening event with respect to viewing a normal control arm, but when viewing a reddened-distorted virtual arm, SCR was comparatively reduced in response to the threat. There was a positive relationship between the reported level of ownership over the distorted and reddened-distorted virtual arms and the level of reported pain/discomfort, but not in the normal control arm. Based on these outcomes, one may postulate that even though observing changes in the morphological characteristics of the virtual body may have an impact on pain perception, these changes are most likely to exert an effect when they are tailored to the specific characteristics of patients with chronic pain and their pain etiology. This idea can be in terms of interactions between body image and pain perception (Matamala-Gomez et al. 2019b, 2021a, b) such that patients with different chronic pain conditions can present different distortions, in terms of body representation, of the painful limb of the body. According to this hypothesis, an intervention using virtual embodiment to modulate such distortions should thus be tailored to each specific group of patients.

A recent study demonstrated that increasing immersiveness by inducing virtual body ownership illusions of virtual avatars may increase the analgesic effect of VR, compared to a VR exposure without using virtual avatars (Hoffman 2021). In this study, the author showed that inducing a sense of embodiment toward an interactive virtual hand substantially augments the analgesic effect of the intervention over a non-embodied condition with minimal environmental interaction. Similarly, a number of studies have shown that increasing the immersiveness of a VR experience through interactive or realistic scenarios that evoke a sense of presence amplifies the analgesic impact of that VR experience in cases of acute pain perception (Al-Ghamdi et al. 2020; Donegan et al. 2020; Hoffman et al. 2004b; Wender et al. 2009). Moreover, since having a virtual body in virtual reality has also been shown to increase the sense of presence in VR (Slater et al. 2010; Slater and Usoh 1993), adding virtual embodiment to the VR experience is likely to have additional analgesic effects. Another crucial component when using virtual body ownership illusions for pain relief is the co-location of the virtual body with the real one (Nierula et al. 2017). In this study, the authors found that to increase pain threshold while applying external heat stimulation with a thermode to the real body, the virtual and real body should be co-located. Further, the authors observed that for the arm, this analgesic effect diminishes when there are more than 30 cm difference between the virtual and real arm (Nierula et al. 2017).

3 Creating Effective Analgesic VR Illusions

As discussed, one of the key elements of the analgesic effect of VR is thought to operate via distraction or divided attention. In theory, the more engaged the subject is in the virtual task, the less "neural bandwidth" is available for processing of nociception and production of pain (Hoffman 2021). So how can we more fully engage our subjects, and immerse them more completely in the virtual scenario? Mel Slater (Slater 2009) describes two constituent parts of an effective "sense of presence" VR illusion that can induce real-life responses as the place illusion (the participant's sense that they are in a real place) and plausibility illusion (the participant's sense that the events taking place in the virtual scenario are actually happening). Of course, the participant *knows* that they are only experiencing an illusion, but *responds* as if the situation and events are real.

To ensure a more effective presence illusion, the sensorimotor contingencies that occur in VR should, as closely as possible, mimic those of reality. In addition, real-world distractions such as external light sources and sounds should be minimized, as well as any communication with the experimenter if present. Creating effective plausibility illusions also requires that the interactions that occur in VR between the subject and the environment have realistic consequences. For example, touching a moving object in VR may cause it to stop or change direction; or perhaps at a more social level, waving hello at someone in VR should cause a response (e.g., they smile or wave back) (Slater 2009).

There is a balance to be struck between making virtual scenarios engaging enough to provide significant distraction, yet realistic enough to be salient and relevant to the patient. Blasting lasers at space monsters on an alien planet might well be hugely entertaining and provide a significant distraction, and therefore analgesia, *at the time*, but once the patient removes the headset and returns to the real world, how much carryover is there from the session? How relevant is such a scenario to normal activities of daily living for the patient? Perhaps one needs to focus on the reasons for trying to induce VR analgesia. For acute pain or for painful procedures, virtual scenarios with highly distractive fantastical elements may indeed be useful (see Hoffman et al. 2000a, b, c; Hoffman et al. 2011; Hoffman et al. 2014). For chronic pain, however, VR interventions that focus on real-world challenges, or those that are particularly relevant for the individual patient, and combined with elements of graded exposure, may be a better option.

For VR interventions that utilize embodiment of a virtual body seen from a first-person perspective, there are a number of factors that make for a stronger illusion. First and foremost, the multisensory stimulation used to induce the illusion must be spatiotemporally congruent or the illusion is easily broken. Visuo-motor congruent stimulation has the additional advantage of producing a sense of agency over the virtual body's movements, which is an important component of embodiment. Finally, whereas spatiotemporal congruency is crucial, avatars do not need to closely match the subject's real body. A "lookalike" avatar may be more easily embodied, and, thanks to customizable photorealistic facial mapping, "lookalike" avatars are easily and readily available; however, it is not essential for embodiment. Indeed, Lugrin and colleagues showed that in fact cartoon-like and machine-like avatars were more readily embodied than more realistic human avatars, perhaps a demonstration of an uncanny valley effect (Lugrin et al. 2015). As discussed previously, embodying non-realistic or fantastic avatars may confer therapeutic benefit (Lugrin et al. 2015; Hoffman et al. 2014, 2011).

Considerations when Using Immersive VR in Patients with Pain

While VR is considered to be low risk (see (Corbetta et al. 2015) for a review on safety), patients in pain often have decreased mobility, tire easily, and have poor concentration spans and therefore need careful consideration when designing, testing, and implementing therapeutic VR treatment. Patients should be positioned comfortably before starting, usually seated with neck or back support, if necessary, but encouraged to move and change position should they feel the need to. For scenarios involving movement, the available range of movement should be considered and either kept within a comfortable range, or patients can be challenged to move beyond a comfortable range, depending on the goals of therapy.

As a general rule of thumb, VR exposure time should be kept to 10–15 min initially, and not extend beyond 20 min with repeated sessions once patients become accustomed to VR (Donegan et al. 2020). They should be encouraged beforehand to immediately discontinue if any symptoms of nausea or dizziness are experienced, since once these symptoms are felt, it is very difficult to get them to settle whilst still in VR. Such symptoms are more likely to occur if there are mismatches between observed visual movement in VR and real-life movement (LaViola 2000). Other ways to reduce nausea include the use of airflow (e.g., with a fan) (D'Amour et al. 2017); slower movements (Kemeny et al. 2017); temporarily narrowing the field of view during head movement (Fernandes and Feiner 2016); and ensuring the interpupillary distance (IPD) is matched to that of the patient (Fulvio et al. 2019).

The virtual task should be set at an appropriate level for the patient. More elderly patients are less likely to be familiar with VR and may be unsure what is required of them. A comprehensive discussion beforehand may be helpful, in which any doubts and questions can be answered. Often, interaction with a virtual therapist in VR, who responds in a realistic way, can be reassuring and can provide motivation for the patient (Matamala-Gomez et al. 2021a; Rehm et al. 2016).

4 Current Trends and Future Directions of IVR in the Field of Pain

Even in its early stages of development, the promise of immersive VR as a tool for mitigating pain through distraction has been clear. Since then, IVR has also become an important tool for understanding and treating pain. What are current trends and potential future developments of the discipline? Which topics are still underexplored and what may become possible only with further technological advancements?

4.1 Recent Developments in IVR and Biosignal Research

Combining biosignal recording, such as electroencephalography or near-infrared spectroscopy, and immersive VR has been successfully implemented in many research areas, including chronic pain research (e.g., Gentile et al. 2020). A novel approach, not fully explored in rehabilitation, is combining an immersive VR experience and non-invasive brain stimulation (NIBS). The two most common NIBS techniques are transcranial direct current stimulation (tDCS) and transcranial magnetic stimulation (TMS). Cassani et al. (2020) have recently reviewed the literature in which NIBS and immersive VR are used for rehabilitation and found two (out of sixteen) studies where NIBS-immersive VR was used for the treatment of neuropathic pain as a result of spinal cord injury, which remains poorly understood and very difficult to treat. It is important to note that both cited studies used an early, non-immersive VR system following the one described by Moseley (2007) and not what we would call immersive VR today. The illusion was based on a video which showed the legs of a gender-matched person walking on a treadmill. To induce the experience of realistic gait perception, a vertical mirror was placed on top of the projected video, so that the own patient's upper part of the body and the walking legs displayed on the screen were aligned in the most realistic position possible, and a synchronized sound of walking shoes was played to increase the realism of the experience. The first study included a 12-week-long intervention and had a 2 (real stimulation/sham stimulation) × 2 (visual illusion/control illusion) design (Soler et al. 2010). The real tDCS with visual illusion condition yielded the largest reduction of all pain subtypes. Results suggest that virtual walking may be a viable treatment for pain after spinal cord injury.

The second study also used a visual illusion of walking to improve neuropathic pain in severe spinal cord injury (Kumru et al. 2013). The study included 5 min of tDCS alone and another 15 min where the stimulation was accompanied by the visual illusion. A 2-week intervention of daily tDCS and visual illusion resulted in an average of 50% pain reduction, which was significantly higher than in the two control conditions.

Despite these promising results, this line of research had received little attention until recent years, which have witnessed the return of NIBS+immersive VR in a new, more advanced form. One example is a study protocol, for two randomized, controlled trials for treating pain in Parkinson's disease patients (González-Zamorano et al. 2021). The studies include a combination of tDCS over the primary motor cortex, action observation, and motor imagery training based on a brain–computer interface using immersive VR. Although the results are not yet known, we may see that the conceptual and technological jump of the last decade opens completely new possibilities for pain research.

Another, almost non-existent approach in this field is a combination of immersive VR and eye tracking (ET), widely used in non-clinical research, such as marketing (Meißner et al. 2019). Although the role of ET in pain modulation has been explored in prior studies (for a review, see Chan et al. 2020), surprisingly only now are we starting to see published research that includes immersive VR + ET paradigms. There are numerous applications of immersive VR + ET. For instance, in chronic pain, it can be used as a measure of attention paid to a specific body part or events in the virtual environment. Moreover, in acute pain, it can be a tool for actively drawing attention to the virtual world to maximize the distraction from a painful procedure. For instance, Al-Ghamdi et al. (2020) used it to test the analgesic effect of interactive vs. non-interactive VR after receiving a thermal pain stimulus. The active condition included a "hands-free" ET-based manipulation of the virtual objects. In the control condition, patients could only passively watch the virtual objects. The study yielded a clear advantage of using the immersive VR + ET technology, as patients reported more fun, stronger presence, and reduced worst pain in this condition, compared to the passive one. As the HMDs with built-in ET become more accessible, we would expect to see a growing number of studies taking advantage of this technology.

4.2 VR for Pain Psychotherapy

Social immersive VR, where users can interact with each other in real time through avatars, is a big part of the VR industry. It is highly attractive because of the full anonymity offered by avatars, combined with an embodied, first-person perspective. It is also important that one can participate in such an experience without leaving home, but still feel "present" with others in cyberspace. This feature is particularly valuable for people with limited mobility or who are isolated, as has been the case in the COVID-19 pandemic. An intriguing study has recently demonstrated the usefulness of social immersive VR for acute pain. Won et al. (2020) showed that interacting in real time in immersive VR with another person can be successfully used for pain distraction, as it boosted thermal pain thresholds in participants. It corroborates the results from the above-mentioned ET study, suggesting that the immersive VR, where patients can interact with objects or others, may be more efficient than passive immersive VR, where patients are alone and not interacting actively.

In sum, immersive VR brings new opportunities for many kinds of group therapy (Cîmpean 2019), and therapy for chronic pain patients is not an exception. So far, to the best knowledge of the authors, no studies have tested such group therapy for pain. Nonetheless, one may predict that such studies will emerge, since various therapeutic groups already exist and are accessible freely through various social VR applications (for a review, see Best et al. 2022). The comparison between an in-person and teletherapy for chronic pain patients based on video-calls suggests a similarly positive effect in both groups (Mariano et al. 2021). It is too early to predict the future popularity and efficiency of an individual or group therapy for pain patients; nonetheless, it seems likely that this line of research will be developed, following spontaneous activity of immersive VR users and an overall growing acceptability among the stakeholders (Dilgul et al. 2021).

4.3 Immersive VR for Cancer Pain, Palliative, and Intensive Care

Immersive VR has also attracted the interest of researchers and practitioners seeking alternatives to opiates and other types of harsh or addictive medications for alleviating cancer-related pain or pain encountered in palliative and intensive care. Some of the first attempts did not yield entirely optimistic results: Laghlam et al. (2021) tested the efficiency of immersive VR for pain reduction during and after drain removal in patients in the intensive care unit. Compared to an inhaled equimolar mixture of N_2O and O_2 (Kalinox®), the immersive VR group was in significantly higher pain immediately after the procedure and equal pain 10 min later, suggesting that the distraction was not effective. However, the virtual experience consisted of a 360° video – thus, it was not interactive and not exactly immersive VR.

Another study investigated 360° videos of nature scenes for cancer patients during their intravenous (IV)/port access. Compared to the first visit, which served as a control, patients reported increased relaxation, feelings of peace, and positive distractions, although no change in pain or stress during the second visit (Scates et al. 2020). Based on these outcomes, it appears that to observe a significant analgesic effect, a more engaging experience may be necessary, especially for more invasive medical procedures. Hurd (2021) analyzed archival data to determine whether using audio-visual immersive VR and morphine is more efficient in inducing analgesia than morphine alone in hospice patients experiencing chronic cancer pain. The author did not find any analgesic effect of the immersive VR experience; however, due to the study limitations (no detailed information on the technical aspects of the immersive VR exactly, potential diversity of experiences, no information of the duration and frequency of the experience, etc.), it is hard to judge what was the reason for this lack of effect. Therefore, further, well-controlled experimental studies are necessary to understand whether and perhaps, how, immersive VR can be implemented in hospice care for palliative patients.

4.4 VR for Pain Diagnosis and Simulation

Among paths still to be explored thoroughly are pain diagnosis in immersive VR. Examples from other disciplines, such as neurology and cognitive psychology, suggest that immersive VR can be a useful tool for diagnosis of a range of medical conditions, including pain diagnosis (Llobera et al. 2012). In a study from Llobera et al. (2013), the authors presented a method that exploits virtual body ownership in combination with a simple brain–computer interface (BCI) based on EEG, while observing virtual movements of the painful limb in a virtual mirror and recording muscle activity (electromyography, EMG) in the corresponding real limb to complement a neurological assessment. The authors found that body ownership induced changes in both the recorded physiological measures and BCI task performance in one case study patient compared to five healthy controls – which suggests that induction of virtual body ownership combined with simple electrophysiological measures could be useful for the diagnosis of patients with neurological conditions.

Another unique possibility offered by immersive VR is a simulation of painful conditions in healthy participants, such as inducing phantom limb pain through avatars with body parts. This line of research may have several applications, such as a better understanding of the underlying mechanisms of pain disorders (Kocur et al. 2020) or an increase in empathy in healthcare employees working with pain patients (Brydon et al. 2021).

5 Conclusions

The present chapter examines the use of VR and more specifically the use of virtual embodiment for pain relief. The current literature in the field highlights the importance of immersiveness and evoking a sense of presence in order to create an effective virtual environment for pain relief. One way of increasing such aspects is by inducing a sense of embodiment within the virtual world via an avatar. Moreover, current studies in the field demonstrate that there is a link between body perception and clinical disorders such as pain, creating a viable therapeutic avenue for virtual embodiment using immersive VR systems. Nevertheless, there is a gap in the scientific literature on the use of immersive VR avatars to treat acute pain (Trost et al. 2021). Hence, robust and suitably powered randomized control trials are needed to further explore the full potential of embodiment technologies such as VR to modulate pain perception. Further investigations aimed at modulating pain perception through an embodied virtual body with larger sample sizes will allow a better understanding of the link between body representation and pain perception. Along these lines, future studies on this topic may make use of brain imaging techniques, which will allow better identification of the neural structures underlying the complex link between modification of body perception and pain.

To date, nearly all studies involving virtual embodiment have targeted treatments designed for patients with chronic pain (Llobera et al. 2013; Matamala-Gomez et al. 2020b, 2018; Pozeg et al. 2017; Solcà et al. 2018). However, while the use of VR generally for acute pain relief is well established, chronic pain management using embodiment in immersive VR in clinical populations requires further study. Rapid development of the VR hardware and software offers continuously new technological solutions which open novel research and treatment methods. This includes between others, combining NIBS and VR for pain mechanisms analysis and diagnosis, group teletherapy for chronic pain, and pain simulation in healthy participants.

References

Al-Ghamdi NA, Meyer WJ, Atzori B, Alhalabi W, Seibel CC, Ullman D, Hoffman HG (2020) Virtual reality analgesia with interactive eye tracking during brief thermal pain stimuli: a randomized controlled trial (crossover design). Front Hum Neurosci 13:467. https://doi.org/10.3389/fnhum.2019.00467

Apps MA, Tsakiris M (2014) The free-energy self: a predictive coding account of self-recognition. Neurosci Biobehav Rev 41:85–97

Arzy S, Overney LS, Landis T, Blanke O (2006a) Neural mechanisms of embodiment: asomatognosia due to premotor cortex damage. Arch Neurol 63(7):1022–1025. https://doi.org/10.1001/archneur.63.7.1022

Arzy S, Thut G, Mohr C, Michel CM, Blanke O (2006b) Neural basis of embodiment: distinct contributions of temporoparietal junction and extrastriate body area. J Neurosci 26(31): 8074–8081. http://www.jneurosci.org/content/26/31/8074.short

Austin PD, Siddall PJ (2021) Virtual reality for the treatment of neuropathic pain in people with spinal cord injuries: a scoping review. J Spinal Cord Med 44(1):8–18. https://doi.org/10.1080/10790268.2019.1575554

Banakou D, Beacco A, Neyret S, Blasco-Oliver M, Seinfeld S, Slater M (2020) Virtual body ownership and its consequences for implicit racial bias are dependent on social context. R Soc Open Sci 7(12):201848

Banakou D, Groten R, Slater M (2013) Illusory ownership of a virtual child body causes overestimation of object sizes and implicit attitude changes. Proc Natl Acad Sci 110 (31):12846–12851

Banakou D, Hanumanthu PD, Slater M (2016) Virtual embodiment of white people in a black virtual body leads to a sustained reduction in their implicit racial bias. Front Hum Neurosci 601

Bani Mohammad E, Ahmad M (2019) Virtual reality as a distraction technique for pain and anxiety among patients with breast cancer: a randomized control trial. Palliat Support Care 17(1):29–34. https://doi.org/10.1017/S1478951518000639

Barrett LF (2017) The theory of constructed emotion: an active inference account of interoception and categorization. Soc Cogn Affect Neurosci 12(1):1–23. https://doi.org/10.1093/scan/nsw154

Barsalou LW (2010) Grounded cognition: past, present, and future. Top Cogn Sci 2(4):716–724

Berlucchi G, Aglioti S (1997) The body in the brain: neural bases of corporeal awareness. Trends Neurosci 20(12):560–564

Best P, Meireles M, Schroeder F, Montgomery L, Maddock A, Davidson G, Galway K, Trainor D, Campbell A, Van Daele T (2022) Freely available virtual reality experiences as tools to support mental health therapy: a systematic scoping review and consensus based interdisciplinary analysis. J Technol Behav Sci 7(1):100–114. https://doi.org/10.1007/s41347-021-00214-6

Birckhead B, Eberlein S, Alvarez G, Gale R, Dupuy T, Makaroff K, Fuller G, Liu X, Yu KS, Black JT, Ishimori M, Venuturupalli S, Tu J, Norris T, Tighiouart M, Ross L, McKelvey K, Vrahas M,

Danovitch I, Spiegel B (2021) Home-based virtual reality for chronic pain: protocol for an NIH-supported randomised-controlled trial. BMJ Open 11(6):e050545. https://doi.org/10.1136/bmjopen-2021-050545

Blanke O, Metzinger T (2009) Full-body illusions and minimal phenomenal selfhood. Trends Cogn Sci 13(1):7–13. https://doi.org/10.1016/j.tics.2008.10.003

Blount RL, Piira T, Cohen LL (2003) Management of pediatric pain and distress due to medical procedures. In: Roberts MC (ed) Handbook of pediatric psychology, 3rd edn. The Guilford Press

Bondin L, Dingli A (2021) Automating pain reduction using biosensors and realtime adaptive VR. In: Musiolik T, Cheok A (eds) Analyzing future applications of AI, sensors, and robotics in society. IGI Global, pp 132–140. https://doi.org/10.4018/978-1-7998-3499-1.ch008

Botella C, Palacios AG, Baños R, Quero S, Breton-Lopez J (2008) Virtual reality in the treatment of pain. J Cyberther Rehabil 1(1):93–100. https://search.proquest.com/docview/622073106?accountid=16562

Brydon M, Kimber J, Sponagle M, MacLaine J, Avery J, Pyke L, Gilbert R (2021) Virtual reality as a tool for eliciting empathetic behaviour in carers: an integrative review. J Med Imaging Radiat Sci 52(3):466–477. https://doi.org/10.1016/j.jmir.2021.04.005

Buckelew SP, Degood DE, Schwartz DP, Kerler RM (1986) Cognitive and somatic item response pattern of pain patients, psychiatric patients, and hospital employees. J Clin Psychol 42(6): 852–860. https://doi.org/10.1002/1097-4679(198611)42:6<852::AID-JCLP2270420603>3.0.CO;2-R

Carter GT, Duong V, Ho S, Ngo KC, Greer CL, Weeks DL (2014) Side effects of commonly prescribed analgesic medications. Phys Med Rehabil Clin N Am 25(2):457–470. https://doi.org/10.1016/j.pmr.2014.01.007

Cassani R, Novak GS, Falk TH, Oliveira AA (2020) Virtual reality and non-invasive brain stimulation for rehabilitation applications: a systematic review. J Neuroeng Rehabil 17(1): 147. https://doi.org/10.1186/s12984-020-00780-5

Chan F, Suen H, Jackson T, Vlaeyen J, Barry T (2020) Pain-related attentional processes: a systematic review of eye-tracking research. Clin Psychol Rev 80:101884. https://doi.org/10.1016/j.cpr.2020.101884

Chirico A, Lucidi F, De Laurentiis M, Milanese C, Napoli A, Giordano A (2016) Virtual reality in health system: beyond entertainment. A mini-review on the efficacy of VR during cancer treatment. J Cell Physiol 231(2):275–287

Cîmpean AI (2019) A pilot study to compare cognitive behavioral therapy with virtual reality vs. standard cognitive behavioral therapy for patients who suffer from cervical cancer. J Evid Based Psychothér 19(1):115–127. https://doi.org/10.24193/jebp.2019.1.7

Corbetta D, Imeri F, Gatti R (2015) Rehabilitation that incorporates virtual reality is more effective than standard rehabilitation for improving walking speed, balance and mobility after stroke: a systematic review. J Physiother 61(3):117–124. https://doi.org/10.1016/j.jphys.2015.05.017

Crombez G, Eccleston C, Baeyens F, Eelen P (1996) Disruptive nature of pain: an experimental investigation. Behav Res Ther 34(11–12):911–918. https://doi.org/10.1016/S0005-7967(96)00058-7

D'Amour S, Bos JE, Keshavarz B (2017) The efficacy of airflow and seat vibration on reducing visually induced motion sickness. Exp Brain Res 235(9):2811–2820. https://doi.org/10.1007/s00221-017-5009-1

Dahlquist L (1999a) Commentary on "Treatments that work in pediatric psychology: Procedure-related pain." J Pediatr Psychol 24(2):153–154. https://psycnet.apa.org/record/1999-13153-010

Dahlquist L (1999b) Pediatric pain management. https://books.google.it/books?hl=it&lr=&id=eIvA9aFUCqUC&oi=fnd&pg=PR11&dq=Pediatric+pain+management,+1999,+Dahlquist&ots=PhK1cb5p81&sig=1PmLxcs8vmLMZp-euAOaZbXy66g

De Oliveira EC, Bertrand P, Lesur MER, Palomo P, Demarzo M, Cebolla A, Banos R, Tori R (2016) Virtual body swap: a new feasible tool to be explored in health and education. In:

Proceedings - 18th symposium on virtual and augmented reality, SVR 2016, pp 81–89. https://doi.org/10.1109/SVR.2016.23
de Preester H, Tsakiris M (2009) Body-extension versus body-incorporation: is there a need for a body-model? Phenomenol Cogn Sci 8(3):307–319. https://doi.org/10.1007/s11097-009-9121-y
De Vignemont F (2011) Embodiment, ownership and disownership. Conscious Cogn 20(1):82–93. https://doi.org/10.1016/j.concog.2010.09.004
Derbyshire SWG, Vogt BA, Jones AKP (1998) Pain and Stroop interference tasks activate separate processing modules in anterior cingulate cortex. Exp Brain Res 118(1):52–60. https://doi.org/10.1007/s002210050254
Di Pietro F, Stanton TR, Moseley GL, Lotze M, Mcauley JH (2015) Interhemispheric somatosensory differences in chronic pain reflect abnormality of the healthy side. Hum Brain Mapp 36(2):508–518. https://doi.org/10.1002/hbm.22643
Diers M, Zieglgänsberger W, Trojan J, Drevensek AM, Erhardt-Raum G, Flor H (2013) Site-specific visual feedback reduces pain perception. Pain 154(6):890–896. https://doi.org/10.1016/j.pain.2013.02.022
Dilgul M, Hickling LM, Antonie D, Priebe S, Bird VJ (2021) Virtual reality group therapy for the treatment of depression: a qualitative study on stakeholder perspectives. Front Virtual Real. https://doi.org/10.3389/frvir.2020.609545
Donegan T, Ryan BE, Swidrak J, Sanchez-Vives MV (2020) Immersive virtual reality for clinical pain: considerations for effective therapy. Front Virtual Real 1(9):9. https://doi.org/10.3389/frvir.2020.00009
Eccleston C, Crombez G (1999) Pain demands attention: a cognitive-affective model of the interruptive function of pain. Psychol Bull 125(3):356–366. https://doi.org/10.1037/0033-2909.125.3.356
Fernandes AS, Feiner SK (2016) Combating VR sickness through subtle dynamic field-of-view modification. IEEE symposium on 3D user interfaces
Flor H (2003) Cortical reorganisation and chronic pain: implications for rehabilitation. J Rehabil Med Suppl 41. https://doi.org/10.1080/16501960310010179
Flor H, Elbert T, Knecht S, Wienbruch C, Pantev C, Birbaumers N, Larbig W, Taub E (1995) Phantom-limb pain as a perceptual correlate of cortical reorganization following arm amputation. Nature 375(6531):482–484. https://doi.org/10.1038/375482a0
Flor H, Braun C, Elbert T, Birbaumer N (1997) Extensive reorganization of primary somatosensory cortex in chronic back pain patients. Neurosci Lett 224(1):5–8. https://doi.org/10.1016/S0304-3940(97)13441-3
Flor H, Nikolajsen L, Jensen TS (2006) Phantom limb pain: a case of maladaptive CNS plasticity? Nat Rev Neurosci 7(11):873–881. https://doi.org/10.1038/nrn1991
Fulvio JM, Ji M, Rokers B (2019) Variability in sensory sensitivity predicts motion sickness in virtual reality. BioRxiv 488817. https://doi.org/10.1101/488817
Gallagher S (2001) The practice of mind. Theory, simulation or primary interaction? J Conscious Stud 8(5–6):83–108
Gentile E, Brunetti A, Ricci K, Delussi M, Bevilacqua V, de Tommaso M (2020) Mutual interaction between motor cortex activation and pain in fibromyalgia: EEG-fNIRS study. PLoS One 15(1): e0228158. https://doi.org/10.1371/journal.pone.0228158
Giummarra MJ, Gibson SJ, Georgiou-Karistianis N, Bradshaw JL (2007) Central mechanisms in phantom limb perception: the past, present and future. Brain Res Rev 54(1):219–232. https://doi.org/10.1016/j.brainresrev.2007.01.009
González-Zamorano Y, Fernández-Carnero J, Sánchez-Cuesta FJ, Arroyo-Ferrer A, Vourvopoulos A, Figueiredo P, Serrano JI, Romero JP (2021) New approaches based on non-invasive brain stimulation and mental representation techniques targeting pain in Parkinson's disease patients: two study protocols for two randomized controlled trials. Brain Sci 11(1):65. https://doi.org/10.3390/brainsci11010065
Gorman JM (2006) Virtual reality: a real treatment option. CNS Spectr 11(1):12–13. https://doi.org/10.1017/S1092852900024081

Graziano MS, Botvinick MM (2002) How the brain represents the body: insights from neurophysiology and psychology. In: Common mechanisms in perception and action: attention and performance XIX, vol 19, pp 136–157

Gutierrez-Maldonado J, Gutierrez-Martinez O, Cabas-Hoyos K (2011) Interactive and passive virtual reality distraction: effects on presence and pain intensity. Annu Rev Cyberther Telemed 9(1)

Gutiérrez-Martínez O, Gutiérrez-Maldonado J, Loreto-Quijada D (2011) Control over the virtual environment influences the presence and efficacy of a virtual reality intervention on pain. Annu Rev Cyberther Telemed 9(1)

Hegedüs G, Darnai G, Szolcsányi T, Feldmann Á, Janszky J, Kállai J (2014) The rubber hand illusion increases heat pain threshold. Eur J Pain 18(8):1173–1181. https://doi.org/10.1002/j.1532-2149.2014.00466.x

Hoffman HG (2021) Interacting with virtual objects via embodied avatar hands reduces pain intensity and diverts attention. Sci Rep 11(1):10672. https://doi.org/10.1038/s41598-021-89526-4

Hoffman HG, Doctor JN et al (2000a) Use of virtual reality for adjunctive treatment of adolescent burn pain during wound care. Pain 85:305–309

Hoffman HG, Doctor JN, Patterson DR, Carrougher GJ, Furness TA (2000b) Virtual reality as an adjunctive pain control during burn wound care in adolescent patients. Pain 85(1-2):305–309. https://doi.org/10.1016/S0304-3959(99)00275-4

Hoffman HG, Patterson DR, Carrougher GJ (2000c) Use of virtual reality for adjunctive treatment of adult burn pain during physical therapy: a controlled study. Clin J Pain 16(3):244–250. https://doi.org/10.1097/00002508-200009000-00010

Hoffman HG, Garcia-Palacios A, Patterson DR, Jensen M, Furness T, Ammons JF (2001) The effectiveness of virtual reality for dental pain control: a case study. CyberPsychol Behav 4(4): 527–535. https://doi.org/10.1089/109493101750527088

Hoffman HG, Richards TL, Coda B, Bills AR, Blough D, Richards AL, Sharar SR (2004a) Modulation of thermal pain-related brain activity with virtual reality: evidence from fMRI. Neuroreport 15(8):1245–1248. https://doi.org/10.1097/01.wnr.0000127826.73576.91

Hoffman HG, Sharar SR, Coda B, Everett JJ, Ciol M, Richards T, Patterson DR (2004b) Manipulating presence influences the magnitude of virtual reality analgesia. Pain 111(1–2):162–168. https://doi.org/10.1016/j.pain.2004.06.013

Hoffman HG, Richards TL, Van Oostrom T, Coda BA, Jensen MP, Blough DK, Sharar SR (2007) The analgesic effects of opioids and immersive virtual reality distraction: evidence from subjective and functional brain imaging assessments. Anesth Analg 105(6):1776–1783. https://doi.org/10.1213/01.ane.0000270205.45146.db

Hoffman HG, Chambers GT, Meyer WJ, Arceneaux LL, Russell WJ, Seibel EJ, Richards TL, Sharar SR, Patterson DR (2011) Virtual reality as an adjunctive non-pharmacologic analgesic for acute burn pain during medical procedures. Ann Behav Med 41(2):183–191. https://doi.org/10.1007/s12160-010-9248-7

Hoffman HG, Meyer WJ, Ramirez M, Roberts L, Seibel EJ, Atzori B, Sharar SR, Patterson DR (2014) Feasibility of articulated arm mounted oculus rift virtual reality goggles for adjunctive pain control during occupational therapy in pediatric burn patients. Cyberpsychol Behav Soc Netw 17(6):397–401. https://doi.org/10.1089/cyber.2014.0058

Holmes NP, Spence C (2004) The body schema and multisensory representation(s) of peripersonal space. Cogn Process 5(2):94–105. https://doi.org/10.1007/s10339-004-0013-3

Hurd A (2021) Integration of morphine and virtual reality: pain management for adult hospice cancer patients. Walden University

Iannetti GD, Mouraux A (2010) From the neuromatrix to the pain matrix (and back). Exp Brain Res 205(1):1–12. https://doi.org/10.1007/s00221-010-2340-1

Indovina P, Barone D, Gallo L, Chirico A, De Pietro G, Giordano A (2018) Virtual reality as a distraction intervention to relieve pain and distress during medical procedures: a comprehensive literature review. Clin J Pain 34(9):858–877

Kemeny A, George P, Merienne F, Colombet F, Mérienne F (2017) New VR navigation techniques to reduce cybersickness. Eng Real Virtual Real. https://doi.org/10.2352/ISSN.2470-1173.2017.3.ERVR-097ï

Kilteni K, Groten R, Slater M (2012) The sense of embodiment in virtual reality. Presence Teleop Virt 21(4):373–387. https://doi.org/10.1162/PRES_a_00124

Kocur M, Graf S, Schwind V (2020) The impact of missing fingers in virtual reality. In: 26th ACM symposium on virtual reality software and technology, pp 1–5. https://doi.org/10.1145/3385956.3418973

Kokkinara E, Slater M, López-Moliner J (2015) The effects of visuomotor calibration to the perceived space and body, through embodiment in immersive virtual reality. ACM Trans Appl Percept (TAP) 13(1):1–22

Krupić D, Žuro B, Corr PJ (2021) Anxiety and threat magnification in subjective and physiological responses of fear of heights induced by virtual reality. Personal Individ Differ 169:109720. https://doi.org/10.1016/j.paid.2019.109720

Kumru H, Soler D, Vidal J, Navarro X, Tormos JM, Pascual-Leone A, Valls-Sole J (2013) The effects of transcranial direct current stimulation with visual illusion in neuropathic pain due to spinal cord injury: an evoked potentials and quantitative thermal testing study. Eur J Pain 17(1):55–66. https://doi.org/10.1002/j.1532-2149.2012.00167.x

Laghlam D, Naudin C, Coroyer L, Aidan V, Malvy J, Rahoual G, Estagnasié P, Squara P (2021) Virtual reality vs. Kalinox® for management of pain in intensive care unit after cardiac surgery: a randomized study. Ann Intensive Care 11(1):74. https://doi.org/10.1186/s13613-021-00866-w

LaViola JJ (2000) A discussion of cybersickness in virtual environments. ACM SIGCHI Bull 32(1):47–56. https://doi.org/10.1145/333329.333344

Lenggenhager B, Tadi T, Metzinger T, Blanke O (2007) Video ergo sum: manipulating bodily self-consciousness. Science 317(5841):1096–1099. https://doi.org/10.1126/science.1143439

Lewis JS, Kersten P, McCabe CS, McPherson KM, Blake DR (2007) Body perception disturbance: a contribution to pain in complex regional pain syndrome (CRPS). Pain 133(1–3):111–119. https://doi.org/10.1016/j.pain.2007.03.013

Llobera J, Gonzalez-Franco M, Perez-Marcos D, Valls-Solé J, Slater M, Sanchez-Vives M (2012) Virtual reality for assessment of patients suffering chronic pain: a case study. Exp Brain Res 225(1):105–117. https://doi.org/10.1007/s00221-012-3352-9

Llobera J, González-Franco M, Perez-Marcos D, Valls-Solé J, Slater M, Sanchez-Vives MV (2013) Virtual reality for assessment of patients suffering chronic pain: a case study. Exp Brain Res 225(1):105–117. https://doi.org/10.1007/s00221-012-3352-9

Longo MR, Betti V, Aglioti SM, Haggard P (2009) Visually induced analgesia: seeing the body reduces pain. J Neurosci 29(39):12125–12130. https://doi.org/10.1523/JNEUROSCI.3072-09.2009

Longo MR, Iannetti GD, Mancini F, Driver J, Haggard P (2012) Linking pain and the body: neural correlates of visually induced analgesia. J Neurosci 32(8):2601–2607. https://doi.org/10.1523/JNEUROSCI.4031-11.2012

Lopez C, Halje P, Blanke O (2008) Body ownership and embodiment: vestibular and multisensory mechanisms. Neurophysiol Clin 38(3):149–161. https://doi.org/10.1016/j.neucli.2007.12.006

Lugrin JL, Latt J, Latoschik ME (2015) Avatar anthropomorphism and illusion of body ownership in VR. In: IEEE virtual reality conference, VR 2015 - proceedings, pp 229–230. https://doi.org/10.1109/VR.2015.7223379

Macaluso E, Maravita A (2010) The representation of space near the body through touch and vision. Neuropsychologia 48(3):782–795. https://doi.org/10.1016/j.neuropsychologia.2009.10.010

Maihöfner C, Handwerker HO, Neundörfer B, Birklein F (2003) Patterns of cortical reorganization in complex regional pain syndrome. Neurology 61(12):1707–1715. https://doi.org/10.1212/01.WNL.0000098939.02752.8E

Malloy KM, Milling LS (2010) The effectiveness of virtual reality distraction for pain reduction: a systematic review. Clin Psychol Rev 30(8):1011–1018. https://doi.org/10.1016/j.cpr.2010.07.001

Mancini F, Longo MR, Kammers MPM, Haggard P (2011) Visual distortion of body size modulates pain perception. Psychol Sci 22(3):325–330. https://doi.org/10.1177/0956797611398496
Mancini F, Longo MR, Canzoneri E, Vallar G, Haggard P (2013) Changes in cortical oscillations linked to multisensory modulation of nociception. Eur J Neurosci 37(5):768–776. https://doi.org/10.1111/ejn.12080
Mancini F, Wang AP, Schira MM, Isherwood ZJ, McAuley JH, Iannetti GD, Sereno MI, Moseley GL, Rae CD (2019) Fine-grained mapping of cortical somatotopies in chronic complex regional pain syndrome. J Neurosci 39(46):9185–9196. https://doi.org/10.1523/JNEUROSCI.2005-18.2019
Mariano TY, Wan L, Edwards RR, Lazaridou A, Ross EL, Jamison RN (2021) Online group pain management for chronic pain: preliminary results of a novel treatment approach to teletherapy. J Telemed Telecare 27(4):209–216. https://doi.org/10.1177/1357633X19870369
Martini M (2016) Real, rubber or virtual: the vision of "one's own" body as a means for pain modulation. A narrative review. Conscious Cogn 43:143–151. https://doi.org/10.1016/j.concog.2016.06.005
Maselli A, Slater M (2013) The building blocks of the full body ownership illusion. Front Hum Neurosci 7:83. https://doi.org/10.3389/fnhum.2013.00083
Matamala-Gomez M, Diaz Gonzalez AM, Slater M, Sanchez-Vives MV (2019a) Decreasing pain ratings in chronic arm pain through changing a virtual body: different strategies for different pain types. J Pain 20(6):685–697. https://doi.org/10.1016/j.jpain.2018.12.001
Matamala-Gomez M, Donegan T, Bottiroli S, Sandrini G, Sanchez-Vives MV, Tassorelli C (2019b) Immersive virtual reality and virtual embodiment for pain relief. Front Hum Neurosci 13:279. https://doi.org/10.3389/fnhum.2019.00279
Matamala-Gomez M, Maselli A, Malighetti C, Realdon O, Mantovani F, Riva G (2020a) Body illusions for mental health: a systematic review. https://doi.org/10.31234/osf.io/mk69x
Matamala-Gomez M, Nierula B, Donegan T, Slater M, Sanchez-Vives MV (2020b) Manipulating the perceived shape and color of a virtual limb can modulate pain responses. J Clin Med 9(2): 291. https://doi.org/10.3390/jcm9020291
Matamala-Gomez M, Bottiroli S, Realdon O, Riva G, Galvagni L, Platz T, Sandrini G, De Icco R, Tassorelli C (2021a) Telemedicine and virtual reality at time of COVID-19 pandemic: an overview for future perspectives in neurorehabilitation. Front Neurol 12:646902. https://doi.org/10.3389/fneur.2021.646902
Matamala-Gomez M, Maselli A, Malighetti C, Realdon O, Mantovani F, Riva G (2021b) Virtual body ownership illusions for mental health: a narrative review. J Clin Med 10:139. https://doi.org/10.3390/jcm10010139
May A (2008) Chronic pain may change the structure of the brain. Pain 137(1):7–15. https://doi.org/10.1016/j.pain.2008.02.034
McCaul KD, Malott JM (1984) Distraction and coping with pain. Psychol Bull 95(3):516–533. https://doi.org/10.1037/0033-2909.95.3.516
Medina J, Coslett HB (2010) From maps to form to space: touch and the body schema. Neuropsychologia 48:645–654. https://doi.org/10.1016/j.neuropsychologia.2009.08.017
Meißner M, Pfeiffer J, Pfeiffer T, Oppewal H (2019) Combining virtual reality and mobile eye tracking to provide a naturalistic experimental environment for shopper research. J Bus Res 100: 445–458. https://doi.org/10.1016/j.jbusres.2017.09.028
Melzack R (1996) Gate control theory: on the evolution of pain concepts. Pain Forum 5(2): 128–138. https://doi.org/10.1016/S1082-3174(96)80050-X
Melzack R (1999a) Pain and stress: a new perspective. In: Psychosocial factors in pain
Melzack R (1999b) From the gate to the neuromatrix. Pain 82(Suppl 1):S121–S126. https://doi.org/10.1016/S0304-3959(99)00145-1
Melzack R (2001) Pain and the neuromatrix in the brain. J Dent Educ 65(12):1378–1382. https://doi.org/10.1002/j.0022-0337.2001.65.12.tb03497.x
Melzack R, Wall PD (1965) Pain mechanisms: a new theory. Science 150(3699):971–979. https://doi.org/10.1126/science.150.3699.971

Merskey H (2002) Clarifying definition of neuropathic pain. Pain 96(3):408–409. https://journals.lww.com/pain/fulltext/2002/04000/Clarifying_definition_of_neuropathic_pain.27.aspx?casa_token=r02BmOE73rMAAAAA:69fDIRxE4XBSlIb5eAGNDlY7Ur4pFa6i_ZUbvNg229QTDclDl5aAihozgA2iO_ePjsCGUNmepVTosFuCLXNeoGI

Moseley GL (2003) A pain neuromatrix approach to patients with chronic pain. Man Ther 8(3):130–140. https://doi.org/10.1016/S1356-689X(03)00051-1

Moseley GL (2007) Using visual illusion to reduce at-level neuropathic pain in paraplegia. Pain 130(3):294–298. https://doi.org/10.1016/j.pain.2007.01.007

Moseley GL, Parsons TJ, Spence C (2008) Visual distortion of a limb modulates the pain and swelling evoked by movement. Curr Biol 18(22):R1047–R1048. https://doi.org/10.1016/J.CUB.2008.09.031

Moseley GL, Gallace A, Spence C (2012) Bodily illusions in health and disease: physiological and clinical perspectives and the concept of a cortical "body matrix.". Neurosci Biobehav Rev 36(1):34–46. https://doi.org/10.1016/j.neubiorev.2011.03.013

Newport R, Preston C (2010) Pulling the finger off disrupts agency, embodiment and peripersonal space. Perception 39(9):1296–1298. https://doi.org/10.1068/p6742

Newport R, Pearce R, Preston C (2010) Fake hands in action: embodiment and control of supernumerary limbs. Exp Brain Res 204(3):385–395. https://doi.org/10.1007/s00221-009-2104-y

Nierula B, Martini M, Matamala-Gomez M, Slater M, Sanchez-Vives MV (2017) Seeing an embodied virtual hand is analgesic contingent on colocation. J Pain 18(6):645–655

Petkova VI, Ehrsson HH (2008) If I were you: perceptual illusion of body swapping. PLoS One 3(12):e3832

Pleger B, Tegenthoff M, Schwenkreis P, Janssen F, Ragert P, Dinse HR, Völker B, Zenz M, Maier C (2004) Mean sustained pain levels are linked to hemispherical side-to-side differences of primary somatosensory cortex in the complex regional pain syndrome I. Exp Brain Res 155(1):115–119. https://doi.org/10.1007/s00221-003-1738-4

Ploghaus A, Tracey I, Gati JS, Clare S, Menon RS, Matthews PM, Rawlins JNP (1999) Dissociating pain from its anticipation in the human brain. Science 284(5422):1979–1981. https://doi.org/10.1126/science.284.5422.1979

Pozeg P, Palluel E, Ronchi R, Solcà M, Al-Khodairy AW, Jordan X, Kassouha A, Blanke O (2017) Virtual reality improves embodiment and neuropathic pain caused by spinal cord injury. Neurology 89(18):1894–1903. https://doi.org/10.1212/WNL.0000000000004585

Preston C, Newport R (2011) Analgesic effects of multisensory illusions in osteoarthritis. Rheumatology 50(12):2314–2315. https://doi.org/10.1093/rheumatology/ker104

Price D (1999) Psychological mechanisms of pain and analgesia. Soc Esp Dolor 8:283–285. http://revista.sedolor.es/pdf/2001_04_10.pdf

Ramachandran VS, Altschuler EL (2009) The use of visual feedback, in particular mirror visual feedback, in restoring brain function. Brain 132(7):1693–1710. https://doi.org/10.1093/brain/awp135

Ramachandran VS, Hirstein W (1998) The perception of phantom limbs. The D. O. Hebb lecture. Brain 121(9):1603–1630. https://doi.org/10.1093/brain/121.9.1603

Ramachandran VS, Brang D, McGeoch PD (2009) Size reduction using mirror visual feedback (MVF) reduces phantom pain. Neurocase 15(5):357–360. https://doi.org/10.1080/13554790903081767

Rehm IC, Foenander E, Wallace K, Abbott J-AM, Kyrios M, Thomas N (2016) What role can avatars play in e-mental health interventions? exploring new models of client–therapist interaction. Front Psychiatry 7:186. https://doi.org/10.3389/FPSYT.2016.00186

Ribu L, Holmen H, Torbjornsen A, Wahl AK, Grottland A, Smastuen MC, Elind E, Bergmo TS, Breivik E, Arsand E (2013) Low-intensity self-management intervention for persons with type 2 diabetes using a mobile phone-based diabetes diary, with and without health counseling and motivational interviewing: protocol for a randomized controlled trial. JMIR Res Protoc 2(2):e34. https://doi.org/10.2196/resprot.2768

Riva G, Mantovani F, Capideville CS, Preziosa A, Morganti F, Villani D, Gaggioli A, Botella C, Alcañiz M (2007) Affective interactions using virtual reality: the link between presence and emotions. Cyberpsychol Behav. https://doi.org/10.1089/cpb.2006.9993

Riva G, Wiederhold BK, Mantovani F (2019) Neuroscience of virtual reality: from virtual exposure to embodied medicine. Cyberpsychol Behav Soc Netw 22(1):82–96. https://doi.org/10.1089/cyber.2017.29099.gri

Rivier C (1995) Luteinizing-hormone-releasing hormone, gonadotropins, and gonadal steroids in stress. Ann N Y Acad Sci 771(1):187–191. https://doi.org/10.1111/j.1749-6632.1995.tb44680.x

Romano D, Maravita A (2014) The visual size of one's own hand modulates pain anticipation and perception. Neuropsychologia 57(1):93–100. https://doi.org/10.1016/j.neuropsychologia.2014.03.002

Salomons TV, Iannetti GD, Liang M, Wood JN (2016) The "pain matrix" in pain-free individuals. JAMA Neurol 73(6):755–756. https://doi.org/10.1001/JAMANEUROL.2016.0653

Scates D, Dickinson JI, Sullivan K, Cline H, Balaraman R (2020) Using nature-inspired virtual reality as a distraction to reduce stress and pain among cancer patients. Environ Behav 52(8): 895–918. https://doi.org/10.1177/0013916520916259

Schneider SM, Workman ML (1999) Effects of virtual reality on symptom distress in children receiving chemotherapy. CyberPsychol Behav 2(2):125–134. https://doi.org/10.1089/cpb.1999.2.125

Schneider SM, Ellis M, Coombs WT, Shonkwiler EL, Folsom LC (2003) Virtual reality intervention for older women with breast cancer. Cyberpsychol Behav 6(3):301–307. https://doi.org/10.1089/109493103322011605

Scuteri D, Matamala-Gomez M, Bottiroli S, Corasaniti MT, De Icco R, Bagetta G, Tonin P (2020) Pain assessment and treatment in dementia at the time of coronavirus disease covid-19. Front Neurol 11:890. https://doi.org/10.3389/fneur.2020.00890

Serino A, Haggard P (2010) Touch and the body. Neurosci Biobehav Rev 34(2):224–236. https://doi.org/10.1016/j.neubiorev.2009.04.004

Serino A, Noel JP, Mange R, Canzoneri E, Pellencin E, Ruiz JB et al (2018) Peripersonal space: an index of multisensory body–environment interactions in real, virtual, and mixed realities. Front ICT 4:31

Sharar SR, Alamdari A, Hoffer C, Hoffman HG, Jensen MP, Patterson DR (2016) Circumplex model of affect: a measure of pleasure and arousal during virtual reality distraction analgesia. Games Health J 5(3):197–202. https://doi.org/10.1089/g4h.2015.0046

Slater M (2009) Place illusion and plausibility can lead to realistic behaviour in immersive virtual environments. Philos Trans R Soc Lond B Biol Sci 364(1535):3549–3557. https://doi.org/10.1098/rstb.2009.0138

Slater M (2014) Grand challenges in virtual environments. Front Robot AI 1. https://doi.org/10.3389/frobt.2014.00003

Slater M (2018) Immersion and the illusion of presence in virtual reality. Br J Psychol 109(3): 431–433. https://doi.org/10.1111/bjop.12305

Slater M, Sanchez-Vives MV (2016) Enhancing our lives with immersive virtual reality. Front Robot AI 3:74. https://doi.org/10.3389/frobt.2016.00074

Slater M, Usoh M (1993) Representations systems, perceptual position, and presence in immersive virtual environments. Presence Teleop Virt 2(3):221–233. https://doi.org/10.1162/pres.1993.2.3.221

Slater M, Pérez Marcos D, Ehrsson H, Sanchez-Vives MV (2008) Towards a digital body: the virtual arm illusion. Front Hum Neurosci 6

Slater M, Spanlang B, Corominas D (2010) Simulating virtual environments within virtual environments as the basis for a psychophysics of presence. In: ACM SIGGRAPH 2010 papers, SIGGRAPH 2010. https://doi.org/10.1145/1778765.1778829

Solcà M, Ronchi R, Bello-Ruiz J, Schmidlin T, Herbelin B, Luthi F, Konzelmann M, Beaulieu JY, Delaquaize F, Schnider A, Guggisberg AG, Serino A, Blanke O (2018) Heartbeat-enhanced immersive virtual reality to treat complex regional pain syndrome. Neurology 91(5):e1–e11. https://doi.org/10.1212/WNL.0000000000005905

Soler MD, Kumru H, Pelayo R, Vidal J, Tormos JM, Fregni F, Navarro X, Pascual-Leone A (2010) Effectiveness of transcranial direct current stimulation and visual illusion on neuropathic pain in spinal cord injury. Brain 133(9):2565–2577. https://doi.org/10.1093/brain/awq184

Stanton TR, Gilpin HR, Edwards L, Moseley GL, Newport R (2018) Illusory resizing of the painful knee is analgesic in symptomatic knee osteoarthritis. PeerJ 6:e5206. https://doi.org/10.7717/peerj.5206

Stasolla F, Matamala-Gomez M, Bernini S, Caffò AO, Bottiroli S (2021) Virtual reality as a technological-aided solution to support communication in persons with neurodegenerative diseases and acquired brain injury during COVID-19 pandemic. Front Public Health 8: 635426. https://doi.org/10.3389/fpubh.2020.635426

Tajadura-Jiménez A, Grehl S, Tsakiris M (2012) The other in me: interpersonal multisensory stimulation changes the mental representation of the self. PLoS One 7(7):14. https://doi.org/10.1371/journal.pone.0040682

Tanja-Dijkstra K, Pahl S, White MP, Andrade J, Qian C, Bruce M, May J, Moles DR (2014) Improving dental experiences by using virtual reality distraction: a simulation study. PLoS One 9(3):e91276. https://doi.org/10.1371/journal.pone.0091276

Themelis K, Newport R (2018) An investigation of contextual factors in the application of multisensory illusions for analgesia in hand osteoarthritis. Rheumatol Adv Pract 2(2):1–10. https://doi.org/10.1093/rap/rky019

Thériault R, Olson JA, Krol SA, Raz A (2021) Body swapping with a Black person boosts empathy: using virtual reality to embody another. Q J Exp Psychol 74(12):2057–2074

Triberti S, Repetto C, Riva G (2014) Psychological factors influencing the effectiveness of virtual reality–based analgesia: a systematic review. Cyberpsychol Behav Soc Netw 17(6):335–345. https://doi.org/10.1089/cyber.2014.0054

Trost Z, France C, Anam M, Shum C (2021) Virtual reality approaches to pain: toward a state of the science. Pain 162(2):325–331. https://doi.org/10.1097/j.pain.0000000000002060

Tsakiris M, Prabhu G, Haggard P (2006) Having a body versus moving your body: how agency structures body-ownership. Conscious Cogn 15(2):423–432. https://doi.org/10.1016/j.concog.2005.09.004

Tsay A, Allen TJ, Proske U, Giummarra MJ (2015) Sensing the body in chronic pain: a review of psychophysical studies implicating altered body representation. Neurosci Biobehav Rev 52: 221–232. https://doi.org/10.1016/j.neubiorev.2015.03.004

Valeriani M, Betti V, Le Pera D, De Armas L, Miliucci R, Restuccia D, Avenanti A, Aglioti SM (2008) Seeing the pain of others while being in pain: a laser-evoked potentials study. NeuroImage 40(3):1419–1428. https://doi.org/10.1016/j.neuroimage.2007.12.056

Watkins LR, Maier SF (2000) The pain of being sick: implications of immune-to-brain communication for understanding pain. Annu Rev Psychol 51:29–57. https://doi.org/10.1146/annurev.psych.51.1.29

Wender R, Hoffman HG, Hunner HH, Seibel EJ, Patterson DR, Sharar SR (2009) Interactivity influences the magnitude of virtual reality analgesia. J Cyber Ther Rehabil 2(1):27–33

Wesslein AK, Spence C, Frings C (2014) Vision affects tactile target and distractor processing even when space is task-irrelevant. Front Psychol 5:84. https://doi.org/10.3389/fpsyg.2014.00084

Wiederhold BK, Wiederhold MD (2012) Managing pain in military populations with virtual reality. Pain syndromes – from recruitment to returning troops. NATO Sci Peace Secur Ser E Hum Soc Dyn 91:75–93

Wiederhold BK, Gao K, Sulea C, Wiederhold MD (2014a) Virtual reality as a distraction technique in chronic pain patients. Cyberpsychol Behav Soc Netw 17(6):346–352. https://doi.org/10.1089/cyber.2014.0207

Wiederhold MD, Gao K, Wiederhold BK (2014b) Clinical use of virtual reality distraction system to reduce anxiety and pain in dental procedures. Cyberpsychol Behav Soc Netw 17(6):359–365. https://doi.org/10.1089/cyber.2014.0203

Won AS, Pandita S, Kruzan KP (2020) Social interaction and pain threshold in virtual reality. Cyberpsychol Behav Soc Netw 23(12):829–845. https://doi.org/10.1089/cyber.2020.0055

Yang S, Chang MC (2019) Chronic pain: structural and functional changes in brain structures and associated negative affective states. Int J Mol Sci 20(13):3130

Virtual Reality for Motor and Cognitive Rehabilitation

Anuja Darekar

Contents

Abstract Virtual Reality (VR) affords clinicians the ability to deliver safe, controlled, task-specific customised interventions that are enjoyable, motivating and engaging. Elements of training in VR comply with principles of learning implicated in new skill acquisition and re-learning skills post-neurological disorders. However, heterogeneity in the description of VR systems and the description and control of 'active' ingredients of interventions (like dosage, type of feedback, task specificity, etc.) have led to inconsistency in the synthesis and interpretation of evidence related to the effectiveness of VR-based interventions, particularly in post-stroke and Parkinson's Disease (PD) rehabilitation. This chapter attempts to describe VR interventions with respect to their compliance with principles of neurorehabilitation,

A. Darekar (✉)
RehabCare Solutions, Pune, Maharashtra, India

Curr Topics Behav Neurosci (2023) 65: 337–370
https://doi.org/10.1007/7854_2023_418
Published Online: 12 April 2023

with the goal of optimising interventions for effective training and facilitation of maximum functional recovery. This chapter also advocates using a uniform framework to describe VR systems to promote homogeneity in literature in order to help in the synthesis of evidence. An overview of the evidence revealed that VR systems are effective in mediating deficits in upper extremity, posture and gait function seen in people post-stroke and PD. Generally, interventions were more effective when they were delivered as an adjunct to conventional therapy and were customised for rehabilitation purposes, in addition to complying with principles of learning and neurorehabilitation. Although recent studies imply that their VR intervention is compliant with principles of learning, only a few explicitly describe how these principles are incorporated as 'active ingredients' of the intervention. Finally, VR interventions targeting community ambulation and cognitive rehabilitation are yet limited and therefore warrant attention.

Keywords Cerebrovascular accidents · Cognition · Parkinson's disease · Rehabilitation · Sensorimotor systems · Virtual reality

Rehabilitation plays an important role in enabling functional independence in daily life and participation in society, post-neurological disorders. Various interventions are tailored to re-train function across multiple sensorimotor and cognitive domains. Virtual reality (VR) technology is a rapidly evolving technological aid used for research and development of interventions for various neurological disorders. This chapter attempts an overview of VR applications used in motor and cognitive rehabilitation post-stroke and Parkinson's Disease (PD).

Section 1 provides an overview of VR technology and basic concepts that inform the design of VR-based rehabilitation systems. Interventions used in neurorehabilitation facilitate improvement in function using principles of learning, such as repetition, task and goal-specific practice, feedback, etc. These principles not only support new skill acquisition, but also guide recovery and re-acquisition of skills in people with neurological deficits (although learning mechanisms may differ between the two populations) (Krakauer 2006; Bermúdez i Badia et al. 2016). Section 2 outlines learning and neurorehabilitation principles that inform formulation of effective interventions, including those utilising VR. This section will also attempt to draw parallels between integral elements of design in VR–human interaction and learning, and also draw attention towards building learning-driven VR interventions for rehabilitation.

Finally, Sects. 3 and 4 provide an overview of the application and effectiveness of VR-based interventions for motor and cognitive rehabilitation in individuals post-stroke and PD, respectively, with reference to concepts described in earlier sections.

1 Overview of VR Systems

VR encompasses a range of technologies comprising hardware and software that artificially generate sensory information (visual, auditory, haptic or olfactory) to create virtual environments (VE) that enable interaction with users in near real-time, and can be sometimes perceived by the users as similar to the real world (Rand et al. 2005; Fung et al. 2006). VR is also a powerful perceptual medium that extracts responses from users in the presence of cognitive certainty that the VE is not real (Slater 2009, 2018).

In the context of neurorehabilitation, these realistic/imaginary VEs are used to create interventions to treat deficits in single or multiple domains while making the rehabilitation process interactive, engaging and oftentimes 'fun'.

1.1 Key Elements of the VR System

The interactive and engaging experience in a VE is influenced by a few key elements (Muhanna 2015):

1. The virtual world: This is the space created by the computer wherein one or more users (participants) interact with the virtual world via graphical representations called avatars in the first-person (1PP) or third-person (3PP).
2. Immersion: Immersion is related to the sensorimotor contingencies (SCs) that the objective characteristics of the VR system can support (Slater 2009, 2018) to facilitate near-normal perception in the VE. A VR system that is able to change its characteristics in response to user action (for instance, change in the rendered image with change in gaze direction) to enable expected and meaningful change in perception is deemed to be more immersive (Slater 2009; Høeg et al. 2021).
3. Interactivity: In VR systems, interactivity enables participants to interact with and modify the VE through various sensors that may enable navigation and/or selection and manipulation of virtual objects (Muhanna 2015). The resultant change in the state of the VE should be conveyed through feedback in real-time thereby lending realism to the VR experience, facilitating the feeling of 'presence'.
4. Feedback: A key element of the VR experience, internal feedback is provided by the user's sensory systems as a direct result of interaction with the VE, and external feedback is provided by the system or an external source during the task or after task-completion.
5. Participants: VR experience can differ based on participant characteristics such as age, gender, cultural experiences and previous exposure to VR. In the context of rehabilitation, participants with neurological deficits may have a different response to the virtual experience than their healthy counterparts owing to sensorimotor and cognitive deficits.

These elements related to the system and the participants together shape presence within the VE. Presence can be simply defined as the 'illusion of being there' (Slater 2018) in the virtual world that provokes perceptual and physiological responses notwithstanding the cognitive knowledge that the experience is not real. Presence may influence the strength of response to the VE and is therefore an important design parameter for VR-based rehabilitation systems. It is important to note that while immersion may influence presence, this relationship is not linear, and other factors including those related to the participants may guide presence.

1.2 Framework for the Description of VR Systems

Rehabilitation literature exhibits considerable heterogeneity in defining VR technologies. This circumstance may have previously contributed to non-definitive conclusions about the effectiveness of VR (Høeg et al. 2021; Tieri et al. 2018). In order to homogenise reporting, researchers are now attempting to outline a consistent framework to describe VR systems. This chapter will use a framework similar to that described by Høeg et al. (Høeg et al. 2021) (outlined below and in Fig. 1) to discuss applications of VR technology in rehabilitation.

Degree/level of immersion: Depending on the sensorimotor contingencies (SCs) they support, VR systems are classified as non-immersive, semi-immersive or fully immersive (Bermúdez i Badia et al. 2016; Levin et al. 2014). Non-immersive systems generally use two-dimensional displays delivered on flat screens (computer screens, TVs, tablets, etc.) that do not provide any depth information.

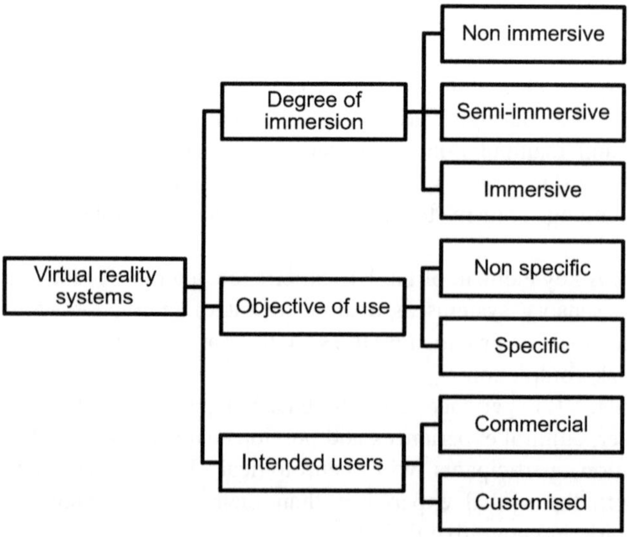

Fig. 1 Framework for description of VR systems

Semi-immersive systems are three-dimensional presentations that utilise stereoscopic projections requiring use of flicker glasses to provide depth information. Both non-immersive and semi-immersive systems have a fixed field of regard (FOR) and a fixed field of view (FOV). Fully immersive systems in contrast provide users with visualisation in all directions (360° FOR) wherein the visual perspective is updated with head movements – for instance, with a head mounted display (HMD). The change in virtual perspective in response to user action is dependent upon the data collected from the sensors that is processed to control and/or modify the VE (Aguilar-Lazcano et al. 2019). Aguilar-Lazcano et al. (2019) describe three types of sensor systems or 'interaction modalities' used to control the VE task. These are (a) vision systems (that use camera sensors such as the Kinect; Microsoft USA); (b) vision complementary systems (that use camera and other additional sensors such as the Leap Motion Controller (LMC; Leap Motion Inc. USA); and c) no-vision systems (that use sensors other than the camera such as haptic interfaces, game consoles or PC components).

Primary objective of usage: VR systems can be classified as 'specific VR systems' (SVR; designed specifically for rehabilitation interventions) and 'non-specific VR systems' (NSVR; designed for healthy adults for non-specific gaming purposes) (Høeg et al. 2021; Maier et al. 2019a).

Intended consumers: VR systems can be 'custom-made' for individual research laboratories to address their area of research. These customised systems are not available easily and are generally costly and cumbersome (Darekar et al. 2015a). Conversely, VR systems can be 'commercial' (called off-the-shelf systems) thereby available easily and are generally cost-effective.

Each of these classifications can be used alone or in combination to define a VR system (see Fig. 2 for examples). For instance, the Wii (Nintendo Co Ltd., Japan), Playstation (Sony Interactive Entertainment, USA) or Xbox Kinect (Microsoft, USA) are all 'non-immersive', 'non-specific', 'commercial' systems; while systems like the Interactive Rehabilitation and Exercise System (IREX; GestureTek, Canada), Vast.Rehab (Brontes Processing, Poland) or Jintronix (Jintronix, Canada) are 'semi-immersive or non-immersive' (depending on the display), 'commercial' but 'specific' systems that are built for rehabilitation. The Cave Automated Virtual Environment (CAVE) (Muhanna 2015) can be described as an 'immersive, specific, commercial' system while the Rutgers Ankle system (Mirelman 2008) can be described as a 'non-immersive, specific, customised' system.

The use of such a framework may help guide researchers to appropriately define their systems and in turn enable effective synthesis of evidence to gauge the effectiveness of interventions (Høeg et al. 2021; Tieri et al. 2018; Maier et al. 2019a).

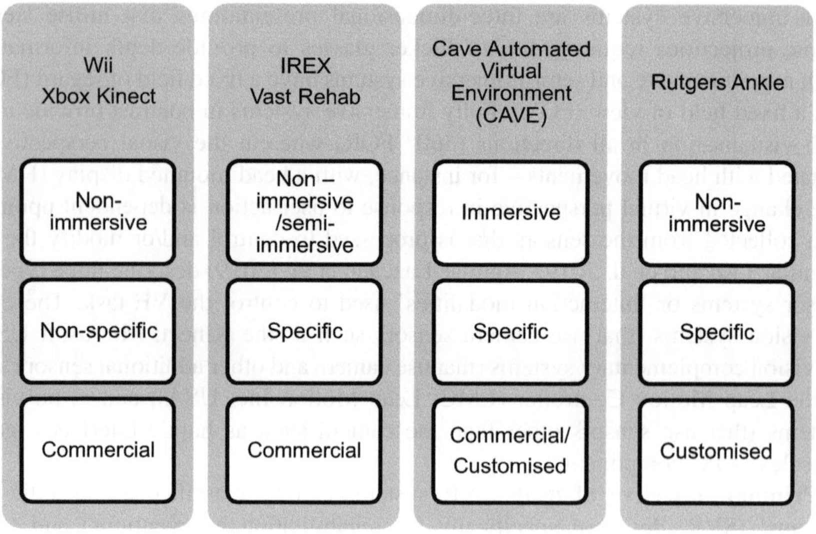

Fig. 2 VR systems' description according to the framework

2 Learning and Neurorehabilitation

2.1 Motor Learning and Skill Acquisition

Learning principles that drive new skill acquisition also influence acquiring skills in people with neurological deficits (Krakauer 2006; Bermúdez i Badia et al. 2016; Kitago and Krakauer 2013; Kleim 2011). Learning facilitates neuroplasticity and cortical reorganisation following a neurological insult/disorder and can be behaviourally correlated to improvement in function (Kleim 2011). Recently, Maier et al. (2019b) have identified the following learning principles that may guide effective neurorehabilitation interventions, namely:

(a) *Massed practice* (prolonged practice with little or no breaks)
(b) *Spaced practice* (providing appropriate rest periods during training)
(c) *Dosage* (duration of single sessions, their frequency, and the total duration over which training is imparted)
(d) *Task-specific practice* (such that internal sensorimotor representation of a skill is learned for effective implementation; this is commonly used for training of activities of daily living (ADL))
(e) *Goal-oriented practice* (wherein the focus is on achieving goals – for instance, reaching for an object – without excessive emphasis on learning individual muscle action or movement patterns)
(f) *Variable practice* (variability within the training sequence or random presentation of variable practice blocks)

(g) *Increasing difficulty* (increasing difficulty levels within a task or graduating to more difficult or complex tasks – graded practice to match the learner's capabilities)
(h) *Multisensory stimulation* (facilitates sensorimotor integration and efficient sensory reweighting, typically achieved in enriched environments (EE), leading to efficient learning)
(i) *Explicit feedback/knowledge of results* (KR; augmented feedback that can be quantitative (success/failure, number or errors, scores, etc.) or qualitative (playback of movement pattern); could be linked to reward mechanisms of the brain that augment learning through positive outcomes (Maier et al. 2019b), and encourage exercise adherence)
(j) *Implicit feedback/knowledge of performance* (KP; information about movement execution or movement kinematics such as feedback about erroneous trunk movement using auditory or visual stimuli)
(k) *Facilitating use of the affected limb* (to promote use of the more affected extremity; important in stroke rehabilitation where the paretic upper limb is not favoured for use leading to learned non-use)
(l) *Action observation* (learning occurs through observation of another person's actions; may lead to internalisation of someone else's movement but also ascribing ownership and agency to body parts that are not one's own – e.g., the rubber hand illusion (Botvinick and Cohen 1998))
(m) *Motor imagery/mental practice* (rehearsal of future movements and action plans; activates the same brain areas as required for actual movement execution; can be useful for individuals with severe motor deficits).

These principles of neurorehabilitation are considered as 'active ingredients' of an intervention – the description and evaluation of which may help optimise the training stimulus and thereby aid efficacious learning to guide behavioural change or return of function.

2.2 VR Can Effectively Facilitate Learning

VR-based rehabilitation is uniquely suited to fulfil many of the above-mentioned principles. First, VEs can provide rich visual scenes that incorporate multimodal sensory stimulation (Keshner and Lamontagne 2021) and feedback (Richards et al. 2018) in order to create enriched environments (EEs) for task-specific training. These EEs enhance presence and task-focussed attention, leading to augmentation of learning.

Further, VR affords the unique opportunity to create 'ecologically valid' VEs that depict everyday scenarios – a kitchen, a street crossing, a train station, etc. (Kizony 2011). Training in such VEs amounts to meaningful, task-specific and goal-oriented practice in a safe, graded, customised and controlled fashion. Real-world training of these tasks (like walking in the community) may not be feasible due to constraints of therapists' time and resources, safety concerns and lack of customisations.

Additionally, through avatars or third-person representations, action observation (Miclaus et al. 2021) or mental practice (Bermúdez i Badia et al. 2016) can also be incorporated in VEs. These factors may facilitate better learning and foster neuroplasticity to obtain lasting behavioural change and functional improvements in clinical populations.

Another feature of VR that is particularly amenable to neurorehabilitation applications is its ability to provide both implicit and explicit feedback (Levin et al. 2014). Implicit feedback is provided in real time through changes in the VE in response to the participants' actions and by the system's ability to detect and flag undesirable movements. Explicit feedback about errors, task-completion time, or success/failure can also be provided in real time or on task completion.

Also, greater engagement with a virtual task resulting from task-focussed attention and enhanced enjoyment due to the game-like features of the VE may result in increased repetition and practice leading to neuroplastic and behavioural changes (Baniña et al. 2020).

Subsequent sections provide an overview of various VR applications and their effectiveness in motor and cognitive rehabilitation in individuals post-stroke and those with PD. Readers should note that these are only brief overviews of otherwise highly specialised areas of research. It should also be noted that the effectiveness of rehabilitation interventions is generally measured within the framework of the International Classification of Functioning (ICF), which describes function across three domains – Body structure/Body function, Activities (including ADLs), and Participation (including fulfilment of societal roles and quality of life (QOL)) (World Health Organization 2001). Most reviews evaluating the effectiveness of VR rehabilitation have also described the impact of VR rehabilitation across these three aforementioned domains. On the other hand, most interventional studies report primarily on body structure/function and activity outcomes (Laver et al. 2017; Saposnik and Levin 2011; Lohse et al. 2014).

3 Stroke

Stroke is the leading cause of motor, sensory and cognitive dysfunction in the adult population resulting in limitations in ADL and participation restriction. Neurorehabilitation is recommended in the acute, sub-acute, as well as chronic phases of stroke (Teasell et al. 2014) to mediate deficits and to facilitate return to function. As described earlier, VR-based neurorehabilitation is guided by motor learning principles to achieve effective re-learning and return to function post-stroke. Indeed, VR has been used extensively for assessment and treatment of post-stroke dysfunctions. This section will provide an overview of its applications in upper limb, postural control (balance) and gait, and cognitive rehabilitation. The framework described earlier will guide usage of terms in Sects. 3 and 4.

3.1 Upper Limb and Hand

Upper extremity (UE) impairments post-stroke manifest as spasticity and weakness in the paretic UE, along with loss of control of movement leading to deficits in UE reach and grasp, difficulty in lifting objects and loss of hand dexterity. The subsequent loss of independence is significant leading to reduction in QOL. For this reason, treating UE dysfunctions have received considerable attention in VR-based rehabilitation, resulting in a substantial number of studies evaluating its effectiveness (Bermúdez i Badia et al. 2016). In fact, this field has seen an almost exponential rise in the number of publications after 2010. Prior to this year, published reviews included more observational studies and only a few randomised controlled trials (RCTs) (Saposnik and Levin 2011). In comparison, recent reviews included only RCTs that surpass the total number of studies using varied study designs, included in earlier reviews (Laver et al. 2017; Mekbib et al. 2020a; Aminov et al. 2018).

Both specific and non-specific VR systems (SVR and NSVR respectively) with different levels of immersion have been used for UE rehabilitation. Customised SVR systems were primarily used in earlier pilot and feasibility studies that tested usability of VR-based interventions for post-stroke rehabilitation (Bermúdez i Badia et al. 2016; Levin et al. 2014; Keshner et al. 2019). These systems, usually customised for research groups, consisted of costly equipment for motion sensing and rendering and display of VEs (Bermúdez i Badia et al. 2016; Darekar et al. 2015a). For instance, Levin et al. (2009) described a VE created to study reaching movements using the Computer Assisted Rehabilitation ENvironments (CAREN; Motek Medical B.V. Netherlands) software and viewed through a HMD (Kaiser XL50). The VE mimicked the physical environment (PE) consisting of two rows of three targets that the participants had to reach in both the PE and VE, so that arm and hand trajectories in both environments could be compared. Arm and hand movements were recorded with an Optotrak Motion Capture System (Northern Digital Inc., Canada). This customised set-up was specifically designed to support the research group's investigation and could not be easily transferred to a clinical set-up. The advent of relatively low-cost off-the-shelf SVR and NSVR systems facilitated the transfer of VR-based rehabilitation from laboratories to clinics. Particularly, clinicians increasingly have used NSVR gaming consoles for UE and balance/gait rehabilitation.

Although both SVR and NSVR systems have been used for training primarily reaching tasks in VR, a larger number of studies have used SVR systems to design their interventions (Maier et al. 2019a; Laver et al. 2017; Mekbib et al. 2020a; Peng et al. 2021; Doumas et al. 2021). For instance, in a meta-analysis, Mekbib et al. (2020a) reported that of the 27 included studies, 21 used SVR systems such as the Virtual Reality Rehabilitation system (VRRS, Khymeia group, Italy), YouGrabber (YouRehab, Switzerland) and others. Also, a higher number of semi-immersive or non-immersive systems have been used. Most of these systems utilised vision-based or vision complementary tracking systems (Aguilar-Lazcano et al. 2019; Doumas et al. 2021; Kim et al. 2020). Thus, in general, semi- or non-immersive, SVR systems utilising vision-based sensor systems are used frequently in UE rehabilitation.

In contrast, studies exhibit heterogeneity in the objectives and usage of the VR system, treatment parameters (dosage, intensity, adaptability, etc.) and outcome measures (Levin et al. 2014; Saposnik and Levin 2011; Lohse et al. 2014; Karamians et al. 2020). With respect to dosage, for instance, intervention times varied from 20 min to 1 h sessions (Kim et al. 2020), delivered over 2–12 weeks (Maier et al. 2019a) for the set of studies considered in this chapter. VR intervention was either dose-matched with the control treatment (CT) or was provided as an adjunct to CT (Laver et al. 2017; Chen et al. 2022).

Similarly, a variety of outcome measures spanning all domains of the ICF framework have been utilised to compare the effectiveness of VR intervention with CT (Subramanian et al. 2020). Subramanian et al. (2020) provide an overview of the various measures used in UE VR training trials based on the ICF framework. As such, the upper extremity section of the Fugl-Meyer Assessment (FMA-UE) (Fugl-Meyer et al. 1975; Gladstone et al. 2002) was the most commonly used to assess UE function (body structure/function domain); while the Box and Block test (BBT) (Mathiowetz et al. 1985; Platz et al. 2005), the Wolf Motor Function test (WMFT) (Wolf et al. 2001; Morris et al. 2001; Fritz et al. 2009), and the Action Research Arm Test (ARAT) (Lyle 1981; McDonnell 2008; Chen et al. 2012) were used to assess the ability to perform UE tasks (activities domain). Participation measures included the Motor activity log (MAL) (van der Lee et al. 2004; Uswatte et al. 2005, 2006) and the Stroke Impact Scale (SIS) (Duncan et al. 1999; Mulder and Nijland 2016) amongst others (participation domain).

Significant improvements in UE impairment/body function (Saposnik and Levin 2011; Lohse et al. 2014; Mekbib et al. 2020a; Aminov et al. 2018) have been reported after VR intervention as compared to CT. Specifically, significantly greater improvements in the FMA-UE have been found after VR training (Saposnik and Levin 2011; Mekbib et al. 2020a; Aminov et al. 2018; Domínguez-Téllez et al. 2020) relative to CT. On the other hand, mixed results have been reported on outcomes in the activities domain. While some studies reported a significant benefit of VR intervention over CT on the upper limb activities (either the BBT alone or on composite scores including the BBT, WMFT and ARAT) (Lohse et al. 2014; Mekbib et al. 2020a; Aminov et al. 2018; Doumas et al. 2021; Karamians et al. 2020), others did not report any differences between VR and CT interventions (Saposnik and Levin 2011; Domínguez-Téllez et al. 2020). Lastly, only a few studies have evaluated the effects of VR intervention over measures of participation (Laver et al. 2017). Of these, some studies report a significant increase in the 'amount of use' scale in the MAL indicating a greater increase in the use of the paretic upper limb post VR intervention (Housman et al. 2009; Subramanian et al. 2013; Jang et al. 2005). In contrast, mixed effects of VR intervention over CT have been found on QOL, with some studies reporting greater improvement in the SIS scale (Chen et al. 2022; Norouzi-Gheidari et al. 2020; Park et al. 2019), while others reporting no additional impact (Saposnik et al. 2016; Kong et al. 2016). These findings indicate that improvement in UE function post VR intervention could foster improvement in participation and to some extent in the QOL.

The above-outlined variability in the reported effectiveness of VR intervention has inspired an alternate way of gauging impact by defining the 'active ingredients' of a therapy (Maier et al. 2019a; Doumas et al. 2021). This approach has led to analysis of system-specific and participant factors. Examples of system-specific factors include SVR vs. NSVR, dosage, active vs. passive intervention for the control group, and the use of guiding principles of neurorehabilitation in VR systems. Examples of participant factors include chronicity of stroke, age, severity of impairment, and more.

Mekbib et al. (2020a) found that a VR training dosage that exceeded 15 h was significantly more effective than dosages involving fewer hours. This finding was partly supported by Laver et al. (2017) but in other studies, dosing did not reliably impact outcomes (Maier et al. 2019a; Karamians et al. 2020). Regardless, an optimally intensive practice along with task specificity and focused attention, necessary for learning, was provided by VR training (Baniña et al. 2020; Broderick et al. 2021; Brunner et al. 2016).

VR interventions using SVR systems provided greater improvement in outcomes than NSVR systems (Maier et al. 2019a; Laver et al. 2017; Lohse et al. 2014; Aminov et al. 2018), although both systems lead to significantly greater improvement in comparison with CT (Aminov et al. 2018; Karamians et al. 2020), thus validating the use of commercial NSVR systems in clinics. The effectiveness of NSVR – and more importantly – SVR systems may be related to their adherence to neurorehabilitation principles, described in Sect. 2.1. Maier et al. (2019a) proposed that the greater the number of neurorehabilitation principles included in the design of the VR intervention, the greater the likely effectiveness of that VR intervention. Generally, SVR systems were superior to NSVR in both effectiveness and use of neurorehabilitation principles to affect learning and plasticity. Maier et al. (2019a) found that NSVR systems focused on three neurorehabilitation principles centred on variable practice, dosage and promoting use of the paretic limb; while more than 50% of studies employing SVR systems used six principles – namely, variable practice, promoting use of paretic limb, implicit feedback, increasing difficulty, task-specific practice and explicit feedback. Another meta-analysis (Doumas et al. 2021) supported these findings by reporting larger effect sizes in favour of VR systems that complied with usage of at least eight out of eleven of the neurorehabilitation principles outlined by Maier et al. (2019a) in comparison with those that used fewer than eight principles. These recent studies are noteworthy as they signify a shift from the technology itself to 'active ingredients' of the technology that can be used to optimise treatment for greater benefit of the patient populations. More importantly, studies have begun reporting on these principles explicitly so that these may further inform VR systems design and interventions (Wüest et al. 2014).

3.2 Postural Control/Balance and Gait

Postural control and gait impairments are common dysfunctions post-stroke leading to increased fall risk (Batchelor et al. 2012), reduced ambulatory activity (Lord et al. 2006) and limitation in activities of daily living and particularly participation due to restricted ambulation outdoors (Lord et al. 2004; Mayo et al. 2002). Therefore, rehabilitation interventions should be geared towards facilitating functional balance, mobility and gait – particularly adaptive gait – required for independent walking in the community (Darekar et al. 2015a). VR systems can be used to create complex, challenging and functional VEs that train postural control and/or walking abilities while introducing various cognitive loads, thereby facilitating improved balance and gait. Posture and gait rehabilitation is an extensive field with more studies focused on harnessing features of VR to evaluate questions related to sensorimotor integration, perception-action and motor control and navigational variables. However, in recent years, the interventional potential of VR has attracted more attention (Keshner et al. 2019).

Non-immersive or semi-immersive VR systems have been used for both posture and gait training, while a few studies delivering gait training have used fully immersive set-ups (Darekar et al. 2015a; Corbetta et al. 2015; de Rooij et al. 2016; Mohammadi et al. 2019; Cano Porras et al. 2018; Luque-Moreno et al. 2015; Palma et al. 2017). However, the effectiveness of one system over the other has not been established.

Studies aimed at improving balance utilised tasks that required quiet standing or using weight shifts, trunk leans or reaching out to objects, while studies aimed at improving mobility and walking used treadmill walking, training components of gait (Darekar et al. 2015a) or 'pre-gait activities' (Bermúdez i Badia et al. 2016; Deutsch and Mirelman 2007) like stepping in place.

For instance, You et al. (2005) used games such as stepping up/down, Sharkbait, and Snowboarding in the IREX system for retraining balance in post-stroke individuals. The stepping up/down game involved performing in-place stepping using hip-knee flexion to climb up a step in the VE; the Sharkbait game involved avoiding sharks and eels by performing weight shifts, stepping, and squatting actions; and finally, the Snowboarding game involved advancing a snowboard by performing weight shifts and stepping. Similarly, the Wii balance board provides games such as Ski Slalom, tilt table, etc. that also involve making weight shifts to accomplish skiing movements and movement of spheres towards their target (Marques-Sule et al. 2021), leading to balance retraining.

For gait training, Fung et al. (2006) used treadmill walking, while Mirelman et al. (2009) used a VR-based robot-assisted task to train one aspect of gait, namely ankle movements, wherein participants were required to perform ankle movements in the seated position to navigate in the VE.

For balance evaluation, most studies used outcomes representing the activity domain of the ICF. The Berg Balance Scale (BBS) (Berg et al. 1992, 1995) and the Timed Up and Go (TUG) (Mathias et al. 1986; Podsiadlo and Richardson 1991;

Flansbjer et al. 2005; Ng and Hui-Chan 2005) tests, used to evaluate functional balance and functional mobility respectively, were the most commonly used measures in this domain. On the other hand, posturography measures that evaluate standing balance (either static or dynamic) while standing on a force plate that records the Centre of Pressure (CoP)*[1] were the most commonly used outcomes in the body structure/function domain. Posturography measures such as bilateral limb-loading symmetry (symmetry of weight distribution under both feet) and relative weight distribution were used to evaluate static balance. Outcomes like CoP path excursion (relative movement of the CoP under either foot) and sway angle (angle between a vertical line from the spatial centre of the supporting feet and a second line connecting from the same point to the subject's centre of gravity) were used to quantify dynamic balance capabilities.

Similarly, outcomes representing body structure/function as well as activity domains were used for gait evaluation. Among these, gait velocity (body structure/function domain) was most commonly used (Darekar et al. 2015a; Cano Porras et al. 2018). In addition, other kinematic variables such as cadence (number of steps taken per minute), step length (distance between heel strike of one leg and the next heel strike of the opposite leg) and stride length (distance between heel strike of one leg and the next heel strike of the same leg), etc. were also used. Outcomes of the activities domain such as Functional Gait Assessment (Wrisley et al. 2004; Thieme et al. 2009; Lin et al. 2010), Dynamic Gait Index (Lin et al. 2010; Jonsdottir and Cattaneo 2007), etc. were less frequently used.

VR-based interventions significantly improved functional balance (BBS scores) (Darekar et al. 2015a; Corbetta et al. 2015; de Rooij et al. 2016; Mohammadi et al. 2019; Cano Porras et al. 2018; Luque-Moreno et al. 2015; Garay-Sánchez et al. 2021; Zhang et al. 2021a; Gibbons et al. 2016; Li et al. 2016; Iruthayarajah et al. 2017) and functional mobility (TUG time) (Darekar et al. 2015a; Laver et al. 2017; Corbetta et al. 2015; de Rooij et al. 2016; Mohammadi et al. 2019; Cano Porras et al. 2018; Zhang et al. 2021a; Li et al. 2016; Iruthayarajah et al. 2017). Some studies reported improvements in static balance reflected in increased bilateral limb-loading symmetry (Yang et al. 2011; Song et al. 2014) in quiet stance and during mobility, and improved weight distribution (Song et al. 2014) in quiet stance. Further, improvement in dynamic balance such as an increased Centre of Pressure (CoP) path excursion under the paretic foot during a sit-to-stand task (Yang et al. 2011) and increased sway angles (Kim et al. 2009) in dynamic tasks was also observed. Thus, both static and dynamic balance showed a greater improvement after VR training than CT. With respect to gait, a large number of studies report an increase in gait speed post VR intervention (Darekar et al. 2015a; Corbetta et al. 2015; de Rooij et al. 2016; Cano Porras et al. 2018; Luque-Moreno et al. 2015; Zhang et al. 2021a;

[1] Centre of Pressure (CoP) is the point location of the vertical ground reaction force vector and is a direct reflection of the control of ankle muscles. In quiet stance, with both feet on the ground, CoP lies between the 2 feet, depending on the weight distribution under each foot. If the weight distribution is asymmetric, CoP shifts towards the foot bearing more weight.

Gibbons et al. 2016; Ghai et al. 2020; Rodrigues-Baroni et al. 2014). Significant improvements in cadence (Zhang et al. 2021a; Gibbons et al. 2016; Ghai et al. 2020), step length (Darekar et al. 2015a; Gibbons et al. 2016) and stride length (Gibbons et al. 2016; Ghai et al. 2020) were also reported.

As compared with UE VR interventions, 'active ingredients' of balance and gait-related VR interventions have not been extensively evaluated. In general, SVR systems were more effective in improving balance and gait than NSVR systems (Mohammadi et al. 2019; Iruthayarajah et al. 2017). One review found that VR interventions led to better outcomes in individuals with chronic stroke than in sub-acute stroke (Gibbons et al. 2016).

Further, although the total VR training dosage varied between studies, it did not seem to significantly influence outcomes (Zhang et al. 2021a). However, an optimal practice intensity – recommended to be at least 10 sessions of VR training (Luque-Moreno et al. 2015) – was required for an intervention to be beneficial. Earlier reviews consisted of a greater proportion of studies using dose-matched CT and VR interventions and reported slightly greater benefits of VR intervention over CT (Darekar et al. 2015a; Corbetta et al. 2015; de Rooij et al. 2016). However, recent meta-analyses and reviews (Laver et al. 2017; Mohammadi et al. 2019; Garay-Sánchez et al. 2021; Iruthayarajah et al. 2017) report a greater benefit when VR intervention was used in conjunction with CT.

Apart from these 'active ingredients', studies have not explicitly evaluated the effect of compliance with specific neurorehabilitation principles. However, some studies have outlined the type of feedback, task specificity, variation or difficulty levels in the description of their VR intervention (Darekar et al. 2015a; Corbetta et al. 2015; Cano Porras et al. 2018), without specific analyses of their effects.

3.3 Adaptive Locomotion

Adaptive locomotion that is safe and effective requires sensorimotor and cognitive processing to meet the increased demands of the task (such as negotiating slopes), and complex environments as encountered in the community consisting of crowds, distractions, obstacles and uneven surfaces (Shumway-Cook et al. 2003; Shumway-Cook et al. 2002). This capability may be compromised in post-stroke survivors, leading to reduced outdoor mobility even 6 months after stroke (Mayo et al. 2002). VR systems have been used to assess adaptive locomotion in complex environments (Fung et al. 2006; Richards et al. 2018) and in tasks related to obstacle avoidance (Darekar et al. 2015b, 2017).

A few studies have also evaluated the effect of training adaptive locomotion in VEs, wherein improvements in the ability to negotiate locomotor challenges (Fung et al. 2006; Richards et al. 2018; Jaffe et al. 2004) have been found post-intervention. For instance, Fung et al. (2006) used a VR-based treadmill locomotor system that incorporated VEs resembling real-life scenarios such as walking in a train station, at an intersection, in a park, and on the beach. Their self-paced treadmill allowed users to adapt their walking speeds in response to task demands. The treadmill was mounted on a modified Stewart platform with 6° of freedom allowing translational and rotational movements to emulate slopes and curbs encountered in the VE, thus providing a near-realistic scenario of walking in the community with its various features. The presence of audio-visual distractions, uneven terrains and slopes as well as obstacles in the VE required users to make physical, temporal and cognitive adjustments for successful negotiation of the task. The authors observed that people with stroke were able to adjust their gait speed in response to the VE and adapt gait to meet the challenges posed via slopes and obstacles. Similarly, Jaffe et al. (2004) used a walking task involving stepping over virtual stationary obstacles in the walking path in comparison with a real-world obstacle-crossing task and observed that people with stroke demonstrated a greater improvement in the obstacle clearance, and an increased gait speed after training in the VR task. These studies indicate the unique advantage of VR gait training that allows users to train for adaptive locomotion required for independence in community ambulation. More importantly, a transfer of VR-based adaptive gait training to real-world community walking has been reported (Mirelman 2008; Yang et al. 2008).

Further, VR training paradigms that have incorporated dual task training have reported improved performance in single as well as dual tasks (Kizony et al. 2010; Kannan et al. 2019; Fishbein et al. 2019). For instance, Kizony et al. (2010) used a walking task that required navigating a virtual shopping market aisle (single task) or shopping for items from a list announced before or during the task while navigating the aisle (dual cognitive task). They found that individuals with stroke could complete the dual task with minimal mistakes and showed a general tendency to increase gait speed while dual tasking.

The potential of VR for assessing and treating impairments of adaptive locomotion is immense (Keshner and Lamontagne 2021). Particularly, adaptive gait training for ambulation in the community is an unmet need in post-stroke rehabilitation that could be effectively addressed with VR technologies, as they enable mimicking environments and situations encountered in the community in a safe and controlled fashion. New treatment paradigms and evidence are thereby expected from this subfield.

Finally, deficits in dual task abilities, as well as the presence of unilateral spatial neglect, can complicate rehabilitation in post-stroke individuals. VR has been used for assessment and treatment of unilateral spatial neglect in particular. This topic, however, is beyond the scope of this chapter and readers are encouraged to refer to work by Lamontagne and colleagues (Keshner and Lamontagne 2021) and detailed reviews (Ogourtsova et al. 2017; Pedroli et al. 2015) for further information.

3.4 Cognition

Cognitive impairment post-stroke occurs in 30–40% of individuals with stroke in one or more cognitive domains, including attention, concentration, memory, spatial awareness, perception, praxis and executive function (EF) (Loetscher et al. 2019). These functions, along with affective and behavioural sequelae of stroke, are strongly associated with QOL in this cohort. Traditional cognitive rehabilitation has limited success in the amelioration of these deficits (Loetscher et al. 2019), thereby necessitating exploration of other treatment avenues, including VR.

The field of VR-based cognitive rehabilitation is recent, but promising. It includes smaller studies used for exploring the feasibility of virtually adapted neuropsychological tests (Ku et al. 2009) and training protocols (Rand et al. 2005; Vourvopoulos et al. 2014; Carelli et al. 2008) along with some larger recent trials (Faria et al. 2020). A meta-analysis by Wiley et al. (Wiley et al. 2020) that included 8 studies with 196 participants did not report any significant effect of VR-based cognitive rehabilitation on global cognition, memory, attention and language. In contrast, a meta-analysis by Zhang et al. (2021b) including 23 trials with 894 participants found significant improvements in EF and visuospatial abilities post VR intervention, revealing considerable heterogeneity in studies included in these reviews. Indeed, some studies evaluated the effect of motor training on cognition, whereas some used SVR systems geared to provide cognitive training only. For instance, Gamito et al. (2017) used VR to mimic daily life activities to train cognitive functions such as working memory (buying several items), visuospatial orientation (finding the way to the minimarket), calculation (retention of digits), selective attention (finding avatars dressed in a particular colour), etc. Studies that used SVR systems for cognitive training in comparison with an active (Cho and Lee 2019; Kim et al. 2011) or a passive (Gamito et al. 2017) control group demonstrated improvements in visuospatial attention and memory after a 4- to 6-week intervention. Some studies also reported improvement in ADL (Cho and Lee 2019; Kim et al. 2011). Thus, in general VR-based cognitive training seems beneficial.

Because cognitive and motor skills are used in conjunction in functional settings, they should be assessed and treated as such (Rizzo et al. 1997; Albani et al. 2002; Cipresso et al. 2014; Kafri et al. 2021). VR offers the opportunity to create these functional environments wherein cognitive and motor demands can be presented simultaneously in a controlled manner, while also enabling accurate measurement of each function (Kafri et al. 2021). In two studies, Faria et al. (2016, 2020) assessed the effectiveness of using a functional city scenario for cognitive training in multiple domains such as memory, attention, visuospatial skills and EF. The VR tasks, which could be customised to accommodate heterogeneous capabilities and needs, required participants to perform multiple errands while wayfinding through the city. The errands were functional everyday tasks such as shopping at a supermarket or going to the bank, etc. After training in this functional VE, an improvement in global cognition (Faria et al. 2016, 2020), attention (Faria et al. 2016, 2020), EF (Faria et al. 2016, 2020) and visuospatial ability (Faria et al. 2020) was observed in comparison with CT (Faria et al. 2016) or customised cognitive training (Faria et al. 2020). The ability of VR systems to provide ecologically valid training scenarios spanning multiple domains of function, should therefore be harnessed more effectively to enable transfer to life-skills that could lead to meaningful functional gains in the post-stroke population.

4 Parkinson's Disease

PD is one of the most prevalent neurodegenerative disorders. Over six million people live with PD as of 2016 (Dorsey et al. 2018). Motor symptoms of PD include bradykinesia (slowness of movement), akinesia, rest tremors, stiffness of limbs and torso and, postural instability (Dorsey et al. 2018). Non-motor symptoms are also common and include fatigue, anxiety and depression (Hunter et al. 2019; Bloem et al. 2021) and other cognitive deficits. These symptoms are progressive in nature and can lead to significant dysfunction in UE function, balance, gait and cognition leading to limitations in ADL, adoption of sedentary lifestyle and reduced QOL.

In addition to pharmacological interventions, exercise and physical therapy are important to mitigate symptoms of PD (Bloem et al. 2021). However, people with PD demonstrate poor adherence to regular exercise programs (Hunter et al. 2019; Ellis et al. 2011) due to many factors, including 'lack of outcome expectation' (low expectation of benefit from exercise) (Hunter et al. 2019; Ellis et al. 2013). Interventions that are tailored to the individual needs and preferences, allow goal-setting, and provide feedback while being enjoyable facilitate adherence to regular exercise (Hunter et al. 2019; Ellis et al. 2013). Because VR-based interventions can allow for customisation, real-time feedback, and goal-setting within the context of engaging movement tasks, they are well-suited therapy tools for individuals with PD (Bloem et al. 2021). VR interventions can also be adapted for cognitive rehabilitation (Larson et al. 2014) in order to address impairments in attention, visuospatial abilities, EF and memory (Aarsland et al. 2021). Indeed, VR systems are used in both assessment and treatment of motor and cognitive functions in PD (Dockx et al. 2016; Triegaardt et al. 2020). Subsequent sections describe applications of VR in the assessment and treatment of various dysfunctions related to PD.

4.1 Upper Extremity

In PD, the ability to reach, grasp and manipulate objects is compromised (Morris 2000), and sequential tasks, such as dressing, grooming and eating are slowed, along with the underscaling of movement trajectories (Morris 2000). In addition, repetitive tasks like finger tapping (Ingram et al. 2021) and dexterity needed for tasks like writing (Ponsen et al. 2008) are affected early on in the disease (Ponsen et al. 2008). However, in contrast to stroke rehabilitation, the use of VR in the assessment and treatment of UE dysfunction in PD is fairly recent. For instance, a recent review by Chau et al. (2021) included only three studies that used VR-based intervention for UE rehabilitation.

Most assessment (Ma et al. 2012; Robles-García et al. 2013; Bank et al. 2018; Oña et al. 2020) and intervention studies (Beek et al. 2019; Ma et al. 2011; Fernández-González et al. 2019; Allen et al. 2017; Cikajlo and Peterlin 2019; Sánchez-Herrera-Baeza et al. 2020) used SVR systems that were either semi- or fully immersive. Most studies utilised the LMC, a small, computer-mouse sized camera that has 2 CCD stereo cameras and 3 Infrared LED for tracking motion of the forearm, wrist and finger with reasonable accuracy (Beek et al. 2019). While semi-immersive systems used the LMC in isolation, immersive systems paired it with HMDs such as the Oculus Rift (Meta Quest, USA).

VR interventions that included reaching, grasping and manipulation of objects led to a greater improvement in UE function and dexterity (as measured by the 9-Hole Peg Test (Earhart et al. 2011), BBT, Purdue Pegboard test, etc.) as compared to CT (Fernández-González et al. 2019; Sánchez-Herrera-Baeza et al. 2020).

Additionally, VR training also unmasked some motor learning deficits in PD. Ma et al. (Ma et al. 2012) trained people with PD in a real vs. virtual task where participants had to reach for and pick a moving ball from a specified contact area on a ramp. The VE used for this task replicated the PE. It was found that training in VR did not improve visuomotor coordination in individuals with PD. In comparison with the physical training task (that improved speed and synchronisation of trunk and arm motions), VR training induced different movement termination strategies that led to termination of arm and trunk motion well before reaching the target, thus revealing deficits in visuomotor coordination in a VE.

Further, it may be difficult for people with PD to transfer learning between contexts (Krebs et al. 2001; Messier et al. 2007) such that outcome measures that are dissimilar to the trained task may not demonstrate any measurable changes (Allen et al. 2017). For instance, Allen et al. (2017) reported that a 12-week home-based SVR training in VE tasks (that required correct timing of responses or rapid movement for task completion) did not improve performance on the 9-hole peg test. Improvement was reported only on trained parameters wherein only improved speed but decreased accuracy was observed post VR training in a horizontal tapping task (Allen et al. 2017). Thus, VR training that required increasing movement speed for success, but did not demand accuracy or penalise errors, led to improvement in only the learned parameter.

Interestingly, Robles-García et al. (2013) used a VR imitation task where tapping the non-affected/least affected hand in the real world was viewed as movement of the affected hand in VR. Increased amplitude of finger tapping on the affected side was observed post-intervention suggesting that imitation of repetitive movements remains intact in individuals with PD. This study also reveals a novel avenue to use mirroring with VR – which may engage a mirror neuron like system (Mekbib et al. 2020b) – to train repetitive movements in PD.

VR thus seems to be a promising tool to assess and train UE function in participants with PD. However, the inquiry into whether VR interventions are beneficial for upper limb rehabilitation in PD is fairly new. Therefore, studies have not yet begun specifying the 'neurorehabilitation principles' guiding VR interventions and their impact on outcomes. Future research may address these important parameters to guide design and implementation of UE VR interventions in PD.

4.2 Postural Control/Balance and Gait

Postural instability along with bradykinesia (cardinal symptoms of PD) leads to increased fall risk (Morris 2000). 45–68% of individuals with PD, including those with mild PD (Ellis et al. 2011; Cattaneo et al. 2019; Lamont et al. 2017), fall frequently. This increased fall risk is contributed to gait hypokinesia and reduced step length while walking (shuffling gait), reduced foot clearance (Morris 2000) in level walking and in obstacle crossing, and the inability to process and respond to high attentional demands – particularly in busy or cluttered areas, or on challenging terrain, etc. (Lamont et al. 2017). In fact, complex tasks and situations may trigger freezing of gait (FOG) that is typically observed with disease progression. Since VR allows testing and training these phenomena in a safe, controlled environment (Canning et al. 2020; Bluett et al. 2019), VR-based interventions are now used in balance and gait rehabilitation in individuals with PD.

Most balance intervention studies in PD have favoured commercially available non-immersive NSVR systems (Triegaardt et al. 2020; Canning et al. 2020) such as Nintendo Wii (the most favoured) (Gandolfi et al. 2017; Liao et al. 2015a; Pompeu et al. 2012; Ribas et al. 2017) and Kinect (Palacios-Navarro et al. 2015). Some studies have used custom-made balance systems that were also non-immersive in nature (Yen et al. 2011; Yang et al. 2016; Shen and Mak 2014). Some studies used semi-immersive systems like Computer Assisted Rehabilitation ENvironments (CAREN) (Calabrò et al. 2020) or NIRVANA (BTS Engineering Corp., USA) (Pazzaglia et al. 2020). Most balance interventions involved tasks that required standing weight shifts, trunk lean for movement or obstacle avoidance, pelvic movements, and reaching with the hand or the foot (Gandolfi et al. 2017; Calabrò et al. 2020; Brachman et al. 2021; Esculier et al. 2012; Feng et al. 2019; Galna et al. 2014; Severiano et al. 2018). Some studies also used pre-gait activities (Mirelman et al. 2013), such as stepping (in place or side steps) or short walks (Shen and Mak 2014; Brachman et al. 2021; Galna et al. 2014; Lee et al. 2015).

Outcome measures representing both body function and activities domain of the ICF framework have been used. The BBS (Activities domain) was the most reported outcome measuring functional balance. VR training was found to be significantly effective in improving BBS scores in comparison with CT (Dockx et al. 2016; Wang et al. 2019; Chen et al. 2020; Sarasso et al. 2021; Lei et al. 2019). Studies also reported improvement in functional mobility assessed using the TUG test (Dockx et al. 2016; Triegaardt et al. 2020; Wang et al. 2019; Chen et al. 2020; Sarasso et al. 2021; Lei et al. 2019). Thus, VR training seems to be effective in improving balance, leading to improvement in the ability to perform functional activities.

Posturography outcomes (body structure/function domain) were also reported to improve with VR interventions. Significant improvements in static balance measures, such as reduced sway velocity (Brachman et al. 2021) and improved root mean square (RMS) of the CoP velocity in the eyes open and closed condition (Esculier et al. 2012), as well as dynamic balance measures such as greater amount and speed of lean in a forward lean task (Brachman et al. 2021), were reported after

VR training as compared with CT. Some studies reported decreased latency of postural responses to perturbations (Shen and Mak 2014; Nuic et al. 2018), indicating improved reactive balance. Improvement in sensory integration and reweighting in response to task demands was also reported (Yen et al. 2011; Liao et al. 2015b). All of these outcomes indicate improvement in static and dynamic balance capabilities in individuals with PD post VR intervention. This may have also led to an improvement in balance confidence reported in some studies (Liao et al. 2015a; Esculier et al. 2012; Lei et al. 2019).

VR balance training was found to be superior to 'passive' (Dockx et al. 2016; Triegaardt et al. 2020; Canning et al. 2020) control interventions. While some studies report an advantage of VR training over dose-matched 'active' control interventions (Dockx et al. 2016; Mirelman et al. 2016), many other studies found no differences between VR and dose-matched active control interventions (Triegaardt et al. 2020; Canning et al. 2020; Pompeu et al. 2012; Yen et al. 2011; García-López et al. 2021), suggesting that VR balance training as an adjunct to CT may be more effective in improving outcomes. Further, VR balance training resulted in greater benefits in individuals with more deficits (Sarasso et al. 2021; Mirelman et al. 2016; Bekkers et al. 2020). Finally, studies that trained individuals primarily on balance tasks also reported improvements in gait parameters such as stride length (Dockx et al. 2016; Triegaardt et al. 2020; Liao et al. 2015a; Wang et al. 2019; Lei et al. 2019) and gait speed (Dockx et al. 2016; Triegaardt et al. 2020; Liao et al. 2015a; Chen et al. 2020; de Melo et al. 2018), suggesting that while task-specific training is important, training that resembles the intended task or trains its sub-components may also lead to improvements (Bermúdez i Badia et al. 2016; Mirelman et al. 2013).

As mentioned above, task-specific gait training is important and VR-based gait training has been found to be beneficial. A large RCT (V-TIME) (Mirelman et al. 2016) evaluated effects of a 6-week task-specific gait training using non-VR treadmill walking (TW) and VR treadmill walking (VR-TW). The custom-made VR gait training program consisted of walking in a VE while responding to challenges like obstacles, multiple pathways and distractors that necessitated step adaptations. The VE imposed a concomitant cognitive load involving attention, planning, dual tasking, sensory processing and response selection. Post-intervention, both training arms reported a reduction in the number of falls, though notably, the fall incidence rate was significantly lower in the VR-TW group than the TW group. The VR-TW group also demonstrated significant improvements in gait variability during level walking and adaptive gait (obstacle crossing), improvements in leading foot clearance during obstacle crossing, as well as balance and gait function and health-related QOL. Some of these effects were preserved at 6-month follow-up. Gait training in complex VEs involving cognitive–motor interaction (CMI) is therefore particularly important since it leads to training of adaptive gait used for community ambulation, an area in which individuals with PD experience severe deficits.

4.3 Cognitive–Motor Interaction

Simultaneous use of motor and cognitive capabilities required in dual tasks can compound impairments in both domains (Wajda et al. 2017) and impose a cost on one or both processes, leading to an increase in the risk of falls in individuals with PD (Kelly et al. 2012). Cognitive dual task deficits are related to impairments in EF, set shifting and attention (Kelly et al. 2012). Further, deterioration in walking during dual tasks may be attributed to a decrease in automaticity in gait due to basal ganglia dysfunction (Redgrave et al. 2010) leading to increased engagement of attentional resources in walking than that seen in normal individuals. This has also been observed in neuroimaging studies demonstrating increased prefrontal cortex activation during steady-state walking in PD (Wajda et al. 2017). However, dual task deficits are known to be responsive to gait and dual task training (Wajda et al. 2017; Kelly et al. 2012). VR tasks that impose a cognitive load on walking have resulted in improvements in dual task performance (such as reduction in falls, decreased gait variability in level walking and in obstacle crossing, and improvement in cognitive function (Mirelman et al. 2011, 2016)). Neuroimaging findings have also revealed an overall reduction of prefrontal activation in simple and complex dual tasks post VR training (Maidan et al. 2018), suggesting post-training neuroplastic changes.

4.4 Freezing of Gait

FOG in PD is characterised by ‘a periodic absence or reduction of forward progression of the feet despite the intention to walk’ (Bluett et al. 2019) and is usually triggered during gait initiation, passing through narrow doorways, turning, dual task conditions and with increased anxiety (Morris 2000). Almost 50% of individuals with PD experience FOG, and it is the most common cause of falls in this population (Bluett et al. 2019). Further, provoking FOGs in real-world scenarios may be dangerous and thus not feasible. VR provides the potential for a safe and controlled environment to trigger FOGs, thereby facilitating assessment and development of training interventions for prevention and adverse event management.

Both semi-immersive and fully immersive systems are used in VR-based FOG assessment. These studies revealed that deficits in set shifting and concurrent motor processing, impaired regulation of automatic behaviour, increased step variability in the immediate steps preceding an FOG event, increased response latency for step initiation and more frequent stop failures may be implicated in PD-FOG (Bluett et al. 2019). Neuroimaging studies that used VR paradigms for FOG assessment have revealed multiple impairments within the motor, cognitive and affective domains (for detailed information, please refer to Bluett et al. (2019) and Canning et al. (2020)). These VR paradigms were designed to evoke behavioural responses to certain stimuli in order to unveil their effect on gait dysfunction in PD (Canning et al. 2020). Canning et al. (2020) described a few paradigms used in FOG research, including immersive VEs that mimic walking under conditions of low versus high threat (e.g., level ground versus walking on a walkway high above the ground) in order to induce anxiety leading to FOG. Another paradigm involved both immersive and non-immersive VR to mimic walking through narrow corridors, doorways or environments with distractions – all of which can provoke FOG. The users could advance in these VEs by walking, stepping in place or operating foot pedals. VR-based assessment thereby provides a safe alternative wherein FOG triggers and response behaviour can be reliably studied in a controlled manner without exposing the participants to adverse events and fall risks.

VR interventions in participants with PD-FOG have reported improvements in motor and cognitive outcomes (Bekkers et al. 2020; Killane et al. 2015). Killane et al. (2015) used a non-immersive SVR to train a time-limited VR maze task incorporating narrow corridors, doorways and turns in the VE. The dual-task condition included addition of a virtual Stroop task. Individuals with PD-FOG demonstrated reduced mean reaction times on the Stroop task, coupled with improved rhythmicity of stepping post-intervention, and reduced instances of freezing episodes during the intervention. Bekkers et al. (2020) in a paradigm similar to that of Mirelman et al. (2016) (refer Sect. 4.2) demonstrated that despite exhibiting more severe impairments, a greater number of falls, and a greater fear of falling relative to PD without FOG, PD-FOG patients showed a similar extent of improvement on measures of postural stability, EF and general mobility post VR intervention. Also, PD-FOG reported a greater reduction in the number of falls post VR intervention (Bekkers et al. 2020) though this outcome was not associated with improvement in balance confidence during the FOG events.

4.5 VR Use for Cognitive Rehabilitation in PD

Cognitive domains including memory, attention, visuospatial abilities and EF may be affected in individuals with PD with cognitive impairment resulting in difficulties in ADL and restriction in participation. Performance on standardised paper and pencil tests may not correspond to cognitive impairments evident in performance of everyday activities (Cipresso et al. 2014), especially in individuals with mild cognitive impairments. These assessments should therefore be based in functional and ecologically valid environments (Rizzo et al. 1997; Albani et al. 2002; Cipresso et al. 2014). Accordingly, VR has been used to evaluate performance in a home environment (Albani et al. 2002) or in a virtual supermarket (Virtual Multiple Errands Test; VMET) (Cipresso et al. 2014). The VMET involves assessment of multimodal cognitive abilities while participants navigate through a virtual supermarket to carry out multiple errands while bound by time and rules of navigation. This test was able to reveal cognitive deficits in people with PD – both in those with previously identified mild cognitive impairments and more importantly, in those without diagnosed cognitive impairments, thus unmasking cognitive deficits that were undetected on conventional neuropsychological tests. This unmasking of deficits that were otherwise undetected by conventional testing methods has been reported in a number of VR-based functional assessments (Ogourtsova et al. 2017; Rizzo et al. 1997; Cipresso et al. 2014).

Studies using VR for cognitive rehabilitation in PD are fairly recent, but provide encouraging insights. Maggio et al. (2018) used a semi-immersive SVR to train attention, spatiotemporal orientation, memory, language fluency and EF in individuals with PD (VR group) in comparison with a dose-matched control group (also individuals with PD) that received non-VR cognitive training. Post-training, the VR group showed greater improvements in EF, global cognition and visuospatial abilities. Zimmerman et al. (2014) compared cognitive outcomes after training with sports games such as Table Tennis, Archery, Swordplay and Air Sports using the Nintendo Wii (VR group) with computer-based cognitive training (CT group). Significant improvements in attention, and greater, but non-significant trends towards improvement in visuo-construction abilities and episodic memory were found in the VR group, suggesting that non-immersive NSVR systems may also be used for cognitive training (Pompeu et al. 2012; Maggio et al. 2018; Alves et al. 2018; Mendes et al. 2012). Thus, both SVR and NSVR systems can prove beneficial for cognitive training in people with PD. However, studies in PD overall have not begun evaluating the impact of 'active ingredients' on outcomes. Future research should therefore focus on identifying and evaluating these factors to optimise training parameters and guide functional recovery.

5 Conclusions

This work provides an overview of the application of VR in motor and cognitive rehabilitation in individuals with stroke and those with PD. The use of VR applications in rehabilitation has been around for more than 20 years and has been found to be effective in UE, balance and gait rehabilitation in both stroke and PD. However, this evidence is mixed (low or moderate effect) owing to the considerable heterogeneity in frameworks of description, objectives and treatment paradigms. The present work attempts to provide an overview of evidence with a common framework that may lead to better understanding of 'active ingredients' of VR intervention and the adoption of which may lead to greater and optimal results. Newer evidence indicates that combining VR training with conventional therapy, using SVR systems to impart training, and attention to compliance to neurorehabilitation principles in treatment design may be most beneficial in improving function. Finally, unmet needs identified in the area of adaptive gait training and cognitive rehabilitation, as well as in evaluation of 'active ingredients' to affect optimal outcomes in these areas should also receive more attention from the rehabilitation research community.

References

Aarsland D, Batzu L, Halliday GM et al (2021) Parkinson disease-associated cognitive impairment. Nat Rev Dis Primers 7(1):47

Aguilar-Lazcano CA, Rechy-Ramirez EJ, Hu HS, Rios-Figueroa HV, Marin-Hernandez A (2019) Interaction modalities used on serious games for upper limb rehabilitation: a systematic review. Games Health J 8(5):313–325

Albani G, Pignatti R, Bertella L et al (2002) Common daily activities in the virtual environment: a preliminary study in parkinsonian patients. Neurol Sci 23(2):s49–s50

Allen NE, Song J, Paul SS et al (2017) An interactive videogame for arm and hand exercise in people with Parkinson's disease: a randomized controlled trial. Parkinsonism Relat Disord 41: 66–72

Alves MLM, Mesquita BS, Morais WS, Leal JC, Satler CE, dos Santos Mendes FA (2018) Nintendo Wii™ versus Xbox Kinect™ for assisting people with Parkinson's disease. Percept Mot Skills 125(3):546–565

Aminov A, Rogers JM, Middleton S, Caeyenberghs K, Wilson PH (2018) What do randomized controlled trials say about virtual rehabilitation in stroke? A systematic literature review and meta-analysis of upper-limb and cognitive outcomes. J Neuroeng Rehabil 15(1):29

Baniña MC, Molad R, Solomon JM et al (2020) Exercise intensity of the upper limb can be enhanced using a virtual rehabilitation system. Disabil Rehabil Assist Technol:1–7

Bank PJM, Cidota MA, Ouwehand PEW, Lukosch SG (2018) Patient-tailored augmented reality games for assessing upper extremity motor impairments in Parkinson's disease and stroke. J Med Syst 42(12):246–246

Batchelor FA, Mackintosh SF, Said CM, Hill KD (2012) Falls after stroke. Int J Stroke 7(6): 482–490

Bekkers EMJ, Mirelman A, Alcock L et al (2020) Do patients with Parkinson’s disease with freezing of gait respond differently than those without to treadmill training augmented by virtual reality? Neurorehabil Neural Repair 34(5):440–449

Berg KO, Wood-Dauphinee SL, Williams JI, Maki B (1992) Measuring balance in the elderly: validation of an instrument. Can J Public Health 83(Suppl 2):S7–S11

Berg K, Wood-Dauphinee S, Williams JI (1995) The balance scale: reliability assessment with elderly residents and patients with an acute stroke. Scand J Rehabil Med 27(1):27–36

Bermúdez i Badia S, Fluet GG, Llorens R, Deutsch JE (2016) Virtual reality for sensorimotor rehabilitation post stroke: design principles and evidence. In: Reinkensmeyer DJ, Dietz V (eds) Neurorehabilitation technology. Springer, Cham, pp 573–603

Bloem BR, Okun MS, Klein C (2021) Parkinson’s disease. Lancet 397(10291):2284–2303

Bluett B, Bayram E, Litvan I (2019) The virtual reality of Parkinson's disease freezing of gait: a systematic review. Parkinsonism Relat Disord 61:26–33

Botvinick M, Cohen J (1998) Rubber hands ‘feel’ touch that eyes see. Nature 391(6669):756

Brachman A, Marszałek W, Kamieniarz A, Michalska J, Pawłowski M, Juras G (2021) Biomechanical measures of balance after balance-based exergaming training dedicated for patients with Parkinson’s disease. Gait Posture 87:170–176

Broderick M, Almedom L, Burdet E, Burridge J, Bentley P (2021) Self-directed exergaming for stroke upper limb impairment increases exercise dose compared to standard care. Neurorehabil Neural Repair 35(11):974–985

Brunner I, Skouen JS, Hofstad H et al (2016) Is upper limb virtual reality training more intensive than conventional training for patients in the subacute phase after stroke? An analysis of treatment intensity and content. BMC Neurol 16(1):219

Calabrò RS, Naro A, Cimino V et al (2020) Improving motor performance in Parkinson’s disease: a preliminary study on the promising use of the computer assisted virtual reality environment (CAREN). Neurol Sci 41(4):933–941

Canning CG, Allen NE, Nackaerts E, Paul SS, Nieuwboer A, Gilat M (2020) Virtual reality in research and rehabilitation of gait and balance in Parkinson disease. Nat Rev Neurol 16(8): 409–425

Cano Porras D, Siemonsma P, Inzelberg R, Zeilig G, Plotnik M (2018) Advantages of virtual reality in the rehabilitation of balance and gait. Neurology 90(22):1017

Carelli L, Morganti F, Weiss PL, Kizony R, Riva G (2008) A virtual reality paradigm for the assessment and rehabilitation of executive function deficits post stroke: feasibility study. Paper presented at: 2008 Virtual Rehabilitation; 25–27 Aug 2008

Cattaneo D, Gervasoni E, Pupillo E et al (2019) Educational and exercise intervention to prevent falls and improve participation in subjects with neurological conditions: the NEUROFALL randomized controlled trial. Front Neurol 10(865)

Chau B, Humbert S, Shou A (2021) Systemic literature review of the use of virtual reality for rehabilitation in Parkinson disease. Fed Pract 38(Suppl 1):S20–S27

Chen H-f, Lin K-c, Wu C-y, Chen C-l (2012) Rasch validation and predictive validity of the action research arm test in patients receiving stroke rehabilitation. Arch Phys Med Rehabil 93(6): 1039–1045

Chen L, Cai G, Weng H et al (2020) The effect of virtual reality on the ability to perform activities of daily living, balance during gait, and motor function in Parkinson disease patients: a systematic review and meta-analysis. Am J Phys Med Rehabil 99(10):917–924

Chen J, Or CK, Chen T (2022) Effectiveness of using virtual reality–supported exercise therapy for upper extremity motor rehabilitation in patients with stroke: systematic review and meta-analysis of randomized controlled trials. J Med Internet Res 24(6):e24111

Cho D-R, Lee S-H (2019) Effects of virtual reality immersive training with computerized cognitive training on cognitive function and activities of daily living performance in patients with acute stage stroke: a preliminary randomized controlled trial. Medicine 98(11)

Cikajlo I, Peterlin PK (2019) Advantages of using 3D virtual reality based training in persons with Parkinson's disease: a parallel study. J Neuroeng Rehabil 16(1):119–119

Cipresso P, Albani G, Serino S et al (2014) Virtual multiple errands test (VMET): a virtual reality-based tool to detect early executive functions deficit in Parkinson's disease. Front Behav Neurosci 8(405)

Corbetta D, Imeri F, Gatti R (2015) Rehabilitation that incorporates virtual reality is more effective than standard rehabilitation for improving walking speed, balance and mobility after stroke: a systematic review. J Physiother 61(3):117–124

Darekar A, McFadyen BJ, Lamontagne A, Fung J (2015a) Efficacy of virtual reality-based intervention on balance and mobility disorders post-stroke: a scoping review. J Neuroeng Rehabil 12(1):46

Darekar A, Lamontagne A, Fung J (2015b) Dynamic clearance measure to evaluate locomotor and perceptuo-motor strategies used for obstacle circumvention in a virtual environment. Hum Mov Sci 40:359–371

Darekar A, Lamontagne A, Fung J (2017) Locomotor circumvention strategies are altered by stroke: I. Obstacle clearance. J Neuroeng Rehabil 14(1):56

de Melo GEL, Kleiner AFR, Lopes JBP et al (2018) Effect of virtual reality training on walking distance and physical fitness in individuals with Parkinson's disease. NeuroRehabilitation 42: 473–480

de Rooij IJ, van de Port IG, Meijer JG (2016) Effect of virtual reality training on balance and gait ability in patients with stroke: systematic review and meta-analysis. Phys Ther 96(12): 1905–1918

Deutsch JE, Mirelman A (2007) Virtual reality-based approaches to enable walking for people poststroke. Top Stroke Rehabil 14(6):45–53

Dockx K, Bekkers EMJ, Van den Bergh V et al (2016) Virtual reality for rehabilitation in Parkinson's disease. Cochrane Database Syst Rev 12

Domínguez-Téllez P, Moral-Muñoz JA, Salazar A, Casado-Fernández E, Lucena-Antón D (2020) Game-based virtual reality interventions to improve upper limb motor function and quality of life after stroke: systematic review and meta-analysis. Games for Health Journal 9(1):1–10
Dorsey ER, Elbaz A, Nichols E et al (2018) Global, regional, and national burden of Parkinson's disease, 1990–2016: a systematic analysis for the Global Burden of Disease Study 2016. Lancet Neurol 17(11):939–953
Doumas I, Everard G, Dehem S, Lejeune T (2021) Serious games for upper limb rehabilitation after stroke: a meta-analysis. J Neuroeng Rehabil 18(1):100
Duncan PW, Wallace D, Lai SM, Johnson D, Embretson S, Laster LJ (1999) The stroke impact scale version 2.0. Evaluation of reliability, validity, and sensitivity to change. Stroke 30(10): 2131–2140
Earhart GM, Cavanaugh JT, Ellis T, Ford MP, Foreman KB, Dibble L (2011) The 9-hole PEG test of upper extremity function: average values, test-retest reliability, and factors contributing to performance in people with Parkinson disease. J Neurol Phys Ther 35(4):157–163
Ellis T, Cavanaugh JT, Earhart GM et al (2011) Factors associated with exercise behavior in people with Parkinson disease. Phys Ther 91(12):1838–1848
Ellis T, Boudreau JK, DeAngelis TR et al (2013) Barriers to exercise in people with Parkinson disease. Phys Ther 93(5):628–636
Esculier JF, Vaudrin J, Bériault P, Gagnon K, Tremblay LE (2012) Home-based balance training programme using Wii Fit with balance board for Parkinsons's disease: a pilot study. J Rehabil Med 44(2):144–150
Faria AL, Andrade A, Soares LI, Badia SB (2016) Benefits of virtual reality based cognitive rehabilitation through simulated activities of daily living: a randomized controlled trial with stroke patients. J Neuroeng Rehabil 13(1):96–96
Faria AL, Pinho MS, Bermúdez i Badia S. (2020) A comparison of two personalization and adaptive cognitive rehabilitation approaches: a randomized controlled trial with chronic stroke patients. J Neuroeng Rehabil 17(1):78
Feng H, Li C, Liu J et al (2019) Virtual reality rehabilitation versus conventional physical therapy for improving balance and gait in Parkinson's disease patients: a randomized controlled trial. Med Sci Monit 25:4186–4192
Fernández-González P, Carratalá-Tejada M, Monge-Pereira E et al (2019) Leap motion controlled video game-based therapy for upper limb rehabilitation in patients with Parkinson's disease: a feasibility study. J Neuroeng Rehabil 16(1):133
Fishbein P, Hutzler Y, Ratmansky M, Treger I, Dunsky A (2019) A preliminary study of dual-task training using virtual reality: influence on walking and balance in chronic poststroke survivors. J Stroke Cerebrovasc Dis 28(11):104343
Flansbjer UB, Holmbäck AM, Downham D, Patten C, Lexell J (2005) Reliability of gait performance tests in men and women with hemiparesis after stroke. J Rehabil Med 37(2):75–82
Fritz SL, Blanton S, Uswatte G, Taub E, Wolf SL (2009) Minimal detectable change scores for the Wolf Motor function test. Neurorehabil Neural Repair 23(7):662–667
Fugl-Meyer AR, Jääskö L, Leyman I, Olsson S, Steglind S (1975) The post-stroke hemiplegic patient. 1. A method for evaluation of physical performance. Scand J Rehabil Med 7(1):13–31
Fung J, Richards CL, Malouin F, McFadyen BJ, Lamontagne A (2006) A treadmill and motion coupled virtual reality system for gait training post-stroke. Cyberpsychol Behav 9(2):157–162
Galna B, Jackson D, Schofield G et al (2014) Retraining function in people with Parkinson's disease using the Microsoft kinect: game design and pilot testing. J Neuroeng Rehabil 11:60–60
Gamito P, Oliveira J, Coelho C et al (2017) Cognitive training on stroke patients via virtual reality-based serious games. Disabil Rehabil 39(4):385–388
Gandolfi M, Geroin C, Dimitrova E et al (2017) Virtual reality telerehabilitation for postural instability in Parkinson's disease: a multicenter, single-blind, randomized, controlled trial. Biomed Res Int 2017:7962826

Garay-Sánchez A, Suarez-Serrano C, Ferrando-Margelí M, Jimenez-Rejano JJ, Marcén-Román Y (2021) Effects of immersive and non-immersive virtual reality on the static and dynamic balance of stroke patients: a systematic review and meta-analysis. J Clin Med 10(19):4473

García-López H, Obrero-Gaitán E, Castro-Sánchez AM, Lara-Palomo IC, Nieto-Escamez FA, Cortés-Pérez I (2021) Non-immersive virtual reality to improve balance and reduce risk of falls in people diagnosed with Parkinson's disease: a systematic review. Brain Sci 11(11):1435

Ghai S, Ghai I, Lamontagne A (2020) Virtual reality training enhances gait poststroke: a systematic review and meta-analysis. Ann N Y Acad Sci 1478(1):18–42

Gibbons EM, Thomson AN, de Noronha M, Joseph S (2016) Are virtual reality technologies effective in improving lower limb outcomes for patients following stroke – a systematic review with meta-analysis. Top Stroke Rehabil 23(6):440–457

Gladstone DJ, Danells CJ, Black SE (2002) The Fugl-Meyer assessment of motor recovery after stroke: a critical review of its measurement properties. Neurorehabil Neural Repair 16(3): 232–240

Høeg ER, Povlsen TM, Bruun-Pedersen JR et al (2021) System immersion in virtual reality-based rehabilitation of motor function in older adults: a systematic review and meta-analysis. Front Virtual Real 2(30)

Housman SJ, Scott KM, Reinkensmeyer DJ (2009) A randomized controlled trial of gravity-supported, computer-enhanced arm exercise for individuals with severe hemiparesis. Neurorehabil Neural Repair 23(5):505–514

Hunter H, Lovegrove C, Haas B, Freeman J, Gunn H (2019) Experiences of people with Parkinson's disease and their views on physical activity interventions: a qualitative systematic review. JBI Evid Synth 17(4)

Ingram LA, Carroll VK, Butler AA, Brodie MA, Gandevia SC, Lord SR (2021) Quantifying upper limb motor impairment in people with Parkinson's disease: a physiological profiling approach. PeerJ 9:e10735

Iruthayarajah J, McIntyre A, Cotoi A, Macaluso S, Teasell R (2017) The use of virtual reality for balance among individuals with chronic stroke: a systematic review and meta-analysis. Top Stroke Rehabil 24(1):68–79

Jaffe DL, Brown DA, Pierson-Carey CD, Buckley EL, Lew HL (2004) Stepping over obstacles to improve walking in individuals with poststroke hemiplegia. J Rehabil Res Dev 41(3A):283–292

Jang SH, You SH, Hallett M et al (2005) Cortical reorganization and associated functional motor recovery after virtual reality in patients with chronic stroke: an experimenter-blind preliminary study. Arch Phys Med Rehabil 86(11):2218–2223

Jonsdottir J, Cattaneo D (2007) Reliability and validity of the dynamic gait index in persons with chronic stroke. Arch Phys Med Rehabil 88(11):1410–1415

Kafri M, Weiss PL, Zeilig G, Bondi M, Baum-Cohen I, Kizony R (2021) Performance in complex life situations: effects of age, cognition, and walking speed in virtual versus real life environments. J Neuroeng Rehabil 18(1):30

Kannan L, Vora J, Bhatt T, Hughes SL (2019) Cognitive-motor exergaming for reducing fall risk in people with chronic stroke: a randomized controlled trial. NeuroRehabilitation 44(4):493–510

Karamians R, Proffitt R, Kline D, Gauthier LV (2020) Effectiveness of virtual reality- and gaming-based interventions for upper extremity rehabilitation poststroke: a meta-analysis. Arch Phys Med Rehabil 101(5):885–896

Kelly VE, Eusterbrock AJ, Shumway-Cook A (2012) A review of dual-task walking deficits in people with Parkinson's disease: motor and cognitive contributions, mechanisms, and clinical implications. Parkinsons Dis 2012:918719

Keshner EA, Lamontagne A (2021) The untapped potential of virtual reality in rehabilitation of balance and gait in neurological disorders. Front Virtual Real 2:641650

Keshner EA, Weiss PT, Geifman D, Raban D (2019) Tracking the evolution of virtual reality applications to rehabilitation as a field of study. J Neuroeng Rehabil 16(1):76

Killane I, Fearon C, Newman L et al (2015) Dual motor-cognitive virtual reality training impacts dual-task performance in freezing of gait. IEEE J Biomed Health Inform 19(6):1855–1861

Kim JHP, Jang SHM, Kim CSP, Jung JHM, You JHP (2009) Use of virtual reality to enhance balance and ambulation in chronic stroke: a double-blind, randomized controlled study. Am J Phys Med Rehabil 88(9):693–701

Kim BR, Chun MH, Kim LS, Park JY (2011) Effect of virtual reality on cognition in stroke patients. Ann Rehabil Med 35(4):450–459

Kim W-S, Cho S, Ku J et al (2020) Clinical application of virtual reality for upper limb motor rehabilitation in stroke: review of technologies and clinical evidence. J Clin Med 9(10)

Kitago T, Krakauer JW (2013) Chapter 8 – motor learning principles for neurorehabilitation. In: Barnes MP, Good DC (eds) Handbook of clinical neurology, vol 110. Elsevier, pp 93–103

Kizony R (2011) Virtual reality for cognitive rehabilitation. In: Cognition, occupation, and participation across the life span: neuroscience, neurorehabilitation, and models of intervention in occupational therapy, 3rd edn. American Occupational Therapy Association, Rockville, pp 143–158

Kizony R, Levin MF, Hughey L, Perez C, Fung J (2010) Cognitive load and dual-task performance during locomotion poststroke: a feasibility study using a functional virtual environment. Phys Ther 90(2):252–260

Kleim JA (2011) Neural plasticity and neurorehabilitation: teaching the new brain old tricks. J Commun Disord 44(5):521–528

Kong K-H, Loh Y-J, Thia E et al (2016) Efficacy of a virtual reality commercial gaming device in upper limb recovery after stroke: a randomized, controlled study. Top Stroke Rehabil 23(5): 333–340

Krakauer JW (2006) Motor learning: its relevance to stroke recovery and neurorehabilitation. Curr Opin Neurol 19(1)

Krebs H, Hogan N, Hening W, Adamovich S, Poizner H (2001) Procedural motor learning in Parkinson's disease. Exp Brain Res 141(4):425–437

Ku J, Lee JH, Han K, Kim SI, Kang YJ, Park ES (2009) Validity and reliability of cognitive assessment using virtual environment technology in patients with stroke. Am J Phys Med Rehabil 88(9)

Lamont RM, Morris ME, Menz HB, McGinley JL, Brauer SG (2017) Falls in people with Parkinson's disease: a prospective comparison of community and home-based falls. Gait Posture 55:62–67

Larson EB, Feigon M, Gagliardo P, Dvorkin AY (2014) Virtual reality and cognitive rehabilitation: a review of current outcome research. NeuroRehabilitation 34:759–772

Laver KE, Lange B, George S, Deutsch JE, Saposnik G, Crotty M (2017) Virtual reality for stroke rehabilitation. Cochrane Database Syst Rev 11(11):CD008349

Lee N-Y, Lee D-K, Song H-S (2015) Effect of virtual reality dance exercise on the balance, activities of daily living, and depressive disorder status of Parkinson's disease patients. J Phys Ther Sci 27(1):145–147

Lei C, Sunzi K, Dai F et al (2019) Effects of virtual reality rehabilitation training on gait and balance in patients with Parkinson's disease: a systematic review. PloS One 14(11):e0224819

Levin MF, Knaut LA, Magdalon EC, Subramanian S (2009) Virtual reality environments to enhance upper limb functional recovery in patients with hemiparesis. Stud Health Technol Inform 145:94–108

Levin M, Weiss P, Keshner E (2014) Emergence of virtual reality as a tool for upper limb rehabilitation. Phys Ther 95

Li Z, Han XG, Sheng J, Ma SJ (2016) Virtual reality for improving balance in patients after stroke: a systematic review and meta-analysis. Clin Rehabil 30(5):432–440

Liao Y-Y, Yang Y-R, Cheng S-J, Wu Y-R, Fuh J-L, Wang R-Y (2015a) Virtual reality–based training to improve obstacle-crossing performance and dynamic balance in patients with Parkinson's disease. Neurorehabil Neural Repair 29(7):658–667

Liao Y-Y, Yang Y-R, Wu Y-R, Wang R-Y (2015b) Virtual reality-based Wii Fit training in improving muscle strength, sensory integration ability, and walking abilities in patients with Parkinson's disease: a randomized control trial. Int J Gerontol 9(4):190–195

Lin JH, Hsu MJ, Hsu HW, Wu HC, Hsieh CL (2010) Psychometric comparisons of 3 functional ambulation measures for patients with stroke. Stroke 41(9):2021–2025
Loetscher T, Potter KJ, Wong D, das Nair R (2019) Cognitive rehabilitation for attention deficits following stroke. Cochrane Database Syst Rev 11
Lohse KR, Hilderman CGE, Cheung KL, Tatla S, Van der Loos HFM (2014) Virtual reality therapy for adults post-stroke: a systematic review and meta-analysis exploring virtual environments and commercial games in therapy. PloS One 9(3):e93318
Lord SE, McPherson K, McNaughton HK, Rochester L, Weatherall M (2004) Community ambulation after stroke: how important and obtainable is it and what measures appear predictive? Arch Phys Med Rehabil 85(2):234–239
Lord SE, Rochester L, Weatherall M, McPherson KM, McNaughton HK (2006) The effect of environment and task on gait parameters after stroke: a randomized comparison of measurement conditions. Arch Phys Med Rehabil 87(7):967–973
Luque-Moreno C, Ferragut-Garcías A, Rodríguez-Blanco C et al (2015) A decade of progress using virtual reality for poststroke lower extremity rehabilitation: systematic review of the intervention methods. Biomed Res Int 2015:342529–342529
Lyle RC (1981) A performance test for assessment of upper limb function in physical rehabilitation treatment and research. Int J Rehabil Res 4(4):483–492
Ma H-I, Hwang W-J, Fang J-J et al (2011) Effects of virtual reality training on functional reaching movements in people with Parkinson's disease: a randomized controlled pilot trial. Clin Rehabil 25(10):892–902
Ma H-I, Hwang W-J, Wang C-Y, Fang J-J, Leong I-F, Wang T-Y (2012) Trunk–arm coordination in reaching for moving targets in people with Parkinson's disease: comparison between virtual and physical reality. Hum Mov Sci 31(5):1340–1352
Maggio MG, De Cola MC, Latella D et al (2018) What about the role of virtual reality in Parkinson disease's cognitive rehabilitation? Preliminary findings from a randomized clinical trial. J Geriatr Psychiatry Neurol 31(6):312–318
Maidan I, Nieuwhof F, Bernad-Elazari H et al (2018) Evidence for differential effects of 2 forms of exercise on prefrontal plasticity during walking in Parkinson's disease. Neurorehabil Neural Repair 32(3):200–208
Maier M, Rubio Ballester B, Duff A, Duarte Oller E, Verschure PFMJ (2019a) Effect of specific over nonspecific VR-based rehabilitation on poststroke motor recovery: a systematic meta-analysis. Neurorehabil Neural Repair 33(2):112–129
Maier M, Ballester BR, Verschure PFMJ (2019b) Principles of neurorehabilitation after stroke based on motor learning and brain plasticity mechanisms. Front Syst Neurosci 13:74–74
Marques-Sule E, Arnal-Gómez A, Buitrago-Jiménez G, Suso-Martí L, Cuenca-Martínez F, Espí-López GV (2021) Effectiveness of Nintendo Wii and physical therapy in functionality, balance, and daily activities in chronic stroke patients. J Am Med Dir Assoc 22(5):1073–1080
Mathias S, Nayak US, Isaacs B (1986) Balance in elderly patients: the "get-up and go" test. Arch Phys Med Rehabil 67(6):387–389
Mathiowetz V, Volland G, Kashman N, Weber K (1985) Adult norms for the box and block test of manual dexterity. Am J Occup Ther 39(6):386–391
Mayo NE, Wood-Dauphinee S, Côté DL, Carlton J (2002) Activity, participation, and quality of life 6 months poststroke. Arch Phys Med Rehabil 83(8):1035–1042
McDonnell M (2008) Action research arm test. Aust J Physiother 54(3):220
Mekbib DB, Han J, Zhang L et al (2020a) Virtual reality therapy for upper limb rehabilitation in patients with stroke: a meta-analysis of randomized clinical trials. Brain Inj 34(4):456–465
Mekbib DB, Zhao Z, Wang J et al (2020b) Proactive motor functional recovery following immersive virtual reality–based limb mirroring therapy in patients with subacute stroke. Neurotherapeutics 17(4):1919–1930
Mendes FAS, Pompeu JE, Lobo AM et al (2012) Motor learning, retention and transfer after virtual-reality-based training in Parkinson's disease – effect of motor and cognitive demands of games: a longitudinal, controlled clinical study. Physiotherapy 98(3):217–223

Messier J, Adamovich S, Jack D, Hening W, Sage J, Poizner H (2007) Visuomotor learning in immersive 3D virtual reality in Parkinson's disease and in aging. Exp Brain Res 179(3):457–474

Miclaus RS, Roman N, Henter R, Caloian S (2021) Lower extremity rehabilitation in patients with post-stroke sequelae through virtual reality associated with mirror therapy. Int J Environ Res Public Health 18(5):2654

Mirelman A (2008) Comparison of robotic-virtual reality lower extremity training with robotic lower extremity training alone for rehabilitation of gait of individuals post-stroke. Diss Abstr Int Sect B Sci Eng 68:7968

Mirelman A, Bonato P, Deutsch JE (2009) Effects of training with a robot-virtual reality system compared with a robot alone on the gait of individuals after stroke. Stroke 40(1):169–174

Mirelman A, Maidan I, Herman T, Deutsch JE, Giladi N, Hausdorff JM (2011) Virtual reality for gait training: can it induce motor learning to enhance complex walking and reduce fall risk in patients with Parkinson's disease? J Gerontol A Biol Sci Med Sci 66(2):234–240

Mirelman A, Maidan I, Deutsch JE (2013) Virtual reality and motor imagery: promising tools for assessment and therapy in Parkinson's disease. Mov Disord 28(11):1597–1608

Mirelman A, Rochester L, Maidan I et al (2016) Addition of a non-immersive virtual reality component to treadmill training to reduce fall risk in older adults (V-TIME): a randomised controlled trial. Lancet 388(10050):1170–1182

Mohammadi R, Semnani AV, Mirmohammadkhani M, Grampurohit N (2019) Effects of virtual reality compared to conventional therapy on balance poststroke: a systematic review and meta-analysis. J Stroke Cerebrovasc Dis 28(7):1787–1798

Morris ME (2000) Movement disorders in people with Parkinson disease: a model for physical therapy. Phys Ther 80(6):578–597

Morris DM, Uswatte G, Crago JE, Cook EW 3rd, Taub E (2001) The reliability of the wolf motor function test for assessing upper extremity function after stroke. Arch Phys Med Rehabil 82(6): 750–755

Muhanna MA (2015) Virtual reality and the CAVE: taxonomy, interaction challenges and research directions. J King Saud Univ Comput Inf Sci 27(3):344–361

Mulder M, Nijland R (2016) Stroke impact scale. J Physiother 62(2):117

Ng SS, Hui-Chan CW (2005) The timed up and go test: its reliability and association with lower-limb impairments and locomotor capacities in people with chronic stroke. Arch Phys Med Rehabil 86(8):1641–1647

Norouzi-Gheidari N, Hernandez A, Archambault PS, Higgins J, Poissant L, Kairy D (2020) Feasibility, safety and efficacy of a virtual reality exergame system to supplement upper extremity rehabilitation post-stroke: a pilot randomized clinical trial and proof of principle. Int J Environ Res Public Health 17(1):113

Nuic D, Vinti M, Karachi C, Foulon P, Van Hamme A, Welter M-L (2018) The feasibility and positive effects of a customised videogame rehabilitation programme for freezing of gait and falls in Parkinson's disease patients: a pilot study. J Neuroeng Rehabil 15(1):31–31

Ogourtsova T, Souza Silva W, Archambault PS, Lamontagne A (2017) Virtual reality treatment and assessments for post-stroke unilateral spatial neglect: a systematic literature review. Neuropsychol Rehabil 27(3):409–454

Oña ED, Jardón A, Cuesta-Gómez A, Sánchez-Herrera-Baeza P, Cano-de-la-Cuerda R, Balaguer C (2020) Validity of a fully-immersive VR-based version of the box and blocks test for upper limb function assessment in Parkinson's disease. Sensors 20(10):2773

Palacios-Navarro G, García-Magariño I, Ramos-Lorente P (2015) A kinect-based system for lower limb rehabilitation in Parkinson's disease patients: a pilot study. J Med Syst 39(9):103

Palma GC, Freitas TB, Bonuzzi GM et al (2017) Effects of virtual reality for stroke individuals based on the international classification of functioning and health: a systematic review. Top Stroke Rehabil 24(4):269–278

Park M, Ko M-H, Oh S-W et al (2019) Effects of virtual reality-based planar motion exercises on upper extremity function, range of motion, and health-related quality of life: a multicenter, single-blinded, randomized, controlled pilot study. J Neuroeng Rehabil 16(1):122

Pazzaglia C, Imbimbo I, Tranchita E et al (2020) Comparison of virtual reality rehabilitation and conventional rehabilitation in Parkinson's disease: a randomised controlled trial. Physiotherapy 106:36–42
Pedroli E, Serino S, Cipresso P, Pallavicini F, Riva G (2015) Assessment and rehabilitation of neglect using virtual reality: a systematic review. Front Behav Neurosci 9(226)
Peng Q-C, Yin L, Cao Y (2021) Effectiveness of virtual reality in the rehabilitation of motor function of patients with subacute stroke: a meta-analysis. Front Neurol 12:639535
Platz T, Pinkowski C, van Wijck F, Kim IH, di Bella P, Johnson G (2005) Reliability and validity of arm function assessment with standardized guidelines for the Fugl-Meyer test, action research arm test and box and block test: a multicentre study. Clin Rehabil 19(4):404–411
Podsiadlo D, Richardson S (1991) The timed "Up and Go": a test of basic functional mobility for frail elderly persons. J Am Geriatr Soc 39(2):142–148
Pompeu JE, Mendes FAS, Silva KG et al (2012) Effect of Nintendo Wii™-based motor and cognitive training on activities of daily living in patients with Parkinson's disease: a randomised clinical trial. Physiotherapy 98(3):196–204
Ponsen MM, Daffertshofer A, Wolters E, Beek PJ, Berendse HW (2008) Impairment of complex upper limb motor function in de novo Parkinson's disease. Parkinsonism Relat Disord 14(3): 199–204
Rand D, Katz N, Shahar M, Kizony R, Weiss PL (2005) The virtual mall: a functional virtual environment for stroke rehabilitation. Annu Rev Cyberther Telemed 3:193–198
Redgrave P, Rodriguez M, Smith Y et al (2010) Goal-directed and habitual control in the basal ganglia: implications for Parkinson's disease. Nat Rev Neurosci 11(11):760–772
Ribas CG, Alves da Silva L, Corrêa MR, Teive HG, Valderramas S (2017) Effectiveness of exergaming in improving functional balance, fatigue and quality of life in Parkinson's disease: a pilot randomized controlled trial. Parkinsonism Relat Disord 38:13–18
Richards CL, Malouin F, Lamontagne A et al (2018) Gait training after stroke on a self-paced treadmill with and without virtual environment scenarios: a proof-of-principle study. Physiother Can 70(3):221–230
Rizzo AA, Buckwalter JG, Neumann U (1997) Virtual reality and cognitive rehabilitation: a brief review of the future. J Head Trauma Rehabil 12(6)
Robles-García V, Arias P, Sanmartín G et al (2013) Motor facilitation during real-time movement imitation in Parkinson's disease: a virtual reality study. Parkinsonism Relat Disord 19(12): 1123–1129
Rodrigues-Baroni JM, Nascimento LR, Ada L, Teixeira-Salmela LF (2014) Walking training associated with virtual reality-based training increases walking speed of individuals with chronic stroke: systematic review with meta-analysis. Braz J Phys Ther 18(6):502–512
Sánchez-Herrera-Baeza P, Cano-de-la-Cuerda R, Oña-Simbaña ED et al (2020) The impact of a novel immersive virtual reality technology associated with serious games in Parkinson's disease patients on upper limb rehabilitation: a mixed methods intervention study. Sensors 20(8):2168
Saposnik G, Levin M (2011) Virtual reality in stroke rehabilitation. Stroke 42(5):1380–1386
Saposnik G, Cohen LG, Mamdani M et al (2016) Efficacy and safety of non-immersive virtual reality exercising in stroke rehabilitation (EVREST): a randomised, multicentre, single-blind, controlled trial. Lancet Neurol 15(10):1019–1027
Sarasso E, Gardoni A, Tettamanti A, Agosta F, Filippi M, Corbetta D (2021) Virtual reality balance training to improve balance and mobility in Parkinson's disease: a systematic review and meta-analysis. J Neurol
Severiano MIR, Zeigelboim BS, Teive HAG, Santos GJB, Fonseca VR (2018) Effect of virtual reality in Parkinson's disease: a prospective observational study. Arq Neuropsiquiatr 76(2): 78–84
Shen X, Mak MKY (2014) Technology-assisted balance and gait training reduces falls in patients with Parkinson's disease: a randomized controlled trial with 12-month follow-up. Neurorehabil Neural Repair 29(2):103–111

Shumway-Cook A, Patla AE, Stewart A, Ferrucci L, Ciol MA, Guralnik JM (2002) Environmental demands associated with community mobility in older adults with and without mobility disabilities. Phys Ther 82(7):670–681

Shumway-Cook A, Patla A, Stewart A, Ferrucci L, Ciol MA, Guralnik JM (2003) Environmental components of mobility disability in community-living older persons. J Am Geriatr Soc 51(3): 393–398

Slater M (2009) Place illusion and plausibility can lead to realistic behaviour in immersive virtual environments. Philos Trans R Soc Lond B Biol Sci 364(1535):3549–3557

Slater M (2018) Immersion and the illusion of presence in virtual reality. Br J Psychol 109(3): 431–433

Song YB, Chun MH, Kim W, Lee SJ, Yi JH, Park DH (2014) The effect of virtual reality and tetra-ataxiometric posturography programs on stroke patients with impaired standing balance. Ann Rehabil Med 38(2):160–166

Subramanian SK, Lourenço CB, Chilingaryan G, Sveistrup H, Levin MF (2013) Arm motor recovery using a virtual reality intervention in chronic stroke: randomized control trial. Neurorehabil Neural Repair 27(1):13–23

Subramanian SK, Cross MK, Hirschhauser CS (2020) Virtual reality interventions to enhance upper limb motor improvement after a stroke: commonly used types of platform and outcomes. Disabil Rehabil Assist Technol:1–9

Teasell RW, Murie Fernandez M, McIntyre A, Mehta S (2014) Rethinking the continuum of stroke rehabilitation. Arch Phys Med Rehabil 95(4):595–596

Thieme H, Ritschel C, Zange C (2009) Reliability and validity of the functional gait assessment (German version) in subacute stroke patients. Arch Phys Med Rehabil 90(9):1565–1570

Tieri G, Morone G, Paolucci S, Iosa M (2018) Virtual reality in cognitive and motor rehabilitation: facts, fiction and fallacies. Expert Rev Med Devices 15(2):107–117

Triegaardt J, Han TS, Sada C, Sharma S, Sharma P (2020) The role of virtual reality on outcomes in rehabilitation of Parkinson's disease: meta-analysis and systematic review in 1031 participants. Neurol Sci 41(3):529–536

Uswatte G, Taub E, Morris D, Vignolo M, McCulloch K (2005) Reliability and validity of the upper-extremity motor activity Log-14 for measuring real-world arm use. Stroke 36(11): 2493–2496

Uswatte G, Taub E, Morris D, Light K, Thompson PA (2006) The motor activity Log-28: assessing daily use of the hemiparetic arm after stroke. Neurology 67(7):1189–1194

van Beek JJW, van Wegen EEH, Bohlhalter S, Vanbellingen T (2019) Exergaming-based dexterity training in persons with Parkinson disease: a pilot feasibility study. J Neurol Phys Ther 43(3)

van der Lee JH, Beckerman H, Knol DL, de Vet HC, Bouter LM (2004) Clinimetric properties of the motor activity log for the assessment of arm use in hemiparetic patients. Stroke 35(6): 1410–1414

Vourvopoulos A, Faria AL, Ponnam K, Badia SBI (2014) RehabCity: design and validation of a cognitive assessment and rehabilitation tool through gamified simulations of activities of daily living. Proceedings of the 11th conference on advances in computer entertainment technology; Funchal, Portugal

Wajda DA, Mirelman A, Hausdorff JM, Sosnoff JJ (2017) Intervention modalities for targeting cognitive-motor interference in individuals with neurodegenerative disease: a systematic review. Expert Rev Neurother 17(3):251–261

Wang B, Shen M, Wang Y-x, He Z-w, Chi S-q, Yang Z-h (2019) Effect of virtual reality on balance and gait ability in patients with Parkinson's disease: a systematic review and meta-analysis. Clin Rehabil 33(7):1130–1138

Wiley E, Khattab S, Tang A (2020) Examining the effect of virtual reality therapy on cognition post-stroke: a systematic review and meta-analysis. Disabil Rehabil Assist Technol:1–11

Wolf SL, Catlin PA, Ellis M, Archer AL, Morgan B, Piacentino A (2001) Assessing Wolf motor function test as outcome measure for research in patients after stroke. Stroke 32(7):1635–1639

World Health Organization (2001) International classification of functioning, disability, and health: ICF [computer program] Version 1.0. World Health Organization, Geneva [2001] ©2001
Wrisley DM, Marchetti GF, Kuharsky DK, Whitney SL (2004) Reliability, internal consistency, and validity of data obtained with the functional gait assessment. Phys Ther 84(10):906–918
Wüest S, van de Langenberg R, de Bruin ED (2014) Design considerations for a theory-driven exergame-based rehabilitation program to improve walking of persons with stroke. Eur Rev Aging Phys Act 11(2):119–129
Yang YR, Tsai MP, Chuang TY, Sung WH, Wang RY (2008) Virtual reality-based training improves community ambulation in individuals with stroke: a randomized controlled trial. Gait Posture 28(2):201–206
Yang SP, Hwang WHM, Tsai YCP, Liu FKM, Hsieh LFM, Chern JSP (2011) Improving balance skills in patients who had stroke through virtual reality treadmill training. Am J Phys Med Rehabil 90(12):969–978
Yang W-C, Wang H-K, Wu R-M, Lo C-S, Lin K-H (2016) Home-based virtual reality balance training and conventional balance training in Parkinson's disease: a randomized controlled trial. J Formos Med Assoc 115(9):734–743
Yen C-Y, Lin K-H, Hu M-H, Wu R-M, Lu T-W, Lin C-H (2011) Effects of virtual reality–augmented balance training on sensory organization and attentional demand for postural control in people with Parkinson disease: a randomized controlled trial. Phys Ther 91(6):862–874
You SH, Jang SH, Kim Y-H et al (2005) Virtual reality-induced cortical reorganization and associated locomotor recovery in chronic stroke: an experimenter-blind randomized study. Stroke 36(6):1166–1171
Zhang B, Li D, Liu Y, Wang J, Xiao Q (2021a) Virtual reality for limb motor function, balance, gait, cognition and daily function of stroke patients: a systematic review and meta-analysis. J Adv Nurs 77(8):3255–3273
Zhang Q, Fu Y, Lu Y et al (2021b) Impact of virtual reality-based therapies on cognition and mental health of stroke patients: systematic review and meta-analysis. J Med Internet Res 23(11): e31007
Zimmermann R, Gschwandtner U, Benz N et al (2014) Cognitive training in Parkinson disease. Neurology 82(14):1219

Virtual Reality Interventions for Mental Health

Oswald D. Kothgassner, Adelais Reichmann, and Mercedes M. Bock

Contents

Abstract Virtual Reality (VR) is a growing field in psychological research and therapy. While there is strong evidence for the efficacy of exposure therapy in VR (VRET) to treat anxiety disorders, new opportunities for using VR to treat mental health disorders are emerging. In this chapter, we first describe the value of VRET for the treatment of several anxiety disorders. Next, we introduce some recent developments in research using VR investigating schizophrenia, neurodevelopmental disorders, and eating disorders. This includes therapeutic strategies beyond VRET, including avatar-based therapies or those combining VR with biofeedback approaches. Although VR offers many convincing advantages,

O. D. Kothgassner (✉) and A. Reichmann
Department of Child and Adolescent Psychiatry, Medical University of Vienna, Vienna, Austria
e-mail: oswald.kothgassner@meduniwien.ac.at

M. M. Bock
Child and Adolescent Psychiatry, Social Psychiatric Services Vienna, Vienna, Austria

Curr Topics Behav Neurosci (2023) 65: 371–388
https://doi.org/10.1007/7854_2023_419
Published Online: 28 April 2023

contraindications in treatment must be considered when implementing VR-supported therapy in clinical practice. Finally, we provide an outlook for future research, highlighting the integration of augmented reality and automation processes in VR environments to create more efficient and tailored therapeutic tools. Further, future treatments will benefit from the gamification approach, which integrates elements of computer games and narratives that promote patients' motivation and enables methods to reduce drop-outs during psychological therapy.

Keywords Biofeedback · Exposure therapy · Mental health · Virtual reality

1 Introduction

The assessment and treatment of mental disorders using Virtual Reality (VR) is a growing field in current psychological research. Within the last decades, rapid technological advances enabled the realization of powerful VR systems for clinical practice. VR is based on the interaction between humans and an immersive computer-generated 3D environment using head-mounted-displays (HMDs; Mühlberger and Pauli 2011). HMDs comprise two screens that present slightly different images to each eye, allowing the viewer to perceive a digitally created 360° environment in full stereoscopic depth. At the same time, sensors continuously capture position and orientation, allowing the system to update the visual environment in response to the user's movements. This is a key element for inducing natural movements in the virtual simulation, giving users the impression of actually being in this virtual environment. This phenomenon – the "experience of being there" (Lombard and Ditton 1997) is an important precondition for the occurrence of emotional reactivity within the virtual simulation (e.g., Felnhofer et al. 2015). Individuals respond on a cognitive, emotional, and physiological level to the virtual environment while interacting with a VR. Further, the experience of another entity in this simulation by using collaborative environments enables social interactions between individuals comparable to real-life situations (Kothgassner et al. 2016, 2019a).

Users can freely explore a virtual environment, interact with it, and be exposed to novel situations (Riva and Mantovani 2012). The latter bears great potential for use in psychological interventions, especially since physical-, time-, and resource-related limitations of typical therapeutic settings can be overcome. This means that patients can train in a safe environment without uncontrollable influences from the therapist. In addition, VR can induce authentic emotional states to work with in a therapeutic setting, as well as environments that are not real or even possible in the real world. VR also allows the integration of continuous objective assessment (physiological, behavioral) throughout the therapy. Therefore, VR has the potential to provide an ecologically valid option for assessment and treatment.

VR promises numerous other benefits. For instance, VR presents a time- and cost-efficient option, thereby overcoming the limitations and restrictions of traditional exposure therapy, with minimal effort. Training repetition (both with and without supervision) is also easy in VR, and patient responses can be observed and captured directly (Mühlberger and Pauli 2011). VR may also be effective for patients who are resistant to conventional psychological or pharmacological therapy (see Gonçalves et al. 2012). Another benefit of this treatment method lies in the ability to create personalized treatment programs. All scenarios and stimuli can be adapted to the needs of the individual patient, serving the need for personalized medicine and therapy (Perna et al. 2018). Further, patients show high acceptance of VR technology in therapy, which might be related to an increased feeling of safety due to the control of the situation and stimuli (Eichenberg 2007). In the following sections, we describe how these unique advantages, afforded by VR, can improve the treatment of various mental health disorders.

2 Anxiety Disorders

Strong evidence for the therapeutic use of VR for mental disorders stems from research on virtual exposure therapy. Exposure to fearful stimuli or situations is a well-established and effective therapeutic tool in Cognitive Behavioral Therapy, especially for anxiety disorders (Carl et al. 2019). Traditional Exposure Therapy is based on Foa and Kozak's (1986) Emotional Processing Theory, which describes the relation between threatening stimuli and the consequent habituation or extinction of emotional responses during continuous exposure with a stimulus. These processes lead to the formation of new memory representations that can override the older fear structures associated with anxious behavior (e.g., Krijn et al. 2004a). This desensitization can be conducted by having the patient repeatedly imagine threatening situations (exposure in-sensu) or confronting the patient with them in a secure real-life setting (exposure in-vivo). However, many patients are not able to adequately imagine a threatening situation, and some situations are not easily accessible (e.g., for fear of flying) or there are ethical restrictions (e.g., after post-traumatic events) on exposure. Therefore, virtual environments can be used to close the gap between imagination and the real world. Confrontation with the threatening situation leads to a correction of inadequate emotional, cognitive, and behavioral patterns (Abramowitz 2013; Foa and Kozak 1986). It is important that the respective virtual environment evoke emotions so that adaptive strategies can be trained in a virtual context (Bordnick et al. 2012; Felnhofer et al. 2015).

2.1 *Specific Phobias*

Specific phobias are intense but irrational fear of a situation, an object, or an animal that causes severe anxiety symptoms. This mental disorder shows a broad range of subtypes (e.g., agoraphobia, arachnophobia, panic disorder, fear of heights, fear of flying). Traditional exposure therapy has been found to be the gold standard for treatment (Wolitzky-Taylor et al. 2008). Specific phobias are the most studied mental disorder regarding the efficacy of VRET to date. In a meta-analysis on specific phobias, Morina et al. (2015) found that standard cognitive behavioral techniques yielded comparable symptom improvements to VRET when both are compared to non-treatment groups. A more recent meta-analysis (Carl et al. 2019) included a subgroup of 12 randomized controlled trials treating specific phobias, comparing VRET to non-treatment controls or in-vivo exposure. They also supported the conclusion that VRET and in-vivo exposure yield comparable outcomes for this subgroup of specific phobias. Moreover, a large effect size in favor of VRET was found in the comparison to non-treatment groups. Importantly, there is evidence that these effects are stable after the therapeutic intervention (e.g., for fear of heights; Emmelkamp et al. 2002; for fear of spiders, Lindner et al. 2020; for dental phobia, Gujjar et al. 2018; and for acrophobia, Krijn et al. 2004b).

VRET in the treatment of anxiety disorders in children and adolescents has not yet been sufficiently investigated (for a brief review see Kothgassner and Felnhofer 2021). Presently, the number of studies (four, including two randomized controlled trials) is insufficient to draw conclusive and generalizable results on treatment success. There are reports from the younger ages regarding school phobia (Gutiérrez-Maldonado et al. 2009) and arachnophobia (St-Jacques et al. 2010), both randomized control trials. Additionally, there are single-arm trials (intervention studies that consist of only one intervention group) for fear of darkness (Severa et al. 2020), as well as public speaking anxiety (Kahlon et al. 2019). Although all studies showed promising results indicating that VRET is effective for children and adolescents, more studies are needed before it can be recommended to be used in clinical practice for this population.

Recent developments include the use of automated VRET interventions; oftentimes these self-help treatments follow a gamification approach and are easy to use. Lindner et al. (2020) showed strong and sustained effects for this approach for treating fear of spiders. This form of intervention could be a practical option for self-help or to augment in-home treatment use to strengthen new therapeutic progress. Examples of VR environments for the treatment of specific phobias are shown in Fig. 1.

Fig. 1 Virtual Reality environments to treat specific phobias (fear of heights, public speaking, social situations in a restaurant). © Beutl and Hlavacs (2011)

2.2 *Social Phobia*

As social phobia – an overwhelming fear of social situations – is one of the most common mental health problems (Kessler et al. 2005) the implementation of VRET in therapeutic interventions of social phobia are of great importance. VR in the treatment of social phobia has the advantage that numerous social situations, involving even more than one person (e.g., dialogs, presentations, job interviews), can be presented in a standardized form and help to extinct the fear of these social situations. Additionally, social skills can be learned and trained in such an environment, which could improve coping with several interpersonal situations (e.g., conflicts, rejections).

There has been much debate about whether virtual characters in VR can induce similar thoughts and feelings as humans in real life. Indeed, VRET to treat social phobia depends on people experiencing anxious responses to virtual characters (Hartanto et al. 2014; Kothgassner et al. 2016) and therefore could change human experiences and behaviors in real-life (Felnhofer et al. 2018; Kothgassner et al. 2017, 2019a). Mechanisms by which people are affected by virtual characters can be explained by the Integrated Process Model of social influence in digital media (for details see Kothgassner and Felnhofer 2020). In support of the model, several studies have shown that, compared to healthy controls, individuals with social phobia tend to higher levels of self-reported anxiety and more psychophysiological arousal during interactions with virtual characters (Kishimoto and Ding 2019; Felnhofer et al. 2019). While Kishimoto and Ding (2019) used a short speech task, the study by Felnhofer et al. (2019) used more complex dialog sequences with virtual characters (e.g., dining in a restaurant, returning an order, talking to a stranger). Interestingly, the latter study found that the social presence (i.e., the experience of another social entity) of individuals with social phobia is significantly higher during the interactions, indicating that people with social phobia share a stronger feeling of a mutual awareness with the other (virtual) person than healthy individuals while interacting in the virtual space.

Summarizing these findings, VR presents a promising tool for psychological assessment of social phobia, although there is still need for validating such tools in future research. It seems that VR also has the potential for effective therapeutic intervention. Klinger et al. (2005) showed that VRET effectively reduced key symptoms of social phobia and increased social functioning. Participants were confronted with social situations in which they had to argue and defend their point of view, hold a speech, initiate contact with unknown people or in which they were observed. Numerous other studies showed efficacy of VRET in reducing social phobia symptoms compared with exposure in-vivo and waiting list controls (e.g., Anderson et al. 2013; Bouchard et al. 2017). One randomized controlled trial (Kampmann et al. 2016) of individuals with social phobia compared VRET with exposure in-vivo as well as with a waiting list control. This study included a broad range of complex social situations (e.g., dining in a restaurant, buying and returning clothes, talking to a stranger, a blind date situation, a job interview) in VR. VRET and exposure in-vivo were significantly more effective than the control condition, but in-vivo showed higher efficacy in the improvement of social phobia symptoms at follow-up. The authors suggest that the superior efficacy of VRET stemmed from the fact that they used stand-alone treatment without cognitive elements typically used in the therapy. Patients were able to use cognitive avoidance strategies during the therapeutic intervention and differences regarding the efficacy of VRET as reported by other studies (e.g., Anderson et al. 2013) could be explained by the inclusion of cognitive elements in the therapy. There is only one study investigating the effects of VRET in children and adolescents with social fears (isolated public speaking anxiety), with promising pre-post evaluation of a single session (Kahlon et al. 2019). Taken together, this indicates a strong need for randomized control trials regarding the use of VRET in social phobia in children and adolescents.

2.3 Panic Disorder and Agoraphobia

Individuals with panic disorder are suffering from panic attacks. Panic attacks are unexpected, but intense episodes of extreme fear and can result in avoidance behavior (e.g., avoiding places where the panic attacks occur). Therefore, people suffering from panic disorder oftentimes present symptoms of agoraphobia, an anxiety of places or situations where they perceive problems to escape. Hence, research on VRET is not limited to specific phobia populations; several investigations in this field have been carried out on its use in panic disorders and agoraphobia. VRET has been shown to be effective and comparable to traditional exposure treatments in several studies (e.g., Meyerbroeker et al. 2013; Vincelli et al. 2003). Vincelli et al. (2003) found that repeated exposure to symptom-inducing virtual environments (elevator, grocery store, subway, public space) reduced anxious and depressive reactions. Furthermore, it has been shown that VRET leads to a rapid decrease in panic attacks, anxiety, and depression. Meyerbroeker et al. (2013) showed similar results, but also stated that regarding the severity of panic disorder, an exposure in-vivo was superior to the usage of VRET in a CBT setting. VRET reduced avoidance behavior, agoraphobic cognitions, and the severity of panic symptoms. The improvements attributed to VRET are reported to be stable over longer follow-up periods subsequent to the intervention, but additional empirical evidence is needed (e.g., Pelissolo et al. 2012; Castro et al. 2008).

2.4 General Anxiety

General anxiety disorder (GAD) is characterized by a condition of extreme worrying about simple everyday situations. People suffering from GAD show often high psychological stress leading to excessive sleeping problems, irritability, exhaustion, or trembling. Beside VRET, a recent development that can be used to reduce general anxiety is the combination of VR environment with biofeedback. In this approach, biological signals, such as heart rate or breathing, can be visually reported to the patient. This happens traditionally through a monitor, but in VR, feedback can be integrated into the virtual environment. Using this technique individuals learn self-regulation of their physiological stress. A recent review (Kothgassner et al. 2022) showed optimistic results from randomized controlled studies in healthy or subclinical individuals but foresees the need of future studies with longer treatment duration and focus on long-term outcomes. One small study (Pallavicini et al. 2009) included patients with general anxiety disorder in a VR-biofeedback trial, comparing them to waitlist controls. They found a significant reduction of GAD symptoms and anxiety after the short treatment. VR-Biofeedback may not be limited to the treatment of general anxiety, but may also be efficacious for the treatment of phobias and psychosis (Kothgassner et al. 2022).

3 Post-Traumatic Stress Disorder

Post-traumatic stress disorder (PTSD) develops usually after an exposure to a traumatic event or threats on the individual's life (e.g., sexual abuse, neglect, car accident). Individuals suffering from PTSD show intrusive thoughts or dreams including flashbacks, avoidance of situations or persons that are related to the trauma, physical sensations, and alertness. The goal of cognitive behavioral interventions is to stabilize the patient and expose the individual with the trauma, after that the trauma will be integrated in the person's history. The treatment of PTSD has been a huge challenge for clinicians because in-vivo exposure often is impossible, for either ethical or practical reasons. In this context, research increasingly focusses on VR-based interventions. Difede et al. (2007), for example, evaluated VRET in patients with PTSD in consequence of the collapse of the World Trade Center buildings following terrorist attacks in September 2001. Participants were all victims, most of whom had responded favorably to Cognitive Behavioral Therapy. Virtual scenarios mimicked the attack, with participants approaching the traumatic situation in a stepwise procedure. The first scenario presented an airplane flying above the World Trade Center. In the second scenario, the building was hit by the airplane. Gradually, further events such as explosions, screaming, or people jumping from the building were added. Results showed that VRET led to a significant decrease in PTSD symptoms.

In a recent meta-analysis, Kothgassner et al. (2019b) compared nine randomized controlled intervention studies assessing the use of VRET in PTSD. The meta-analysis aimed at evaluating the effects of VRET on PTSD symptom severity, as well as anxiety and depressive symptoms. Compared to waitlist controls, patients in the VRET condition showed significant decreases in PTSD and depressive symptoms. In comparison to active controls (mainly in-sensu exposition) there was only a small, non-significant effect of VRET. Most of the studies included assessed VRET in military staff with post-traumatic symptoms. Only two studies evaluated the effects of VRET in patient groups other than military staff. Difede et al. (2007) included firefighters, paramedics, and civilians who were direct victims of the terrorist attacks on September 2001, and Cárdenas-López et al. (2015) assessed VRET in criminal-associated traumata in Mexican civilians. It has to be concluded that efficacy of VRET as treatment for patients suffering from PTSD is not yet generalizable, because only a few studies in non-military personnel have been conducted (Kothgassner et al. 2019b). Participants in the included studies were mostly male military staff or veterans, with an age span from early adulthood to geriatric patients. Studies including women are especially needed, since the female gender is one of the main risk factors for diverse trauma types (Brewin et al. 2000).

4 Schizophrenia

Schizophrenia is a severe mental disorder involving a range of symptoms in cognition, behavior, and emotions. It is usually divided into positive symptoms (e.g., hallucinations, paranoid ideations, disorganized thinking) and negative symptoms (e.g., reduced cognitive and executive functions). Further, individuals with schizophrenia show problems in social situations that are mostly associated with schizophrenia symptoms (e.g., paranoid ideations). Beside VRET, the current research focusses on new therapeutic options to treat psychosis and schizophrenia using VR. Treatments target hallucinations and paranoid ideations, as well as social skills and cognitive functions.

4.1 Hallucinations and Paranoid Ideations

One study (Craig et al. 2018) addressed acoustic hallucinations in schizophrenic patients. They adapted the voice of a virtual character to the sound and pitch of the hallucinated voice. Patients switched between hearing their own voice and that of the virtual characters in order to allow for controlled exposure in a safe therapeutic setting. This setting can be installed by doing this in familiar places with conscientious preparation. There was an improvement in delusional beliefs and auditory hallucinations in patients after the training. In order to augment standard treatment of psychotic disorders, VR-based Cognitive Behavioral Therapy (VR-CBT) might support the enhancement of social involvement and reduction of paranoid ideations (Pot-Kolder et al. 2018). VR-CBT uses exposure to various virtual environments involving social situations and cues (e.g., café, public transport, grocery store), where the therapist can vary the number of virtual characters and their characteristics. Therapeutic work in the respective virtual environment supports coping with various social and sensory stimuli. Individualization of treatment is achieved by adapting reactions of the virtual characters (e.g., neutral, hostile, eye contact) to the patient and their paranoid anxiety. Furthermore, pre-recorded voice notes of the patient might be included in virtual interactions. Overall, studies show significant decreases in paranoid ideations, anxiety, and negative affective states (Pot-Kolder et al. 2018). However, improved functioning with respect to basic paranoid tendency and safety behaviors in social interactions within groups was not observed (Pot-Kolder et al. 2018).

4.2 Social Skills

Social Skills Training (SST) has proven to be an effective method to promote social abilities (Bustillo et al. 2001). In a randomized controlled trial, Park et al. (2011) applied VR roleplay trainings using a virtual environment and virtual characters. Patients with psychotic disorders showed greater motivation for SST after training with VR-SST. Moreover, the VR-SST group improved their conversational skills more than the traditional treatment group, although VR-SST was inferior to conventional SST in teaching of vocal and nonverbal abilities. Park et al. (2011) argue that virtual characters cannot mimic therapists in characteristics of interpersonal interaction, such as closeness, muscle tone, or modulated eye contact. A main advantage of VR-SST is that it can more closely approximate everyday life. Rus-Calafell et al. (2014) developed a multilevel SST-VR training program employing positive and negative reinforcement to promote the progressive acquisition of social skills. Strengths of this VR program are random presentation of virtual faces, interactions, and characters, enabling diversification of social situations. Therapists observe patients in real time and vary virtual environments and virtual characters accordingly. If necessary, the program can be paused so that the therapist and patient can directly discuss specific situations or behaviors. A pilot study showed significant increases in social functioning (less social discomfort, less social anxiety, less avoidance behavior, increased social interaction) after the VR training. In this study, patients reported satisfaction with the perceived benefits of the treatment and the performance of the psychologist, as well as a high acceptance of the VR system (Rus-Calafell et al. 2014).

5 Neurodevelopmental Disorders

5.1 Autism Spectrum Disorder

Several studies evaluating VR technologies in the treatment of patients suffering from Autism Spectrum Disorder (ASD) have been conducted (see review by Mesa-Gresa et al. 2018). Due to the key symptoms of ASD resulting in deficits in social interaction, it is not surprising that most VR treatments for ASD focus on social skills training (Mesa-Gresa et al. 2018). One example is a VR approach to improve social and emotional adaption of school-aged children with ASD, using VR learning scenarios with different social events and situations (e.g., morning routines; waiting to use bathroom, following library rules, food is sold out). The authors (Ip et al. 2016) report significant improvements in emotion expression and self-regulation, as well as social reciprocity after the training. Another study, Kandalaft et al. (2013), developed a VR-training protocol to improve social cognition. Specifically, social competences, social cognitions, and psychosocial functioning of adults with ASD were trained in a variety of social situations (e.g., a virtual job interview). They observed marked increases in theory of mind and in emotion recognition, as well as an enhancement of social and job-related skills. Relatedly, Chen et al. (2016) focus on the training of emotion recognition capacities with promising improvements using virtual hints in storybooks for children. In this study, augmented reality was used to integrate aspects of the real-life environment and the VR. Other studies have focussed on skills of daily living (e.g., shopping, Adjorlu et al. 2017; driving, Wade et al. 2016). VR training can be a powerful tool to improve socio-emotional skills and support for daily living, especially in children and adolescents with ASD.

5.2 Attention Deficit and Hyperactivity Disorders

People with attention deficit and hyperactivity disorder (ADHD) show three key symptoms - attention problems, hyperactivity, and impulsivity - that can be addressed by cognitive training. These cognitive training environments are transferable to VR, for example, to treat Schizophrenia (LaBarbera et al. 2016). Cho et al. (2002) used a virtual classroom to train children and adolescents with attention deficits. Participants had to work on tasks of selective attention; for example upon random presentation of yellow, red, and purple flags, they were asked to respond only to purple flags. The authors found an overall enhancement in attention after training, whereby especially attention span and focus increased. Further, the congruence between experience and behavior in a virtual environment and in comparable real-life situations indicates that VR might also be a useful tool in the diagnosis of mental health. Advances in the field of psychological assessment included VR as a valuable tool, especially since standard test batteries or computer testing systems often have the disadvantage of limited ecological validity. Already in the mid-2000s, research showed that attention deficit and hyperactivity disorders (ADHD) can be reliably assessed by the use of an immersive virtual classroom (e.g., Parsons et al. 2007; Pollak et al. 2009). Results showed that distractions in a virtual classroom (e.g., cars passing by) influenced performance during a learning task just as in the real-life situation. Additionally, children exhibited a similar number of omission and commission errors in a virtual classroom and in a comparable real-life school situation.

6 Eating Disorders

There are several potential targets in the therapy of eating disorders, where VR could play a key role in therapy. One of the primary aspects in patients suffering from eating disorders (such as Anorexia Nervosa) is body image distortion (Cesa et al. 2013), which might be directly addressed with VR. In one approach, the set-up of the virtual environment might be such that the patient encounters situations typically inducing comparison of their body with others or related to relapse (e.g., Swimming Pool, Restaurant, Supermarket, Kitchen; Riva et al. 2006). In other applications, patients can vary the height and width of their virtual body dynamically to "wear" different body shapes. Even in the early days of VR, Riva et al. (1999) applied VRET to patients (e.g., eating situations, food) suffering from anorexia nervosa and showed that body awareness increased, and body dissatisfaction decreased with treatment. Cesa et al. (2013) assessed the long-term effects of VRET in overweight patients who engaged in binge eating, finding that patients showed weight reductions after 1 year. Bingeing stopped during treatment and resumed after treatment, although bingeing attacks were less frequent. In virtual environments, patients were confronted with situations typically inducing sustained bingeing, or relapse into this problematic behavior (e.g., grocery store, public swimming pool, restaurant). Similar results were found by Riva et al. (2006) who showed that overweight patients presented with decreased body image distortion, less avoidance behaviors as well as more adaptive coping strategies after treatment. Moreover, Paslakis et al. (2017) used a novel VRET scenario to treat compulsive exercising in VR (e.g., virtual jogging). They showed that the urge for physical activity decreased after a VR training session, suggesting that this approach could be used to reduce compulsive physical activity in anorexia nervosa (as an example for such VR setup see Fig. 2).

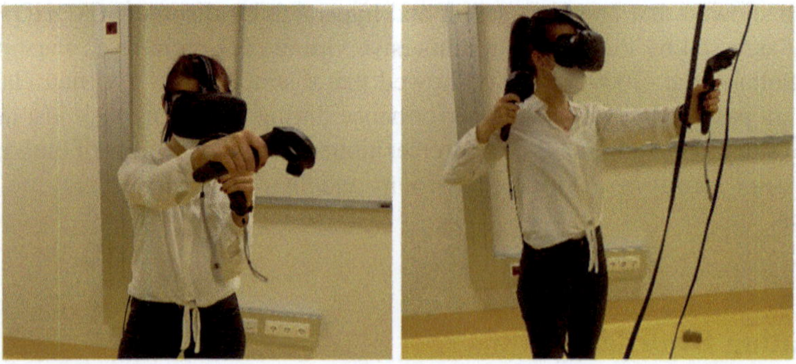

Fig. 2 Virtual Reality Setup with a Head-Mounted Display to interact with the virtual environment

7 Contraindications and Limitations for the Use of Virtual Reality

While VR shows great promise in the treatment of a number of psychological disorders, there are nonetheless a number of medical conditions (e.g., migraine, headaches, seizures, cardiovascular diseases) that might contraindicate the use of VR (Gorini and Riva 2008). VR in patients suffering from psychotic disorders needs to be well planned and organized, since virtual realities might lead to irritation and drifting off into psychotic symptoms (Repetto and Riva 2011). Some individuals report on vertigo and nausea during or after the use of VR. This phenomenon is known as "cybersickness." There is still a debate on how this phenomenon is related to age and if children are more or less affected than adults (Arns and Cerney 2005; Garcia-Palacios et al. 2007; St-Jacques et al. 2010). In some people, a slow introduction to the use of VR is supportive to avoid this adverse event. Further, it is important to consider the risk of depersonalization and derealization during the exposure with VR environments. This is particularly of interest for the treatment of individuals with high preexisting dissociation levels as it might occur in mental disorders such as PTSD or psychosis and might lower the patients' sense of presence in physical reality (Aardema et al. 2010). However, newer studies state that the technological advances – predominantly in the field of VR graphics – could possibly reduce the risk of dissociation during VR experiences and discuss a positive correlation between cybersickness and reported dissociation levels, which should be further investigated in future studies (Mondellini et al. 2021).

Despite the success of VRET, the development and adjustment of VR settings for clinical settings is still expensive and highly challenging, because clinicians can effort hardware, but there is a significant hurdle in getting access to proper software solutions. There are only few commercial clinical applications available, mainly restricted without the possibility to adapt the simulation. Especially, certified products are rare and most clinicians do not have the ability to develop these VR simulations on their own. Another issue is that newer studies (Kothgassner et al.

2021) report that stress and anxiety inducing scenarios might habituate differently in real-life and virtual environments over time. This could be a problem for therapy plans and should be considered when working with VR-based therapy tools.

8 Future Perspectives

Further advances in VR technology will allow for further developments in treatments, especially the integration of augmented reality and automation processes in VR settings to create more efficient therapy tools. Machine learning, advanced sensors, and developments for more user-friendly software applications would fulfill the potential for the realization of personalized therapy approaches in mental health (Perna et al. 2018). Generally, there are many studies that show that VRET is a powerful tool to treat anxiety disorders and it is very promising regarding the treatment of PTSD (Carl et al. 2019; Kothgassner et al. 2019b). However, evidence is lacking regarding the efficacy of VRET in the population of children and adolescents. Additionally, some VR applications developed for the treatment of mental disorders lack a clear evidence base (e.g., virtual social skills training, VR-biofeedback, cognitive training), although preliminary results are promising (e.g., Kothgassner et al. 2022; Mesa-Gresa et al. 2018). A future challenge for therapeutic VR is the development of software applications and the resources to develop efficient designs for tools beyond VRET, especially for developments integrating a gamification approach. Integration of VR into a holistic narrative via merging therapeutic programs and computer games seems especially promising. Gamification tools in therapeutic interventions are of high stimulative nature and include a well-designed story outline that constantly promotes patients' motivation, which might serve as an ideal platform for future treatment. These programs also offer numerous self-administered treatment contents aside from given instructions, which might increase treatment efficiency.

Take-Home Messages

- Virtual Reality Exposure Therapy is highly effective in specific phobias, including social anxiety. However, future developments need more trials to accurately estimate their efficacy.
- There is no sufficient evidence to use VR-based therapies in children and adolescents. More research trials are warranted to investigate VR therapies in this population.
- Research and clinical practice should consider potential contraindications of VR in the treatment of mental health issues (e.g., cybersickness, dissociation).
- Augmented reality for VRET, the integration of automation processes for highly efficacious treatments and gamification approaches to reduce drop-out rates in therapies, as well as the inclusion of more objective measures to assess the efficacy of the therapeutic intervention are important to move forward with VR as a therapeutic tool.

References

Aardema F, O'Connor K, Côté S, Taillon A (2010) Virtual reality induces dissociation and lowers sense of presence in objective reality. Cyberpsychol Behav Soc Netw 13(4):429–435

Abramowitz JS (2013) The practice of exposure therapy: relevance of cognitive-behavioral theory and extinction theory. Behav Ther 44(4):548–558. https://doi.org/10.1016/j.beth.2013.03.003

Adjorlu A, Høeg ER, Mangano L, Serafin S (2017) Daily living skills training in virtual reality to help children with autism spectrum disorder in a real shopping scenario. In: 2017 IEEE international symposium on mixed and augmented reality (ISMAR-adjunct), IEEE, pp 294–302

Anderson PL, Price M, Edwards SM, Obasaju MA, Schmertz SK, Zimand E, Calamaras MR (2013) Virtual reality exposure therapy for social anxiety disorder: a randomized controlled trial. J Consult Clin Psychol 81(5):751

Arns LL, Cerney MM (2005) The relationship between age and incidence of cybersickeness among immersive environment users. IEEE proceedings virtual reality, pp 267–268

Bordnick PS, Traylor AC, Carter BL, Graap KM (2012) A feasibility study of virtual reality-based coping skills training for nicotine dependence. Res Soc Work Pract 22(3):293–300

Bouchard S, Dumoulin S, Robillard G, Guitard T, Klinger E, Forget H et al (2017) Virtual reality compared with in vivo exposure in the treatment of social anxiety disorder: a three-arm randomised controlled trial. Br J Psychiatry 210(4):276–283

Brewin CR, Andrews B, Valentine JD (2000) Meta-analysis of risk factors for posttraumatic stress disorder in trauma-exposed adults. J Consult Clin Psychol 68(5):748–766

Bustillo JR, Lauriello J, Horan WP, Keith SJ (2001) The psychosocial treatment of schizophrenia: an update. Am J Psychiatry 158(2):163–175. https://doi.org/10.1176/appi.ajp.158.2.163

Cárdenas-López G, de la Rosa-Gómez A, Durán-Baca X, Bouchard S (2015) Virtual reality PTSD treatment program for civil victims of criminal violence. In: Cipresso P, Seriono S (eds) Virtual reality: technologies, medical applications and challenges. Nova Science, New York, pp 269–289

Carl E, Stein AT, Levihn-Coon A, Pogue JR, Rothbaum B, Emmelkamp P, Asmundson GJG, Carlbring P, Powers MB (2019) Virtual reality exposure therapy for anxiety and related disorders: a meta-analysis of randomized controlled trials. J Anxiety Disord 61:27–36. https://doi.org/10.1016/j.janxdis.2018.08.003

Castro W, Pitti CT, Bethencourt JM, de la Fuente J, Gracia R (2008) The effects of a treatment based on the use of virtual reality exposure and cognitive-behavioral therapy applied to patients with agoraphobia. Int J Clin Health Psychol 8(1):5–22

Cesa GL, Manzoni GM, Bacchetta M, Castelnuovo G, Conti S, Gaggioli A et al (2013) Virtual reality for enhancing the cognitive behavioral treatment of obesity with binge eating disorder: randomized controlled study with one-year follow-up. J Med Internet Res 15(6):e2441

Chen CH, Lee IJ, Lin LY (2016) Augmented reality-based video-modeling storybook of nonverbal facial cues for children with autism spectrum disorder to improve their perceptions and judgments of facial expressions and emotions. Comput Hum Behav 55:477–485

Cho BH, Ku J, Jang DP, Kim S, Lee YH, Kim IY et al (2002) The effect of virtual reality cognitive training for attention enhancement. Cyberpsychol Behav 5(2):129–137

Craig TK, Rus-Calafell M, Ward T, Leff JP, Huckvale M, Howarth E, Emsley R, Garety PA (2018) AVATAR therapy for auditory verbal hallucinations in people with psychosis: a single-blind, randomised controlled trial. Lancet Psychiatry 5(1):31–40. https://doi.org/10.1016/S2215-0366(17)30427-3

Difede J, Cukor J, Jayasinghe N, Patt I, Jedel S, Spielman L et al (2007) Virtual reality exposure therapy for the treatment of posttraumatic stress disorder following September 11, 2001. J Clin Psychiatry 68(11):1639–1647

Eichenberg C (2007) Einsatz von “virtuellen Realitäten” in der Psychotherapie. Psychotherapeut 52(5):362–367

Emmelkamp PM, Krijn M, Hulsbosch AM, De Vries S, Schuemie MJ, van der Mast CA (2002) Virtual reality treatment versus exposure in vivo: a comparative evaluation in acrophobia. Behav Res Ther 40(5):509–516

Felnhofer A, Kothgassner OD, Schmidt M, Heinzle AK, Beutl L, Hlavacs H, Kryspin-Exner I (2015) Is virtual reality emotionally arousing? Investigating five emotion inducing virtual park scenarios. Int J Hum Comput Stud 82:48–56

Felnhofer A, Kafka JX, Hlavacs H, Beutl L, Kryspin-Exner I, Kothgassner OD (2018) Meeting others virtually in a day-to-day setting: investigating social avoidance and prosocial behavior towards avatars and agents. Comput Hum Behav 80:399–406

Felnhofer A, Hlavacs H, Beutl L, Kryspin-Exner I, Kothgassner OD (2019) Physical presence, social presence, and anxiety in participants with social anxiety disorder during virtual cue exposure. Cyberpsychol Behav Soc Netw 22(1):46–50

Foa EB, Kozak MJ (1986) Emotional processing of fear: exposure to corrective information. Psychol Bull 99(1):20–35. https://doi.org/10.1037/0033-2909.99.1.20

Garcia-Palacios A, Botella C, Hoffman H, Fabregat S (2007) Comparing acceptance and refusal rates of virtual reality exposure vs. in vivo exposure by patients with specific phobias. Cyberpsychol Behav 10(5):722–724

Gonçalves R, Pedrozo AL, Coutinho ESF, Figueira I, Ventura P (2012) Efficacy of virtual reality exposure therapy in the treatment of PTSD: a systematic review. PloS One 7(12):e48469

Gorini A, Riva G (2008) Virtual reality in anxiety disorders: the past and the future. Expert Rev Neurother 8(2):215–233

Gujjar KR, Van Wijk A, Sharma R, De Jongh A (2018) Virtual reality exposure therapy for the treatment of dental phobia: a controlled feasibility study. Behav Cogn Psychother 46(3): 367–373

Gutiérrez-Maldonado J, Magallón-Neri E, Rus-Calafell M, Peñaloza-Salazar C (2009) Virtual reality exposure therapy for school phobia. Anuario de Psicología 40(2):223–236

Hartanto D, Kampmann IL, Morina N, Emmelkamp PG, Neerincx MA, Brinkman WP (2014) Controlling social stress in virtual reality environments. PloS One 9(3):e92804

Ip HH, Wong SW, Chan DF, Byrne J, Li C, Yuan VS et al (2016) Virtual reality enabled training for social adaptation in inclusive education settings for school-aged children with autism spectrum disorder (ASD). In: International conference on blended learning. Springer, Cham, pp 94–102

Kahlon S, Lindner P, Nordgreen T (2019) Virtual reality exposure therapy for adolescents with fear of public speaking: a non-randomized feasibility and pilot study. Child Adolesc Psychiatry Ment Health 13(1):47. https://doi.org/10.1186/s13034-019-0307-y

Kampmann IL, Emmelkamp PM, Hartanto D, Brinkman W, Zijstra BJ, Morina N (2016) Exposure to virtual social interactions in the treatment of social anxiety disorder: a randomized controlled trial. Behav Res Ther 77:147–156

Kandalaft MR, Didehbani N, Krawczyk DC, Allen TT, Chapman SB (2013) Virtual reality social cognition training for young adults with high-functioning autism. J Autism Dev Disord 43(1): 34–44

Kessler RC, Berglund P, Demler O, Jin R, Merikangas KR, Walters EE (2005) Lifetime prevalence and age-of-onset distributions of DSM-IV disorders in the National Comorbidity Survey Replication. Arch Gen Psychiatry 62(6):593–602

Kishimoto T, Ding X (2019) The influences of virtual social feedback on social anxiety disorders. Behav Cogn Psychother 47(6):726–735

Klinger E, Bouchard S, Légeron P, Roy S, Lauer F, Chemin I, Nugues P (2005) Virtual reality therapy versus cognitive behavior therapy for social phobia: a preliminary controlled study. Cyberpsychol Behav 8(1):76–88

Kothgassner OD, Felnhofer A (2020) Does virtual reality help to cut the gordian knot between ecological validity and experimental control? Ann Int Commun Assoc 44(3):210–218

Kothgassner OD, Felnhofer A (2021) Lack of research on efficacy of virtual reality exposure therapy (VRET) for anxiety disorders in children and adolescents: a systematic review. Neuropsychiatr 35(2):68–75. https://doi.org/10.1007/s40211-020-00349-7

Kothgassner OD, Felnhofer A, Hlavacs H, Beutl L, Palme R, Kryspin-Exner I, Glenk LM (2016) Salivary cortisol and cardiovascular reactivity to a public speaking task in a virtual and real-life environment. Comput Hum Behav 62:124–135

Kothgassner OD, Griesinger M, Kettner K, Wayan K, Völkl-Kernstock S, Hlavacs H et al (2017) Real-life prosocial behavior decreases after being socially excluded by avatars, not agents. Comput Hum Behav 70:261–269

Kothgassner OD, Goreis A, Kafka JX, Kaufmann M, Atteneder K, Beutl L et al (2019a) Virtual social support buffers stress response: an experimental comparison of real-life and virtual support prior to a social stressor. J Behav Ther Exp Psychiatry 63:57–65

Kothgassner OD, Goreis A, Kafka JX, Van Eickels RL, Plener PL, Felnhofer A (2019b) Virtual reality exposure therapy for posttraumatic stress disorder (PTSD): a meta-analysis. Eur J Psychotraumatol 10(1):1654782. https://doi.org/10.1080/20008198.2019.1654782

Kothgassner OD, Goreis A, Glenk LM, Kafka JX, Pfeffer B, Beutl L et al (2021) Habituation of salivary cortisol and cardiovascular reactivity to a repeated real-life and virtual reality Trier social stress test. Physiol Behav 242:113618

Kothgassner OD, Goreis A, Bauda I, Ziegenaus A, Glenk LM, Felnhofer A (2022) Virtual reality biofeedback interventions for treating anxiety. Wien Klin Wochenschr 134:49–59. https://doi.org/10.1007/s00508-021-01991-z

Krijn M, Emmelkamp PM, Olafsson RP, Biemond R (2004a) Virtual reality exposure therapy of anxiety disorders: a review. Clin Psychol Rev 24(3):259–281

Krijn M, Emmelkamp PM, Biemond R, de Ligny CDW, Schuemie MJ, van der Mast CA (2004b) Treatment of acrophobia in virtual reality: the role of immersion and presence. Behav Res Ther 42(2):229–239

LaBarbera D, Francomono A, Sideli L, La Cascia C, La Paglia F (2016) Virtual reality environments to rehabilitation attention deficits in schizophrenic patients. Annu Rev Cyberther Telemed 14:143–148

Lindner P, Miloff A, Bergman C, Andersson G, Hamilton W, Carlbring P (2020) Gamified, automated virtual reality exposure therapy for fear of spiders: a single-subject trial under simulated real-world conditions. Front Psych 11:116

Lombard M, Ditton T (1997) At the heart of it all: the concept of presence. J Comput-Mediat Commun 3(2):JCMC321

Mesa-Gresa P, Gil-Gómez H, Lozano-Quilis JA, Gil-Gómez JA (2018) Effectiveness of virtual reality for children and adolescents with autism spectrum disorder: an evidence-based systematic review. Sensors 18(8):2486

Meyerbroeker K, Morina N, Kerkhof GA, Emmelkamp PMG (2013) Virtual reality exposure therapy does not provide any additional value in agoraphobic patients: a randomized controlled trial. Psychother Psychosom 82(3):170–176

Mondellini M, Mottura S, Guida M, Antonietti A (2021) Influences of a virtual reality experience on dissociation, mindfulness, and self-efficacy. Cyberpsychol Behav Soc Netw 24(11):767–771

Morina N, Ijntema H, Meyerbröker K, Emmelkamp PM (2015) Can virtual reality exposure therapy gains be generalized to real-life? A meta-analysis of studies applying behavioral assessments. Behav Res Ther 74:18–24

Mühlberger A, Pauli P (2011) Virtuelle realität in der psychotherapie. PiD-Psychotherapie im Dialog 12(02):143–147

Pallavicini F, Algeri D, Repetto C, Gorini A, Riva G (2009) Biofeedback, virtual reality and mobile phones in the treatment of Generalized Anxiety Disorder (GAD): a phase-2 controlled clinical trial. J Cyber Ther Rehabil 2(4):315–327

Park K-M, Ku J, Choi S-H, Jang H-J, Park J-Y, Kim SI, Kim J-J (2011) A virtual reality application in role-plays of social skills training for schizophrenia: a randomized, controlled trial. Psychiatry Res 189(2):166–172. https://doi.org/10.1016/j.psychres.2011.04.003

Parsons TD, Bowerly T, Buckwalter JG, Rizzo AA (2007) A controlled clinical comparison of attention performance in children with ADHD in a virtual reality classroom compared to standard neuropsychological methods. Child Neuropsychol 13(4):363–381

Paslakis G, Fauck V, Röder K, Rauh E, Rauh M, Erim Y (2017) Virtual reality jogging as a novel exposure paradigm for the acute urge to be physically active in patients with eating disorders: implications for treatment. Int J Eat Disord 50(11):1243–1246

Pelissolo A, Zaoui M, Aguayo G, Yao SN, Roche S, Ecochard R et al (2012) Virtual reality exposure therapy versus cognitive behavior therapy for panic disorder with agoraphobia: a randomized comparison study. J Cyberther Rehabil 5(1):35–43. Rnal Clin Health Psychol 8(1): 5–22

Perna G, Grassi M, Caldirola D, Nemeroff CB (2018) The revolution of personalized psychiatry: will technology make it happen sooner? Psychol Med 48(5):705–713

Pollak Y, Weiss PL, Rizzo AA, Weizer M, Shriki L, Shalev RS, Gross-Tsur V (2009) The utility of a continuous performance test embedded in virtual reality in measuring ADHD-related deficits. J Dev Behav Pediatr 30(1):2–6

Pot-Kolder RMCA, Geraets CNW, Veling W, van Beilen M, Staring ABP, Gijsman HJ, Delespaul PAEG, van der Gaag M (2018) Virtual-reality-based cognitive behavioural therapy versus waiting list control for paranoid ideation and social avoidance in patients with psychotic disorders: a single-blind randomised controlled trial. Lancet Psychiatry 5(3):217–226. https://doi.org/10.1016/S2215-0366(18)30053-1

Repetto C, Riva G (2011) From virtual reality to interreality in the treatment of anxiety disorders. Neuropsychiatry 1(1):31

Riva G, Mantovani F (2012) From the body to the tools and back: a general framework for presence in mediated interactions. Interact Comput 24(4):203–210

Riva G, Bacchetta M, Baruffi M, Rinaldi S, Molinari E (1999) Virtual reality based experiential cognitive treatment of anorexia nervosa. J Behav Ther Exp Psychiatry 30(3):221–230

Riva G, Bacchetta M, Cesa G, Conti S, Castelnuovo G, Mantovani F, Molinari E (2006) Is severe obesity a form of addiction?: Rationale, clinical approach, and controlled clinical trial. Cyberpsychol Behav 9(4):457–479

Rus-Calafell M, Gutiérrez-Maldonado J, Ribas-Sabaté J (2014) A virtual reality-integrated program for improving social skills in patients with schizophrenia: a pilot study. J Behav Ther Exp Psychiatry 45(1):81–89. https://doi.org/10.1016/j.jbtep.2013.09.002

Severa M, Saez B, Mir JM (2020) Feasibility of a virtual reality program to treat children with fear of darkness with nonexpert therapists. Revista de Psicología Clínica con Ninos y Adolescentes 7(2):16–21

St-Jacques J, Bouchard S, Bélanger C (2010) Is virtual reality effective to motivate and raise interest in phobic children toward therapy?: a clinical trial study of in vivo with in virtuo versus in vivo only treatment exposure. J Clin Psychiatry 71(07):924–931. https://doi.org/10.4088/JCP.08m04822blu

Vincelli F, Anolli L, Bouchard S, Wiederhold BK, Zurloni V, Riva G (2003) Experiential cognitive therapy in the treatment of panic disorders with agoraphobia: a controlled study. Cyberpsychol Behav 6(3):321–328

Wade J, Zhang L, Bian D, Fan J, Swanson A, Weitlauf A et al (2016) A gaze-contingent adaptive virtual reality driving environment for intervention in individuals with autism spectrum disorders. ACM Trans Interact Intell Syst 6(1):1–23

Wolitzky-Taylor KB, Horowitz JD, Powers MB, Telch MJ (2008) Psychological approaches in the treatment of specific phobias: a meta-analysis. Clin Psychol Rev 28(6):1021–1037

MIX
Papier aus verantwortungsvollen Quellen
Paper from responsible sources
FSC® C105338

FSC
www.fsc.org

If you have any concerns about our products,
you can contact us on
ProductSafety@springernature.com

In case Publisher is established outside the EU,
the EU authorized representative is:
Springer Nature Customer Service Center GmbH
Europaplatz 3, 69115 Heidelberg, Germany

Printed by Libri Plureos GmbH
in Hamburg, Germany